I0055289

Bioenergy: Biomass to Biofuels

Bioenergy: Biomass to Biofuels

Edited by Nina Goldman

SYRAWOOD
PUBLISHING HOUSE

New York

Published by Syrawood Publishing House,
750 Third Avenue, 9th Floor,
New York, NY 10017, USA
www.syrawoodpublishinghouse.com

Bioenergy: Biomass to Biofuels
Edited by Nina Goldman

© 2019 Syrawood Publishing House

International Standard Book Number: 978-1-68286-759-4 (Hardback)

This book contains information obtained from authentic and highly regarded sources. Copyright for all individual chapters remain with the respective authors as indicated. All chapters are published with permission under the Creative Commons Attribution License or equivalent. A wide variety of references are listed. Permission and sources are indicated; for detailed attributions, please refer to the permissions page and list of contributors. Reasonable efforts have been made to publish reliable data and information, but the authors, editors and publisher cannot assume any responsibility for the validity of all materials or the consequences of their use.

Trademark Notice: Registered trademark of products or corporate names are used only for explanation and identification without intent to infringe.

Cataloging-in-Publication Data

Bioenergy : biomass to biofuels / edited by Nina Goldman.
      p. cm.
Includes bibliographical references and index.
ISBN 978-1-68286-759-4
1. Biomass energy. 2. Fuel. I. Goldman, Nina.
TP339 .B59 2019
662.88--dc23

TABLE OF CONTENTS

PREFACE

Bioenergy is a form of renewable energy that is derived from biological sources. Biomass is a plant or plant-based material that is used directly via combustion to produce heat or indirectly after conversion into biofuel. Such conversion of biomass to biofuel is achieved by different thermal, chemical and biochemical methods. Biomass fuels are usually a by-product or residue of forestry, farming or animal husbandry. Various solid and sewage biomass are used as sources of energy. Solid biomass includes manure, garden waste and crop residues. Some agricultural products like corn, soya bean, sugarcane, jatropha, sugar beet, etc. are specifically grown for biofuel production across the world. Some examples of biofuels are bioethanol, biodiesel, biofuel gasoline, etc. This book aims to equip students and experts with the advanced topics and upcoming concepts in the field of bioenergy. As this field is emerging at a fast pace, this book will help the readers to better understand the concepts of biomass and biofuels. This book will help new researchers by foregrounding their knowledge in this domain.

After months of intensive research and writing, this book is the end result of all who devoted their time and efforts in the initiation and progress of this book. It will surely be a source of reference in enhancing the required knowledge of the new developments in the area. During the course of developing this book, certain measures such as accuracy, authenticity and research focused analytical studies were given preference in order to produce a comprehensive book in the area of study.

This book would not have been possible without the efforts of the authors and the publisher. I extend my sincere thanks to them. Secondly, I express my gratitude to my family and well-wishers. And most importantly, I thank my students for constantly expressing their willingness and curiosity in enhancing their knowledge in the field, which encourages me to take up further research projects for the advancement of the area.

Editor

Additional supporting evidence for significant iLUC emissions of oilseed rape biodiesel production in the EU based on causal descriptive modeling approach

ANIL BARAL and CHRIS MALINS

The International Council on Clean Transportation, 1225 Eye St., NW Suite 900, Washington, DC 20005, USA

Abstract

Agro-economic modeling studies have shown that indirect land-use change (iLUC) emissions of first-generation biofuels can be significant, reducing or eliminating the climate change mitigating potential of these fuels. Recognizing this, proposed amendments to the European Union's Renewable Energy Directive (RED) would require reporting iLUC emissions of biofuels. The objective of this paper was to provide additional evidence of the iLUC emissions of oilseed rape (OSR) biodiesel using a noneconomic modeling approach called the causal descriptive (CD) model. The CD model originally developed by E4tech (A Causal Descriptive Approach to Modelling the GHG Emissions Associated with the Indirect Land Use Impacts of Biofuels, 2010, E4tech, London, UK) is one of the first noneconomic modeling approaches used for estimating indirect land-use change (iLUC). Using the E4tech CD modeling framework, we refine assumptions for key parameters such as yields in marginal land, displacement of OSR oil by palm oil, land availability for OSR expansion in the EU, imports of OSR from Canada and Ukraine, and palm oil expansion on peatland and thereby estimate iLUC GHG emissions for a likely scenario (Central Scenario). We find GHG emissions of OSR biodiesel to be 57 g CO_2 eq./MJ for the Central Scenario. To capture the possible range of iLUC GHG emissions, we calculate iLUC GHG emissions by changing assumptions for the Central Scenario and land-use emission factors. We find that GHG emissions of OSR biodiesel may vary from 18 to 101 CO_2 eq./MJ. The results provide additional evidence supporting the previous conclusions derived from agro-economic modeling studies that iLUC emissions of food-based biofuels can be expected to be significant compared to potential savings. Hence, to achieve meaningful GHG reductions from biofuel use and avoid policy failure, it is important that the EU should take concrete policy action to target support for biofuels toward those with the lowest expected iLUC emissions.

Keywords: biofuels, causal descriptive modeling, climate change mitigation, EU biofuel policy, indirect land-use change, oilseed rape biodiesel

Introduction

Oilseed rape (OSR) has traditionally been grown as a rotation crop to produce rapeseed oil to meet vegetable oil demand in the EU, and for export. However, OSR demand in the EU has grown to meet the demand for biofuels as required by the Renewable Energy Directive (RED), likely driving increased production of OSR and increased vegetable oil imports (Malins, 2011a,b). The RED aims to achieve 10% renewable energy in transport by 2020, whereas the Fuel Quality Directive (FQD) requires 6% GHG reduction from road transport by 2020. Currently, the FQD and RED are being amended with regard to the role of iLUC factors and support for advanced biofuels. There are differing positions on these issues by the

European Parliament and the European Council, and a final position is in the process of negotiation by the European institutions.

A position adopted by the EU Parliament in 2013 stated that biofuels from food-based crops shall be no more than 6% in the transport sector and iLUC emissions should be counted toward the FQD target, but not when assessing sustainability criteria under the FQD/RED (Lucas, 2013). In contrast, the position reached by the European Council in June 2014 would adjust the threshold for crop-based biofuels to 7% and require that iLUC emissions be used only for reporting purposes by the European Commission, not for any compliance purpose. It is noted that biofuel policies in the US such as the Renewable Fuel Standard and California's Low Carbon Fuel Standard require that iLUC emissions be incorporated in life cycle GHG emissions in determining biofuels' eligibility as renewable fuels and estimating GHG reduction credits.

Correspondence: Anil Baral, 10243 Roberts Common Lane, Burke, VA 22015, USA e-mail: baralsab@gmail.com

In the absence of iLUC factors, OSR biodiesel is expected to be an important contributor to meeting the targets – for example, Bauen *et al.* (2010) estimated that 41% of the biodiesel demand (23 billion liters) in 2020 would be met by OSR biodiesel. However, the increased utilization of OSR for biodiesel will lead to indirect land-use change (iLUC), as rapeseed oil consumed by the biodiesel industry results in OSR area expanding, other crops being displaced and/or rapeseed oil being diverted from vegetable oil markets and replaced by other oils. Since 2008, partial equilibrium models such as AGLINK/COSIMO, ESIM, and Food and Agricultural Policy Research Institute (FAPRI) and general equilibrium models such as GTAP and IFPRI-MIRAGE have been used to estimate iLUC and/or associated greenhouse gas (GHG) emissions. For example, Laborde (2011) used a global computable general equilibrium (CGE) model to estimate that the land-use change emissions of rapeseed biodiesel are 54–55 g CO_2e/MJ effectively making it a net emitter of GHG emissions in comparison with diesel after including direct GHG emissions. For canola biodiesel, a related food-based biodiesel, the California Air Resources Board reports iLUC emissions of 41.6 g MJ^{-1}, supporting the evidence that iLUC emissions of food-based biodiesel can be high (California Air Resources Board, 2014). An iLUC model comparison study carried out by Edwards *et al.* (2010) found significant indirect land-use change GHG emissions for various types of biodiesel ranging from about 50 g CO_2e/MJ to 350 g CO_2e/MJ. The iLUC models considered for comparison included both the general and partial equilibrium models – Global Trade Analysis Project (GTAP), LEITAP, AGLINK-COSIMO, and FAPRI models.

In addition, a simple bottom-up modeling approach known as the causal descriptive (CD) model has been used as an alternative method to economic models to estimate iLUC GHG emissions (Bauen *et al.*, 2010). According to Bauen *et al.* (2010), iLUC emissions of OSR biodiesel can be in the range of 17–35 g CO_2e/MJ. In contrast, a case study for rapeseed biodiesel in France carried out by Akhurst *et al.* (2011) using the CD model showed a negative iLUC emission factor of 0.16. This is primarily due to an assumption of a relatively high yield increase of 1.5% per year.

Whereas in economic models, it can often be difficult to understand the predicted responses to increasing biofuel demand because of the complex interactions between various economic sectors, the CD model is relatively easy to follow, transparent (in the sense that the major market effects are explicitly specified), and allows for a quick evaluation of several 'what if' scenarios by changing assumptions and input parameters. For example, Bauen *et al.* (2010), referred to as the 'E4tech study' henceforth, have used this method to estimate iLUC

GHG emissions and analyze 'what if' scenarios for several biofuels including OSR biodiesel, palm oil biodiesel, and wheat ethanol. While the E4tech study includes several scenarios, used to test the impact of various assumptions, there are some equally (or possibly more) reasonable possible scenarios that are left out from this analysis. In addition, there has been criticism surrounding the use of some optimistic assumptions such as high yields on abandoned and newly converted land that led to lower iLUC GHG emissions (Searchinger, 2010; Marelli *et al.*, 2011).

The objective of this study was to estimate iLUC emissions of rapeseed biodiesel under various scenarios by updating/revising input data used by the E4tech study, mainly pertaining to yields and land-use changes and using reasonable assumptions for various scenarios.

Materials and methods

Overview of CD modeling

The CD methodology is a simple bottom-up approach that maps out a chain of significant causes and effects in response to additional biofuel demand. It quantifies these causes and effects based on a set of assumed parameters based on some combination of historical trends, future projections, and expert opinions and stakeholder inputs. These expert opinions and stakeholder inputs are often corroborated by the literature review.

A CD model is a spreadsheet model; it is intended that it should be easy to track down the assumptions made and input data used, and thus, the model is more easily accessible to most stakeholders than are economic models. This aspect might be considered appealing to regulators and policy analysts. On the other hand, given simplifications compared to the real economy such as ignoring various potential economic feedback mechanisms, these models are unable to capture comprehensively the economy-wide impacts of meeting additional biofuel demand. One major drawback is that the CD methodology does not take into account the prices of commodities in mapping out the cause and effect relationships. Part of CD modeling is to identify a limited set of regions in which responses to increased biofuel demand are expected to happen. CD modeling will therefore work best when it is reasonable to assume that the market impacts of biofuels demand will be predominantly felt in a small number of regions and commodity markets. If we believe the world market is more integrated, with responses spread across a wide set of countries, CD modeling is a less useful simplification.

The baseline in the CD model is a scenario without a new demand for additional biofuel, that is, in the absence of biofuel policy. For its baseline, a CD model uses projections of the future demand, supply, and international trade of feedstocks obtained from economic models (such as FAPRI) or simpler regression analyses. This is coupled with a market analysis to examine the entry of new feedstocks and coproducts and likely substitutions. A biofuel scenario incorporates policy changes, projecting additional demand for biofuel into the future.

The CD methodology can be used to perform a retrospective analysis in which one looks into land-use changes that occurred in a given period and assigns them to biofuels and other economic drivers. The ICONE study (Nassar et al., 2010) that analyzes the indirect land-use impacts of Brazilian sugarcane is one example. However, in most cases, this approach has been utilized for predictive analysis, that is, to estimate how much of the land-use change at a future date, for instance 2020, is triggered by biofuel demand in that year (Bauen et al., 2010).

Illustrative example

To facilitate a basic understanding of how a CD model works, an illustration for a biofuel from a notional crop A is provided below.

The starting point for building a CD model is to determine the additional biofuel demand in a biofuel policy scenario. Next, the total feedstock (crop A) needed to produce a given amount of biofuel is calculated, that is, demand for crop A. In the E4tech CD modeling, there are three primary ways the additional demand for feedstock can be met: (1) increased crop yield, (2) crop area expansion, and (3) increased feedstock import. Additionally, biofuel demand could partly be met by reduced food consumption or displacement of rapeseed oil from nonbiofuel vegetable oil markets.

An increase in yield results in no further land impact, but crop land expansion to grow more crop A causes land-use change. There can be several possibilities for how land-use change might occur. It is possible that the existing crop A is diverted to biofuel production leading to its reduced export. The demand for crop A in other countries could be met by expanding crop area and/or increasing yields in those countries or by importing crop A from countries with surplus production; for example, African countries import wheat from Australia when European wheat exports to Africa decline due to its use for biofuel production.

Alternatively, the demand for crop A can be simply met by bringing more new area under cultivation by converting either abandoned land or natural land to cropland.

In addition, crop A may displace crop B on the existing land and the demand for crop B may in turn be met by a combination of area expansion and a yield increase.

As more biofuel is produced, there is an increase in the supply of the economically valuable coproducts such as distillers dried grain with solubles, oil meal, etc. (Malins et al., 2014). These coproducts can find uses in products such as animal feeds, thereby displacing crop-based products and in doing so reduce the overall demand for other crops. In Fig. 1, coproduct C displaces crop D, which avoids the requirement for land to grow the displaced crop D. In other words, coproducts offset to some extent the land-use impact of biofuel production. However, if crop D itself was associated with two coproducts (in this case E and F) and coproduct C only displaces coproduct E, then the demand for crop D declines but the unmet demand for other coproduct F would need to be met in some other way, potentially through growing another crop. One such example is that if wheat DDGS replaces soybean meal, the demand for soybean declines, presumably leading to its decreased production. This means that production of soy oil, an important coproduct from soybean, would also decline.

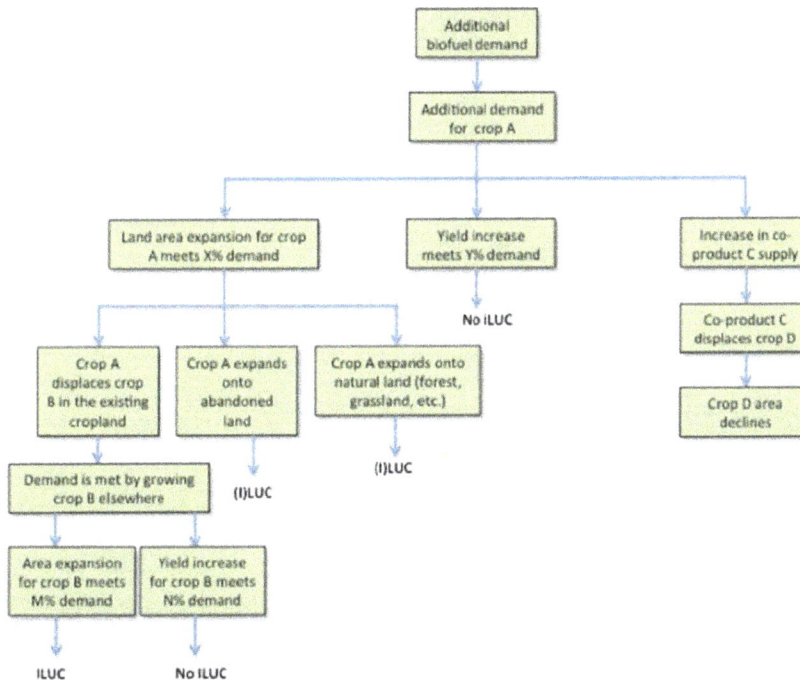

Fig. 1 Illustration of a CD model.

Hence, other vegetable oil needs to be produced to make up the unmet demand of soy oil.

An important distinction that may be drawn here is that a biofuel coproduct may replace either a 'determinant' coproduct or a 'dependent' coproduct. A determinant coproduct is a major coproduct that is the primary determinant of the value of a crop. A dependent coproduct is a product with less value, the production of which only marginally affects the overall crop value. It is presumed for CD modeling that only changes in demand for the determinant coproduct will affect overall production of a crop. If a biofuel coproduct displaces a determinant coproduct (e.g., soy meal) causing the whole crop no longer to be produced, this means that the dependent coproduct (e.g., soy oil) will no longer be produced. In the case of a biofuel coproduct displacing a dependent coproduct, the displaced dependent coproduct may substitute yet another product in the market, or in some cases could be simply discarded if it has no alternative economic use.

In some cases, coproducts may also displace non-land-based products. For example, electricity can be produced from lignin, which is a coproduct of cellulosic ethanol production; this displaces some electricity from the grid and thereby avoids GHG emissions which otherwise would occur.

CD modeling involves making an assumption about what happens at each stage of this cascade of product displacements (either production increase or another layer of displacement). Production changes can come from yields or from area increases. Area increase can come from natural land, or by displacing some other crop, requiring an additional displacement assessment. Once the net land-use change areas in various countries/region are identified, they are multiplied by the respective land-use emission factors to estimate iLUC GHG emissions.

The main steps in the CD methodology to estimate iLUC GHG emissions are as follows:

1. Determine the land-use requirement in the baseline, that is, in the absence of biofuel demand,
2. Determine feedstock demand in a biofuel scenario,
3. Determine market responses to meet the feedstock demand,
4. Estimate land-use impacts in various regions/countries for different market responses, and
5. Multiply land-use change area with emission factors and estimate net iLUC GHG emissions.

Model, data, and assumptions

The central biofuel scenario developed for this study considers OSR biodiesel based primarily on the framework, data, and assumptions provided by the E4tech study, modifying them, where appropriate, based on insight and reasonable assumptions gathered from the literature review. Following the E4tech study, in the Central Scenario, we assume that rapeseed biodiesel meets 41% of the total biodiesel demand (23 billion liters) in 2020. In the baseline, while some portion of rapeseed is used for rapeseed biodiesel, no further rapeseed is assumed to be used for biodiesel after 2008.

To meet 41% of the biodiesel demand in the Central Scenario in 2020, the EU requires 7.2 million tons (Mt) of additional production of OSR. In the EU, OSR is used as a break crop grown in rotation with cereals such as wheat and barley. In the E4tech study, it is assumed that this additional demand is met by a combination of harvested area expansion for OSR in Europe, yield increase, increased imports from Ukraine and Canada, and displacement of OSR from the vegetable oil market as shown in the cause and effect diagram (Fig. 2).

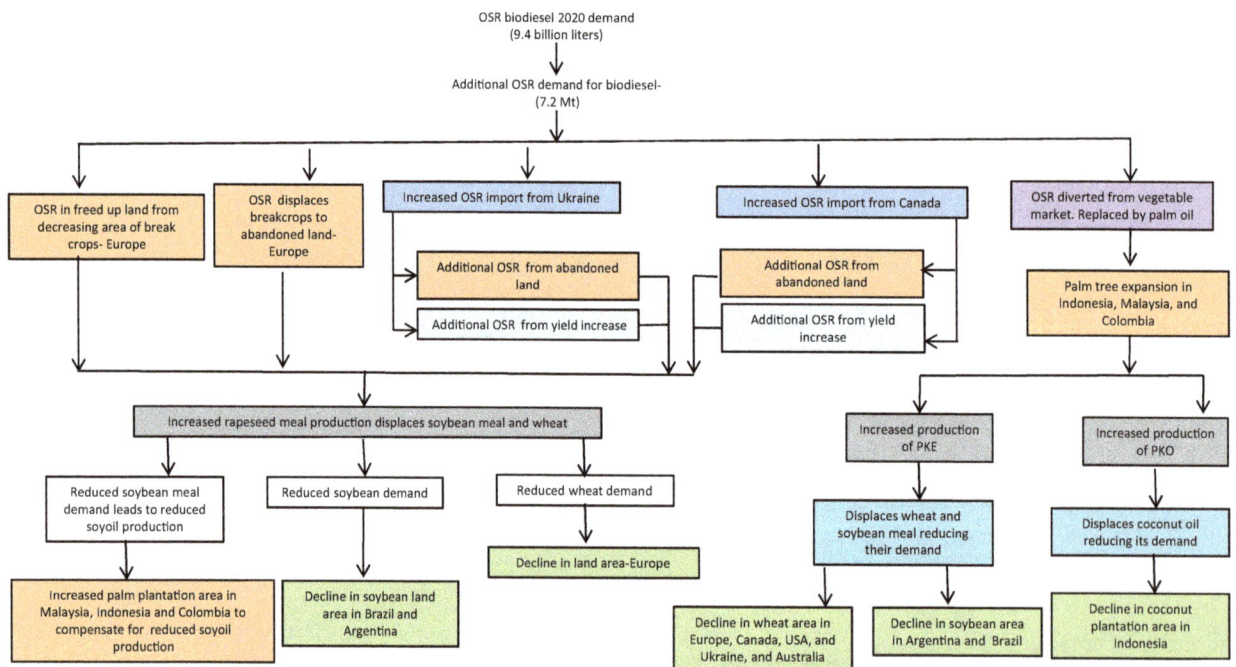

Fig. 2 Cause and effect relationships in response to meeting OSR biodiesel demand of 9.4 billion liters.

The key assumptions and data directly adopted from the E4tech study without any modification are discussed below.

Harvested area expansion in the EU. Following the E4tech study, we assume that harvested area for cereals will decline, and harvested area of OSR as a rotation crop will not increase. There are two possibilities for area expansion. First, OSR is grown on land that becomes free due to a decline in the production of other break crops such as peas and mustard (based on historical trends from 2000 to 2008). Secondly, due to its higher economic value, OSR displaces less economical break crops such as lupin and beans. This creates additional land for OSR expansion. The displaced break crops are in turn grown on abandoned cropland. Thus, the abandoned cropland foregoes carbon sequestration, as it would otherwise have reverted to natural land (forest, shrub land, savanna, grassland, wetlands, etc.).

As in the E4tech study, we assume that there is no OSR yield increase in either the baseline or Central Scenario (biofuel) in the EU from 2008 to 2020. The OSR yield in 2020 for the Central Scenario and baseline is 3.6 tonnes ha^{-1} (t ha^{-1}). The E4tech study notes that the EU is poised to eliminate some pesticide application practices in the coming years, and anticipates that this will cancel out any potential yield gains. However, yield increases are assumed for OSR exporting countries – Canada and Ukraine. Following the E4tech study, we assume that there are yield increases of 37% in Ukraine and 9% in Canada in response to an increase in OSR biodiesel demand compared to the baseline in 2020.

Imports from Ukraine and Canada. As shown in Fig. 2, the increased supply of OSR from Ukraine results from a combination of yield increase and additional production in abandoned cropland. It has been argued that abandoned land is available in plenty in Ukraine after the fall of the Soviet Union in the 1990s (Bauen *et al.*, 2010). Following the E4tech study, we assume that the 2020 OSR yield would be 2.83 t ha^{-1} in the biofuel scenario compared to 2.06 t ha^{-1} in the baseline, an increase of 37%.

Likewise, the increased supply of OSR from Canada results from a combination of yield increase and area expansion. We assume that the OSR yield in 2020 for the Central Scenario would be 2.16 compared to 1.99 t ha^{-1} in the baseline in 2020, an increase of 0.17 t ha^{-1} or 9%.

Coproduct impact. Besides the abovementioned land-use impacts that result from an increase in OSR demand, there is a positive land-use impact from the displacement of soybean meal and wheat in animal feed by rapeseed meal, a coproduct of OSR biodiesel production. Displacement of soybean meal by rapeseed meal leads to reduced demand for soybean, which in turn leads to avoided land use in Brazil and Argentina. Likewise, wheat displacement leads to reduced wheat demand in Europe, resulting in avoided land use in Europe. We do not take into account other coproducts such as glycerol, which do not displace land-based products and hence do not offset the land-use impact.

However, there is a countervailing impact of reduced soybean demand such that it leads to an increase in land-use impact. Reduced soybean demand implies reduced production of soy oil. The demand from soy oil has to be met from other sources, in this case by palm oil. The increased demand for palm oil leads to an increase in palm plantation area in Indonesia and Malaysia causing an increase in GHG emissions.

This land-use impact of palm expansion is in turn tempered by land use avoided by the use of coproducts of palm oil production – palm kernel extract (PKE) and palm kernel oil (PKO). PKE can be used as animal feed ingredient. In this study, we assume that PKE use in animal feed leads to displacement of wheat and soybean meal, resulting in reduced demand for wheat in Europe and soybean in Brazil and Argentina. This implies reduced wheat area in Europe and reduced soybean area in Brazil and Argentina. Likewise, it is assumed that PKO will displace coconut oil, thereby reducing its demand. It is further assumed that this displacement of coconut oil leads to a reduction in coconut plantation area in Indonesia with avoided GHG emissions.

Likewise, other key assumptions and data directly adopted from the E4tech study without any modification are as follows:

- We use average yields as opposed to marginal yields. There are several factors that contribute to marginal yields. Hence, it is very difficult to evaluate the impact of a particular factor on marginal yields.

- To avoid double counting, any reductions in GHG emissions achieved from carbon mitigation policies already in place, such as Reducing Emissions from Deforestation and Forest Degradation (REDD), are included in the baseline. This means that if a coproduct substitution leads to the avoidance of deforestation, it will not be counted as avoided deforestation due to forest/biodiversity protection policies already in place.

- Land-use emission factors are based on Winrock land conversion data and MODIS satellite data, which were used by US EPA (USEPA, 2009) to calculate emission factors for land conversion and reversion. These data were validated by comparing them with actual land-use change data obtained from aircraft surveys and other satellite data (Bauen *et al.*, 2010). In one cases, MODIS tended to misclassify land-use categories and was corrected using Monte Carlo analysis.

- Coproduct substitution rates are adopted from Lywood *et al.* (2009), while in one scenario, substitution ratios suggested by Joint Research Centre, EUCAR & CONCAWE (2008), are used.

On the other hand, the key assumptions of the E4tech study that we modified for the Central Scenario and the reasons for doing so are provided below.

Lower yields on abandoned cropland. We assume that OSR yields on abandoned cropland or freed-up land are 20% lower than the average yield on existing cropland. We postulate that abandoned cropland is likely to have lower yields and the size of abandoned land in the EU would be small due to the following reasons:

1. Crop yields vary, among regions and even within a single farm. When land goes out of production, it is logical to assume that the least productive land would be abandoned first.
2. The E4tech report argues that cereal cultivation in the EU decreases by about 2.6 million hectares (Mha) between 2008 and 2020 primarily because of a yield increase in cereals like wheat. However, this figure is a result of assumptions that no additional cereals are used for biofuels beyond 2008 in the baseline, and the corresponding area has been subtracted from the FAPRI baseline projection. This is not true, and even in the absence of biofuel policy, some additional cereals are produced beyond 2008. For example, FAPRI estimates that harvested wheat area only declines by 0.8 Mha as opposed to 2.2 Mha between 2008 and 2020 (FAPRI, 2009).

On the other hand, the OECD-FAO baseline predicts that the total EU cultivation area for oilseed and cereals combined will increase (OECD & FAO, 2009) even though cereal area declines. This implies that there may be no abandoned land at all, or only some abandoned land if we assume a high yield increase.

Therefore, unlike in the E4tech study, we assume that only small part of the OSR demand is met by expansion of OSR production onto freed-up land from declining harvested areas of break crops. Most of the OSR demand for biodiesel is met by imports from Ukraine and Canada and displacement of OSR from the vegetable oil market by palm oil. It is to be noted that even a 20% lower yield in abandoned land can be an optimistic assumption as inferred from the weighted-average-national wheat yield in abandoned land which comes out at 67% of EU average when land abandoned between 1997 and 2007 is considered.

Imports of OSR from Ukraine and Canada. The ESIM model, which is regarded as the best existing model for EU agriculture production and markets, estimates that only 18% of biodiesel in 2022 comes from extra oilseed production in the EU (Bauen *et al.*, 2010); the remaining demand (80%) is met with either renewable diesel from cellulosic feedstock (30%) and vegetable oil imports (50%) or vegetable oil imports only.

For the Central Scenario, we assume that 20% of the biodiesel demand is met from the domestic production, and 58% of the demand is met by importing rapeseed meal from Canada and Ukraine. We assume conservatively that the remaining 22% demand is met by the displacement of OSR oil by palm oil from the EU vegetable market as discussed later in this section.

Displacement of OSR oil from the vegetable oil market by palm oil. Citing historical trends in the EU vegetable oil market, the E4tech study argues that OSR oil is unlikely to be displaced out of the EU food market by palm oil.

However, due to its low price and fungibility, palm oil is very competitive in the vegetable oil market and may likely be

a significant or even the main substitute for OSR oil in the EU despite some differences in physical properties. This is consistent with the observation that palm oil imports have increased steadily in line with biodiesel demand and that OSR production has not been sufficient to meet the vegetable oil demand in the EU (Malins, 2011a,b). To account for the possibility that an increase in the demand for biodiesel could divert OSR oil from the vegetable oil market to biofuels and be replaced by palm oil (Schmidt & Weidema, 2007), we assume that 22% of rapeseed demand for biofuels is met with palm oil. The 22% displacement figure for OSR was also used in the IFPRI-MIRAGE modeling (Laborde, 2011). However, no displacement of OSR out of Ukrainian food markets is assumed for the Central Scenario, which is consistent with the E4tech study.

Peatland conversion of 33%. For most scenarios, the E4tech study assumes that 5% of oil palm expansion occurs on peatland in Malaysia and Indonesia. However, the available evidence indicates that at least one-third of oil palm expansion is likely to occur on peatland (Marelli *et al.*, 2011). Between 2003 and 2008, oil palm cultivation area increased by 600 thousand hectares (Kha) in Malaysia. Most of the expansion occurred in Sarawak State and will likely continue to do so in the future. There is no land conservation policy in Sarawak State, which has the largest peatland area in Malaysia.

In the case of Indonesia, a study by Hooijer *et al.* (2006) showed that 25% of land concessions granted for oil palm plantations were in Sumatra and Papua. Although not all concessions were used for oil palm plantations, the share of oil palm on peat may expand, accounting for up to 50% of palm oil plantations in the future. Therefore, high peatland conversion is more probable than low or no peatland conversion. Hence, peatland conversion of 33% is assumed, while acknowledging that the current peatland conversion rate can be higher than 33%.

To sum up, the Central Scenario is created by modifying the assumptions of the E4tech study by considering lower availability of abandoned land in the EU for OSR expansion, 20% lower yields on abandoned cropland, imports of OSR from Canada and Ukraine, high peatland conversion for palm oil, and 22% of OSR demand for biodiesel being met through the displacement of OSR oil from the vegetable market by palm oil.

Sensitivity analysis

To get a sense of the likely range of GHG emissions associated with OSR biodiesel, we further performed a sensitivity analysis around this Central Scenario. For the sensitivity analysis, scenarios A to F are created by changing one assumption/parameter at a time to assess the sensitivities toward land-use emissions factors, the extent of OSR oil displacement from vegetable oil markets, coproduct displacement ratios, yield increase, and the impact of reduced food consumption (Table 1).

As land-use emission factors are important, we outline below some of the criticisms of the emission factors used by the E4tech study and how an alternate set of emission factors were compiled for a sensitivity run for this study. It is to be noted here that the emission factors take into account forgone carbon sequestration. For example, if abandoned land is used for crop production, it forgoes an opportunity to sequester carbon by natural increases in levels of vegetation and soil carbon.

The E4tech study uses the land-use emission factors used by US EPA (2009), which are based on Winrock International land-use data. Winrock land-use reversion factors for forest might be low due to an accounting error in the dataset. Instead of providing a value for the 'carbon accumulation rate for the first 20 years' following land conversion, the Winrock International accidently set a value of zero in Europe (Marelli et al., 2011). In addition, changes in below-ground biomass from both land reversion and conversion are not considered in calculating change in carbon stocks. The Winrock land-use data for cropland reversion is based on the satellite data of natural land in 2006 and that was cropland in 2001. This means that the dataset only captures the intermediate land cover classes as it would take several years to establish mature forests with significant canopy. Hence, forest areas and carbon accumulation rates are underestimated (Edwards et al., 2010). For the Central Scenario, we did not correct for the error in forest C accumulation rate in the Winrock dataset, which means that in this respect, these iLUC emissions are conservative estimates.

A separate set of emission factors were compiled by the International Council on Clean Transportation (ICCT), which consider changes in soil and vegetative carbon. Carbon stocks were taken from the Harmonized World Soil Database (FAO/IIASA/ISRIC/ISSCAS/JRC, 2009). The WWF Terrestrial Ecoregions Map was used to map biomes on to the world soil map. This allows us to calculate average soil carbon concentrations for each biome in each country. For vegetation emissions, IPCC Tier 1 default values were used to calculate total vegetative carbon above ground. Above-ground dry biomass stocks were multiplied by (1+ root: shoot ratio) to account for below-ground vegetation biomass. Dead litter C stocks were added to live vegetation C stocks for total vegetation C per hectare. Total C for forests systems was multiplied by 0.9 for developed regions (USA, Europe, Pacific Developed) and by 0.96 for developing regions (all other regions) to calculate C emissions, accounting for harvested wood products (Searle & Malins, 2011). For reversion emission factors, sequestration was calculated by multiplying above-ground forest growth from the IPCC Tier I default values (1+ root: shoot ratio) to account for

below-ground vegetation growth. The land extension coefficients from Searchinger et al. (2008) were then applied to each climate/ecological system for each region, matching E4tech regions with Searchinger regions as appropriately as possible. Total C emissions were summed across climate/ecological system for each region. The emission factors compiled by the ICCT are summarized in Table 2.

Results

Figure 3 illustrates the iLUC emissions for the Central Scenario and highlights how modifying some of the underlying assumptions in the Central Scenario can alter iLUC emissions. Based on reasonable assumptions backed by the available literature, we find that the iLUC central estimate of OSR biodiesel to be 57 g CO_2e/MJ. This value is in close agreement with 54–55 g CO_2e/MJ reported by Laborde (2011) for OSR biodiesel using IFRI-MIRAGE, a general equilibrium model. Alternative scenario analysis shows that under certain conditions and assumptions, iLUC emissions can vary from 18 g CO_2e/MJ to 101 g CO_2e/MJ.

To better understand the major market responses that contribute significant iLUC GHG emissions in the Central Scenario, a decomposition of iLUC emissions by market responses is provided in Fig. 3. As can be seen in Fig. 3, OSR expansion due to demand for OSR biodiesel, and indirect palm expansion in response to the displacement of OSR oil from the vegetable markets by palm oil and the shortage of soy oil from reduced soybean production are responsible for indirect land-use change emissions. OSR expansion contributes 53 g CO_2e/MJ, whereas palm expansion contributes 54 g CO_2/MJ of land-use change emissions. Palm expansion is assumed to occur in Indonesia, Malaysia, and Colombia, whereas soybean expansion is assumed to occur in Brazil and Argentina. However, these land-use change emissions are offset by avoided land-use emissions from wheat, soybean, and coconut oil displacements by three coproducts – rapeseed meal, PKO, and PKE. Total offset emissions are −50 g CO_2e/MJ with

Table 1 Additional scenarios and assumptions for sensitivity runs

Scenario A	Same as the Central Scenario except no OSR displacement by palm from the EU vegetable oil market
Scenario B	Central Scenario except 80% OSR displacement by palm from the EU vegetable oil market.
Scenario C	Central Scenario except land-use emissions factors compiled by the ICCT instead of the E4tech emission factors
Scenario D	Central Scenario except JEC displacement ratios for rapeseed meal
Scenario E	Central Scenario except annual OSR yield improvement of 1.0% in the central scenario as opposed to no yield increase in the baseline. This leads to lower OSR imports from Ukraine.
Scenario E	Central Scenario with a consideration of the impact of reduced food consumption on land savings. It has been postulated that diverting food-based crop to fuel drives up the price of crop and hence causes a reduction in food consumption. It is assumed that reduced food consumption accounts for 20% of the OSR biodiesel demand, although a preliminary analysis of the existing agro-economic models suggests that the impact can be even higher.

Table 2 Weighted average emission factors for land conversion and reversion

Country	Weighted avg. emission factors compiled by the ICCT- (t C ha^{-1}, over 30 years)		Weighted avg. emission factors- E4tech study (t C ha^{-1} over 30 years)	
	Conversion	Reversion	Conversion	Reversion
Argentina	90	48	22	22
Australia	50	29	26	
Brazil	88	48	59	59
Canada	41	16		30
Colombia	106	48	69	
EU	112	46	35	29
Indonesia	175 (384)	96	75 (229)	
Malaysia	175 (384)		74 (226)	
Ukraine	91	46	25	41
USA	56	31		32

Data in parentheses represent emission factors accounting for 33% of oil palm expansion on peatland.

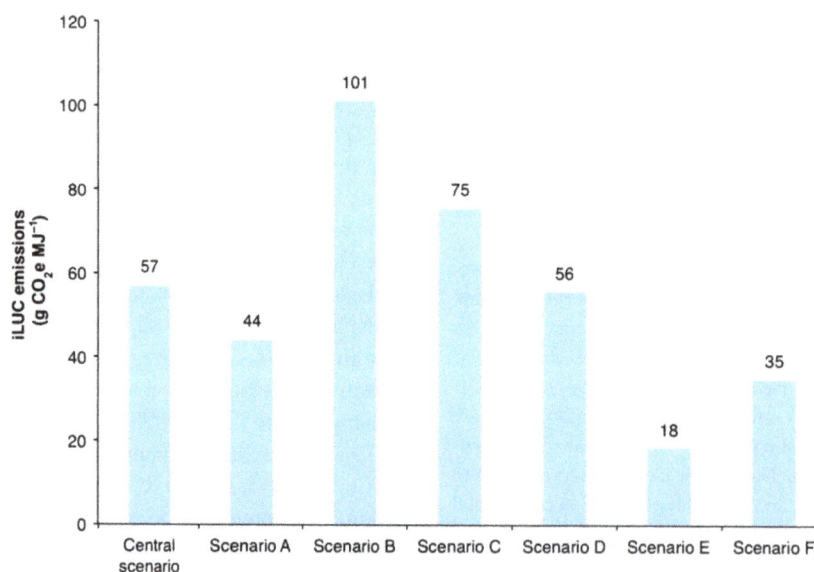

Fig. 3 iLUC emissions of OSR biodiesel for various scenarios.

the majority of offset emissions coming from soybean displacement followed by coconut oil displacement. Soybean displacement leads to avoided land use in Brazil and Argentina, whereas coconut displacement leads to avoided land use in Indonesia. Wheat displacement leads to avoided land use in the EU, Ukraine, Canada, US, and Australia. The higher emissions from palm expansion are due to a reasonable assumption that at least 33% of oil palm expansion occurs on peatland (Hooijer *et al.*, 2006) and that 22% OSR oil demand is met by palm oil substitution for OSR oil. As scenarios A (no OSR oil displacement) and B (80% OSR displacement) suggest, the extent of OSR oil displacement by palm oil can have a significant impact on iLUC emissions despite including GHG credits from the palm oil coproducts PKO and PKE. Moreover, the yield increase

has the most significant impact on iLUC emissions. Following the E4tech study, in the Central Scenario, we assume there to be no yield increase in response to an increase in biodiesel demand in the EU. Akhurst *et al.* (2011) assumed a 1.5% annual yield improvement of OSR in France in the baseline. Here, to assess the likely impact of a yield increase on iLUC emissions, an annual yield improvement of 1.0% is assumed in the Central Scenario (Scenario E) in addition to yield increases already assumed for Ukraine and Canada. In this case, GHG emissions decrease significantly to 18 g CO_2 eq./MJ (Fig. 3) as most of the demand for OSR is met by yield increase without requiring significant additional land area. This suggests that increasing crop yields can be one of the strategies for mitigating iLUC emissions.

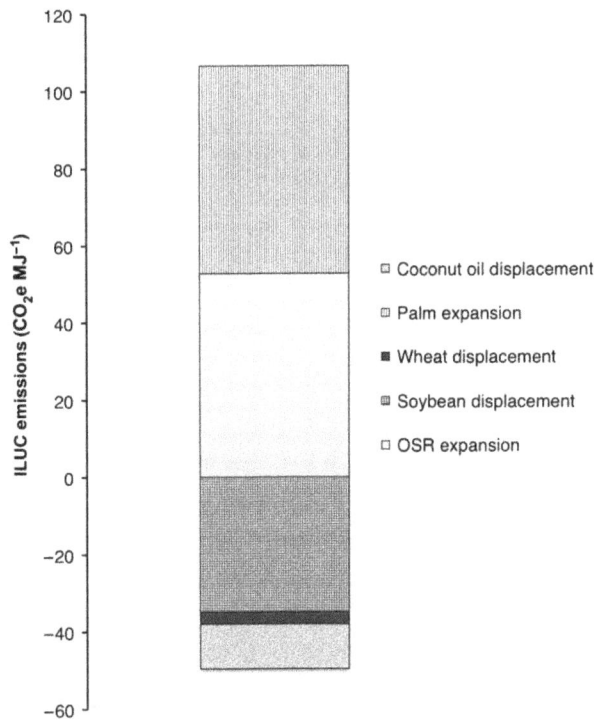

ILUC emissions (CO_2e MJ^{-1})

- Coconut oil displacement
- Palm expansion
- Wheat displacement
- Soybean displacement
- OSR expansion

Fig. 4   Major causal factors (market responses) contributing to iLUC emissions of rapeseed biodiesel.

The use of the land-use emission factors compiled by the ICCT increase iLUC GHG emissions by 32% (Scenario C). This underscores the fact that E4tech emissions factors may have been underestimated and highlights the need for more accurate assessments of emissions from various land types in general. As shown in Table 2, the emission factors for Ukraine, Malaysia, and Argentina are higher than those used in the E4tech study.

Using OSR displacement ratios of oilseed rape meal from JEC (2008), which entails a lower displacement of soybean meal (0.4) than wheat (0.5), does not result in an appreciable reduction in GHG emissions (Scenario D).

Agro-economic iLUC models often acknowledge that increased biofuel demand leads to price increases, resulting in reduced food consumption; this unconsumed crop may then be used for biofuel production, avoiding further cropland expansion. This is a desirable outcome in terms of iLUC, but not socially. To reflect this likelihood, in Scenario F, we assume that increased OSR prices lead to reduced food consumption that could meet 20% of OSR demand for biodiesel. This results in a 37% reduction in GHG emissions relative to the Central Scenario. Such a high reduction is attributed to the assumption that reduced food consumption eliminates the need to divert 20% of OSR from vegetable markets, thereby avoiding significant GHG emissions from palm expansion on peatland. The E4tech study purposefully did not consider the impact of reduced food consumption.

Discussion

The CD methodology is a simple bottom-up approach that allows one to identify the cause and effect links in the biofuel production chain and estimate the associated impacts on land-use change. It helps us analyze how, and to what extent, different market-mediated responses impact iLUC. This can be useful in determining appropriate actions at global/regional/country levels to avoid/minimize the magnitude of iLUC and associated emissions.

However, accurately estimating iLUC emissions requires a clear understanding of future crop yields, regions where land-use change occurs, trade balances, types of land that will be brought in for cultivation, displacement ratios of coproducts, the fate of abandoned land (reversion), and more importantly carbon stocks of the various land types. The E4tech study correctly acknowledges that it is a challenging task to accurately predict even a range of likely iLUC factors, let alone a single iLUC factor, when we consider uncertainties in assumptions about future projections and estimates of carbon stocks of various land cover types. This is particularly because it would be prohibitively costly to capture all the causes and effects and to model all possible scenarios for a given biofuel.

By capturing broader economy-wide impacts, agro-economic models avoid the second limitation but are still subjected to uncertainties with regard to future yield increases, regions where biofuel expansion occurs, types of land conversion, and land-use emission factors.

These problems can be addressed to some extent by refining the major assumptions contributing to iLUC based on the available literature, particularly the most recent studies that shed new light on likely land-use changes and emissions. This study is an attempt in this direction. The major assumptions we analyzed are as follows: yield increase, availability of abandoned land, the extent of peatland conversion, impact of reduced food consumption, the fungibility of palm oil and other vegetable oil, and emission factors and OSR imports from Canada and Ukraine. Based on the available evidence, we considered 20% lower yields on land where additional OSR production occurs, 33% expansion of oil palm production on peatland, 58% of the OSR demand being met by OSR imports from Ukraine and Canada, and 22% of the OSR demand being met by OSR oil displacement by palm oil.

Based on these considerations, we find that the iLUC GHG emissions of OSR biodiesel in Europe are 57 g CO_2e/MJ for the Central Scenario. For all the scenarios analyzed in this study, iLUC GHG emissions vary from 18 to 101 g CO_2 eq./MJ. On the other hand, Bauen et al.

(2010) found a range of 17–35 g CO_2 eq./MJ for various scenarios analyzed for OSR biodiesel.

We would like to emphasize that we have obtained these higher estimates despite assuming several optimistic assumptions of the E4tech study. For example, as in the E4tech study, we make a simplistic assumption that PKO displaces only coconut oil in Indonesia. In reality, PKO competes with many edible oils including palm stearate. Moreover, coconut oil is also supplied by other countries such as the Philippines, which have little tropical forest and peatland. Hence, the assumption of only coconut displacement in Indonesia tends to overestimate credits and contribute to lower iLUC emissions. Likewise, a 37% yield increases in Ukraine compared to the baseline and the use of average EU yield rather than marginal yields are optimistic assumptions. Moreover, by not correcting for the Winrock dataset for accidental zeros, the iLUC emissions for the Central Scenario might be underestimated.

It is not clear that CD modeling offers a more useful reflection of the likely indirect land-use change emissions from biofuels than economic modeling. It is in the nature of the economic models that they capture more subtle market interactions than CD modeling is able to include, and these effects may well be important. Notwithstanding these caveats, the results support the hypothesis that OSR biodiesel may offer little or no GHG saving compared to diesel, considering the default non-iLUC emissions of OSR biodiesel of 22 g CO_2e/MJ as reported in the RED. Further careful examination of assumptions, parameters, and input data used in the model will help to improve confidence in CD estimates of iLUC emissions of OSR biodiesel.

Nonetheless, the results show that incorporation of iLUC factors in the RED is justified to ensure that use of biofuels help achieve the intended climate mitigation goals (Malins, 2013). Hence, it is imperative that the EU should strongly consider transitioning from the proposed reporting-only iLUC emission requirement to actually incorporating iLUC emissions in the life cycle emissions of biofuels for compliance purpose under the FQD and RED.

Acknowledgements

We would like to acknowledge the generous financial support from the ClimateWorks Foundation. We would like to thank an anonymous reviewer for his/her constructive feedback and comments.

References

Akhurst M, Kalas N, Woods J (2011) *Applying causal descriptive modelling techniques to the assessment of the potential indirect land use change impact of French grown oilseed rape biodiesel: case study* (Draft report). Imperial College London and LCA[works].

Bauen A, Chudziak C, Vad K, Watson P (2010) *A Causal Descriptive Approach to Modelling the GHG Emissions Associated with the Indirect Land Use Impacts of Biofuels.* E4Tech, London, UK.

California Air Resources Board (2014) Low Carbon Fuel Standard Re-Adoption Indirect Land Use Change (iLUC) Analysis. September 29, 2014 Workshop, Sacramento, CA. http://www.arb.ca.gov/fuels/lcfs/lcfs_meetings/092914iluc-prestn.pdf (accessed 15 December 2014).

Edwards R, Mulligan D, Marelli L (2010) *Indirect Land Use Change from Increased Biofuels Demand: Comparison of Models and Results for Marginal Biofuels Production from Different Feedstocks).* EC Joint Research Centre, Ispra.

FAO/IIASA/ISRIC/ISSCAS/JRC (2009) *Harmonized world soil database (version 1.1).* Available at: http://www.iiasa.ac.at/Research/LUC/External-World-soil-database/HTML/ (accessed 9 August 2014).

FAPRI (2009) FAPRI 2009 U.S. and world agricultural outlook. Available at: http://www.fapri.org/outlook/2009/ (accessed 9 August 2014).

Hooijer A, Silvius M, Wösten H, Page S (2006) *PEAT-CO2, Assessment of CO2 Emissions from Drained Peatlands in SE Asia.* Delft Hydraulics, HD Delft, Netherlands.

Joint Research Centre, EUCAR & CONCAWE (JEC) (2008) *Well-to-wheels analysis of future automotive fuels and powertrains in the European context.* (version 3).

Laborde D (2011) *Assessing the land use change consequences of European biofuel policies* (Specific Contract No SI2.580403 Implementing Framework Contract No TRADE/07/A2). Washington DC:IFPRI.

Lucas P (2013) EP Vote on ILUC Proposal, the (Long) Road to Sustainable Biofuels. Environment and Energy Affairs, 20 September 2013.

Lywood W, Pinkney J, Cockerill SAM (2009) Impact of protein concentrate coproducts on net land requirement for European biofuel production. *GCB Bioenergy*, **1**, 346–359.

Malins C (2011a) Vegetable oil markets and the EU biofuel mandate. Briefing paper, International Council on Clean Transportation, Washington, DC.

Malins C (2011b) *IFPRI-MIRAGE 2011 modelling of indirect land use change: Briefing on report for the European Commission Directorate General for Trade.* Available at: http://www.theicct.org/sites/default/files/publications/ICCT_IFPRI-iLUC-briefing_Nov2011-1.pdf (accessed 15 December 2014).

Malins C (2013) A model-based quantitative assessment of the carbon benefits of introducing iLUC factors in the European Renewable Energy Directive. *GCB Bioenergy*, **5**, 639–651.

Malins C, Searle S, Baral A (2014) *A Guide for the Perplexed to the Indirect Effects of Biofuels Production.* International Council on Clean Transportation, Washington, DC.

Marelli L, Mulligan D, Edwards R (2011) *Critical Issues in Estimating ILUC Emissions: Outcomes of an Expert Consultation 9–10 November 2010, Ispra (Italy).* EU, Luxemborg.

Nassar AM, Antoniazzi LB, Moreira MR, Chiodi L, Harfuch L (2010) *An Allocation Methodology to Assess GHG Emissions Associated with Land Use Change.* Institute for International Trade Negotiations, Sao Paulo.

OECD, FAO (2009) *OECD-FAO Agricultural Outlook 2009–2018.* Available at: http://www.agri-outlook.org/dataoecd/2/31/43040036.pdf (accessed 11 December 2011).

Schmidt JH, Weidema BP (2007) Shift in the marginal supply of vegetable oil. *International Journal of Life Cycle Assessment*, **13**, 235–239.

Searchinger T (2010) *Comments of Tim Searchinger on draft ILUC analysis of rapeseed and palm, biodiesel presented by E4tech in February 2010.* Available at: http://www.endsreport.com/docs/20100428b.pdf (accessed 15 December 2014).

Searchinger T, Heimlich R, Houghton RA *et al.* (2008) Use of US croplands for biofuels increases greenhouse gases through emissions from land-use change. *Science*, **319**, 1238–1240.

Searle S, Malins C (2011) *Estimates of carbon storage in wood products following land clearing* (Working paper 2011-4). Available at: http://www.theicct.org/sites/default/files/publications/ICCT_carbon_storage_in_wood_products_August_2011.pdf (accessed 15 December 2014).

USEPA (2009) *Draft Regulatory Impact Analysis: Changes to Renewable Fuel Standard Program.* US Environmental Protection Agency, Washington, DC.

2

Valorization of Sida (*Sida hermaphrodita*) biomass for multiple energy purposes

NICOLAI DAVID JABLONOWSKI[1], TOBIAS KOLLMANN[1,2], MORITZ NABEL[1], TATJANA DAMM[3], HOLGER KLOSE[3], MICHAEL MÜLLER[4], MARC BLÄSING[4], SÖREN SEEBOLD[4], SIMONE KRAFFT[1], ISABEL KUPERJANS[5], MARKUS DAHMEN[5] and ULRICH SCHURR[1]

[1]*Forschungszentrum Jülich GmbH, Institute of Bio- and Geosciences, IBG-2: Plant Sciences, 52428 Jülich, Germany, [2]Project Management Jülich, Renewable Energies – EEN, 52425 Jülich, Germany, [3]RWTH Aachen University, Institute of Botany and Molecular Genetics IBMG, Worringerweg 3, 52074 Aachen, Germany, [4]Forschungszentrum Jülich GmbH, Institute of Energy and Climate Research, IEK-2: Microstructure and Properties of Materials, 52428 Jülich, Germany, [5]FH Aachen, Aachen University of Applied Sciences, Institut NOWUM-Energy, Heinrich-Mussmann-Str. 1, 52428 Jülich, Germany*

Abstract

The performance and biomass yield of the perennial energy plant *Sida hermaphrodita* (hereafter referred to as Sida) as a feedstock for biogas and solid fuel was evaluated throughout one entire growing period at agricultural field conditions. A Sida plant development code was established to allow comparison of the plant growth stages and biomass composition. Four scenarios were evaluated to determine the use of Sida biomass with regard to plant development and harvest time: (i) one harvest for solid fuel only; (ii) one harvest for biogas production only; (iii) one harvest for biogas production, followed by a harvest of the regrown biomass for solid fuel; and (iv) two consecutive harvests for biogas production. To determine Sida's value as a feedstock for combustion, we assessed the caloric value, the ash quality, and melting point with regard to DIN EN ISO norms. The results showed highest total dry biomass yields of max. 25 t ha^{-1}, whereas the highest dry matter of 70% to 80% was obtained at the end of the growing period. Scenario (i) clearly indicated the highest energy recovery, accounting for 439 288 MJ ha^{-1}; the energy recovery of the four scenarios from highest to lowest followed this order: (i) $\gg$ (iii) $\gg$ (iv) > (ii). Analysis of the Sida ashes showed a high melting point of >1500 °C, associated with a net calorific value of 16.5–17.2 MJ kg^{-1}. All prerequisites for DIN EN ISO norms were achieved, indicating Sida's advantage as a solid energy carrier without any post-treatment after harvesting. Cell wall analysis of the stems showed a constant lignin content after sampling week 16 (July), whereas cellulose had already reached a plateau in sampling week 4 (April). The results highlight Sida as a promising woody, perennial plant, providing biomass for flexible and multipurpose energy applications.

Keywords: biogas feedstock, biogenic energy source, biomass, flexible biomass application, lignin source, perennial energy crop, plant development, Sida hermaphrodita, solid fuel

Introduction

In terms of a growing bio-based economy, a sustainable plant biomass supply is becoming a major challenge to meeting the demands. The increased share of bio-based energy carriers is challenging both society and energy supply systems. In Germany, only the share of renewable energy of total electricity production reached 25.8% in 2014 and shall further increase in the future to 40–45% until 2025 (Bundesministerium für Wirtschaft und Energie, 2015). Because biomass can be stored and is therefore ready for use on demand as a reliable sup-

ply source, energy production from biomass plays an important role in the variety of renewable energy sources, such as water, wind, and solar power (Carroll & Somerville, 2009; Graham-Rowe, 2011).

To cover the demand of plant biomass, biomass production should not compete with crop production traditionally used for food and feed (Graham-Rowe, 2011; Voigt et al., 2012). The food vs. fuel controversy must be avoided by directly searching for nonfood plant species and by indirectly avoiding the use of highly valuable crop land to obviate land-use conflicts (Schröder et al., 2008). Therefore, plants characterized by high-yield, low-nutrient demand, and a valuable biological composition for a variety of bioeconomic applications should be emphasized. The perennial nature of such plants could decrease the energy input for its produc-

Correspondence: Dr Nicolai David Jablonowski
e-mails: n.d.jablonowski@fz-juelich.de
n.d.jablonowski@daad-alumni.de

tion because annual field preparation and plant establishment becomes unnecessary. Additionally, such a plant biomass production system may even exhibit additional benefits for ecosystem services and soil protection, following the idea of a 'low input–high output system' in terms of energy investments and additional benefits (Blanco-Canqui, 2010). However, perennial cropping systems reduce the ability of farmers to react to sudden changes on the market. To tackle these relevant aspects of biomass production for energy purposes, we investigated the perennial plant *Sida hermaphrodita* (L.) Rusby (hereafter referred to as Sida), also known as Virginia mallow or Virginia fanpetals. From 1930 until 2000, some research, mostly published in Russian or Polish language, was conducted on this plant with regard to its propagation and establishment (Spooner *et al.*, 1985). The native North American species Sida was introduced to Poland during the 1950s as a fodder and fiber source, but it developed further as a promising energy crop (Spooner *et al.*, 1985; Borkowska *et al.*, 2009). However, in the recent past, Sida attracted attention again as a promising plant for bioenergy production (Borkowska & Wardzinska, 2003; Borkowska & Molas, 2012; Barbosa *et al.*, 2014; Nabel *et al.*, 2014). Along with *Miscanthus*, another perennial, high-yield energy plant, Sida recently attracted attention due to its wood-like, high-yield biomass (Borkowska & Molas, 2012, 2013). Besides its high biomass yield, Sida is also highly attractive for pollinators due to its long flowering period and therefore has high ecosystem service values. Due to the numerous shoots per plant, the biomass yield of Sida is higher compared to that of currently used energy plants like corn (Slepetys *et al.*, 2012). In addition to soil fertility, precipitation, and climate conditions, the organic dry matter yield depends on the age of the plant and the time of harvest and generally varies from 9.6 to 19.7 t ha^{-1} (Slepetys *et al.*, 2012; Borkowska & Molas, 2013).

While Sida is commonly used as a solid fuel for combustion, the first biogas batch tests with Sida showed a potential of 435 Ndm^3 kg^{-1} organic dry matter (oDM) from silage made from a biomass harvest in July, suggesting that Sida is also useful as a substrate for biogas production (Oleszek *et al.*, 2013). Methane obtained from biogas production plays a major role as an energy carrier from biogenic resources. To date, approximately 7800–8000 operating biogas facilities in Germany are producing 27.6–29.0 billion MWh, while a total of 49.1 TWh were produced from biomasses, of which biogas and solid fuels contributed with 59.1% and 24.2%, respectively (Deutscher Bauernverband, 2015; FNR, 2015). Anaerobic batch tests on Sida biomass revealed methane concentrations of 280 to 293 Ndm³ kg^{-1} oDM (Hartmann & Haller, 2014). However, the determined biogas poten-

tial and methane yield of Sida biomass are lower compared with major energy crops like corn (Schattenhauer & Weiland, 2006), even though this varies with the time of harvest. First attempts with chemical pretreatment have been carried out to increase the biological availability and to improve the resulting biogas yield of Sida as a biogas feedstock (Michalska *et al.*, 2012).

To date, Sida has not yet been evaluated as a feedstock for flexible, demand-driven energy applications by analyzing its cell wall composition and energy yield with regard to the plant development stage and time of harvest. A flexible use of the Sida biomass would allow operators to react to market changes using the biomass either as feedstock for solid fuel and biogas production or for industrial applications.

In this study, we evaluated the overall performance, biomass, and energy yield of *Sida hermaphrodita* (L.) Rusby at weekly harvest intervals throughout an entire growing period at agricultural field conditions in the second year after field establishment. The aim of this study was to develop a plant growth and development code allowing a general determination of the best Sida harvesting time: (i) with regard to the maximum biomass yield depending on the plant development stage; (ii) to determine the energy value in terms of biogas production and solid fuel energy content with regard to its verification in accordance with the German industry standard DIN EN ISO 17225-7:2014-09 for solid biogenic fuels (DIN-EN-ISO-17225-7, 2014); and (iii) to obtain a biomass comprising the most suitable cell wall composition enabling specific utilization and upscaling for technical applications. Additionally, we determined that the ash content, composition, and melting point of Sida were important parameters when it is used as a solid fuel in accordance with the aforementioned DIN norm.

To determine the best energy use of Sida biomass and identify the highest energy output, four application scenarios were evaluated in terms of solid fuel and biogas applications. The four tested scenarios were as follows: (i) one harvest for solid fuel only; (ii) one harvest for biogas production only; (iii) one harvest for biogas production, followed by a subsequent harvest of the regrown biomass for solid fuel; and (iv) two consecutive harvests for biogas production.

Material and methods

Experimental site and setup

The experimental site was located in Mersch, Germany (100 m o. NN, 6°22′34 east and 50°57′50 north using ETRS89). The site was exposed to an annual mean temperature in 2014 of 11.5 °C, with a minimum temperature of −5.1 °C, a maximum temperature of 34.7 °C, and an annual precipitation of 801 mm during the time of the experiment. The soil was Orthic Luvisol consisting of 5.6% sand, 79.0% silt, 15.4% clay, with pH 6.2 ($CaCl_2$), containing 2.4% C_{org}, 32 mg kg^{-1} P_2O_5, 18 mg kg^{-1} K_2O, and 9 mg kg^{-1} Mg. The experiment was established in May 2013 using seedlings of the perennial mallow plant *Sida hermaphrodita* (L.) Rusby in the BBCH-Sida plant development stage 12–13 (Table S1). Seedlings were precultivated in the greenhouse at controlled conditions from March until the beginning of the field experiment in May in 2013, using compostable pots made of peat (pot type 30023092, 8 × 8 cm square; Jiffy®, Moerdijk, The Netherlands). A total of 436 individual seedlings were planted in the arable field soil in a plant distance of 0.5 m and a row distance of 0.75 m, in a total area of 165 m^2, equaling 2.7 plants m^{-2}. The herbicide glyphosate (RoundUp, 4 L ha^{-1}) was applied prior planting and in-between the plants and rows during the plant establishment phase to control weed growth. No fertilizers were applied prior and throughout the entire experiment.

Determination of plant development stages: BBCH-Sida code

Plant physiological traits of *Sida* were monitored throughout the experiment. To allow a future estimation of desired harvest times assuming different biomass application scenarios, we analyzed and described the plant development stages for Sida in a modified BBCH code. Our numeric BBCH-Sida code was developed following previous BBCH codes for other plant species created by Hack *et al.* (Hack *et al.*, 1992). The BBCH-Sida code is provided in full length in the Supporting Information of this manuscript (Table S1). The first given number of the numeric BBCH-Sida code 0 (i.e., seed germination) to 9 (aging process, senescence) describes the macrostages of the plant. The second number 0–9 describes the microstages within the macrostages, for example, number of branches. For the assessment of a Sida stand, it must be considered that at least 50% of the plants display the respective development stage. Macrostages, such as leaf or side branch development, may emerge simultaneously. In this case, only the higher development stage is being considered. Our numeric BBCH-Sida code was developed with regard to its application on both the agricultural (stand/plant population) and detailed laboratory level (single plants). Emphasis is attributed to the timely application of pesticides particularly during the establishment phase and to potential harvest during the growing period.

Biomass harvest and preparation

The evaluation of biomass yield started in the second year (2014) after establishment when the mean plant height reached BBCH-Sida development stage BBCH 17 (approximately 38 cm height). To determine the total yield development, biomass of five individual plants was harvested weekly, following a totally randomized design, by cutting the stems approximately 7 cm above ground. All replicate plants were separated into stems and leaves including side branches. All leaves and stems for each individual plant were shredded (<2 cm, Viking AE 1180 E) and dried at 85 °C until weight constant. For further

handling, the dried Sida biomass was milled (<1 mm, Retsch SM 200) and homogenized. Subsamples were additionally milled to powder using a Retsch MM 400 for subsequent elemental and calorific analysis. To determine the optimal biomass use for energy purposes, the following harvest scenarios were applied: (i) Plants were grown throughout the vegetation period and were harvested at BBCH-Sida stage 98 to determine its calorific value when used as a solid fuel only (sample F1, Table 1). (ii) Plants were harvested only once at BBCH-Sida stage 91 to determine its biogas potential only at a late plant development stage (sample B2, Table 1). (iii) To determine the biogas potential of green Sida biomass, randomly selected plants were firstly harvested at BBCH-Sida development stage BBCH 55 and a dry matter content of 16.5% (sample B1.1, Table 1). The freshly shredded biomass was compressed in 60 L PE bins for subsequent ensilaging for 12 weeks. Subsequently, the regrown biomass was harvested as dried biomass at the end of the entire vegetation period at BBCH-Sida 98 to evaluate its calorific value as a solid fuel (sample F2, Table 1). (iv) To determine the maximum biogas potential of green Sida biomass, plants were firstly processed in accordance with scenario (iii) (sample B1.2, Table 1). The regrown biomass of the previously harvested plants was harvested again in BBCH-Sida development stage 71, exhibiting a dry matter content of 28.5%, to evaluate the biogas potential of the regrown fresh plant material (sample B1.2, Table 1).

Biomass harvests for solid fuel application were aimed at obtaining high dry matter content. Harvest times for scenario (iii) and (iv) were chosen to allow a potential supply of alternative biogas feedstock before fresh biomass of other feedstock sources (e.g., maize silage) would be available. Harvest in October for scenario (ii) and (iv) was chosen to obtain a high biomass yield and to avoid late shooting and frost damage of the plants during winter. The two repeated harvests in scenario (iii) and (iv) were meant to evaluate the added energy value of Sida biomass as a feedstock for additional solid fuel and biogas, respectively.

The described harvest scenarios are summarized in Table 1. In every scenario, a total of 10 plants were randomly harvested as biological replicates.

Evaluation of calorific values, ash content, composition, and behavior

Subsamples of the pulverized biomass were compressed into pellets and analyzed for their higher heating value in five biological replicates employing a Parr calorimeter Type 6200, in accordance with DIN 51900-3:2005-01 (DIN-51900-3, 2003). Simultaneously, the ash content, composition, and behavior were analyzed using subsamples of the milled biomass in triplicates. For the investigation on ash melting behavior, the samples were oxidized at 550 °C in platinum crucibles under constant air flow in a muffle furnace for 24 h at constant temperature. The heating rate of the samples was 5 K min^{-1}. The samples were weighed after ashing and ground in a mortar to ensure the homogeneity of the ash samples. X-ray powder diffraction (XRD) was used to identify crystalline compounds of the ashes using a Siemens D500 powder diffractometer. The

Table 1 Overview of the four different Sida harvest and biomass evaluation scenarios to determine the optimum energy usage of Sida biomass

Scenario	1. Harvest: purpose/sample #	BBCH-Sida development stage/Date of sampling	Dry matter content	2. Harvest: purpose/sample #	BBCH-Sida development stage/Date of sampling	Dry matter content
(i)	Solid fuel/F1	BBCH 98/15.01.2015	75.9%	–	–	–
(ii)	Biogas/B2	BBCH 91/15.10.2014	36.1%	–	–	–
(iii)	Biogas/B1.1	BBCH 55/12.06.2014	16.5%	Solid fuel/F2	BBCH 98/15.01.2015	76.3%
(iv)	Biogas/B1.1	BBCH 55/12.06.2014	16.5%	Biogas/B1.2	BBCH 71/15.10.2014	28.5%

Table 2 Comparison of Sida biomasses silages prepared for biogas tests

	B1.1	B1.2	B2
Time of harvesting	12.06.2014	15.10.2014	15.10.2014
Type of ensilage	In barrel	In barrel	In barrel
Time of ensilage	12 weeks	12 weeks	12 weeks
Dry matter after ensilage	20.2%	26.0%	30.8%
Organic dry matter after ensilage	18.0%	23.6%	28.1%

fusibility of the ashes was determined by hot-stage microscopy according to DIN 51730:2007 09 (DIN-51730, 1998).

The higher heating value of the Sida biomass was determined employing the above-mentioned calorimeter. Because the Sida biomass exhibited a residual moisture content, the net calorific value (i.e., lower heating value) was determined mathematically using the preceding equation (Equation 1: Equation for the calculation of the higher heating value of moist biomass).

$$q_{FM} = q_{TS} * \left(\frac{DM}{100}\right) - 0.02443 * (1 - DM)$$

The higher heating value (q_{TS}) is given in MJ kg^{-1}, the dry matter (DM) in %, and the value 0.02443 is the enthalpy of water vaporization at constant pressure and a temperature of 25 °C, given in MJ kg^{-1}. At harvest, the DM content of the Sida biomass for the solid fuel scenario (samples F1 and F2) accounted for 76%; however, from the literature, DM values of 85–89% were reported (Stolarski et al., 2013). Therefore, the corrected heating value was calculated with a dry matter content of 88% and a residual moisture of 12%. For the overall calculation of the energy yield per ha, the resulting heating value was multiplied with a corrected fresh biomass yield, that is, the obtained dry matter yield plus 12% residual moisture. The determination of the dry and organic dry matter was performed using a furnace (Heratherm, Thermo Scientific, USA) and a muffle furnace (N100/14, Nabertherm, Germany), according to DIN EN 12880:2001 (DIN-EN-12880, 2001).

Elemental analysis

Prior elemental analysis all Sida biomass samples were oven dried at 85 °C and subsequently milled (<1 mm, Retsch SM 200) and homogenized. Subsamples were additionally powdered employing a ball mill (Retsch MM400). Al, Ca, Cr,

Fe, K, Mg, P, S, and Si were measured and quantified using ICP-OES. As, Cd, Cu, Hg, and Pb were determined using ICP-MS due to a lower detection limit. To do so, subsamples of 100 mg were diluted and decomposed in a mixture of 3 mL HNO$_3$ and 2 mL H$_2$O$_2$ in a microwave. Subsequently, 1 mL of HF was added. The samples were adjusted to a volume of 14 mL and were measured in a dilution of 1 : 10. For C, H, N, and O determination, triplicate samples of approximately 2.5 and 2 mg, respectively, were oxidized employing an element analyzer (Vario EL Cube, Elementar). The relative standard deviations for the abovementioned methods were for elemental contents of >1 ± 3%, and for elemental contents of <0.1 ± 20%. The total Cl content in the Sida biomass was analyzed in accordance with DIN EN 15408 (DIN-EN-15408, 2011).

Evaluation of biogas potential

The calculated biogas production of Sida was measured in eudiometer batch test systems derived from the specification of the German DIN 38 414-8:2006-03 (DIN-38414-8, 1985). Silage of Sida biomass from scenario (ii), (iii), and (iv) (harvested in June and October) was prepared from shredded and homogenously mixed Sida biomass that was highly compressed into 60-L PE bins. As inoculum, active digestate originating from a commercially operating biogas plant fed with maize silage was used. Prior analysis, the dry matter of the digestate and the Sida silage was determined at 105 °C. The inoculum and the Sida silage was homogeneously mixed in 1 L Shott Duran glass bottles, and the tests were carried out at mesophilic conditions at 37 °C in the dark using a temperature-controlled water bath. The gas volume was measured daily, and the ambient air temperature and air pressure were monitored for calculation to norm conditions. The test ended when the relative gas yield after 24 h was below 1% of the total produced gas volume. Each Sida biomass silage was tested in four replicates. The detailed information about the different silages is shown in Table 2.

Analysis of Sida lignocellulosic residues

Extraction and analysis procedures were modified according to Foster et al. (Foster et al., 2010a). We further ground 70–73 mg dried homogenized Sida leaf, side branch, and stem material to a fine powder using a M 400 mill (Retsch, Haan, Germany) with a frequency of 30 s^{-1} for 2 min (leafy material) or 10 min (stem material). Plant cell wall residues were isolated by washing once with 70% (v/v) ethanol solution and four times with chloroform/methanol (1 : 1; v/v) solution and once with 1 mL acetone solution collecting the pellet every time at 20 000 g (Foster *et al.*, 2010a). The pellet was dried under air flow at room temperature. Starch was removed by enzymatic digestion with α-amylase (3 U) and amyloglucosidase (1.5 U) (Megazyme, Bray, Ireland). The remaining de-starched, alcohol-insoluble residues (d-AIR) were washed four times with water, once with acetone, and then subsequently dried. All analyses were performed with approximately 2 mg d-AIR. The matrix polysaccharide composition was determined by extraction and hydrolysis with 2 M trifluoroacetic acid (Foster *et al.*, 2010a). We collected 100 μL of TFA, which was evaporated under air flow and the remaining pellet was dissolved in water. Single sugar analysis was performed according to Voiniciuc *et al.* (Voiniciuc *et al.*, 2015) using high-performance anion exchange chromatography with pulsed amperometric detection (HPAEC-PAD). For the separation of monosaccharides, a CarboPac PA20 column was used with a flow rate of 0.5 mL min^{-1} and was equilibrated with 2 mM NaOH for 10 min before sample injection. Neutral sugars were separated with 2 mM NaOH over a time course of 18 min. Afterward, 550 mM NaOH was used for 10 min to separate uronic acids. The column was rinsed finally with 800 mM NaOH for 10 min. Monosaccharide amounts were normalized to an internal standard and quantified using standard calibration curves of the different monosaccharides.

The crystalline cellulose content was determined after hydrolyzing and removing the noncrystalline cellulose with acetic and nitric acid (Updegraff, 1969). The remaining crystalline cellulose residues were hydrolyzed with 72% (w/v) sulfuric acid. Carbohydrate content was determined using anthrone assay (Scott & Melvin, 1953).

Lignin content was determined according to Foster *et al.* (Foster *et al.*, 2010b) using the acetyl bromide spectrophotometric method. Different amounts (0.1–0.7 mg) of Kraft-Lignin (Sigma-Aldrich, Seelze, Germany) were used as a standard.

Results

Biomass productivity

The evaluation of the plant performance and biomass yield was conducted in the second year after the establishment of the field trail. Because the biomass yield of a perennial energy plant such as Sida is generally relatively low in the first year, the second year after plant establishment was chosen to allow the development of the BBCH-Sida code and to obtain reliable biomass yield estimations.

The established BBCH-Sida code allowed a detailed characterization of the respective plant development stage at each weekly harvest and enables future comparison of Sida plants irrespective of the climate or the geographic conditions of the experiment (Table S1). The analysis of the Sida cell wall composition at each harvest allows a further, general estimation of the plant BBCH-Sida development stages and the associated plant biomass characteristics. This enables a detailed,

demand-driven application of the biomass at the ideal plant development stage and application purpose.

Sida plants started to regrow in late March 2014 of the second year and increased rapidly in biomass yield, accounting for a maximum of 25 t ha^{-1} dry matter total biomass equivalents for both stems and leaves in sampling week 22 at a BBCH-Sida development stage 71 (Fig. 1). At peak biomass production, the share accounted for approximately 15 t ha^{-1} of stem biomass only (Fig. 1). The subsequent decrease of dry matter biomass yield is attributed to the loss of the plant leaves, resulting in a final dry biomass yield of 15 t ha^{-1} of dried woody stems at the end of the experiment (Fig. 1). In return, the dry matter content increased continuously throughout the growing period (Fig. 2). The highest dry matter content was reached at the termination of the vegetation period and at the harvesting of the dead standing dried biomass in January 2015, accounting for more than 70% dry matter content. The increase of the dry matter content is associated with a loss of leaves and a relative increase of lignin and cellulose in the side branches (Fig. 4). Until the termination of the experiment, the dry matter content increased to 90% until March 2015 (data not shown).

Calorific evaluation and ash characterization

The higher heating value of the Sida biomass used as a solid fuel was equally high for both samples, irrespective of a single or double biomass harvest (sample F1 vs. F2), accounting for approximately 19.5 MJ kg^{-1} (Table 3). However, due to the much higher biomass yield after a single biomass harvest, sample F1 resulted in a 3.6 higher total energy yield compared with the sample F2 (Table 3).

To determine whether Sida biomass meets the prerequisites of a solid fuel in accordance with DIN EN ISO 17225-7:2014-09, we analyzed the heating and calorific value, water and ash content, and the content of N, S, and Cl, among other elements (DIN-EN-ISO-17225-7, 2014). Table 4 shows the elemental analysis from Sida biomass, both harvested in January 2015 at a BBCH-Sida development stage 98.

The crystalline compounds in the two resulting ashes of both samples (F1 and F2) identified by XRD are given in Table 5. The variance of the measurement is 10% (relative). The ash composition of the two Sida samples is very similar, and both are strongly enriched with alkaline earth metal and alkali metal compounds. The main phases are calcium carbonate, fairchildite, and calcium hydroxyapatite, which together count for 87% to 88% of the weight of the ash. Earth alkali silicates and oxides were found only in minor amounts. Sulfur and chlorine compounds have not been found because of the low sulfur and chlorine content of the raw material (Table 4).

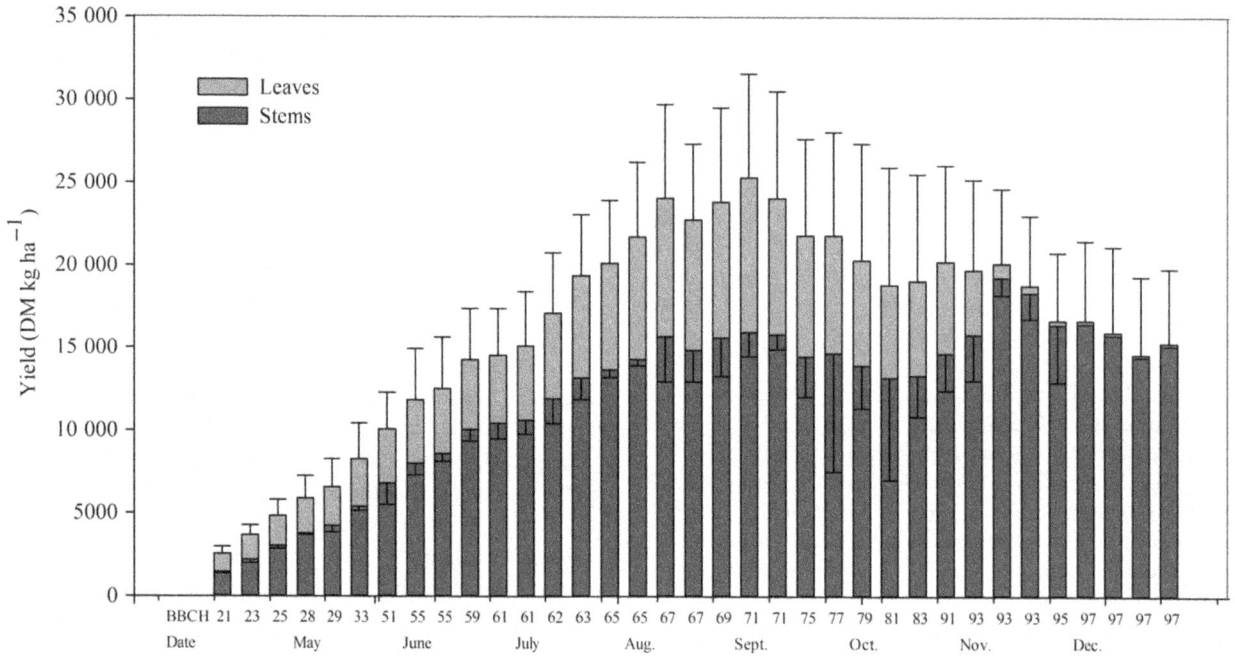

Fig. 1 Share of biomass development (dry mass basis) of stems and leaves over the entire growing period; values are presented in a 5th order moving average; bars indicate the standard error of $n = 5$ biological replicates.

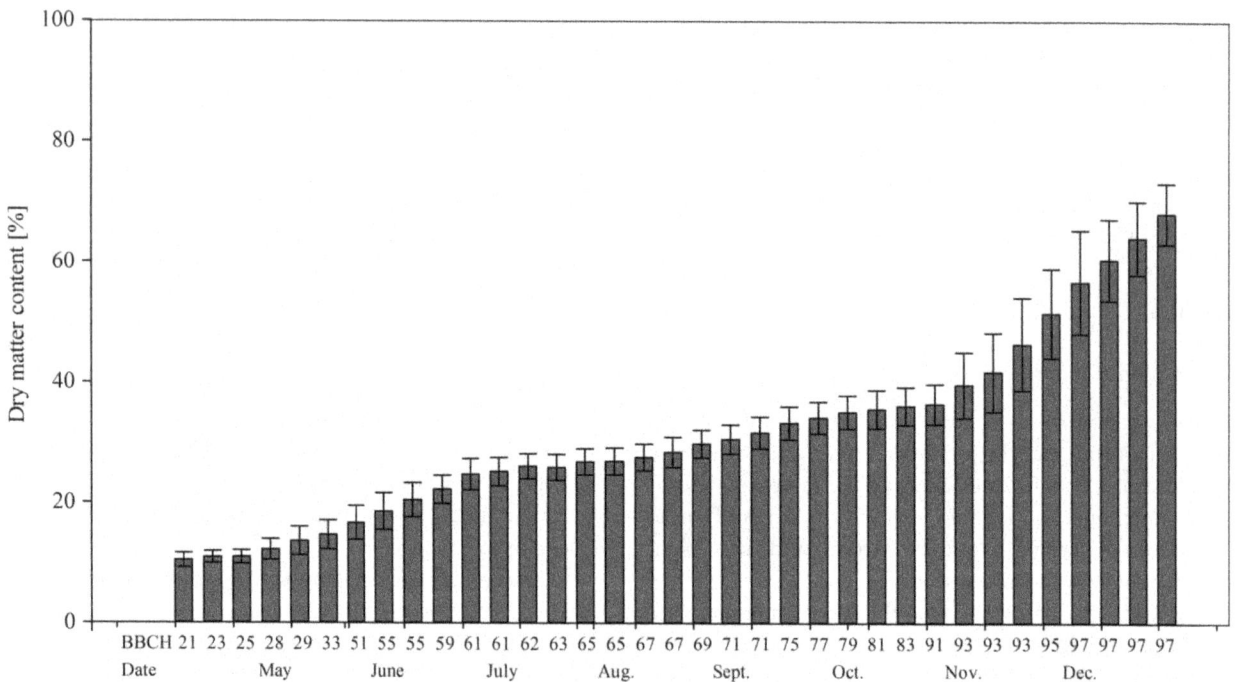

Fig. 2 Development of the total dry matter content in the Sida biomass throughout the entire growing period; values are presented in a 5th order moving average; bars indicate the standard error of $n = 5$ biological replicates.

Biogas production

The parameters of Sida after ensiling showed an increased organic dry matter (oDM) and dry matter content over time. Sample B2, the biomass resulting from a single harvest in October at BBCH-Sida development stage 98, showed the highest dry matter and organic dry matter content, accounting for 30.8% and 28.1%, respectively (Table 2).

Table 3 Calculated values for the energy yield of Sida biomass used as solid fuel

Parameter	One harvest: sample F1	Second harvest: sample F2	Unit
Corrected fresh biomass (FM_c)	26.43	7.11	t ha^{-1}
Corrected dry matter content (DM_c)	88	88	%
Higher heating value	19.21 ± 0.18	19.61 ± 0.27	MJ kg^{-1}
	5.34 ± 0.05	5.45 ± 0.08	kWh kg^{-1}
Net calorific value (q_{FM})	16.62 ± 0.16	16.96 ± 0.24	MJ kg^{-1}
	4.70 ± 0.04	4.79 ± 0.07	kWh kg^{-1}
Energy yield (EE_{ha})			
Total Higher heating value ha^{-1} (q_{TSha})	446 794 ± 4186	122 696 ± 1689	MJ ha^{-1}
Total Net calorific value ha^{-1} (q_{FMha})	439 266 ± 4229	120 586 ± 1706	MJ ha^{-1}

Table 4 Overview of the DIN EN ISO requirements of non-wood biomasses used for energy purposes as solid fuels. Elemental composition of Sida biomass harvested for solid fuel purposes in January at BBCH-Sida development stage 98. The relative standard deviations for all elements were for elemental contents of >1% ±3% and for elemental contents of <0.1% ±20%. The total Cl content in the Sida biomass was analyzed in accordance with DIN EN 15408 (DIN-EN-15408, 2011)

Characteristics	Unit	DIN EN ISO 17225-7: 2014-09	Sida-biomass samples F1	F2
Water content	m-%	12 ≤ 12	12	
Ash content	m-%	6.0 ≤ 6	2.99	2.69
Gross density	g cm^{-3}	0.9 ≥ 0.9		
Net Calorific	MJ kg^{-1}	14.5 ≥ 14.5	16.62	16.96
value	kWh kg^{-1}	4.0 ≥ 4.0	4.62	4.71
Nitrogen, N	m-%	1.5 ≤ 1.5	<0.26	<0.26
Sulfur, S	m-%	0.20 ≤ 0.20	0.028	0.024
Chlorine, Cl	m-%	0.10 ≤ 0.10	0.030	0.053
Arsen, As	mg kg^{-1}	≤1	<0.05	<0.05
Cadmium, Cd	mg kg^{-1}	≤0.5	0.44	0.38
Chrome, Cr	mg kg^{-1}	≤50	<50	<50
Copper, Cu	mg kg^{-1}	≤20	2.25	2.09
Lead, Pb	mg kg^{-1}	≤10	0.34	0.28
Mercury, Hg	mg kg^{-1}	≤0.1	<0.1	<0.1
Nickel, Ni	mg kg^{-1}	≤10	2.51	1.43
Zinc, Zn	mg kg^{-1}	≤100	<50	<50

Table 5 Composition of the ash in weight-% obtained after an oxidation of the Sida biomass at 550 °C

Compound	Sample F1	Sample F2
$CaCO_3$	49	55
$K_2Ca(CO_3)_2$	21	17
$Ca_{10}(PO_4)_6(OH)_2$	17	16
MgO	4	7
$CaMgSi_2O_6$	4	2
SiO_2	2	2
$Ca(OH)_2$	2	2
CaO	1	1

The data in Table 6 show much higher organic dry matter per hectare in sample B2 collected solely in October (BBCH-Sida 91), compared with samples B1.1 and B1.2, which represents two harvests from the same plants at different times and plant development stages (June and October, BBCH-Sida 55 and BBCH-Sida 77). The biggest specific biogas yield was found for sample B1.1. The given methane concentration was calculated in accordance with literature values based on other plant biomasses using a mean value for calculation of 53.5%.

Energy yield evaluation

Allowing a comparison of the total obtained energy values for the Sida biomasses for both biogas and solid fuel feedstock scenarios, we calculated the energy yields into fuel oil equivalents per ha, assuming 36 MJ per L fuel oil (Table 7). Scenario (i), solid fuel obtained from a single harvest of the dried biomass at termination of the plant growing period (harvest in January), yielded by far the highest energy, accounting for 440 GJ, equaling 12 202 L of fuel oil. The energy yield of scenario (iii) – the first biomass harvest used for biogas production, and a subsequent second harvest of the biomass used for solid fuel – accounted for approximately 212 GJ. Scenario (iv) yielded 135 GJ when Sida biomass was harvested two consecutive times to be used as biogas feedstock, whereas scenario (ii), the one late harvest for

The batch test of the biomass resulting from the three harvests of Sida showed different maximum biogas production rates and the time needed (kinetics) to reach the given maxima (Fig. 3). The first biomass harvested in June at BBCH-Sida development stage 55 (sample B1.1) reached an average specific biogas maximum of 420 L kg^{-1} oDM. The second harvested biomass (sample B1.2) and the biomass of the single harvest of B2, both collected in October at BBCH-Sida stage 91, yielded a similar specific maximum of 260 L biogas kg^{-1} oDM. Changes in the kinetic of the biogas production of the three tested samples can be observed from the fourth day (Fig. 3).

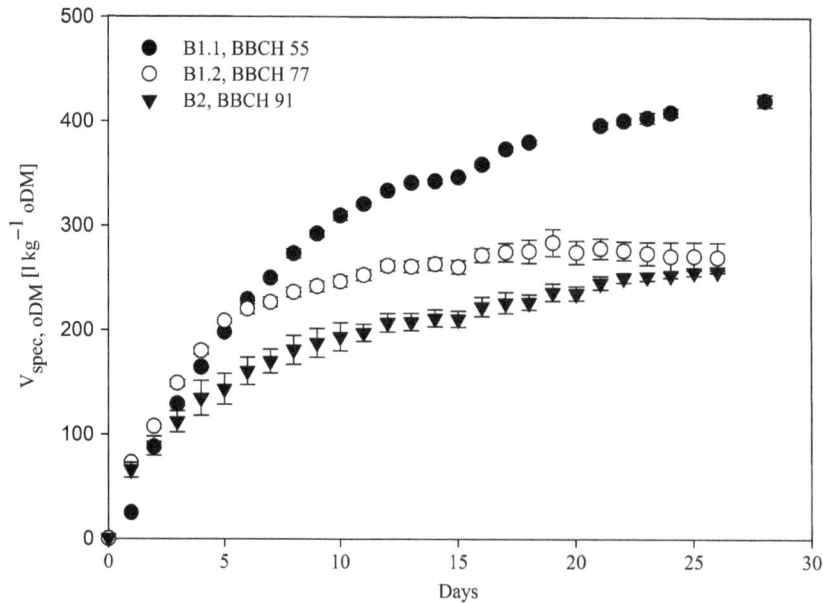

Fig. 3 Specific norm biogas yields from batch tests using biomass silage originating from different harvest dates and Sida plant development stages. B1.1 is biomass from a first harvest at BBCH-Sida development stage 55 (June); B1.2 is biomass from a second harvest of B1.1 plants at BBCH-Sida development stage 77 (October); B2 is biomass from a single harvest at BBCH-Sida development stage 91 (October).

Table 6 Characteristics of the three Sida biomasses harvested at different growth stages and times tested as a feedstock for biogas production used as a silage; EE is energy yield per hectare; oDM is organic dry matter

Parameter	B1.1	B1.2	B2
Organic dry matter (t ha^{-1})	12.5 ± 0.2	9.4 ± 0.2	19.0 ± 0.4
Ensilage loss (t ha^{-1})	1.1 ± 0.4	0.9 ± 0.3	1.7 ± 0.6
Methane content (%)	53.5 ± 1.5	53.5 ± 1.5	53.5 ± 1.5
Biogas yield (m^3_N t^{-1} oDM)	419.5 ± 26.6	269.3 ± 14.3	256.3 ± 1.3
CH$_4$ (m^3_N t^{-1} oDM)	204.2 ± 16.3	131.1 ± 9.5	124.8 ± 6.2
CH$_4$ (m^3_N ha^{-1})	2542.7 ± 202.7	1235.0 ± 89.4	2367.1 ± 117.5
EE (MJ ha^{-1})	91537.2 ± 7292.2	44460.0 ± 3218.4	85215.6 ± 4230.0

Table 7 Total energy yield from the four studied Sida biomass energy use scenarios; $n = 5$ biological replicates, calculated per ha

Scenario	(i)	(ii)	(iii)	(iv)
	1. Harvest solid fuel F1	1. Harvest biogas B2	1. Harvest biogas B1.1	1. Harvest biogas B1.1
EE$_{ha}$ (MJ)	–	–	91 022 ± 7292	91 022 ± 7292
	–	–	2. Harvest solid fuel F2	2. Harvest biogas B1.2
EE$_{ha}$ (MJ)	439 288 ± 4229	85 215 ± 4230	120 760 ± 1706	44 460 ± 3218
Total (MJ ha^{-1})	439 288 ± 4229	85 215 ± 4230	211 782 ± 8998	135 482 ± 10510
Total (l$_{oil}$ ha^{-1})	12202.4 ± 117.5	2367.1 ± 117.5	5882.8 ± 249.9	3763.4 ± 291.9

biogas only, yielded the comparably lowest energy yield of 85 GJ.

Lignocellulose analysis

To characterize Sida lignocellulose, we monitored the crystalline cellulose content, TFA-soluble matrix polysaccharides (MPS), and AcBr soluble lignin (ABSL) during the entire growing period (Foster et al., 2010b). Two different tissues of the plants, that is, main stem and leaves including side branches, were analyzed. ABSL in stem tissue rose from 8% to an almost constant level of 17% from week 20 at BBCH-Sida development stage 67 onwards. The crystalline cellulose content rose from 32% to 50% of d-AIR from the second to the eighth week, that is, BBCH-Sida development stage 17–33, respectively, and remained constant. The amount of TFA-soluble MPS varied between 12% and 20% of d-AIR within the analyzed time period. In the leaf fraction, ABSL was constant at approximately 5% until week 16, that is, BBCH-Sida 63, and was elevated subsequently up to 15% in week 32, equaling BBCH-Sida 93. Crystalline cellulose varied between 17% and 25% within the first 16 weeks and rose afterward up to 42% in week 32, describing the BBCH-Sida development stages 20–63, and 93, respectively. TFA-soluble MPS was low in weeks 2 and 4 (8% to 10%, BBCH-Sida 17 and 23) and remained constant between 17% and 20% until week 32 (BBCH-Sida 93).

Discussion

Biomass production

In our study, we evaluated the biomass growth and development of Sida throughout an entire growing period at real agricultural field conditions. Even though Borkowska & Molas (2013) reported that Sida biomass yield significantly increases in the first 4 years after plantation establishment, our values accounted for approximately 20 t ha^{-1} biomass dry weight when harvested in November at a BBCH-Sida development stage 93. Already 1 year after the plants' establishment, this value even exceeded the reported values from earlier studies by a factor of 1.5–1.6 when the plants were harvested in November as given in the referenced studies (Borkowska & Molas, 2012, Borkowska & Molas, 2013). This could be explained by the favorable field environment and conditions, but it clearly shows the high possible biomass yields of Sida even in the second year after plant establishment when grown at beneficial conditions. However, when used as a solid fuel, a high dry matter content is preferred to increase the energy yield of the biomass and to avoid costly drying processes prior to storage (Borkowska & Molas, 2013). Therefore, we recommend a biomass harvest of the dried Sida stems only in early spring, prior to the regrowth of the plants, because the dried stem material is free of leaves (Borkowska & Molas, 2013) and possesses the highest content of lignin and cellulose, as described below.

The high standard error in the biomass yield throughout the experimental growing period is attributed to the high variation in the phenotype of the Sida plants because this plant is a wild type and varies strongly in its individual biomass yield. An attempt to breed the plants may result in a more homogeneous plant growth and overall appearance with similar biomass yields.

Furthermore, considering Sida as a promising lignocellulosic biomass plant for energy applications and molecular breeding activities may improve future applications as a second generation bioenergy crop (Allwright & Taylor, 2015).

Published data on Sida have not sufficiently considered the time of harvest and the respective plant development stage, which makes detailed comparison between the data rather vague. The established BBCH-Sida development code, given in detail in the Supporting Information (Table S1), allows an estimation of the biomass composition irrespective of the location and environmental influences. This will allow future comparisons between the harvested Sida biomasses and their target-oriented application as a feedstock for energy or other industrial applications, for example, fibers for the paper industry or raw material for chemical applications (Bogusz et al., 2015; Grande et al., 2015).

Cell wall composition

Sida lignocellulose was monitored over the whole growth period by determining a set of standard parameters, which enabled a clear discrimination of primary growing tissue (primary cell walls) and adult tissue (secondary cell walls) characterized by higher ABSL and crystalline cellulose values. Also, the change in the composition of the TFA-soluble MPS in, for example, the leaf fraction, indicates the formation of more adult tissue between the weeks 16 and 32, that is, BBCH-Sida development stage 63 and 93, respectively (Table S1, Supporting information). The overall development of secondary cell walls and therefore the increasing amounts of cellulose and lignin are in line with previous observations (Borkowska et al., 2009; Borkowska & Molas, 2012). However, a direct comparison of distinct lignin values is often difficult due to differences in the determination method and unknown cultivation conditions. Within this study, ABSL determination was used to enable a high throughput measurement. Nevertheless, the observed lignin content in fully developed plants is comparable to previous studies using other methods like Klason determination (Michalska et al., 2012). Determined cellulose and MPS levels are in line with previous observations (Michalska et al., 2012) although measured cellulose in our study reflects only the crystalline part and not holocellulose, suggesting an overall high crystallinity in Sida cellulose fibers. The correlation of the established BBCH-Sida code with the biochemical properties of the plant material enables a targeted harvest strategy for a tailored utilization of Sida biomass.

Energy yield

Solid fuel. The analysis of the Sida biomass revealed its high potential as a solid fuel meeting all guidelines in accordance with DIN EN ISO 17225-7:2014-09 (Table 4) (DIN-EN-ISO-17225-7, 2014). As shown in our analysis, the energy content of the Sida biomass accounted for approximately 19 MJ kg^{-1} dry mass. This measured value is in accordance with the value used in a previous study for the calculation of combustion heat (Borkowska *et al.*, 2009) and is further in line compared with values obtained from *Miscanthus* as another important perennial energy crop (Baxter *et al.*, 2014). The overall energy yield for Sida as a solid fuel accounts for a calorific value of 446 GJ ha^{-1} (net calorific value: 440 GJ ha^{-1}) in its second year after the establishment of the field experiment. As previously reported, the Sida yield increased even during the fourth year after establishment (Borkowska & Molas, 2013). The energy value obtained in our study is even 22% higher than the values of an earlier study from the fourth to the sixth year after plant establishment (Borkowska & Molas, 2012), and even 168% higher when compared with a 4-year average energy yield of Sida cultivated in a light soil (Borkowska *et al.*, 2009). Because no fertilizers were applied in our study, these higher values might be attributed to the higher silt and C_{org} content as well as a higher soil pH value among other environmental factors that were beneficial for the overall plant performance at our field trials.

Even though samples of F1 showed slightly higher values of heavy metals such as Cd, Cu, Pb, and Ni compared with F2 samples, the determined values were still meeting the DIN guidelines. Even though Sida plants express the ability to accumulate heavy metals, the detected values are low and must be attributed to the longer growing period of F1 biomass (Borkowska & Wardzinska, 2003). However, our data were significantly lower compared with data from Sida grown on municipal sewage sludge compost and high-calcium brown coal ash (Krzywy-Gawronska, 2012), and were in accordance with the requirements of DIN EN ISO 17225-7:2014-09 to be classified as a biogenic solid fuel (DIN-EN-ISO-17225-7, 2014).

The presence of Cl and S in solid biomass used for combustion is crucial due to their high corrosion potential in the furnaces. The values of Cl and S in the analyzed Sida biomass were below the maximum values as specified in the norm DIN EN ISO 17225-7:2014-09 for solid biomass fuels, making Sida biomass a promising candidate as a sustainable energy carrier for combustion (DIN-EN-ISO-17225-7, 2014).

Ash analysis

The aim of the investigation was the characterization of the ash composition and the determination of the ash melting temperature of the promising solid fuel Sida. The amount of ash after biomass combustion accounted for 2.7% to 3.0%, which is lower than the 3.6% value reported in another study on Sida (Michalska *et al.*, 2012), which might be due to a different development stage of the used plant biomass. However, the obtained ash values are in the same range as described for numerous wood and woody biomasses (Vassilev & Baxter, 2010) and for *Miscanthus*, which is also considered to be a valuable energy crop used for combustion (Baxter *et al.*, 2012). The analysis of the Sida ashes revealed that both samples did not melt at 1500 °C, which is the maximum temperature achievable in the hot-stage microscope that we used. The explanation for the high melting temperature for Sida ash is in principle based on the amount of compounds with a high melting temperature (Misra *et al.*, 1993; Wang & Dibdiakova, 2014). This is due to the high content of $CaCO_3$, which will decompose to CaO with a melting temperature of 2580 °C. The decomposition of the carbonate starts between 650 and 900 °C. Additionally, $K_2Ca(CO_3)_2$ will decompose, increasing the amount of the high melting compounds. Furthermore, low-melting potassium silicate compounds were not formed due to the insignificant amount of silica in the ash. As shown, the chemical composition of Sida is favorable for combustion in comparison to herbaceous biomass because the mineral content with a high melting point is much higher, leading to a comparably higher melting point of the ash. Therefore, problems related to ash melting, for example, slagging or bed agglomeration, should be less significant for Sida, as shown in this study. In general, the ash behavior of Sida is related more to the woody biomass used for combustion, assuming that a substitution of wood burning systems with Sida biomass seems to be worthwhile.

Fig. 4 Lignocellulose analysis of harvested Sida plants. AcBr soluble lignin (ABSL), TFA-soluble matrix polysaccharides (MPS), and crystalline cellulose content are depicted. The values represent the mean of 5 biological replicates. After week 32, Sida plants were defoliated; therefore, no leaves were harvested from that point. Stem tissue lignification is completed in week 20 (approximately 17% of d-AIR), whereas lignin level rose until defoliation (approximately 14% of d-AIR) in the leaf fraction. Bars indicate the standard deviation of 5 biological replicates.

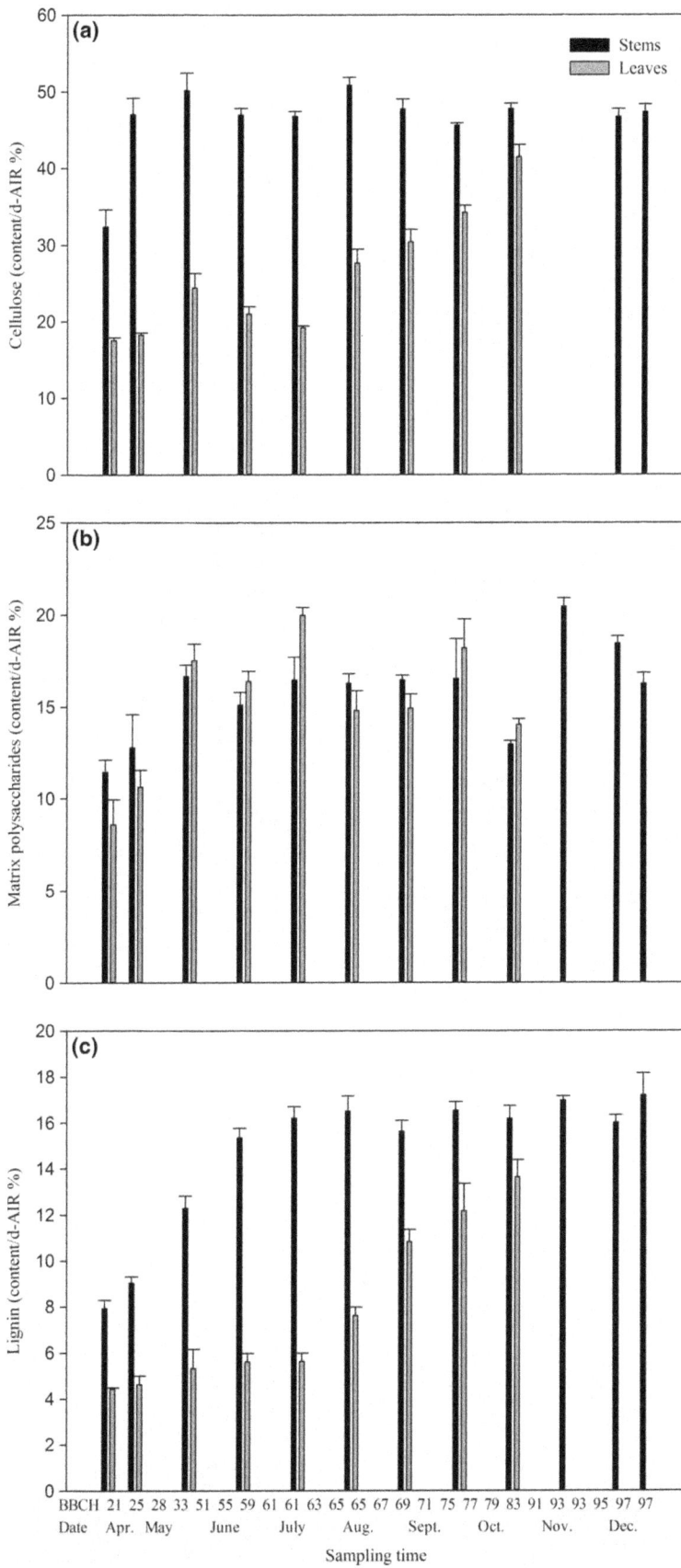

Biogas production

To determine whether the energy yield of Sida biomass could be improved by numerous harvests for both solid fuels and biogas feedstock, we investigated the Sida biomasses from two different development stages of the plants in batch tests. The first cut of Sida in June at a development stage of BBCH-Sida 55 (sample B1.1) displayed a dry matter content of 20% consisting of approximately 90% organic dry matter, which is comparable with grass silage (Michalska *et al.*, 2012). The second cut of the biomass at development stage 77 (sample B1.2) showed a higher organic content due to an increased lignification of the biomass. The sample B2 obtained from a single harvest for biogas production at development stage 91 showed the highest organic content. In the initial stages of the Sida plant's development, a high amount of short-molecule biomass like cellulose is developed, leading to more complex organic compounds like hemicellulose and lignin to achieve stability (Hendriks & Zeeman, 2009). Therefore, the different organic plant tissue composition can be associated with the higher biogas yield of B1.1 sampled at a much earlier plant development stage compared to the biogas yield from samples B1.2 and B2 harvested in October at a development stage of 77 and 91, respectively (Fig. 4). Lignin as a complex polymer consists of three different phenolic monomers that are difficult to hydrolyze for microorganisms and appear to be an inhibitor for the biogas production process, limiting the digestibility (Hendriks & Zeeman, 2009).

As indicated, sample B1.1 contains the least amount of lignin, cellulose, and hemicellulose and shows the highest biogas production rate after 25 days. In comparison, sample B2 showed the highest lignin, cellulose, and hemicellulose content, which resulted in an inhibited biogas production rate. As shown by Brown *et al.* (Brown *et al.*, 2012) using different lignocellulosic biomasses at liquid anaerobic digestion, methane production has an inverse linear relationship with the lignin content (Pokój *et al.*, 2015). As further shown by Pokój *et al.*, Sida silage tested as a biogas feedstock produced from biomass harvested at the same development stage as used in our study (BBCH-Sida development stage 55, i.e., flowering phase) showed a lignin removal efficiency of approximately 45% and an overall organics removal rate of approximately 65% (Pokój *et al.*, 2015). However, as concluded in the study by Pokój *et al.*, lignin content is an inadequate criterion for estimating the methane production.

As further demonstrated in a study by Michalska *et al.*, (2015), chemical and enzymatic pretreatment of Sida biomass resulted in an increased biogas production of 316 L kg^{-1} total solids, equaling a methane content of

200 L. Interestingly, our results revealed a biogas production from Sida silage of approximately 420 L kg^{-1} organic dry matter for sample B1.1 and approximately 269 and 256 L kg^{-1} organic dry matter for samples B1.2 and B2, respectively. These values correspond to a methane content of 204, 131, and 124 L kg^{-1} organic dry matter, respectively, within the same time of incubation. These values also correspond to a previous study using fresh Sida biomass for biogas production at mesophilic conditions (Dêbowski *et al.*, 2012). Our results give evidence that a Sida biomass harvest at plant BBCH-Sida development stage 55 following a thorough homogenization and ensilaging of the biomass makes Sida biomass a promising candidate for biogas feedstock.

Combining our results of the overall measured energy yields from Sida biomass used as a solid fuel and biogas feedstock from the various developments stages, we saw that the four energy-use scenarios clearly indicated the highest energy recovery for scenario (i) (solid fuel) of 439 288 MJ ha^{-1}. The energy recovery of the four scenarios from most to least is as follows: scenario (i) $\gg$ (iii) (first biomass used for biogas, subsequent harvest used as solid fuel: 211 782 MJ ha^{-1}) $\gg$ (iv) (two times biomass harvest at development stage 55 and 77, respectively, used for biogas: 135 482 MJ ha^{-1}) $>$ (ii) (one harvest for biogas only at plant development stage 91: 85 215 MJ ha^{-1}). Although scenario (iii) resulted in approximately half the energy yield of scenario (i), the flexible application of the Sida biomass as biogas feedstock when harvested at the BBCH-Sida stage 55 is an added value. How much the energy yields could be increased with regard to earlier or later harvests at different plant development stages needs further investigation. At this point, it remains unknown to what extent intermediate harvests of Sida biomass may result in sustainable production and continuous yield in subsequent years due to the interruption of the natural plant development. However, harvest of the dried Sida biomass at BBCH-Sida 98–99 allows both a maximum energy yield and reduced impact on the living plants, which results in a sustainable supply of Sida biomass over years (Borkowska *et al.*, 2009; Borkowska & Molas, 2012). The calculated energy yields for the Sida biomass do not consider necessary energy investments for plant establishment, maintenance, and harvest of the biomass; therefore, given energy values represent the possible energy yield depending on the utilization scenario obtained from our measurements.

Acknowledgements

We gratefully acknowledge the help and assistance by Lucy Harrison and the funding of the OrCaCel project by the Bioeconomy Science Center (BioSC). The scientific activities of the BioSC were financially supported by the Ministry of Innovation, Science, and Research within the framework of the NRW Strategieprojekt BioSC (No. 313/323-400-002 13).

References

Allwright MR, Taylor G (2015) Molecular breeding for improved second generation bioenergy crops. *Trends in Plant Science*, **21**, 43–54.

Barbosa DBP, Nabel M, Jablonowski ND (2014) Biogas-digestate as nutrient source for biomass production of *Sida Hermaphrodita, Zea Mays* L. and *Medicago sativa* L. *Energy Procedia*, **59**, 120–126.

Baxter XC, Darvell LI, Jones JM, Barraclough T, Yates NE, Shield I (2012) Study of *Miscanthus x giganteus* ash composition – variation with agronomy and assessment method. *Fuel*, **95**, 50–62.

Baxter XC, Darvell LI, Jones JM, Barraclough T, Yates NE, Shield I (2014) Miscanthus combustion properties and variations with Miscanthus agronomy. *Fuel*, **117**, 851–869.

Blanco-Canqui H (2010) Energy crops and their implications on soil and environment. *Agronomy Journal*, **102**, 403–419.

Bogusz A, Oleszczuk P, Dobrowolski R (2015) Application of laboratory prepared and commercially available biochars to adsorption of cadmium, copper and zinc ions from water. *Bioresource Technology*, **196**, 540–549.

Borkowska H, Molas R (2012) Two extremely different crops, Salix and Sida, as sources of renewable bioenergy. *Biomass and Bioenergy*, **36**, 234–240.

Borkowska H, Molas R (2013) Yield comparison of four lignocellulosic perennial energy crop species. *Biomass and Bioenergy*, **51**, 145–153.

Borkowska H, Wardzinska K (2003) Some effects of Sida hermaphrodita R. Cultivation on sewage sludge. *Polish Journal of Environmental Studies*, **12**, 119–122.

Borkowska H, Molas R, Kupczyk A (2009) Virginia fanpetals (sida hermaphrodita rusby) cultivated on light soil; height of yield and biomass productivity. *Polish Journal of Environmental Studies*, **18**, 563–568.

Brown D, Shi J, Li Y (2012) Comparison of solid-state to liquid anaerobic digestion of lignocellulosic feedstocks for biogas production. *Bioresource Technology*, **124**, 379–386.

Bundesministerium für Wirtschaft und Energie (2015) *Bruttostromerzeugung in Deutschland 2014*.

Carroll A, Somerville C (2009) Cellulosic biofuels. *Annual Review of Plant Biology*, **60**, 165–182.

Dębowski M, Dudek M, Zieliński M (2012) Effectiveness of methane fermentation of virginia fanpetals (*Sida hermaphrodita* Rusby) under mesophilic conditions. *Ecological Chemistry and Engineering*, **19**, 1445–1453.

Deutscher Bauernverband (2015) *Situationsbericht 2014/2015 des deutschen Bauernverbandes*.

DIN-38414-8 (1985) German standard methods for the examination of water, waste water and sludge; sludge and sediments (group S); determination of the amenability to anaerobic digestion (S 8).

DIN-51730 (1998) Prüfung fester Brennstoffe Bestimmung des Asche-Schmelzverhaltens. *DIN 51730*.

DIN-51900-3 (2003) Prüfung fester und flüssiger Brennstoffe Bestimmung des Brennwertes mit dem Bomben- Kalorimeter und Berechnung des Heizwertes Teil 3: Verfahren mit adiabatischem Mantel. *DIN 51900-3*.

DIN-EN-12880 (2001) Charakterisierung von Schlämmen - Bestimmung des Trockenrückstandes und des Wassergehalts.

DIN-EN-15408 (2011) Solid recovered fuels - Methods for the determination of sulphur (S), chlorine (Cl), fluorine (F) and bromine (Br) content.

DIN-EN-ISO-17225-7 (2014) Solid biofuels – Fuel specifications and classes – Part 7: Graded non-woody briquettes.

FNR (2015) Basisdaten Bioenergie Deutschland 2015.

Foster CE, Martin TM, Pauly M (2010a) Comprehensive compositional analysis of plant cell walls (Lignocellulosic biomass) part II: carbohydrates. *Journal of Visualized Experiments*, 10–13.

Foster CE, Martin TM, Pauly M (2010b) Comprehensive compositional analysis of plant cell walls (lignocellulosic biomass) part I: lignin. *Journal of Visualized Experiments : JoVE*, 2–5.

Graham-Rowe D (2011) Agriculture: beyond food versus fuel. *Nature*, **474**, S6–S8.

Grande PM, Viell J, Theyssen N, Marquardt W, Domínguez de María P, Leitner W (2015) Fractionation of lignocellulosic biomass using the OrganoCat process. *Green Chemistry*, **17**, 3533–3539.

Hack H, Bleiholder H, Buhr L, Meier U, Schnock-Fricke U, Weber E, Witzenberger A (1992) Einheitliche Codierung der phänologischen Entwicklungsstadien mono- und dikotyler Pflanzen-Erweiterte BBCH-Skala, Allgemein. *Nachrichtenblatt des deutschen Pflanzenschutzdienstes*, **44**, 265–270.

Hartmann A, Haller J (2014) Silphie und Co als Biogassubstrat – Erste Ergebnisse aus dem Dauerkulturanbau. *TFZ-Merkblatt: 14PHm004*, **49**.

Hendriks ATWM, Zeeman G (2009) Pretreatments to enhance the digestibility of lignocellulosic biomass. *Bioresource Technology*, **100**, 10–18.

Krzywy-Gawronska E (2012) The effect of industrial wastes and municipal sewage sludge compost on the quality of virginia fanpetals (SIDA HERMAPHRODITA RUSBY) biomass Part 1. Macroelements content and their uptake dynamics. *Polish Journal of Chemical Technology*, **14**, 2.

Michalska K, Miazek K, Krzystek L, Ledakowicz S (2012) Influence of pretreatment with Fenton's reagent on biogas production and methane yield from lignocellulosic biomass. *Bioresource Technology*, **119**, 72–78.

Michalska K, Bizukojć M, Ledakowicz S (2015) Pretreatment of energy crops with sodium hydroxide and cellulolytic enzymes to increase biogas production. *Biomass and Bioenergy*, **80**, 213–221.

Misra MK, Ragland KW, Baker AJ (1993) Wood ash composition as a function of furnace temperature. *Biomass and Bioenergy*, **4**, 103–116.

Nabel M, Bueno D, Barbosa P, Horsch D, Jablonowski ND (2014) Energy crop (Sida hermaphrodita) fertilization using digestate under marginal soil conditions: a dose-response experiment. *Energy Procedia*, **16**, 14000.

Oleszek M, Matyka M, Lalak J, Tys J, Paprota E (2013) Characterization of Sida hermaphrodita as a feedstock for anaerobic digestion process. *Journal of Food, Agriculture & Environment*, **11**, 1839–1841.

Pokój T, Bułkowska K, Gusiatin ZM, Klimiuk E, Jankowski KJ (2015) Semi-continuous anaerobic digestion of different silage crops: VFAs formation, methane yield from fiber and non-fiber components and digestate composition. *Bioresource Technology*, **190**, 201–210.

Schattenhauer A, Weiland P (2006) Handreichung Biogasgewinnung und – nutzung. *Fachagentur Nachwachsende Rohstoffe e.V.*, 29–31.

Schröder P, Herzig R, Bojinov B et al. (2008) Bioenergy to save the world. Producing novel energy plants for growth on abandoned land. *Environmental Science and Pollution Research International*, **15**, 196–204.

Scott TA, Melvin EH (1953) Determination of dextran with anthrone. *Analytical Chemistry*, **25**, 1656–1661.

Slepetys J, Kadziuliene Z, Sarunaite L, Tilvikiene V, Kryzeviciene A (2012) Biomass potential of plants grown for bioenergy production. *International Scientific Conference: Renewable Energy and Energy Efficiency*, 66–72.

Spooner DM, Cusick AW, Hall GF, Baskin JM (1985) Observation on the distribution and ecology of *Sida hermaphrodita* (L.) Rusby (Malvaceae). *Sida*, **11**, 215–225.

Stolarski MJ, Szczukowski S, Tworkowski J, Krzyzaniak M, Gulczyński P, Mleczek M (2013) Comparison of quality and production cost of briquettes made from agricultural and forest origin biomass. *Renewable Energy*, **57**, 20–26.

Updegraff DM (1969) Semimicro determination of cellulose in biological materials. *Analytical Biochemistry*, **32**, 420–424.

Vassilev SV, Baxter D, Andersen LK, Vassileva CG (2010). An overview of the chemical composition of biomass. *Fuel*, **89**, 913–933.

Voigt TB, Lee OK, Kling GJ (2012) Perennial herbaceous crops with potential for biofuel production in the temperate regions of the USA. *CAB Reviews*, **7**, 45–57.

Voiniciuc C, Schmidt MH-W, Berger A et al. (2015) MUCI10 produces Galactoglucomannan that maintains pectin and cellulose architecture in Arabidopsis seed mucilage. *Plant Physiology*, **169**, 403–420.

Wang L, Dibdiakova J (2014) Characterization of ashes from different wood parts of Norway spruce tree. *Chemical Engineering Transactions*, **37**, 37–42.

3

Comparing methods for measuring the digestibility of miscanthus in bioethanol or biogas processing

SUSANNE FRYDENDAL-NIELSEN[1], UFFE JØRGENSEN[1], MAIBRITT HJORTH[2], CLAUS FELBY[3] and RENÉ GISLUM[4]

[1]Department of Agroecology, Aarhus University, Blichers Allé 20, PO Box 50, 8830 Tjele, Denmark, [2]Department of Engineering, Aarhus University, Hangøvej 2, 8200, Aarhus N, Denmark, [3]Department of Geosciences and Natural Resource Management, University of Copenhagen, Rolighedsvej 23, 1958 Frederiksberg C, Denmark, [4]Department of Agroecology, Aarhus University, Forsøgsvej 1, 4200 Slagelse, Denmark

Abstract

Lignocellulosic biomass is a candidate for future renewable energy resources. Choice of optimum biomass types and biological conversion techniques requires well-founded assessment of the digestibility determining the conversion efficiency. The aim of this study was to investigate and evaluate the digestibility of miscanthus samples that were tested using three methods: 3,5-dinitrosalicylic acid assay (DNS), anaerobic batch digestion test, and high-throughput pretreatment and hydrolysis method, including a grinding and hydrothermal pretreatment prior to the analysis (HTPH). The miscanthus samples were expected to have different digestibilities due to maturity stage, dry matter content and the implementation of extrusion as a mechanical pretreatment. The results of the DNS and the biogas batch test methods were highly correlated (R^2 between 0.75 and 0.92), but not with the results of the HTPH method. The DNS and biogas batch test showed that digestibility differed between samples, probably due to the degree of lignification and content of soluble sugars. For the HTPH method, the digestibility for biorefining was the same irrespective of the variation in the other analyses. The HTPH method had higher biomass use efficiency, closely followed by the biogas batch test running for 91 days on the mechanically pretreated biomass. The HTPH method provided information on the overall quantity of carbohydrates that can be made available from a given biomass. Additionally, DNS and biogas batch test visualize the variation in digestibility between biomass types caused by lignification and particle. The study concludes that the choice of evaluation method for miscanthus will depend on the bioenergy conversion method used and that important information on the interaction between physio-chemical pretreatment and biological accessibility of the biomass can be obtained by comparing the methods. This information will enable sound decisions on the future choice of bioenergy conversion technologies.

Keywords: 3,5-dinitrosalicylic acid assay, energy crop, enzymatic saccharification, harvest time, hydrolysis, methane yield

Introduction

The transition from using fossil to renewable resources has advanced the research on the processing of renewable biomass resources into fuels and chemicals. However, the need for a parallel supply of food and fuels has shown that the biomass supply should preferably be based on nonedible lignocellulosic biomass. A number of lignocellulosic residues and crops are available or have high yields and as such are attractive feedstocks (Bentsen & Felby, 2012), but compared to starch-based feedstocks lignocellulose is highly recalcitrant and thus more difficult to process. Utilizing the carbon from such recalcitrant lignocellulosic structures for energy carriers can provide renewable and storable carbon-based fuels for the transportation sector which are difficult to produce from other renewable sources such as the sun, wind and hydropower. The main component of biorefining is the sustainable processing of lignocellulose-derived sugars in a cascade of processes transforming them into a spectrum of biobased products and fuels (Jungmeier *et al.*, 2013; Parajuli *et al.*, 2015). In this context, digestibility, that is the ease by which the processed biomass is converted into fermentable sugars, has been the subject of a number of studies related to the biorefining process for bioethanol or biogas production (Hendriks & Zeeman, 2009; Weiland, 2010; Lindedam *et al.*, 2012). Both bioethanol and biogas are important bulk products in biorefining processes and are consequently the focus of the present study. Bioethanol is a liquid fuel which can be directly blended

Correspondence: Susanne Frydendal-Nielsen
e-mail: sufn@agro.au.dk

into petrol, whereas biogas (methane) can replace natural gas used in heat and power plants, for transportation or as a feedstock in the petrochemical industry. As a fuel, bioethanol has an advantage over biogas in that it can easily be incorporated in the existing transportation systems and partly or fully substitute fossil fuels. The technology of biogas production is established and can be implemented in simple set-ups for single households or small communities in developing countries (Katuwal & Bohara, 2009) or as part of bigger complex plants feeding a larger gas grid (Berglund & Borjesson, 2006). Biorefineries that convert lignocellulosic biomass to bioethanol are technically more complex and have only recently been established on a commercial level (Somerville et al., 2010). Moreover, the biomass retention time in the bioethanol process is usually shorter than in the biogas process – an important factor determining the capacity of the processing plant. The more readily the lignocellulosic biomass can be degraded by enzymes or microorganisms, the higher the rate of conversion of lignocellulosic carbon into energy carriers such as biogas and bioethanol (Hendriks & Zeeman, 2009). Higher digestibility and conversion rates will decrease the retention time as well as the capacity of the process plants by producing more energy per production unit.

Digestion of recalcitrant biomass is performed by complex interactions of enzymes and microorganisms, and many factors affect the digestibility. The factors can be divided into chemical factors, such as content and structure of the polymers cellulose, hemicellulose and lignin, and physical factors, such as particle size and surface area. Furthermore, the organization of the polymers in the cell wall matrix further affects the availability and digestibility of the cellulose and hemicellulose as particularly the lignin presents a physical barrier to microorganisms and enzymes (Fu et al., 2011).

Plant species vary in their suitability for biological conversion, that is due to variations in digestibility (Karp & Shield, 2008; Somerville et al., 2010), just as the digestibility varies between the components of a given plant (Hayes, 2013; Zhang et al., 2014b). Miscanthus is the focus of this study because this perennial, rhizomatous, C_4 crop is one of the highest-yielding energy crops in Europe (Hastings et al., 2009) which combines with a low environmental impact (Hamelin et al., 2012). The composition of leaves and stems from miscanthus differs (Hodgson et al., 2010; Hayes, 2013; Wahid et al., 2015b), and the lower stem part of miscanthus has a higher lignin content than more juvenile plant parts (Huyen et al., 2010; Hayes, 2013). The composition of the plant will change with the maturing of the plant from autumn to late winter by, for example, increasing the lignin content (Jørgensen, 1997; Hayes, 2013) and decreasing the ash content (Lewandowski et al., 2003).

The lignified fibres and crystalline cellulose are difficult to degrade biologically, but the crystallinity does not per se affect the digestibility (Caulfield & Moore, 1974). Decreasing the crystalline particle size and increasing the surface area of the cellulosic biomass are known to improve the enzymatic accessibility of the biomass and the hydrolysis of the lignocellulosic content (Caulfield & Moore, 1974; Hendriks & Zeeman, 2009; Surendra & Khanal, 2015). Lignocellulosic biomass is recalcitrant to degradation, which means that lignocellulosic biomass, due to its low digestibility, needs a chemical or physical pretreatment to saccharification in the biorefining process (Hendriks & Zeeman, 2009).

As the conversion of biomass is controlled by a number of factors and their interactions, an integrated approach is needed when evaluating digestibility. However, normally just one methodology is used to evaluate biomass convertibility: for example, measuring the convertibility of biomass to ethanol using different thermo-chemical pretreatments (Jørgensen et al., 2007), measuring the amount of sugars enzymatically hydrolysed from the biomass (Wahid et al., 2015a) or using biogas batch tests (Wahid et al., 2015b). In the current study, three methods were used in parallel as the combined information was hypothesized to provide additional information. The three methods included a biogas batch test measuring the methane yield, an enzymatic saccharification followed by quantification of total sugars using a spectrophotometer, and a hydrothermal pretreatment followed by enzymatic saccharification and quantification of glucan, xylan and arabinan using HPLC (Zhang et al., 2014a). Table 1 shows the details and a comparison of the methods.

The overall objective of this study was to evaluate the digestibility of biomass from miscanthus using three different methods and measuring the conversion into either biogas (methane) or fermentable sugars. The biomass samples were chosen to cover the biological variation induced by different harvest times and the altered particle size caused by mechanical extrusion.

Materials and methods

Biomass

Miscanthus (Miscanthus × giganteus) was established in field experiments at Aarhus University in Foulum, Denmark (56.49N, 9.55E), in 2001, and the field has been maintained and harvested annually during winter with the last harvest in 2012. Biomass for this experiment was harvested early, intermediate and late in the harvest season 2013–2014 using a forage harvester. Samples were stored at −18 °C before further analysis.

Table 1 Comparison of experimental conditions for the three methods

	Biogas batch test	DNS	HTPH
Condition of input biomass	As received	As received	Dried (40 °C) and ground
Experimental pretreatment	None	None	Hydrothermal (190 °C, 10 min)
Enzymes	Inoculum	Cellulase	Cellulase
		Mannase	β-Glucosidases
			Hemicellulase
Temperature	35 °C	50 °C	50 °C
Time	91 days	72 h	72 h
Measured variables	Methane yield	Reducing sugars	Glucan
	Methane concentration		Xylan
	in the gas		Arabinan

After thawing at room temperature, the samples were shredded twice (Untha RS 30 4-2, Kuchl, Austria) through a 4-cm sieve to eliminate long stem and leaf parts. The majority of particles were then shorter than 4 cm with an average of approximately 1 cm.

Harvest time changes both the composition and dry matter content of biomass, and to differentiate the effects, the late harvest was randomly subdivided into three samples. One sample was kept at its original dry matter concentration, while the dry matter content of the other two samples was modified by the addition of water to achieve low, medium and high (the original) dry matter concentrations. Water and biomass were blended using a concrete mixer, and the biomass and water mixture was then left for 24 h at 4 °C to let the water saturate the biomass. Samples from early and intermediate harvests were used in their original dry matter concentrations.

Mechanical pretreatment

Subsamples were taken from all biomass types and pretreated mechanically by extrusion to introduce more variation into the digestibility. The extrusions were performed by a corotating twin-screw extruder (PSHJ-65, Xinda Corporation, Jiangyin, Jiangsu, China, barrel length: 2.84 m) with 340 mm of kneading followed by reverse kneading (56 mm). The kneading causes a particle size reduction, and the reverse kneading forces the biomass to change direction, resulting in a build-up of biomass inside the barrel; more force is needed to move the compact biomass and the close contact with the barrel and extruder screw increases the effect due to the build-up, and the barrel heats up due to the friction. The temperature measured was 20 °C at the feeding point and increased steadily to 100–105 °C after 1.42 m. In the rest of the extruder, the temperature was stable (±5 °C), and the last element of the barrel was cooled down to avoid evaporation of water, which would bias the mass balance.

3,5-Dinitrosalicylic acid assay – sugar availability measure

Enzymatic saccharification using 3,5-dinitrosalicylic acid (DNS) assay (Selig *et al.*, 2008) and spectrophotometric quantifications of the total amount of sugars enzymatically hydrolysed from the biomass (Adney & Baker, 2008) were carried out. Hydrolysis was performed in 100-ml blue-cap bottles. Biomass corresponding to 3.5 g dry matter was mixed with 0.928 g cellulase (Celluclast 1.5L; Novozymes, Bagsværd Denmark) and 0.208 g mannanase (Novozym 51054; Novozymes), and to prevent unwanted growth of microorganisms, 1 ml 2% sodium azide was added. To this was added a 50 ml citrate buffer (pH = 4.8) and finally demineralized water to obtain a total of 100 g. Hydrolysis ran for 72 h at 50 °C in a shaking incubator (185 rpm). All samples including enzyme and cellulose controls were analysed in triplicate.

The quantification of sugars was performed by DNS assay. The bottles were shaken and a sample was extracted and centrifuged, after which an aliquot was extracted from the supernatant and diluted (1 : 9 and 3 : 7 in the samples with high and low sugar yields, respectively) with demineralized water. Glucose standards (concentration of 0 to 1.00 g glucose l^{-1}) were made and analysed with the samples to obtain a standard curve for evaluation of the samples. One millilitre of 3,5-dinitrosalicylic acid solution was added to the diluted samples, and colour was developed during 5 min of boiling (Miller, 1959). The samples were homogenized by shaking, and 280 μl was extracted and transferred to a microtiter plate for spectrophotometric measurement at 538 nm. All measurements were corrected for blanks, and the sugar contents were calculated based on the standard curve.

Biogas batch test

The anaerobic biogas production was measured in a batch test using inoculum from Bånlev biogas plant, Aarhus, Denmark. The plant primarily uses pig manure as a feedstock, but also deep litter, slaughterhouse residues and industrial wastes (mainly lipids). The inoculum was stored in mesophilic conditions (35 °C) for 14 days to halt the biogas production from the inoculum. The properties of the inoculum after storage were 3.06% dry matter and 1.62% volatile solids. A 200-g inoculum and biomass mixture (in a substrate/inoculum ratio of 1 : 1 based on volatile solids) was put in 1-l glass bottles. All biomass types were analysed in triplicate and the experiment also included a control containing inoculum only. The bottles were closed with butyric rubber stoppers and a metal seal. Oxygen was removed from the bottle by flushing with N_2 for 2 min, and the bottles were shaken and stored in an incubator under mesophilic conditions (35 °C). During the following 91 days, the volume of produced biogas (mixture of methane and carbon dioxide) was frequently recorded using an acidified water displacement method, and the methane-to-carbon dioxide ratio was measured using a gas chromatograph (Gas Chromatograph System 7890A, Agilent Technologies, Santa Clara, CA, USA, with Agilent Technologies GC sampler 80) equipped with a thermal conductivity detector and a flame photometric detector, and the methane yield was corrected for the methane yield from the inoculum.

High-throughput pretreatment and hydrolysis method

The high-throughput pretreatment and hydrolysis set-up (HTPH) for evaluation of the conversion efficiency of the biomass into sugars was set up at University of Copenhagen as described by Zhang *et al.* (2014a) including an automated plant material grinding and dispensing set-up from Labman Automation Ltd (Stokesley, North Yorkshire, UK). The biomass was ground to powder and 0.028 g was dispensed into a 96-well aluminium plate, 422 μl sodium citrate buffer (pH = 4.8) was added and the plate was closed with a Teflon plate and a clamp. The plate was heated to 190 °C for 10 min, acting as a hydrothermal pretreatment. Subsequently, the plate was cooled, enzymes (Cellic Ctec, 20 FPU) were added and the hydrolysis ran for 72 h at 50 °C in a shaking incubator (600 rpm; Heidolph Titramax, Schwabach, Germany). The samples were filtered and the contents of glucan, xylan and arabinan measured using a Dionex Ultimate HPLC system (Sunnyvale, CA, USA). A schematic overview and comparison of the three methods can be found in Table 1.

Calculations and statistical analysis

Data on glucan, xylan and arabinan from the HTPH analysis and sugar contents from the DNS analysis were analysed using ANOVA and linear models to evaluate the effect of harvest time, pretreatment (with or without extrusion of biomass) and their interaction. When no interaction was significant, the significance of the fixed effects was analysed using the 'general linear hypotheses' of the 'multcomp' package. The effect of three dry matter concentrations at the late harvest was evaluated with a linear model that contained the three dry matter levels of the late harvest, pretreatment (with or without mechanical pretreatment of biomass) and their interaction. The analyses of data with interaction were computed using the least squares means of the 'lsmeans' package in R. All statistic computations were performed using R version 3.1.3.

The methane yield was calculated in normalized litres (l_N) by correction to 0 °C and 1.013 bar and converted into g CH_4 and g CO_2 as a percentage of biomass using the ideal gas law and the methane-to-carbon dioxide ratio measured in the biogas batch test.

Results

3,5-Dinitrosalicylic acid assay – sugar availability measure

DNS assay was able to discriminate between pretreated and nonpretreated samples and only the early harvest yielded significantly more sugar than the other four biomass types (Fig. 1a). Comparing only samples from different harvest dates (all original dry matter concentration) or only late harvest with different dry matter contents, the amount of total enzymatically hydrolysed sugars from the DNS method depended significantly on the dry matter concentration (of the late harvests), the harvest time (comparing biomass types with their original dry matter concentration) and the use of pretreatment. The results show that the more lignified, mature miscanthus is more inaccessible to the enzymes.

However, the mechanical pretreatment was able to reduce the particle size and break the fibre structures whereby the sugar yields were higher than from the nonpretreated miscanthus samples.

High-throughput pretreatment and hydrolysis method

A significant positive effect on glucan yield from the mechanical pretreatment was found on the two biomass types (intermediate harvest, and late harvest with medium dry matter content). In contrast, there was a significant negative effect of the mechanical pretreatment on the late harvest with a high dry matter content. Where samples were not pretreated, there was no effect of biomass. For the pretreated samples with a high dry matter content, the glucan yield of the late harvest was significantly lower than early, intermediate and late harvest with medium dry matter concentrations. The HTPH analysis, on the other hand, revealed no clear pattern for the xylan yield from treatment, dry matter concentration or harvest time (Table 2 and Fig. 1b).

Biogas batch test

Pretreatment resulted in a significant positive effect on all biomass samples at all sampling dates from day 24. The early harvest gave the highest methane yield and methane + CO_2 (biogas) yield compared to all other investigated samples ($P < 0.05$). The pretreatment had a positive effect at all harvests (Fig. 1c,d). The later the biomass harvest, the lower the methane yield, starting at day 18 of the biogas batch test measurements, with significant differences among early, intermediate and late harvests with the original dry matter concentration.

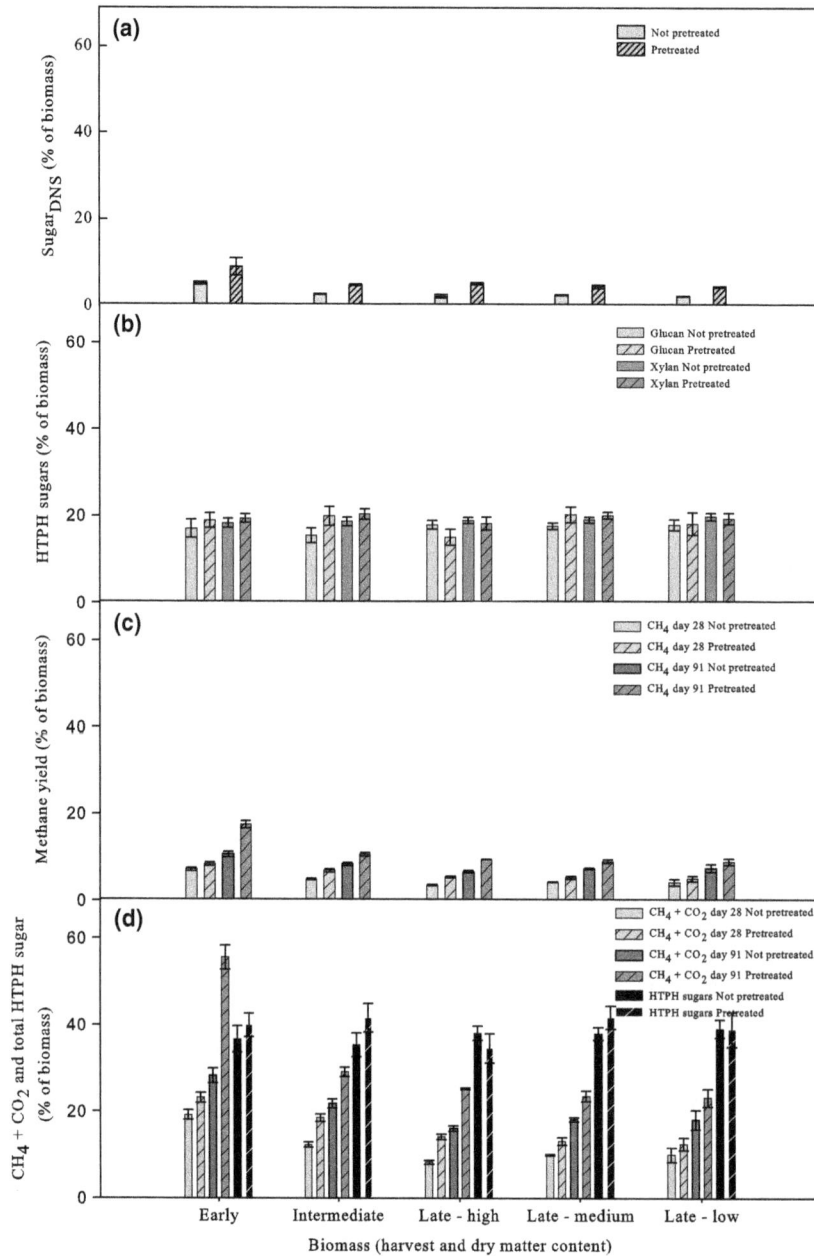

Fig. 1 Yield of enzymatically hydrolysed sugars, methane and biogas at different harvest times and dry matter contents (late harvest). (a) Total amount of sugars hydrolysed in the DNS experiment as a percentage of the biomass; amount of (b) glucan (light grey) and xylan (dark grey) hydrolysed by the HTPH method; (c) methane (CH_4) yield after 28 days (light grey) and 91 days (dark grey) and (d) sum of produced biogas (methane and carbon dioxide) after 28 (light grey) and 91 (dark grey) days and sum of hydrolysed sugars from the HTPH method (black), shown as percentages of the biomass. Hatched bars show extruded biomass whereas bars not hatched show the biomasses that are not pretreated.

The artificial modification of dry matter concentration had no effect on the methane yield of the late harvest for either the extruded or untreated biomass. The effects of harvest time and pretreatment on methane yields at day 91 showed significant interactions (Table 2).

Correlation between methods

The DNS method and methane data correlated well, the correlation coefficient (R^2) being 0.92 at day 91 of methane yield measurement and 0.82 at day 28 (Table 3). The coefficient for earlier methane yields was

Table 2 Significance levels of experimental parameters (dry matter concentration (DM), extrusion, harvest and interactions) for total sugar content (DNS), glucan, xylan and arabinan measured after high-throughput pretreatment and hydrolysis and methane (CH_4) yield after 28 and 91 days. The upper part compares biomass types from the late harvest with different dry matter contents (Late-low, Late-medium and Late-high) and in the lower part are biomass types with the original dry matter content compared (Early, Intermediate and Late-high)

	DNS sugars	Glucan	Xylan	Arabinan	CH_4 day 28	CH_4 day 91
DM	***	0.12	0.18	***	0.60	***
Extrusion	***	*	0.12	**	***	***
DM*Extrusion	0.15	**	0.11	0.08	0.09	***
Harvest	***	0.24	0.25	***	***	***
Extrusion	***	0.12	0.13	*	***	***
Harvest × Extrusion	0.31	**	0.11	*	0.10	***

Indicated P-values: ***$P < 0.001$, **$P < 0.01$ and *$P < 0.05$.

Table 3 Correlation coefficient (R^2) of DNS sugars, glucan, xylan, arabinan, methane yield and $CO_2 + CH_4$ yield, the latter two after 28 and 91 days

	DNS	Glucan	Xylan	Arabinan
Glucan	0.01			
Xylan	0.00	0.76		
Arabinan	0.89	0.12	0.01	
Methane day 28	0.82	0.04	0.00	0.77
Methane day 91	0.92	0.05	0.00	0.91
$CO_2 + CH_4$ day 28	0.81	0.04	0.00	0.75
$CO_2 + CH_4$ day 91	0.91	0.05	0.00	0.90

0.75 for day 18, and increasing until the end of the experiment at day 91, although data are only shown for day 28 and 91. Contrastingly, the correlations between DNS, biogas or methane and glucan or xylan from the HTPH method were poor (Table 3). Arabinan from the HTPH method showed a high correlation with DNS ($R^2 = 0.89$), methane at day 28 and 91 ($R^2 = 0.77$ and 0.91, respectively) and biogas at day 28 and 91 ($R^2 = 0.75$ and 0.90, respectively), whereas the correlations with glucan and xylan from the HTPH method were poor ($R^2 = 0.12$ and 0.01, respectively).

Discussion

The DNS and biogas batch test methods were able to document the expected change in digestibility of biomass types due to a reduction in particle size (Caulfield & Moore, 1974; Hjorth *et al.*, 2011; Surendra & Khanal, 2015). Both the DNS and the biogas batch methods showed a significant positive effect of pretreatment, and the glucan yield of HTPH revealed a difference between the pretreatments for three of the five biomass types. Moreover, DNS and methane yields decreased with later harvests, as also shown by Wahid *et al.* (2015b). In contrast, the sugar yield from HTPH was not related to harvest times, which is inconsistent with the findings of

Hayes (2013) who modelled an increased yield of ethanol per tonne of miscanthus when harvests from November to April were compared.

The starting condition for the HTPH method is dried and ground biomass, which is likely to change its digestibility as particle size influences digestibility (Caulfield & Moore, 1974), and this processing will minimize the physical difference between pretreated and nonpretreated biomass. Secondly, HTPH quantifies the sugars hydrolysed by enzymes after opening of the fibres by hydrothermal pretreatment, redistributing lignin and hemicellulose and even removing hemicellulose (Jørgensen *et al.*, 2007; Hendriks & Zeeman, 2009), and in this way hiding the physical difference between the pretreated and nonpretreated biomass. The DNS method, on the other hand, analysed the raw miscanthus samples with intact fibre structures only exposed to cutting and possibly extrusion. The hydrothermal pretreatment was very effective in degrading fibres of green and mature lignified biomass (Fig. 1b). The result was a 3–15-fold higher sugar yield compared to DNS sugar yields (Fig. 1a and d).

The DNS and HTPH methods both involve enzymatic hydrolysis of the biomass, although different enzymes were used for the two methods whereas duration and temperature were similar (Table 1). The inoculum used in the biogas batch test was a liquid consisting of a cocktail of cellulytic and other enzymes and microbes adapted to the environment in the biogas tank. This cocktail is able to degrade the substrates normally available in the inoculum, providing energy sources for the methanogens on which methane yield depends. The inoculum in this experiment was adapted to industrial and slaughterhouse waste and lignocellulosic material (wheat straw from deep litter). The microbiota therefore needed some time to adapt to the new miscanthus feedstock, which resulted in a slight delay in biogas production compared to what would be expected if the inoculum had already been adapted to the substrate.

The methane yields in the current study were 50–123 l_N CH_4 kg^{-1} VS at day 28 and 95–258 l_N CH_4 kg^{-1} VS at day 91, which was lower than those found by Wahid et al. (2015b) of approximately 200 and 300 l_N CH_4 kg^{-1} VS at days 30 and 90, respectively. The biomass converted by Wahid et al. (2015b) was green and the inoculum used was better adapted to lignocellulosic material (Moset et al., 2015), explaining the higher methane yields. The biogas batch test and DNS method were apparently able to quantify the digestibility of the existing fibre composition and to discriminate among different pretreatments.

Another important difference between the methods was that the DNS and HTPH methods measure sugars from which a theoretical bioethanol or biogas output can be calculated. The biogas batch test, on the other hand, quantifies the output of the energy carrier, methane. Comparing the total output from the HTPH method (sum of the sugars) and the biogas batch test (sum of methane and CO_2) reveals that the HTPH technique facilitates a more efficient conversion of the biomass (Fig. 1d). A 100% fermentation of the hydrolysed sugars into ethanol is impossible to obtain and consequently the proportional ethanol yield from the biomass is expected to be slightly smaller than the proportional sugar yield from the biomass (Hayes & Hayes, 2009).

The DNS and biogas batch test results are well correlated (Table 3) and similar to the findings of Wahid et al. (2015a) after 28 days of methane production (R^2=0.70). This underlines that enzymes used in the DNS method and the mixture of microbiota and enzymes in the inoculum of the biogas batch test have a comparative ability of degrading the biomass. By contrast, methane, biogas and sugar yields from the DNS were not correlated to glucan or xylan (Table 3) due to the difference between the methods (Table 1).

The HTPH set-up is designed to evaluate biomass for the biorefining to ethanol in the Inbicon plant design (Larsen et al., 2012). In Inbicon, hydrothermal pretreatment is chosen because of and efficiency. Modification of the HTPH pretreatment such as lowering the temperature or duration of the treatment could allow a range of physiologically different samples to be produced from which the biomass type (e.g. harvest time) with the lowest pretreatment requirements (and thus lowest costs) relative to its energy yield could be identified. Alternatively, at a given price you could choose the most cost-effective pretreatment for a given biomass to fit the bill.

The coherence between the DNS and the biogas batch test results suggests that the DNS method can be used as a rapid test method, as it allows comparisons of relative methane yields of a certain biomass in only 72 h compared to the 91 days for the batch test. DNS may

also be a more accurate method as the exact methane yields from batch tests will depend on the variation in inoculum.

The HTPH is a model system for the biorefining, and our results revealed that the variation in quality of miscanthus for enzymatic hydrolysis was not important if a severe hydrothermal pretreatment was applied and that the convertibility did not depend on, for example, harvest time. For biogas production, on the other hand, the quality of miscanthus was highly dependent on the timing of the harvest.

In conclusion, the methane measurements from the biogas batch test and the DNS method were highly correlated (R^2 = 0.75 at day 18 of the biogas batch test and increasing at every sampling date to R^2 = 0.92 on day 91) because both methods are based on similar physical biomass input conditions. The HTPH method measures glucan and xylan which did not correlate well with the results of the biogas batch test or the DNS method due to the hydrothermal pretreatment prior to the enzymatic hydrolysis in HTPH. The hydrothermal pretreatment very efficiently breaks and opens the lignocellulosic structures and the differences between the miscanthus samples are thereby eliminated. As a result, HTPH provides information on the overall quantity of carbohydrates that can be made available from a given biomass. Additionally, the DNS and biogas batch test methods visualize the variation in direct digestibility between biomass types caused by lignification and particle size, dependent on harvest time, extrusion and dry matter concentration. Thus, a combination of methods (e.g. DNS plus HTPH) can provide a more complete picture of potential and actual accessibility of lignocellulosic material for biological conversion. This is helpful when choosing the most cost-efficient combination of biomass production (e.g. harvest time), pretreatment (e.g. hydrothermal or extrusion) and final conversion method.

References

Adney B, Baker J (2008) Measurement of Cellulase Activities. Laboratory Analytical Procedure NREL, NREL/TP-510-42628, 1–8.

Bentsen NS, Felby C (2012) Biomass for energy in the European Union – a review of bioenergy resource assessments. *Biotechnology for Biofuels*, **5**, 25–34.

Berglund M, Borjesson P (2006) Assessment of energy performance in the life-cycle of biogas production. *Biomass and Bioenergy*, **30**, 254–266.

Caulfield DF, Moore WE (1974) Effect of varying crystallinity of cellulose on enzyme hydrolysis. *Wood Science*, **6**, 375–379.

Fu C, Mielenz JR, Xiao X et al. (2011) Genetic manipulation of lignin reduces recalcitrance and improves ethanol production from switchgrass. *Proceedings of the National Academy of Sciences of the United States of America*, **108**, 3803–3808.

Hamelin L, Jørgensen U, Petersen BM, Olesen JE, Wenzel H (2012) Modelling the carbon and nitrogen balances of direct land use changes from energy crops in Denmark: a consequential life cycle inventory. *Global Change Biology Bioenergy*, **4**, 889–907.

Hastings A, Clifton-Brown J, Wattenbach M, Mitchell CP, Stampfl P, Smith P (2009) Future energy potential of Miscanthus in Europe. *Global Change Biology Bioenergy*, **1**, 180–196.

Hayes DJM (2013) Mass and compositional changes, relevant to biorefining, in *Miscanthus* × *giganteus* plants over the harvest window. *Bioresource Technology*, **142**, 591–602.

Hayes DJ, Hayes MHB (2009) The role that lignocellulosic feedstocks and various biorefining technologies can play in meeting Ireland's biofuel targets. *Biofuels Bioproducts & Biorefining-Biofpr*, **3**, 500–520.

Hendriks ATWM, Zeeman G (2009) Pretreatments to enhance the digestibility of lignocellulosic biomass. *Bioresource Technology*, **100**, 10–18.

Hjorth M, Granitz K, Adamsen APS, Møller HB (2011) Extrusion as a pretreatment to increase biogas production. *Bioresource Technology*, **102**, 4989–4994.

Hodgson EM, Fahmi R, Yates N *et al.* (2010) Miscanthus as a feedstock for fast-pyrolysis: Does agronomic treatment affect quality? *Bioresource Technology*, **101**, 6185–6191.

Huyen TLN, Remond C, Dheilly RM, Chabbert B (2010) Effect of harvesting date on the composition and saccharification of *Miscanthus* × *giganteus*. *Bioresource Technology*, **101**, 8224–8231.

Jørgensen U (1997) Genotypic variation in dry matter accumulation and content of N, K and Cl in Miscanthus in Denmark. *Biomass and Bioenergy*, **12**, 155–169.

Jørgensen H, Kristensen JB, Felby C (2007) Enzymatic conversion of lignocellulose into fermentable sugars: challenges and opportunities. *Biofuels Bioproducts & Biorefining-Biofpr*, **1**, 119–134.

Jungmeier G, Hingsamer M, van Ree R (2013) Biofuel-driven Biorefineries -A selection of the most promising biorefinery concepts to produce large volumes of road transportation biofuels by 2025. In: *IEA Bioenergy, Task 42 Biorefinery*. Joanneum Research Forschungsgesellschaft mbH, Graz, Austria, pp. 38.

Karp A, Shield I (2008) Bioenergy from plants and the sustainable yield challenge. *New Phytologist*, **179**, 15–32.

Katuwal H, Bohara AK (2009) Biogas: A promising renewable technology and its impact on rural households in Nepal. *Renewable & Sustainable Energy Reviews*, **13**, 2668–2674.

Larsen J, Haven MO, Thirup L (2012) Inbicon makes lignocellulosic ethanol a commercial reality. *Biomass and Bioenergy*, **46**, 36–45.

Lewandowski I, Clifton-Brown JC, Andersson B *et al.* (2003) Environment and harvest time affects the combustion qualities of Miscanthus genotypes. *Agronomy Journal*, **95**, 1274–1280.

Lindedam J, Andersen SB, Demartini J *et al.* (2012) Cultivar variation and selection potential relevant to the production of cellulosic ethanol from wheat straw. *Biomass and Bioenergy*, **37**, 221–228.

Miller GL (1959) Use of dinitrosalicylic acid reagent for determination of reducing sugar. *Analytical Chemistry*, **31**, 426–428.

Moset V, Al-Zohairi N, Møller HB (2015) The impact of inoculum source, inoculum to substrate ratio and sample preservation on methane potential from different substrates. *Biomass and Bioenergy*, **83**, 474–482.

Parajuli R, Dalgaard T, Jørgensen U *et al.* (2015) Biorefining in the prevailing energy and materials crisis: a review of sustainable pathways for biorefinery value chains and sustainability assessment methodologies. *Renewable and Sustainable Energy Reviews*, **43**, 244–263.

Selig M, Weiss N, Ji Y (2008) Enzymatic Saccharification of Lignocellulosic Biomass. Laboratory Analytical Procedure NREL, NREL/TP-510-42629, 1–5.

Somerville C, Youngs H, Taylor C, Davis SC, Long SP (2010) Feedstocks for lignocellulosic biofuels. *Science*, **329**, 790–792.

Surendra KC, Khanal SK (2015) Effects of crop maturity and size reduction on digestibility and methane yield of dedicated energy crop. *Bioresource Technology*, **178**, 187–193.

Wahid R, Hjorth M, Kristensen S, Møller HB (2015a) Extrusion as pretreatment for boosting methane production: effect of screw configurations. *Energy & Fuels*, **29**, 4030–4037.

Wahid R, Nielsen SF, Hernandez VM, Ward AJ, Gislum R, Jørgensen U, Møller HB (2015b) Methane production potential from Miscanthus sp.: effect of harvesting time, genotypes and plant fractions. *Biosystems Engineering*, **133**, 71–80.

Weiland P (2010) Biogas production: current state and perspectives. *Applied Microbiology and Biotechnology*, **85**, 849–860.

Zhang H, Fangel JU, Willats WGT, Selig MJ, Lindedam J, Jørgensen H, Felby C (2014a) Assessment of leaf/stem ratio in wheat straw feedstock and impact on enzymatic conversion. *Global Change Biology Bioenergy*, **6**, 90–96.

Zhang H, Thygesen LG, Mortensen K, Kadar Z, Lindedam J, Jørgensen H, Felby C (2014b) Structure and enzymatic accessibility of leaf and stem from wheat straw before and after hydrothermal pretreatment. *Biotechnology for Biofuels*, **7**, 74–84.

4

Agronomic factors in the establishment of tetraploid seeded *Miscanthus × giganteus*

ERIC K. ANDERSON[1], DOKYOUNG LEE[1], DAMIAN J. ALLEN[2,3] and THOMAS B. VOIGT[1]

[1]*Department of Crop Sciences, University of Illinois at Urbana-Champaign, Urbana, IL 61801, USA*, [2]*Mendel BioEnergy Seeds, Mendel Biotechnology Inc., 3935 Point Eden Way, Hayward, CA 94545, USA*, [3]*Department of Agronomy, Purdue University, West Lafayette, IN 47907, USA*

Abstract

To meet US renewable fuel mandates, perennial grasses have been identified as important potential feedstocks for processing into biofuels. Triploid *Miscanthus × giganteus*, a sterile, rhizomatous grass, has proven to be a high-yielding biomass crop over the past few decades in the European Union and, more recently, in the United States. However, high establishment costs from rhizomes are a limitation to more widespread plantings without government subsidies. A recently developed tetraploid cultivar of *M. × giganteus* producing viable seeds (seeded miscanthus) shows promise in producing high yields with reduced establishment costs. Field experiments were conducted in Urbana, Illinois from 2011 to 2013 to optimize seeded miscanthus establishment by comparing seeding rates (10, 20, and 40 seeds m^{-2}) and planting methods (drilling seeds at 38 and 76 cm row spacing vs. hydroseeding with and without premoistened seeds) under irrigated and rainfed conditions. Drought conditions in 2011 and 2012 coincided with stand establishment failure under rainfed conditions, suggesting that seeded miscanthus may not establish well in water-stressed environments. In irrigated plots, hydroseeding without premoistening was significantly better than hydroseeding with premoistening, drilling at 38 cm and drilling at 76 cm with respect to plant number (18%, 54%, and 59% higher, respectively), plant frequency (13%, 30%, and 40% better, respectively), and the rate of canopy closure (18%, 33%, and 43% faster, respectively) when averaged across seeding rates. However, differences in second-year biomass yields among treatments were less pronounced, as plant size partially compensated for plant density. Both hydroseeding and drilling at rates of 20 or 40 seeds m^{-2} appear to be viable planting options for establishing seeded miscanthus provided sufficient soil moisture, but additional strategies are required for this new biomass production system under rainfed conditions.

Keywords: biomass, cellulosic feedstock, hydroseeding, irrigation, *Miscanthus × giganteus*, perennial grass, planting method, seeding rate

Introduction

Lignocellulosic bioenergy feedstock production is anticipated to expand as mandated volumes of second-generation biofuels in the United States continue to increase (EISA, 2007; Schnepf & Yacobucci, 2013). Reduced greenhouse gas emissions are expected in the production of cellulosic feedstocks compared with conventional fossil fuels and ethanol produced from corn (*Zea mays* L.) (Davis *et al.*, 2012). The majority of current research in the United States focuses on feedstock conversion to liquid transportation biofuel, though it is more widely used to produce heat and power, particularly in the European Union (EU) (Murphy *et al.*, 2013). While current efforts to produce cellulosic biofuels at commercial scale involve crop and forest residues, dedicated biomass crops such as perennial grasses are expected to comprise the single largest contribution to cellulosic biofuels feedstock supplies by 2022 (USDOE, 2011).

Miscanthus × giganteus Greef and Deuter ex Hodkinson and Renvoize (Greef *et al.*, 1997; Hodkinson & Renvoize, 2001) is a perennial, warm-season, rhizomatous grass originating in East Asia that has been evaluated extensively in the EU, and more recently in the United States, as a high-yielding bioenergy feedstock. The triploid (3×) hybrid of *M. sinensis* Andersson (2×) and *M. sacchariflorus* (Maxim.) Benth. (4×) (Linde-Laursen, 1993; Rayburn *et al.*, 2009; Sacks *et al.*, 2013), commonly referred to as the 'Illinois' clone of *M. × giganteus* in the United States, comprises most of the miscanthus grown for bioenergy commercially and for research purposes in the EU and the United States (Głowacka *et al.*, 2013).

Correspondence: Thomas B. Voigt
e-mail: tvoigt@illinois.edu

It is also the only variety currently allowed for miscanthus plantings in Project Areas 2–5 under the Biomass Crop Assistance Program (USDA, 2012).

Being triploid, *M. × giganteus* cv. Illinois does not produce viable seeds (Slomka *et al.*, 2012) and is planted by rhizomes or plugs (typically a rhizome-derived nursery-propagated plantlet). Establishment costs are high for clonal miscanthus compared with seeded grasses (Lewandowski *et al.*, 2003; Christian *et al.*, 2005). Christian *et al.* (2005) estimated a delivered cost of rhizomes or plants to be $0.29–0.57 (€0.25–0.5) and the cost of establishment to be $1725–3450 (€1500–3000) per hectare, although no planting rate was given with this estimate. Current delivered cost estimates in Illinois for *M. × giganteus* cv. Illinois are $0.05 and $0.24 per rhizome and plug, respectively (personal observation). Assuming a goal of 17 300 plants ha^{-1}, establishment costs, not including labor and inputs, would be $1235 and $4150 ha^{-1} from rhizome and plug, respectively. For comparison, seeding costs for switchgrass (*Panicum virgatum*) are $99–741 ha^{-1} based on current market prices, not including labor and inputs and assuming a cost of $22–66 kg^{-1} pure live seeds (PLS) depending on cultivar and a seeding rate of 4.5–11.2 kg PLS ha^{-1}. Cheaper current establishment costs with switchgrass have to be weighed against lower expected yields in many geographic regions compared with *M. × giganteus* cv. Illinois, approximately a 50% penalty in Illinois (Arundale *et al.*, 2014).

Recommendations vary for seeding rates of various perennial grasses. In a study investigating seeding rates for switchgrass and big bluestem (*Andropogon gerardii*), Vogel (1987) found that rates greater than 200 PLS m^{-2} (roughly 3 kg PLS ha^{-1}) were not necessary when planting into firm seed beds in the spring in Nebraska, USA on silty clay loam. Other seeding rate recommendations for switchgrass have included 2.2–4.5 kg PLS ha^{-1} (Foster *et al.*, 2013), 2.2–11.2 kg PLS ha^{-1} (Parrish *et al.*, 2008), 3.4 kg PLS ha^{-1} (Vassey *et al.*, 1985), 5.6–6.7 kg PLS ha^{-1} (Teel *et al.*, 2003), and 9.0–11.2 kg PLS ha^{-1} (Wolf & Fiske, 1995). In general, the recommended seeding rate is 3–10 kg ha^{-1}, which is approximately equivalent to 200–400 PLS m^{-2}. With a typical target of 10–20 plants m^{-2}, 25 plants m^{-2} is considered as excellent stand (Lee *et al.*, 2014). Christian *et al.* (2005) used 500 PLS m^{-2} in a field study investigating conventional sowing methods of establishing *M. sinensis* at the Rothamstad research farm in England on silty clay loam, and observed 5–26 plants m^{-2}, an establishment rate of 1–5%.

While several methods are used to plant grass seed, drilling and hydroseeding are of particular interest and used in this research. In the only previous report on direct-seeding, Christian *et al.* (2005) recorded a 63%

and 40% reduction in seedling emergence with broadcasted pelleted and unpelleted *M. sinensis* seed, respectively, compared with drilling at 1 cm depth. In drilling, seeds are dropped from a box into small furrows made by disk openers or chisels that are followed by packer wheels that firm the soil over the seeds (Houk, 2009). Seeding depth and rates are adjustable to ensure proper seed placement in the furrow and planting uniformity (Houk, 2009). Drilling seeds normally leads to good seed-to-soil contact.

In hydroseeding, seeds are mixed with water, and often with pulp fiber mulch and fertilizer, and the suspended slurry is sprayed on the prepared soil (Beard, 1973). Commonly used to plant turfgrass seed, hydroseeding works well on slopes or rocky sites difficult to cultivate, but since the seeds are on the soil surface, there needs to be adequate rainfall or irrigation for germination and establishment (Beard, 1973). Hydroseeding has been successfully used to plant switchgrass on reclaimed mine sites (Keene & Skousen, 2010). Soaking seeds to initiate germination before planting, in combination with hydroseeding, has been used to minimize the time between planting and seedlings accessing soil water through root growth (Young *et al.*, 1977). At anticipated average air temperatures during planting of approximately 20 °C, it was expected that at least 6–8 days would be required for germination and emergence based on laboratory studies of *M. sinensis* (Clifton-Brown *et al.*, 2011; Christian, 2012) and *M. × giganteus* (Panter, 2010), which highlights the challenge of maintaining a moist environment throughout germination for shallow- and surface-planted seed.

The potential benefits of seeding vs. planting with plugs or rhizomes include: lower cost of propagules; simpler logistics of storing, handling and transporting propagules; lower cost and greater speed of planting; lower cost of 'overplanting' to avoid future gaps in the field; decreased time for production of new cultivars; and increased genetic diversity compared with the clonal system. The potential risks with planting from seeds include: higher susceptibility to establishment failure in dry surface soils; increased weed competition and reduced herbicide options during establishment; and increased invasiveness potential of non-native species. Clonal *M. × giganteus* has been shown to have a low risk of invasiveness in the United States according to the Australian Weed Risk Assessment model (Barney & DiTomaso, 2008; Gordon *et al.*, 2011); however, it is uncertain whether a nonsterile variety of *M. × giganteus* would have as low an invasive potential (Quinn *et al.*, 2010, 2012).

A tetraploid variety of *M. × giganteus* (hereafter referred to as seeded miscanthus) has been developed that produces viable seeds and achieves yields comparable to the triploid 'Illinois' variety (Sacks *et al.*, 2013). The germplasm was acquired, further developed, and patented by Mendel Biotechnology, Inc. (Power-Cane™ Miscanthus, Mendel Biotechnology, Inc., Hayward, CA, USA) as a seeded miscanthus variety grown for bioenergy (Sacks *et al.*, 2013).

The goals of the current research were to answer the following questions with respect to establishment of seeded miscanthus: (i) Is irrigation necessary during the year of establishment? (ii) What seeding rate will result in adequate stand establishment and the highest biomass yields at the end of the second growing season? (iii) What planting methods are optimal for establishing seeded miscanthus and achieving maximum biomass yield at the end of the second growing season?

Materials and methods

A field study was conducted from 2011 to 2013 in Urbana, Illinois (40°3′58.647″, −88°11′32.2938″) on Dana silt loam (fine-silty, mixed, superactive, mesic Oxyaquic Argiudolls). Prior to initiation of the experiment, the field was in a corn-soybean (*Glycine max* L.) rotation with soybean as the prior crop. The field was tilled to a depth of 10 cm with a rotary tiller prior to planting each year. A soil test was conducted to determine soil nutrient baseline. Soil organic matter was 2.5%, pH was 6.8, cation exchange capacity was 12.2 cmol+ kg^{-1}, K was 131 mg kg^{-1}, and Bray-1 P was 13 mg kg^{-1}. No fertilizer was applied to the field throughout the experiment. Weather data were also collected for Urbana, IL for the duration of the experiment (Fig. 1).

The experiment was conducted as a split plot within a randomized complete block design with irrigation as the main plots and a factorial combination of seeding methods and seeding rates as the subplots. The experiment was replicated three times in space and twice in time. Irrigated plots were watered only during the year of planting. Initially, irrigation was conducted with a traveling water cannon (T180 Ag-Rain Water-Reel, Kifco, Inc., Havana, IL, USA) when the soil surface was dry. An irrigation system was installed on 25 July, 2011 with sprinkler heads (P5R impact sprinkler heads, Rain Bird Corporation, Azusa, CA, USA) spaced 9.1 m apart with head-to-head coverage to ensure more uniform watering. Thereafter, plots were irrigated at least once a day if the volumetric water content of the soil dropped to 16% at a depth of 2 cm until the miscanthus was approximately 5 cm tall and thereafter at a depth of 10 cm. Soil moisture sensors (WaterScout SM 100, Spectrum Technologies, Aurora, IL, USA) were installed in both irrigated and rainfed plots, and data were collected during the year of establishment.

Plots were planted 7–9 June, 2011 and 23–24 May, 2012. Seeding rates were 10, 20, and 40 seeds m^{-2} (0.1, 0.2 and 0.4 kg ha^{-1}, respectively). These relatively low rates were used due to limited seed availability and to observe a density response. Planting methods included drilling at 38 cm (Drill 38) and 76 cm (Drill 76) row spacing, and hydroseeding with dry (Hydro) and premoistened (HydroPM) seed. For HydroPM

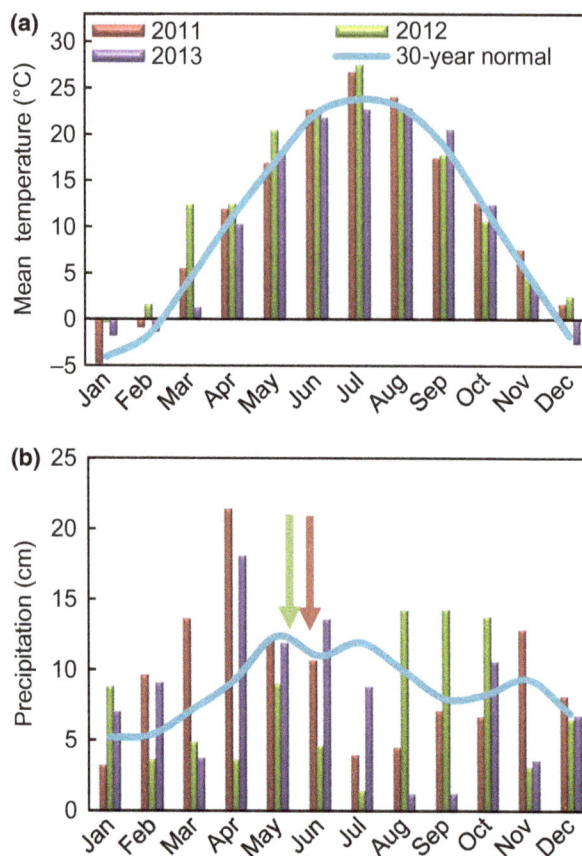

Fig. 1 Air temperature (a) and precipitation (b) conditions from 2011 to 2013 with 30-year normals for Urbana, IL. Arrows represent planting dates for the first (red) and second (green) replications.

plots, seeds were moistened until the first few seeds germinated, then stored at 4 °C for 2–3 days until planting. Plot size was 4.6 m wide by 6.1 m long which allowed for 6 and 12 rows of plants per plot for Drill 38 and Drill 76 treatments, respectively. Drilled seeds were planted at approximately 1.0 cm which was in the range of recommended planting depths for switchgrass (Vassey *et al.*, 1985; Teel *et al.*, 2003; Berti & Johnson, 2013) and *M. sinensis* (Christian *et al.*, 2005). The width of the plot drill (Great Plains 3P606NT, Great Plains Ag, Salina, KS, USA) mounted with a cone seed distributer (Wintersteiger, Ankeny, IA, USA) was 2.1 m and half of the seeds were planted in each of two passes for each plot. Seeds for the six plots of a given hydroseeding method and seeding rate were combined into one batch in the hydroseeder (T60 HydroSeeder, Finn Corporation, Fairfield, OH, USA) and enough water and mulch were added to cover all plots. The seed-mulch slurry was applied with a hand-held nozzle spraying 75 l min^{-1} with 150 l applied to each plot. In 2012, all plots were rolled with a packing drum 1 day after hydroseeding to improve seed-to-soil contact.

Miscanthus seeds of an initial variety of PowerCane™ Miscanthus (Mendel Biotechnology, Inc., Hayward, CA, USA) were obtained for this experiment. Three tetraploid F1 plants resulting from crosses between *M. sinensis* (diploid) and *M. sacchariflorus* (tetraploid) were selected, vegetatively increased, and allowed to cross-pollinate (Sacks *et al.*, 2013), and the F2 seeds from these crosses were collected and pooled for this experiment. Seeds were cleaned with palea and ciliate lemma removed (Smith & Barney, 2014). Seeds used in drill treatments were pelleted to increase accuracy of metering of these small seeds during planting. Seed weights were 1.1 mg seed^{-1} for unpelleted and 12 mg (11 : 1 loading) and 27 mg seed^{-1} (25 : 1 loading) for pelleted seeds in 2011 and 2012, respectively. Pellet sizes were 2.5 mm (9× the size of unpelleted) and 3.5 mm diameter (13× the size of unpelleted) for 2011 and 2012 plantings, respectively. These were significantly larger than the pelleted *M. sinensis* seeds used by Christian *et al.* (2005) at 3.8 mg seed^{-1} (5 : 1 loading). Cleaned seeds were used for hydroseeding without pelleting. There were no germination differences between pelleted and nonpelleted seeds (both 96.5%) when tested per Kim *et al.* (2012) in a growth chamber germination test. This high germination rate results in seed number being approximately equal to PLS, and so the former is reported throughout.

Preemergence herbicides were applied and incorporated with tillage prior to planting. The herbicide mixture included 0.21 kg ai ha^{-1} mesotrione (Callisto®, Syngenta Crop Protection, Inc., Greensboro, NC, USA) + 1.68 kg ai ha^{-1} atrazine (AAtrex®, Syngenta Crop Protection, Inc.) + 0.42 kg ai ha^{-1} quinclorac (Paramount®, BASF Corporation, Research Triangle Park, NC, USA). This herbicide mixture was shown to be safe for preemergence application on seeded miscanthus in a greenhouse experiment (E. Anderson, unpublished data). A postemergence application of 1.69 kg ai ha^{-1} acetochlor + 0.67 kg ai ha^{-1} atrazine (Harness® Xtra, Monsanto Company, St. Louis, MO, USA) + 0.42 kg ai ha^{-1} quinclorac was made after the miscanthus had grown to approximately 30 cm tall on 19 July, 2011 and 10 July, 2012. Quinclorac had already been shown to be safe on seedling miscanthus (E. Anderson unpublished data) and Harness® Xtra is labeled for use on miscanthus at least 5 cm tall (Anonymous, 2010). Herbicides with foliar and residual activity were also applied at the beginning of the second year of each replication. A tank mix of 1.69 kg ai ha^{-1} acetochlor + 0.67 kg ai ha^{-1} atrazine + 0.53 kg ae ha^{-1} 2,4-D ester (Radar™ LV, Growmark, Inc., Bloomington, IL, USA) was applied on 28 March and 18 May, 2012 and on 9 April, 2013, minimizing recruitment of any new miscanthus seedlings from remnant planted or *in situ* produced seed.

Stand counts were taken 1, 2, 4, and 8 weeks after emergence (WAE) by counting all plants in each plot. At 8 WAE, plot frequency measurements were also taken using a modified frequency grid (Vogel & Masters, 2001). With a goal of 24 710 plants ha^{-1}, or 0.4 m^2 plant^{-1}, a PVC grid was fabricated with four quadrants each 0.4 m^2. The grid was placed 1.0 m to the left and 0.5 m back from the front left corner of each plot, and the number of quadrants containing at least one plant was recorded. The grid was then flipped end-over-end seven times for a total of 32 quadrants per plot. The sum of all quadrants containing a plant was divided by 32 to obtain a percentage of area containing a plant in an optimal uniform stand. Stand and frequency counts were also taken at the beginning

of the second growing season on 26 April, 2012 and 14 May, 2013. Spring plant frequency data were then used to calculate plant density.

Canopy closure during the second growing season was estimated by measuring the interception of photosynthetically active radiation (PAR) for each plot. A ceptometer (AccuPar LP-80, Decagon Devices, Inc., Pullman, WA, USA) with 80 PAR sensors spaced 1 cm apart was placed at ground level to record light penetration at five locations within each plot – 1 m^2 in from each corner and in the center, situated parallel with the front of each plot. One above-canopy PAR reading was taken for each plot with the ceptometer. Measurements were taken within 3 h of solar noon each day when skies were clear. Measurements were taken between early June and mid-September five times in 2012 and six times in 2013. Canopy closure was estimated to be when 75% of PAR was intercepted. The number of Julian days to canopy closure was then estimated from a plot of percent light interception over time. If a plot never reached 75% PAR interception, it was assigned a value of 365 days.

Aboveground biomass harvest was conducted after a killing frost at the end of the second growing season. Killing frosts occurred on 10 November, 2012 and 21 October, 2013, and harvests were conducted on 30 November, 2012 and 15 November, 2013. Miscanthus fresh weight yields were then determined by harvesting an area approximately 2.4 m wide by 4.8 m long in the center of each plot with a plot combine (Wintersteiger Cibus S harvester, Wintersteiger Inc., Salt Lake City, UT, USA, mounted with a C1200 forage chopper, Maschinenfabrik KEMPER GmbH & Co. KG, Stadtlohn, Germany) at a height of 10 cm. Subsamples of approximately 0.5 kg harvested biomass were directly collected from the combine and dried at 60 °C for a minimum of 5 days and weighed to determine percent dry matter.

Normality of the residuals and equality of the variances were evaluated using a boxplot of the residuals and a plot of the residuals against their predicted values in SAS software (SAS 9.2, SAS Institute Inc., ©2002–2008, Cary, NC, USA). Data were analyzed in the MIXED procedure in SAS software at $\alpha = 0.05$ with irrigation, seeding method, seeding rate and their interactions as fixed effects and year and spatial replication as random effects. Pair-wise differences from the MIXED procedure were used to create letter groupings of similar means using the PDMIX800 macro (Saxton, 1998) in SAS software. Pearson product–moment correlation was used to identify significant correlations among all dependent variables using the CORR procedure in SAS software.

Results

Air temperatures did not deviate markedly from the 30-year normals for most months during the experiment (Fig. 1a). In 2011, July was warmer than average, and March in 2012 was much warmer than normal as were May and July. Both 2011 and 2012 were unusually dry with 2012 being severely impacted by drought in Urbana (Fig. 1b). Although precipitation was normal the month after planting in 2011, total rainfall was 16 cm below normal from July through October. Total precipitation from July 2011 to July 2012 was 37 cm below normal and 16 cm below normal in the 2 months following planting in 2012. Near-normal rates of precipitation between planting and first signs of emergence occurred in both planting years. However, rainfall was below normal in the 2 months following emergence, particularly in 2012.

Under these postemergence drought conditions, stand establishment failure occurred in rainfed plots in both establishment years. Robust weed control was observed (Figure S1), eliminating this potentially confounding variable. In 2011, plots in one block of the rainfed treatment at the end of the field that was lowest in elevation had reasonably good stand and plant frequency measurements (fall densities ranged from 6% to 97% for individual plots with an average of 44% when averaged across planting methods and seeding rates), and data were collected on those plots. However, plots in the other two blocks in 2011 and all blocks in 2012 had less than 20 plants, with most plots having no living plants (Figure S1). Therefore, data collection was discontinued from rainfed plots and irrigation was removed from the model. Only data from irrigated plots will be presented.

Because little winter kill was observed, and stand counts and plant frequency measurements taken in the fall were strongly correlated with those taken the following spring ($r = 0.95$, $P < 0.0001$ and $r = 0.89$, $P < 0.0001$, respectively), only spring counts are presented. Seeding rate, planting method, and their interaction term were all highly significant for plant count (Table 1). The magnitude of the response to seeding rate was greater with hydroseeding treatments, particularly HydroPM, than with either drilling method for plant count (Fig. 2a), thus explaining the significance of the interaction term. At 40 seeds m^{-2}, Hydro and HydroPM produced 60 000 plants ha^{-1} whereas both Drill 38 and Drill 76 produced just half that number. Hydro at 20 seeds m^{-2} produced 45% and 50% more plants than Drill 76 and Dill 38, respectively, at 40 seeds m^{-2}.

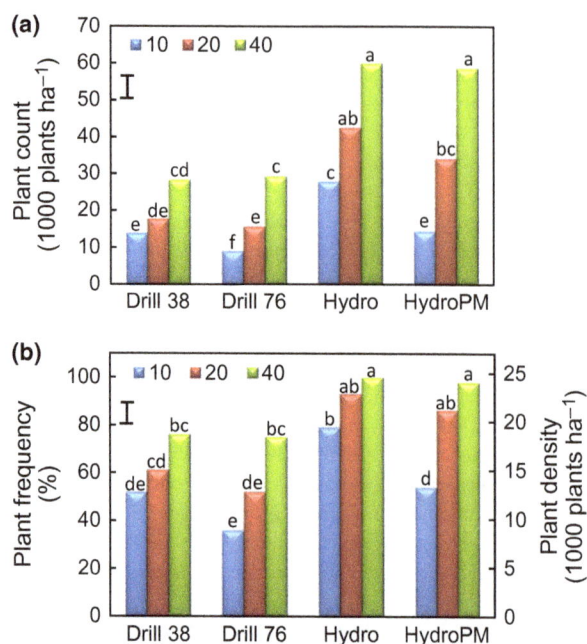

Fig. 2 Seeded miscanthus plant counts (a) and plant frequency (b) measurements taken in the spring following the year of establishment, pooled across two planting years. Plant density was calculated based on plant frequency grid data and is shown for ease of interpretation. Seeding rates included 10, 20, and 40 seeds m^{-2}. Planting methods included hydroseeding without (Hydro) and with (HydroPM) premoistening seeds and drilling at 38 cm (Drill 38) and 76 cm (Drill 76) row spacing. Letter groupings denote differences among treatments at $\alpha = 0.05$. Bars represent SEM at $\alpha = 0.05$.

Seeding rate and planting method were highly significant for plant frequency, a measure of stand uniformity defined as the percentage of 0.4 m^2 quadrats containing at least one plant (Table 1). Though Hydro and HydroPM at 40 seeds m^{-2} resulted in the highest stand uniformity (100% and 98%, respectively), they were not significantly different from the same planting methods at 20 seeds m^{-2} (Fig. 2b). Hydro at 10 seeds m^{-2} and both drill methods at 40 seeds m^{-2} all resulted in a plant frequency of over 75%.

Table 1 P-values ($\alpha = 0.05$) from ANOVA for second-year irrigated seeded miscanthus biomass yield components, pooled across two replicate planting seasons*

Source of variation	Plant count	Plant frequency†	Canopy closure‡	Biomass yield
Seeding rate (SR)	<0.0001	<0.0001	<0.0001	0.0157
Planting method (PM)	<0.0001	<0.0001	<0.0001	0.0872
SR * PM	0.0043	0.3327	0.2515	0.1893

*Seeding rates included 10, 20, and 40 seeds m^{-2}. Planting methods included hydroseeding with and without premoistening seeds and drilling at 38 and 76 cm row spacing. Plant count and plot frequency were measured in the spring, canopy closure during the summer, and biomass yield following a killing frost, all during the year following establishment.
†Plant frequency as measured by percentage of 0.4 m^2 quadrats with at least one plant.
‡Canopy closure as estimated by the number of Julian days to achieve 75% PAR interception.

Seeding rate and planting method were highly significant for canopy closure in the second growing season (Table 1). Julian days to canopy closure were moderately and strongly negatively correlated with plant count and frequency ($r = -0.70$, $P < 0.0001$ and $r = -0.83$, $P < 0.0001$, respectively). Hydro and HydroPM at 40 seeds m^{-2} resulted in the shortest time to canopy closure (172 and 173 Julian days, respectively); however, these were not significantly different from Drill 38 at 40 seeds m^{-2}, Hydro and HydroPM at 20 seeds m^{-2}, or Hydro at 10 seeds m^{-2} (Fig. 3). Hydro plots achieved canopy closure 82 days earlier than Drill 76 plots when averaged across seeding rates. When averaging hydroseeding methods together, canopy closure occurred 56 days earlier compared with drilling when averaged across drill spacing and seeding rates.

Seeding rate was significant at $\alpha = 0.05$ and planting method was significant at $\alpha = 0.10$ for aboveground biomass yield (Table 1). Second-year harvested biomass moisture content was 27% for the 2011 seeding and 34% for the 2012 seeding when averaged across planting methods and seeding rates. Biomass yield was weakly correlated with plant count and frequency ($r = 0.47$, $P < 0.0001$ and $r = 0.52$, $P < 0.0001$, respectively) and weakly negatively correlated with time to canopy closure ($r = -0.59$, $P < 0.0001$). Within-treatment variability was relatively high (1.7 Mg ha^{-1} standard error of the differences in means) (Fig. 4), despite the inclusion of

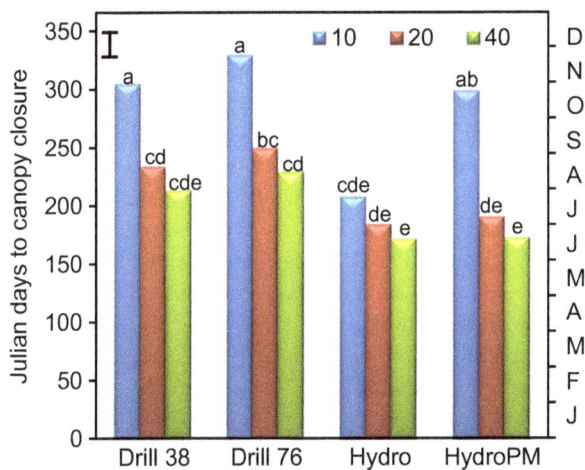

Fig. 4 Aboveground seeded miscanthus biomass yield harvested after a killing frost the year after establishment. Seeding rates included 10, 20, and 40 seeds m^{-2}. Planting methods included hydroseeding without (Hydro) and with (HydroPM) premoistening seeds and drilling at 38 cm (Drill 38) and 76 cm (Drill 76) row spacing. Letter groupings denote differences among treatments at $\alpha = 0.05$. Bar represents SEM at $\alpha = 0.05$.

two temporal and three spatial replicates. Although HydroPM at 20 seeds m^{-2} produced the highest yield, it was not significantly different from any planting methods at 40 seeds m^{-2}, Hydro and Drill 76 at 20 seeds m^{-2}, or Hydro at 10 seeds m^{-2}. Only the yields with the two drill methods at 10 seeds m^{-2} were significantly lower than any planting method at 40 seeds m^{-2} at the end of the second growing season.

Discussion

Seeded miscanthus was successfully established and second-year yields were similar to those for $M. \times gigan$-$teus$ cv. Illinois (Miguez $et\ al.$, 2008). However, limited soil moisture conditions associated with drought during the first 2 months from planting could be challenging for successful stand establishment for the second-year biomass.

Seed emergence was delayed until 3 weeks after planting under irrigated field conditions with generally less than 30% emergence and survival achieved. Even in 2011 when precipitation was near normal during the month following planting, establishment failed in most nonirrigated plots coincident with above-average temperatures and drought-like conditions in the following 2 months. Since the optimum temperature for germination with this miscanthus germplasm is roughly 28 °C (X. Duan, Mendel Biotechnology Inc., personal communication), future research will be needed to identify ideal planting timings in different regions to ensure adequate soil temperatures while taking advantage of spring precipitation. Clifton-Brown $et\ al.$ (2011) observed high temperature requirements of >25 °C in germination of $M.$ $sinensis$, and concluded that direct-seeding of this

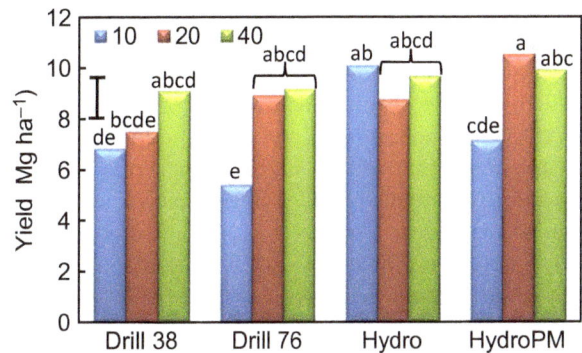

Fig. 3 Rate of canopy closure in seeded miscanthus as estimated by 75% PAR interception during the year after establishment. Seeding rates included 10, 20, and 40 seeds m^{-2}. Planting methods included hydroseeding without (Hydro) and with (HydroPM) premoistening seeds and drilling at 38 cm (Drill 38) and 76 cm (Drill 76) row spacing. Months are displayed on the secondary y-axis for ease of interpretation. Letter groupings denote differences among treatments at $\alpha = 0.05$. Bar represents SEM at $\alpha = 0.05$.

potential biomass crop is currently impossible for much of Northern Europe. While this experiment required 6–25% of seeds to grow to achieve the target density of 24 700 plants ha^{-1} under nonwater-limiting conditions, exploring the impact of higher seeding rates for rainfed seeded miscanthus would also be an appropriate avenue to explore due to the observed lower survival rate under rainfed conditions.

The observed difficulty in establishing *M. × giganteus* from seed without irrigation, even with excellent selective weed control, is reassuring with respect to concerns over unintentional invasive spread outside of crop production. Smith & Barney (2014) recently demonstrated that this seeded *M. × giganteus* emerged in multiple habitats, particularly where there was little competition, but these seedlings subsequently suffered 99.9% mortality within 6 months. Counter-intuitively, establishment requiring very precise conditions which only exist in a miscanthus farmers' field would be an ideal scenario, if they can be created cost effectively. Using only miscanthus seeds that have been cleaned (i.e. palea and ciliate lemma removed) may provide such a scenario (Smith & Barney, 2014).

Under irrigated conditions, hydroseeding was generally more successful than drilling seeds with respect to stand count, plant frequency, and rate of canopy closure. This may be due to a level of protection from seed and seedling desiccation afforded by the hydroseeding mulch, which was generally thick enough to only cover the soil surface. Laboratory germination testing in this study and results from previous field work (E. Anderson unpublished data) did not show a difference between pelleted and unpelleted miscanthus seed, so it is unlikely that differences in germination were related to pelleting the seeds rather than the planting method. Variability in planting depth with the seed drill, combined with small seed size, may also have resulted in seeds being buried too deep for successful emergence. Broadcasting was discounted from this study due to a 63 and 40% reduction in seedling emergence with broadcasted pelleted and unpelleted *M. sinensis* seed, respectively, compared with drilling at 1 cm depth (Christian *et al.*, 2005). However, broadcasting miscanthus seeds with a sand or similar medium followed by rolling could minimize this variability in depth; future research is needed to determine the effectiveness of this planting method. Wolf & Fiske (1995) and Monti *et al.* (2001) noted that rolling following planting was important in establishing switchgrass. Future analyses are also needed to determine the economic feasibility of these various establishment protocols under different landscapes, soil types and geoclimatic regimes.

Plant frequency measurement protocols and analyses were based on our goal of 24 700 plants ha^{-1}

(10 000 plants ac^{-1}). This goal was achieved with all planting methods at 40 seeds m^{-2}, hydroseeding (Hydro and HydroPM) at 20 seeds m^{-2} and Hydro at 10 seeds m^{-2}, with irrigation. Previous recommendations for *M. × giganteus* cv. Illinois plantings suggested a planting rate of 17 215 propagules ha^{-1} (6 970 propagules ac^{-1}) (Pyter *et al.*, 2009). Our goal of 24 700 plants ha^{-1}, compared with 17 215 propagules ha^{-1}, was because the ultimate size and rate of spread of the seeded type were unknown. All planting methods and seeding rates in the current experiment were successful in meeting the goal of 17 215 plants ha^{-1}, except Drill 38 at 10 seeds m^{-2}, Drill 76 at 10 and 20 seeds m^{-2}, and HydroPM at 10 seeds m^{-2}. These findings correlate well with those treatments that resulted in biomass yields that were not significantly different from the highest yields in this study (Fig. 4).

The earliest canopy closure occurred in mid-June and early July during the second growing season, on average, in hydroseeded plots at 40 and 20 seeds m^{-2}, respectively. Hydro at 10 seeds m^{-2} and Drill 38 at 40 seeds m^{-2} plots achieved canopy closure in late July followed by Drill 38 at 20 seeds m^{-2} and Drill 76 at 40 seeds m^{-2} in mid-August. The benefits of early canopy closure include minimizing weed competition (Bullock *et al.*, 1988; Knezevic *et al.*, 2003), minimizing water loss due to evaporation from the soil surface (Chavez *et al.*, 2008), and maximizing PAR interception and thus potential yield (Andrade *et al.*, 2002; Balkcom *et al.*, 2011).

Observed seeded miscanthus yields of approximately 10 Mg ha^{-1} at the end of the second growing season (Fig. 4) are consistent with those seen with triploid *M. × giganteus* cv. Illinois (Miguez *et al.*, 2008). Planting method and seeding rate impacts on second-year aboveground biomass yield of seeded miscanthus are statistically detectable (Table 1; Fig. 4), but substantially less pronounced than stand count, uniformity (Fig. 2), and canopy closure (Fig. 3). This demonstrates that seeded miscanthus has the potential to partially compensate with increased individual plant size, in response to lower plant density by the end of the second growing season. In a meta-analysis of triploid *M. × giganteus* (Miguez *et al.*, 2008), second year yields responded to plant density <40 000 plants ha^{-1}, but these were not statistically detectable after the establishment phase as yields plateau (e.g., 4 years, Clifton-Brown *et al.*, 2011). Foster *et al.* (2013) found that seedling number in switchgrass increased from 2.2 to 11.2 kg PLS ha^{-1}, but biomass yields were not significantly different at the end of the establishment year at two sites in southern Oklahoma, USA. This growth plasticity can be seen in decreasing number of tillers and leaves per plant in response to higher plant density in switchgrass

(Sanderson & Reed, 2000). A reduction in yield in response to excessive plant density is expected, but this is less apparent for biomass yield than grain yield (e.g., corn; (Alessi & Power, 1974). Such supra-optimum plant densities have yet to be identified in miscanthus and were not observed in this study; however with the expected reduction in propagule costs of seeded miscanthus, this becomes a relevant question. Despite this, since reseeding can increase production cost by 36% which is not recovered by increased yield (Perrin *et al.*, 2008), it may be advantageous to err on higher initial seeding rates.

Biomass yield for Hydro at 10 seeds m^{-2} was higher than Hydro at the two higher seeding rates although the differences were not significant. There are several sources of variability that could account for this unpredicted result. First, the plasticity of seeded miscanthus probably caused plots with lower plant density to produce biomass yields higher than were expected based on seeding rate. Second, the seeds for this experiment were from a segregating population, and it is possible that even with thorough seed mixing that this treatment may have ended up with a greater proportion of seeds from higher yielding plants. Finally, blocking in the east–west direction probably did not account for all in-field variability as there was also a north–south yield gradation in the 2012–2013 field. It is possible that plot randomization resulted in Hydro plots with 10 seeds m^{-2} being located in areas of the field with higher fertility.

Establishment of perennial grass from seed under dry conditions can be challenging. It is unlikely that grass crops grown for biofuels will be grown under irrigation due to the relatively brief period of greatest need during the establishment year and the high cost of an irrigation system compared with the price of biomass. In this initial investigation in which water was either limiting or non-limiting, results clearly showed that adequate soil moisture was necessary for longer than 4 weeks after planting under conditions extant during this experiment. In Illinois, *M. × giganteus* cv. Illinois is commonly planted from late March through mid-May using rhizomes or plugs. Moisture from snow melt and rainfall is commonly available to these spring plantings, and even during dry springs, both rhizomes and plugs are able to access soil moisture and become established due to the ~10 cm planting depth of rhizomes and the ~12 cm-long plug root mass. In fact, plugs planted during May 2011 (approximately 1 month before this study's 2011 seeding) in an unirrigated field at the same research farm established well (personal observations). Future research with varying levels and timings of irrigation will be needed to better determine the water needs and optimum geographic regions for establishing this seeded crop.

With irrigation, both hydroseeding and seed drilling at 20 or 40 seeds m^{-2} resulted in adequate seeded miscanthus stand establishment to ensure second-year biomass yields on par with *M. × giganteus* cv. Illinois. Further research is needed to identify regions and improved agronomics to enable miscanthus direct seeding without irrigation.

Acknowledgements

This research was funded by Mendel Biotechnology Inc., the Crop Sciences Department at the University of Illinois at Urbana-Champaign, and the Energy Biosciences Institute. We would like to thank Allen Parrish, Callan Beeson, Santanu Thapa, Collin Reeser, Chris Rudisill, and Tim Mies for technical assistance. Thanks to Bill Keyser of John Deere Landscaping in Champaign, IL for his assistance with irrigation system design. We are also grateful to Xianming Duan (Mendel Biotechnology Inc.) for providing the seeds and advice on miscanthus seed physiology.

References

Alessi J, Power JF (1974) Effects of plant population, row spacing, and relative maturity on dryland corn in the Northern Plains. I. Corn forage and grain yield. *Agronomy Journal*, **66**, 316–319.

Andrade F, Calvino P, Cirilo A, Barbieri P (2002) Yield responses to narrow rows depend on increased radiation interception. *Agronomy Journal*, **94**, 975–980.

Anonymous (2010) *Harness Xtra Supplemental Label, for Weed Control in Miscanthus & Other Non-Food Perennial Bioenergy Crops*. Monsanto Company, St. Louis, MO.

Arundale RA, Dohleman FG, Heaton EA, McGrath JM, Voigt TB, Long SP (2014) Yields of Miscanthus × giganteus and Panicum virgatum decline with stand age in the Midwestern USA. *Global Change Biology Bioenergy*, **6**, 1–13.

Balkcom KS, Satterwhite JL, Arriaga FJ, Price AJ, Van Santen E (2011) Conventional and glyphosate-resistant maize yields across plant densities in single- and twin-row configurations. *Field Crops Research*, **120**, 330–337.

Barney JN, DiTomaso JM (2008) Nonnative species and bioenergy: are we cultivating the next invader? *BioScience*, **58**, 64–70.

Beard JB (1973) *Turfgrass: Science and Culture*. Prentice-Hall Inc, Englewood Cliffs, NJ.

Berti MT, Johnson BL (2013) Switchgrass establishment as affected by seeding depth and soil type. *Industrial Crops and Products*, **41**, 289–293.

Bullock D, Nielsen R, Nyquist W (1988) A growth analysis comparison of corn grown in conventional and equidistant plant spacing. *Crop Science*, **28**, 254–258.

Chavez JL, Neale CMU, Prueger JH, Kustas WP (2008) Daily evapotranspiration estimates from extrapolating instantaneous airborne remote sensing ET values. *Irrigation Science*, **27**, 67–81.

Christian EJ (2012) Seed development and germination of Miscanthus Sinensis. Graduate Theses and Dissertations, Paper 12880, p. 75. Iowa State University, Ames.

Christian DG, Yates NE, Riche AB (2005) Establishing *Miscanthus sinensis* from seed using conventional sowing methods. *Industrial Crops and Products*, **21**, 109–111.

Clifton-Brown J, Robson P, Sanderson R, Hastings A, Valentine J, Donnison I (2011) Thermal requirements for seed germination in miscanthus compared with switchgrass (*Panicum virgatum*), reed canary grass (*Phalaris arundinaceae*), maize (*Zea mays*) and perennial ryegrass (*Lolium perenne*). *Global Change Biology Bioenergy*, **3**, 375–386.

Davis SC, Parton WJ, Del Grosso SJ, Keough C, Marx E, Adler PR, DeLucia EH (2012) Impact of second-generation biofuel agriculture on greenhouse-gas emissions in the corn-growing regions of the US. *Frontiers in Ecology and the Environment*, **10**, 69–74.

EISA (2007) Energy Independence and Security Act of 2007 (EISA), Pub. L. No. 110-140, 121 Stat. 1492, 1783-84 (Dec. 19, 2007), codified at 42 U.S.C. §17381.

Foster JL, Guretzky JA, Huo C, Kering MK, Butler TJ (2013) Effects of row spacing, seeding rate, and planting date on establishment of switchgrass. *Crop Science*, **53**, 309–314.

Głowacka K, Clark LV, Adhikari S *et al.* (2013) Genetic variation in *Miscanthus × giganteus* and the importance of estimating genetic distance thresholds for differentiating clones. *Global Change Biology Bioenergy*, doi: 10.1111/gcbb.12166.

Gordon DR, Tancig KJ, Onderdonk DA, Gantz CA (2011) Assessing the invasive potential of biofuel species proposed for Florida and the United States using the Australian weed risk assessment. *Biomass and Bioenergy*, 35, 74–79.

Greef JM, Deuter M, Jung C, Schondelmaier J (1997) Genetic diversity of European Miscanthus species revealed by AFLP fingerprinting. *Genetic Resources and Crop Evolution*, 44, 185–195.

Hodkinson TR, Renvoize S (2001) Nomenclature of Miscanthus × giganteus (Poaceae). *Kew Bulletin*, 56, 759–760.

Houk MJ (2009) Conservation planting methods for native and introduced species. Plant Materials Technical Note No. 14. United State Department of Agriculture Natural Resources Conservation Service. Available at: http://www.nrcs.usda.gov/Internet/FSE_PLANTMATERIALS/publications/lapmctn9048.pdf. (accessed 24 February 2014).

Keene T, Skousen J (2010) Switchgrass production potential on reclaimed surface mines in West Virginia. In: *Proceedings of the 2010 National Meeting, American Society for Mining and Reclamation, 5–11 June 2010.* American Society for Mining and Reclamation, Pittsburgh, PA. Available at: http://anr.ext.wvu.edu/r/download/77088. (accessed 24 February 2014).

Kim SM, Rayburn AL, Voigt T, Parrish A, Lee DK (2012) Salinity effects on germination and plant growth of prairie cordgrass and switchgrass. *Bioenergy Research*, 5, 225–235.

Knezevic S, Evans S, Mainz M (2003) Row spacing influences the critical timing for weed removal in soybean (*Glycine max*). *Weed Technology*, 17, 666–673.

Lee DK, Parrish AS, Voigt T (2014) Switchgrass and giant miscanthus agronomy. In: *Engineering and Science of Biomass Feedstock Production and Provision* (eds Shastri Y, Hansen A, Rodriguez L, Ting KC), pp. 37–60. Springer Science + Business Media, New York.

Lewandowski I, Scurlock JMO, Lindvall E, Christou M (2003) The development and current status of perennial rhizomatous grasses as energy crops in the US and Europe. *Biomass and Bioenergy*, 25, 335–361.

Linde-Laursen I (1993) Cytogenetic analysis of *Miscanthus-giganteus*, an interspecific hybrid. *Hereditas*, 119, 297–300.

Miguez FE, Villamil MB, Long SP, Bollero GA (2008) Meta-analysis of the effects of management factors on Miscanthus × giganteus growth and biomass production. *Agricultural and Forest Meteorology*, 148, 1280–1292.

Monti A, Venturi P, Elbersen HW (2001) Evaluation of the establishment of lowland and upland switchgrass (*Panicum virgatum* L.) varieties under different tillage and seedbed conditions in northern Italy. *Soil and Tillage Research*, 63, 75–83.

Murphy F, Devlin G, McDonnell K (2013) Miscanthus production and processing in Ireland: an analysis of energy requirements and environmental impacts. *Renewable and Sustainable Energy Reviews*, 23, 412–420.

Panter DM (2010) Breeding and commercializing miscanthus as a biofuels crop for the future. In: *Proceedings of the Southeast Bioenergy Conference 2010*, Tifton, GA. 1–23. Available at: http://www.sebioenergy.org/2010/PDF_10/August_4/Small_Auditorium/1.30-3.00/Panter,%20DM%20-%20Breeding%20and%20Commercializing%20Miscanthus%20as%20a%20Biofuels%20Crop%20%28Final%29.pdf. (accessed 24 February 2014).

Parrish DJ, Fike JH, Bransby DI, Samson R (2008) Establishing and managing switchgrass as an energy crop. *Forage and Grazinglands*, doi: 10.1094/FG-2008-0220-01-RV.

Perrin R, Vogel K, Schmer M, Mitchell R (2008) Farm-scale production cost of switchgrass for biomass. *Bioenergy Research*, 1, 91–97.

Pyter R, Heaton E, Dohleman F, Voigt T, Long S (2009) Agronomic experiences with Miscanthus × giganteus in Illinois, USA. *Methods in Molecular Biology*, 581, 41–52.

Quinn LD, Allen DJ, Stewart JR (2010) Invasiveness potential of *Miscanthus sinensis*: implications for bioenergy production in the United States. *Global Change Biology Bioenergy*, 2, 310–320.

Quinn LD, Stewart JR, Yamada T, Toma Y, Saito M, Shimoda K, Fernandez FG (2012) Environmental tolerances of *Miscanthus sinensis* in invasive and native populations. *Bioenergy Research*, 5, 139–148.

Rayburn AL, Crawford J, Rayburn CM, Juvik JA (2009) Genome size of three *Miscanthus* species. *Plant Molecular Biology Reporter*, 27, 184–188.

Sacks EJ, Jakob K, Gutterson NI (2013) High biomass *Miscanthus* varieties. *United States Plant Patent Application Publication*, 13/513, 173, 1–24.

Sanderson MA, Reed RL (2000) Switchgrass growth and development: water, nitrogen, and plant density effects. *Journal of Range Management*, 53, 221–227.

Saxton AM (1998) A macro for converting mean separation output to letter groupings in Proc Mixed. In: *Proceedings 23rd SAS Users Group Intl., Nashville, TN.* SAS Institute, Cary, NC. 1243–1246.

Schnepf R, Yacobucci BD (2013) Renewable fuel standard: overview and issues. Congressional Research Service Report for Congress. 14 March, 2013. Available at: http://www.fas.org/sgp/crs/misc/R40155.pdf. (accessed 24 February 2014).

Slomka A, Kuta E, Plazek A *et al.* (2012) Sterility of Miscanthus × giganteus results from hybrid incompatibility. *Acta Biologica Cracoviensia Series Botanica*, 54, 113–120.

Smith LL, Barney JN (2014) The relative risk of invasion: evaluation of *Miscanthus × giganteus* seed establishment. *Invasive Plant Science and Management*, 7, 93–106.

Teel A, Barnhart S, Miller G (2003) Management guide for the production of switchgrass for biomass fuel in Southern Iowa. Iowa State University Extension, Ames, IA. Available at: http://www.extension.iastate.edu/publications/pm1710.pdf. (accessed 24 February 2014).

USDA (2012) Biomass crop assistance program project areas listing. Available at: http://www.fsa.usda.gov/FSA/webapp?area=home & subject=ener & topic=bcap-pjt-bloc. (accessed 24 February 2014).

USDOE (2011) *US Billion Ton Update: Biomass Supply for a Bioenergy and Bioproducts Industry.* Perlack RD, Stokes BJ (Leads), ORNL/TM-2011/224. Oak Ridge National Laboratory, Oak Ridge, TN.

Vassey TL, George JR, Mullen RE (1985) Early-spring, mid-spring, and late-spring establishment of switchgrass at several seeding rates. *Agronomy Journal*, 77, 253–257.

Vogel KP (1987) Seeding rates for establishing big bluestem and switchgrass with preemergence atrazine applications. *Agronomy Journal*, 79, 509–512.

Vogel KP, Masters RA (2001) Frequency grid - A simple tool for measuring grassland establishment. *Journal of Range Management*, 54, 653–655.

Wolf DD, Fiske DA (1995) Planting and managing switchgrass for forage, wildlife, and conservation. Virginia Cooperative Extension Publication 418–013. Available at: http://pubs.ext.vt.edu/418/418-013/418-013.html. (accessed 24 February 2014).

Young J, Kay B, Evans R (1977) Accelerating germination of common bermudagrass for hydroseeding. *Agronomy Journal*, 69, 115–119.

5

Biochar amendment of soil improves resilience to climate change

ROGER T. KOIDE[1], BINH T. NGUYEN[2], R. HOWARD SKINNER[3], CURTIS J. DELL[3], MATTHEW S. PEOPLES[4], PAUL R. ADLER[3] and PATRICK J. DROHAN[2]

[1]Department of Biology, Brigham Young University, Provo, UT 84602, USA, [2]Department of Ecosystem Science and Management, The Pennsylvania State University, University Park, PA 16802, USA, [3]Pasture Systems and Watershed Management Research Unit, USDA-ARS, University Park, PA 16802, USA, [4]Department of Plant Science, The Pennsylvania State University, University Park, PA 16802, USA

Abstract

Because of climate change, insufficient soil moisture may increasingly limit crop productivity in certain regions of the world. This may be particularly consequential for biofuel crops, many of which will likely be grown in drought-prone soils to avoid competition with food crops. Biochar is the byproduct of a biofuel production method called pyrolysis. If pyrolysis becomes more common as some scientists predict, biochar will become more widely available. We asked, therefore, whether the addition of biochar to soils could significantly increase the availability of water to a crop. Biochar made from switchgrass (*Panicum virgatum* L.) shoots was added at the rate of 1% of dry weight to four soils of varying texture, and available water contents were calculated as the difference between field capacity and permanent wilting point water contents. Biochar addition significantly increased the available water contents of the soils by both increasing the amount of water held at field capacity and allowing plants to draw the soil to a lower water content before wilting. Among the four soils tested, biochar amendment resulted in an additional 0.8–2.7 d of transpiration, which could increase productivity in drought-prone regions or reduce the frequency of irrigation. Biochar amendment of soils may thus be a viable means of mitigating some of the predicted decrease in water availability accompanying climate change that could limit the future productivity of biofuel crops.

Keywords: available soil water, biochar, bioenergy, climate change, field capacity, pressure plate, soil texture, switchgrass, transpiration, wilting point

Introduction

Soil moisture limits plant productivity in many areas of the world (Seneviratne *et al.*, 2010). In some regions, this limitation may be exacerbated by predicted climate change as a consequence of elevated temperature, reduced rainfall, changing seasonal rainfall patterns, or reduced availability of irrigation water (Schlenker *et al.*, 2007; Vano *et al.*, 2010; O'Neill & Dobrowolski, 2011). Insufficient soil moisture may become particularly problematic for biofuel crops; to reduce competition with food crops, biofuel crops may have to be planted preferentially on sites that are not highly productive for food crops (Vuichard *et al.*, 2009; Cai *et al.*, 2011; Niblick *et al.*, 2013), where some of the challenges include shallow and excessively draining soils. Improving the availability of water in these challenging soils will be of great importance to biofuel production.

An increase in the soil available water content is possible by their amendment with hydrogels, polymers that can hold water at hundreds of times their own weight (Abedi-Koupai *et al.*, 2008). However, the cost and longevity (Frantz *et al.*, 2005) of such materials make their use at the field-scale very unlikely. In this contribution, we ask whether amendment of soils with biochar can similarly increase available water content. In contrast to hydrogels, biochar (one of a continuum of thermal conversion products of organic materials) may be far more temporally stable in soil and, because it is a byproduct of the production of biofuels via pyrolysis, its application may become cost-effective as syngas and bio-oil production increase (Laird, 2008). Considerable interest already exists in the amendment of soils with biochar because it can stably sequester significant amounts of carbon (Nguyen *et al.*, 2014) and, under certain circumstances, significantly improve crop yield (Spokas *et al.*, 2012).

Correspondence: Roger T. Koide
e-mail: rogerkoide@byu.edu

Some research indicates that amendment of coarse-textured soils with biochar increases their capacity to absorb water (Laird *et al.*, 2010; Karhu *et al.*, 2011; Novak *et al.*, 2012; Novak & Watts, 2013). The ability to absorb water is important in situations, where excess drainage leads to loss of soil or nutrients. But simply holding more water does not necessarily result in more water being available to the plant, particularly if the water is held so tightly that the plant has no access to it. Thus, in situations in which water stress limits crop yield, the effect of biochar on soil water-holding capacity per se is less relevant than the effect of biochar on the amount of water that is actually available to the plant, the so-called available water content. There is frequently a positive relationship between soil available water content and plant yield, particularly in arid climates (Wong & Asseng, 2006; Lawes *et al.*, 2009).

Available water content is defined as the difference in water content held at field capacity and at the permanent wilting point of the plant. Increased available water content due to biochar amendment can result from either improved water content at field capacity or reduced water content at permanent wilting point, or both. Most biochars made from plant materials have a high porosity and surface area (Downie *et al.*, 2009) and thus a large capacity to hold water at field capacity (Glaser *et al.*, 2002). It is not clear what to expect from biochar near the wilting point. Therefore, the current study was conducted to examine the effects of switchgrass derived biochar on the water content of soils at both field capacity and the wilting point. We determined available water content using a bioassay. While it is common to use the pressure plate apparatus for this purpose, we showed that it may not be appropriate for soils containing biochar. Because soil texture is known to influence soil hydraulic properties, we added biochar to four soils varying significantly in texture. Finally, because biochar is significantly less dense than mineral soil, biochar addition to soil can appear to increase the water content of soil on a weight basis merely by decreasing its density. Therefore, we determined available soil water content on the basis of soil volume rather than weight.

Materials and methods

Soils

The effect of biochar on the water-holding properties of soils is likely dependent on several factors including soil texture (and thus inherent water-holding capacity). Therefore, we chose to study four soils that varied significantly in texture. Soils were collected from four sites across the Appalachian Plateau and Ridge and Valley Physiographic Provinces in Pennsylvania, United States, which are currently part of a regional study on the use of biochar to produce switchgrass (*Panicum virgatum* L.) in soils from sites that are marginally productive for row crops. Two soils, the Edom (collected at 77°54′32.848″W 40°36′56.727″ N) and Wharton (78°9′41.034″W 41°1′55.388″N) silty clay loams, are from sites that frequently experience extended periods of excessive wetness. The other two soils, the Morrison sandy loam (77°53′55.5″W 40°49′34.062″N) and the Weikert loam (77°53′ 38.94″W 40°39′8.106″N) are from excessively drained sites. Soils were sampled in September 2011 from the plow layer (A over Ap, or Ap horizon) to a depth of 0.15 m. Seven to ten positions at each site were sampled, and the samples were combined. All soils were air-dried in a greenhouse and passed through a 2 mm sieve. Soil C concentration was determined using an Elementar Vario-Max instrument on soils that were finely ground using mortar and pestle. Soil pH was measured with a standard glass electrode using a 1 : 1 soil:distilled water (w : v) ratio following a 30 min equilibration. The soils are further described in Table 1.

To determine the moisture release characteristics of the soils, we used pressure plate extractors (Soilmoisture Equipment Corp, Santa Barbara, CA, USA) to assess the relationship between tension and water content for the four soils. Gravimetric water content was determined at tensions of −0.010, −0.033, −0.100, −0.300, −0.500, −1.00, and −1.50 MPa (Dane & Hopman, 2002). Briefly, rubber rings (5 cm diameter by 1 cm high) were placed on porous, ceramic plates of the appropriate bubbling pressure and filled with 2 mm sieved, air-dried soils. The ceramic plates were placed in plastic pans and a sufficient quantity of deionized water was added to cover the surface of the plates. Plates were left in the pans overnight to allow soil to saturate by capillary action. Once the soils were saturated, the ceramic plates were transferred to pressure chambers. The desired pressure was exerted on each chamber until outflow of water from the pressure chamber was no longer observed. This usually required at least several hours, but the chamber was

Table 1 Mean (SEM) of some characteristics of the four soils used in this study. $n = 4$

Soil series	Sand (%)	Silt (%)	Clay (%)	Total C* (%)	pH†
Edom silty clay loam (Typic Hapludalf)	6 (0.1)	62 (0.5)	32 (0.3)	3.8 (0.3)	6.0 (0.08)
Weikert loam (Lithic Dystrudept)	44 (0.7)	40 (0.6)	16 (0.3)	4.0 (0.2)	6.0 (0.03)
Morrison sandy loam (Ultic Hapludalf)	61 (0.2)	27 (0.2)	12 (0.1)	1.7 (0.1)	5.9 (0.02)
Wharton silty clay loam (Aquic Hapludult)	15 (0.3)	58 (0.3)	27 (0.1)	4.1 (0.6)	5.4 (0.03)

*Total C is assumed to be equivalent to organic C at the given pH values.
†In water.

allowed to equilibrate in each case for 24 h. In any case, the criterion for equilibrium was no further water loss. Upon reaching equilibrium, the plates were removed from the chambers and the gravimetric water contents of soil samples were determined from the wet weights and dry weights (assessed after heating to 105 °C for 24 h). There were three replicates at each tension for each soil.

Biochar

Switchgrass is a perennial grass, and a potentially important bioenergy crop (Parrish & Fike, 2005). It has been grown successfully on sites that are marginally productive for row crops (Keshwani & Cheng, 2009). Biochar was produced from shoots of harvested switchgrass (*P. virgatum var.* Cave-In-Rock) by the torrefaction facility at North Carolina State University (Raleigh, NC, USA). The switchgrass had been produced by Ernst Conservation Seeds (Meadville, PA, USA). Approximately 4.5 dry weight units of air-dry grass shoots were used to produce one dry weight unit of biochar. Biochar production consisted of several steps. First, the switchgrass biomass was preheated to about 100 °C in a hopper before being transferred to and held in the torrefaction chamber by constant stirring with an auger for 1–1.5 min. The torrefaction chamber had a low oxygen environment and was held between 375 and 475 °C. The biochar was then cooled in an exit auger for about 3 min, achieving a near ambient temperature (35 °C). Finished biochar was hydrated to 60% water content by weight to prevent self-heating. After allowing it to stand for 24 h to reach hydration/oxidation equilibrium with the atmosphere, the biochar was packaged and shipped to our research facility in Pennsylvania. Biochar pH was measured with a standard glass electrode using a 1 : 20 biochar:distilled water (w : v) ratio following a 30 min equilibration. Ash content was determined by weighing the residue after heating to 550 °C for 4 h. Total C and N concentrations were determined using an Elementar Vario-Max elemental analyzer. Bulk density was calculated from the dry weight (105 °C) of known volumes.

The mean (se) biochar bulk density was approximately 0.083 (0.004) g cm^{-3} ($n = 9$), the ash content was 10.2 (0.9)% of dry weight ($n = 3$), the C concentration was 74.6 (1.2)% of dry weight ($n = 9$), the N concentration was 1.16 (0.02)% of dry weight ($n = 9$), the C : N ratio was 64.8 (2.1, $n = 9$), and the pH in water was 9.5 (0.1, $n = 16$). Only 3% of the biochar (by weight) was >2.0 mm diameter, while 93% was in the 0.05–2.0 mm fraction, and 4% was <0.05 mm.

The procedure used to determine the moisture release curves of the soils was also used for whole biochar pieces, biochar that was finely ground, and ground biochar mixed with whole biochar pieces (1 : 1 weight ratio), with three replicates at each tension.

Soil-biochar mixtures

Soils were compared with mixtures of soil with 1% biochar by weight. To prepare the mixtures, each soil was mixed with 1% biochar by weight, equivalent to approximately 10 tonne ha^{-1} to a depth of approximately 15 cm. Although the biochar was

1% of soil dry weight, it comprised approximately 10–13% by volume depending on the soil and experiment (see Results).

Available water content

We calculated available water content as the difference in water content of the media when allowed to drain by gravity to field capacity, and at the permanent wilting point, as in Meyer & Green (1980). Four subsamples from each of four unamended soils, biochar and the soil-biochar mixtures (1% biochar by weight) were taken for measurement of water content at field capacity using the Büchner funnel method (Veihmeyer & Hendrickson, 1949) with some modifications. A 186 ml Nalgene Büchner funnel was used for the measurement with Whatman filter paper No. 2 to cover the funnel to prevent soil loss during drainage. A 53.1 cm^3 PVC ring (5.2 cm diameter × 2.5 cm height) was placed on the filter. Fifty to seventy grams of medium, depending on the medium, were weighed and packed into the PVC ring to the bulk densities given in Table 2. To saturate the soil, the removable top portion of the funnel was then placed overnight in a plastic tray containing distilled water to the level of half the ring's height. On the following day, the top portion of the funnel was attached to the bottom portion and allowed to stand for 48 h to permit free water drainage (Sarkar & Haldar, 2005). The top of the funnel was covered with parafilm to prevent evaporation. The media in the rings were collected for determination of water content by weighing both before and after overnight drying at 105 °C.

A completely randomized experiment with four replicates was designed to determine medium water content at the permanent wilting point. The soils, biochar and their mixtures were packed into 1.67-l standard plastic pots, tapping the pot vigorously on a hard surface on five or six separate occasions during the filling process To compact the medium to the bulk densities given in Table 2. The pots were then randomly placed in plastic trays, and 3–5 cm of water was added to the trays to moisten the media. Maize (*Zea mays* L.) seeds (Pioneer PO891AM1 Des Moines, IA, USA) were pregerminated on moist filter paper. Seedlings were transplanted into the pots (five seedlings per pot) and given Hoagland's nutrient solution (Machlis & Torrey, 1956) daily to maintain rapid growth. All trays were placed randomly in one growth chamber with a constant 50% relative humidity and a thermal cycle of 30/22 °C day/night with a 14 h photoperiod. The maximum photosynthetically active radiation flux density was approximately 700 μmol m^{-2} s^{-1} at plant height. Approximately 2 weeks after transplanting, watering was stopped to allow the plants to wilt. To prevent the media from losing water by direct evaporation, the drainage holes at the bottom of each pot were covered with polyethylene, and the surface of the soil, biochar or soil-biochar mixture was covered with a 2.5 cm layer of horticultural perlite. To determine the permanent wilting point, we started to track transpiration rates as soon as water was withheld, before the leaves wilted (data not shown). Transpiration rates were determined using a steady state porometer (LI-1600, LI-COR, Lincoln, NE, USA) for several days, 2 h after the initiation of the light period on a recently fully expanded leaf. When

Table 2 Mean bulk densities and volumetric proportions of biochar in the soil-biochar mixtures used in the wilting point and field capacity determinations of water content. $n = 4$

Soil	Biochar*	Wilting point		Field capacity	
		Bulk density (g cm^{-3})	Volumetric proportion biochar	Bulk density (g cm^{-3})	Volumetric proportion biochar
Edom	no	1.05	0.00	1.07	0.00
Edom	yes	0.94	0.10	0.97	0.11
Wharton	no	1.23	0.00	1.07	0.00
Wharton	yes	1.11	0.12	1.01	0.11
Weikert	no	1.05	0.00	1.01	0.00
Weikert	yes	0.94	0.10	0.99	0.11
Morrison	no	1.37	0.00	1.31	0.00
Morrison	yes	1.17	0.12	1.22	0.13
Experimental SEM		0.01	0.001	0.01	0.001

*1% of dry weight.

transpiration rate was nil, the pot was transferred to a dark growth chamber to allow leaf water potential to equilibrate with soil water potential. After four hours in the dark, the plants were visually examined to determine if the leaves had regained turgor. Pots with leaves that recovered turgor were placed back in the lighted growth chamber. Pots with leaves that had not regained turgor had leaf water potential measured using a pressure chamber. Three subsamples of the medium in the pot were taken for water content determination by weighing the subsamples before and after 105 °C overnight drying. Leaf water potentials at the wilting point were less than −1.9 MPa for all soil × biochar treatment combinations (including 100% biochar). The average was −2.2 MPa (SE = 0.5). Volumetric water content (g cm^{-3}) was calculated by multiplying bulk density by gravimetric water content.

We calculated the available water content of biochar, soils, and soils mixed with biochar as the difference in water content between field capacity and wilting point.

Statistical analysis

Available water content. There were four replicate measurements of water content at field capacity and four replicates at the permanent wilting point. Available water content is the difference between field capacity and permanent wilting point water contents. To determine the means and standard deviations for the differences in each treatment combination (four soils with and without biochar), frequency distributions of the differences were constructed in the following way. For a given treatment combination, the differences were calculated for all possible combinations of the four replicate water contents at wilting point and the four replicate water contents at field capacity. These 16 differences were then randomly sampled (four at a time) 1000 times with replacement. The 1000 means of the four drawn samples were then taken to bootstrap a frequency distribution of the differences, and the mean and standard deviation were determined. Single factor analyses of variance were performed according to Cohen (2002), and means were separated using the 95% confidence intervals.

Field capacity and permanent wilting point water contents. The interactive effects of four soils and two biochar treatments (0% or 1% by weight) on field capacity and wilting point water contents were determined using the analysis of variance procedures in the Statgraphics programs (STSC, 1991).

Results

As expected based on soil textural properties, the four soils exhibited very different moisture release properties (Fig. 1). At any given tension, the sandiest soil (Morrison) had the lowest water content, while the most clayey soil (Edom) had the highest water content. The soils of intermediate texture (Weikert and Wharton) had intermediate moisture release curves.

Whole biochar particles (size range given above) exhibited a moisture release curve that suggested, contrary to expectations, that the least amount of water was

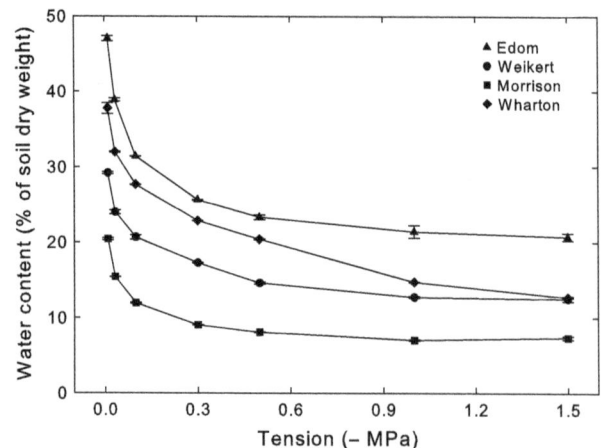

Fig. 1 Moisture release curves for the four soils (unmixed with biochar). Error bars are ±1 SEM. In some cases, errors are smaller than symbols. $n = 3$.

held at lower tensions (-0.033 and -0.100 MPa) and the most at higher tensions (-0.300 to -1.50 MPa) (Fig. 2), suggesting that the pressure plate method may not be useful in describing the moisture release characteristics of this particular biochar. We hypothesized that for whole biochar pieces in the pressure plate apparatus, hydraulic connection was lost either among the individual pieces of biochar, or between the pieces of biochar and the ceramic plate. To test that hypothesis, we used biochar that had been ground to readily pass through a 1 mm sieve using a mortar and pestle as we felt this material would not suffer from the loss of hydraulic connection. We subsequently refer to this as powdered biochar. When powdered biochar was used, the moisture release curve was far closer to what might be expected, with ever-decreasing water contents as tensions increased. When whole and powdered biochar was mixed 1 : 1 (w : w), the moisture release curve was intermediate. This would only be possible if some water were not draining from the whole biochar, suggesting that some loss of hydraulic connectivity does occur in the pressure plate apparatus with whole biochar. Thus, we chose an alternative method to determine available water content of soil and soil-biochar mixtures.

The available water content of biochar, calculated from field capacity and permanent wilting point water contents, was significantly greater than for any of the soils (Fig. 3). The soils themselves varied significantly ($P < 0.05$) in available water content, blending this particular biochar into the soils at 1% of dry weight significantly ($P < 0.05$) increased available water content of all the soils, and the magnitude of the effect of the biochar depended ($P < 0.05$) on the soil (Fig. 3). The increases in available water content due to this biochar were 0.0157, 0.0165, 0.0263, and 0.0549 g cm^{-3} for the Edom, Wharton, Weikert, and Morrison soils, respectively.

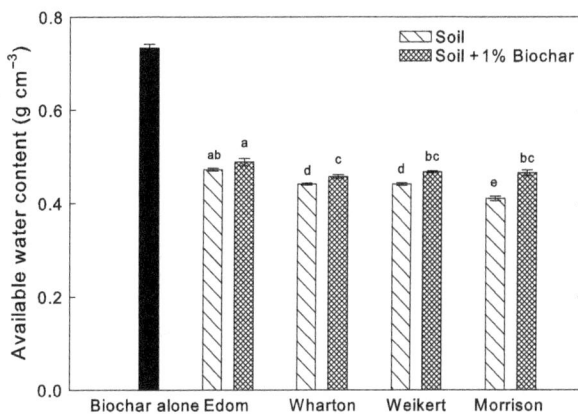

Fig. 3 Mean available water content (g cm^{-3}) of soils and soils mixed with biochar (1% by weight). Available water content was calculated as the difference between water content at field capacity and water content at permanent wilting point. Error bars are ± 1 SEM. $n = 4$. Different letters indicate significantly different means according to 95% confidence intervals.

The available water contents of the soils were influenced by mixture with this biochar in two ways. The biochar significantly ($P < 0.0005$) increased the water content at field capacity. The average increase for the four soils was 0.011 g cm^{-3}, although the effect of the biochar depended ($P < 0.005$) on the soil (Fig. 4a). For example, the biochar effect was large for the Morrison soil (0.0298 g cm^{-3}) and small for Wharton soil (0.0002 g cm^{-3}). The biochar also significantly ($P < 0.0001$) decreased the water content at the permanent wilting point, indicating that, prior to wilting, the plants were able to extract more water from soil when it was mixed with the biochar. Overall, the decrease in water content at permanent wilting due to the biochar amendment was 0.017 g cm^{-3}, (Fig. 4b). In the case of permanent wilting point water content, there was no significant interaction ($P = 0.413$) between soil and biochar amendment.

Discussion

Our goal was to determine the effect of the switchgrass biochar on soil available water content. We utilized four soils that differed significantly in their moisture release characteristics. The addition of the biochar at the rate of 1% by weight (equivalent to between 10% and 13% biochar by volume, or 10 tonne ha^{-1}) resulted in significant increases in available water content.

The increase in available soil water content due to the biochar amendment was biologically significant. This is readily illustrated using the results for the Morrison soil. The biochar amendment increased the available water content of this soil by 0.0551 g cm^{-3}. If the biochar were tilled in to a depth of 15 cm in the field,

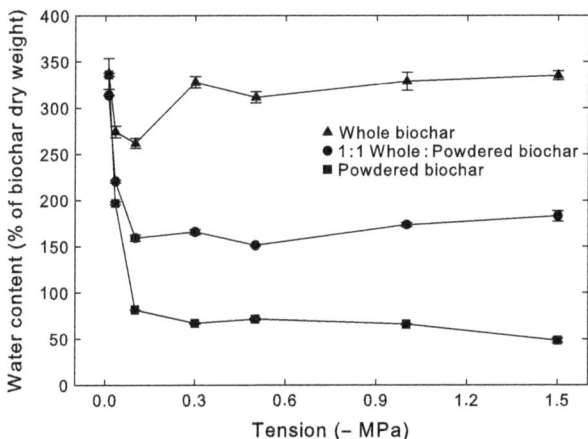

Fig. 2 Moisture release curves of whole and powdered biochars, and a 1 : 1 mixture of the two. Error bars are ± 1 SEM. In some cases, errors are smaller than symbols. $n = 3$.

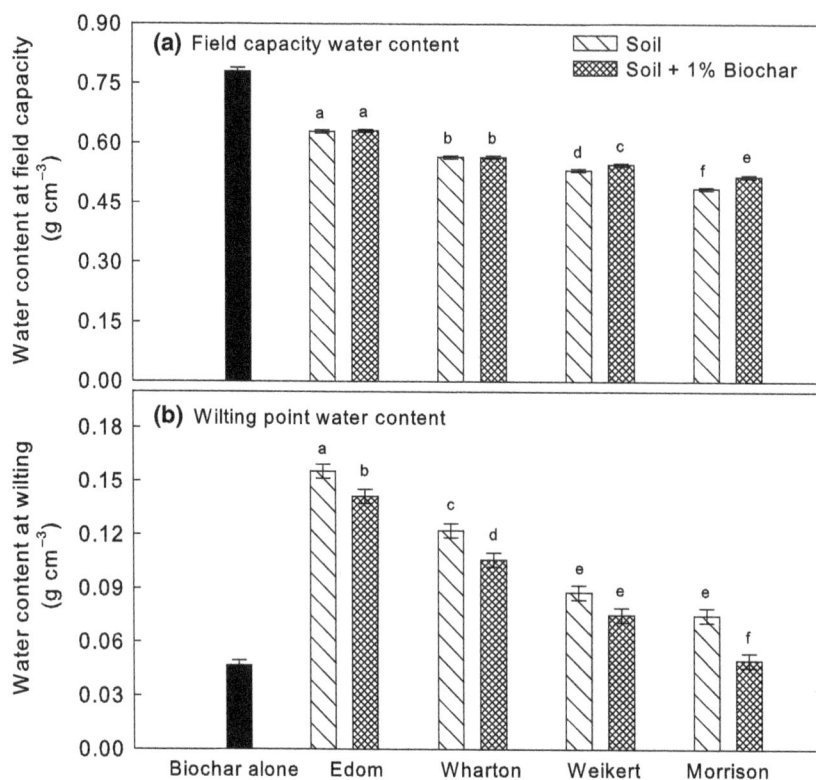

Fig. 4 Water content at field capacity (a) and at wilting point (b) for biochar, the four soils, and the four soils mixed with biochar (1% by weight). Error bars are ±1 SEM. $n = 4$. Different letters indicate significantly different means according to 95% confidence intervals.

which is a common depth for tillage implements, the increase in available soil water in the top 15 cm of soil would amount to 8270 g m^{-2} (0.0551 g cm^{-3} × 150 000 cm^3 m^{-2}). Maximal transpiration rates from a mature switchgrass canopy in Pennsylvania during the summer have been documented from eddy covariance measurements to be approximately 3 mm d^{-1} or 3000 g m^{-2} (Skinner & Adler, 2010). Thus, for a single drying event the increase in available soil water in the Morrison soil due to biochar amendment would amount to 2.7 additional days of full transpiration. Even for the Edom soil, for which the biochar had the least effect on available water content, the biochar amendment would provide approximately 0.8 additional days of full transpiration. This suggests that if water shortage occurred frequently enough, soils amended with this biochar might allow plants to have a significant advantage in terms of photosynthesis (Seneviratne et al., 2010) as well as lessening plant stress, consistent with the relationship between available water content and crop yield in arid climates (Wong & Asseng, 2006; Lawes et al., 2009). In some regions of the United States, there may be both more frequent and longer lasting periods of drought (Cook et al., 2004). The resilience of agricultural systems to that kind of climate change may be improved by

addition to the soil of biochars with similar properties to the one used in this study. While many have considered biochar amendments to be primarily a means of sequestering C in soils over the long term (Nguyen et al., 2014), they may have the additional benefit of making more water available to the crop. In this study, the biochar amendment increased soil available water content in two ways. It increased soil water content at field capacity and allowed plants to draw the soil to lower water content before wilting.

Several studies have investigated the effects of biochar amendment on soil hydraulic properties but, for a variety of reasons, most of them are difficult to interpret. A common difficulty stems from expressing water content on a gravimetric basis, i.e. g water per g soil (see Ulyett et al., 2014, Novak & Watts, 2013; Briggs et al., 2012; Novak et al., 2012; Karhu et al., 2011; Laird et al., 2010). That is problematic when dealing with biochar because, owing to its very low bulk density, when added to soil it significantly increases the volume of a given weight. Thus, because a gram of soil and a gram of soil-biochar mixture are not equivalent in volume, the very basis for the comparison is called into question.

The works by Pereira et al. (2012) and Abel et al. (2013) are exceptional in that they characterized available water on a volumetric basis. As did we, they both found that addition of biochar to soil increased the water available to a plant, the difference between water content at field capacity and at the wilting point. We found that the biochar effect on available water was more pronounced in a coarsely textured soil than in the more finely textured soils as did Abel *et al.* (2013). That stands to reason as fine-textured soils may already have high available water content mainly due to the increased water content at field capacity. Abel *et al.* (2013) found that biochar made from *Zea mays* (whole shoots) increased soil available water mainly because it increased soil water content at field capacity. In contrast, we found that our switchgrass biochar both increased water content at field capacity and caused the soils to retain less water at the wilting point. This difference may have to do with distinct hydraulic properties of different biochars or different soils.

Methodology has also made it difficult to assess the effect of biochar on water availability from soils. For most soils, the pressure plate method is adequate to determine both field capacity (frequently estimated at −0.033 MPa) and wilting point (frequently estimated at −1.5 MPa) water contents. We found, however, that our switchgrass biochar did not produce the expected shape of the moisture release curve, particularly at the greater tensions (toward the wilting point), possibly because of the loss of hydraulic connectivity among biochar particles and/or between the biochar particles and the porous ceramic plate. Mixing whole biochar particles in a matrix of powdered biochar produced a moisture release curve that had a shape closer to what was expected. Thus, it is possible that in a matrix of soil the moisture release curve of biochar-soil mixtures would be acceptable. However, there is no way to determine that. Instead, we determined wilting point water content based on a bioassay. In the field, after all, it is the ability of the plants to extract water from the soil that will determine how much is available for their use.

We conclude that addition of our switchgrass biochar to a variety of soils, from the fine-textured Edom and Weikert soils, to the coarse-textured Morrison and Wharton soils, significantly increased available water content, even with as little as 1% biochar by weight (as in this study). While we cannot yet speak to the economics of this approach, biochar amendment may be a viable means of mitigating current water shortages on drought-prone soils and future water shortages accompanying climate change.

Acknowledgements

We acknowledge support from the Sustainable Bioenergy Research Program of the USDA National Institute of Food and Agriculture (# 2011-67009-20072), Brigham Young University, Penn State University and the USDA-ARS. Many thanks are given to the expert staff at the USDA-ARS Pasture Systems and Watershed Management Research Unit. Significant improvement of the manuscript was made possible by two anonymous reviewers.

References

Abedi-Koupai J, Sohrab F, Swarbrick G (2008) Evaluation of hydrogel application on soil water retention characteristics. *Journal of Plant Nutrition*, **31**, 317–331.

Abel S, Peters A, Trinks S, Schonsky H, Facklam M, Wessolek G (2013) Impact of biochar and hydrochar addition on water retention and water repellency of sandy soil. *Geoderma*, **202–203**, 183–191.

Briggs C, Breiner J, Graham R (2012) Physical and chemical properties of Pinus ponderosa charcoal: implications for soil modification. *Soil Science*, **177**, 263–268.

Cai X, Zhang X, Wang D (2011) Land availability for biofuel production. *Environmental Science & Technology*, **45**, 334–339.

Cohen B (2002) Calculating a factorial ANOVA from means and standard deviations. *Understanding Statistics*, **1**, 191–203.

Cook ER, Woodhouse CA, Eakin CM, Meko DM, Stahle DW (2004) Long-term aridity changes in the western United States. *Science*, **306**, 1015–1018.

Dane JH, Hopman JW (2002) Pressure plate extractor. Section 3.3.2.4. In: *Methods of Soil Analysis: Part 4-Physical Methods* (eds Dane JH, Topp GC), pp. 688–690. Soil Science Society of America, Madison, WI, USA.

Downie D, Crosky A, Munroe P (2009) Physical properties of biochar. In: *Biochar for Environmental Management: Science and Technology* (eds Lehmann J, Joseph S), pp. 13–31. Earthscan, London.

Frantz J, Locke J (2005) Actual performance versus theoretical advantages of polyacrylamide hydrogel throughout bedding plant production. *HortScience*, **40**, 2040–2046.

Glaser B, Lehmann J, Zech W (2002) Ameliorating physical and chemical properties of highly weathered soils in the tropics with charcoal - a review. *Biology and Fertility of Soils*, **35**, 219–230.

Karhu K, Mattila T, Bergström I, Regina K (2011) Biochar addition to agricultural soil increased CH₄ uptake and water holding capacity – Results from a short-term pilot field study. *Agriculture, Ecosystems & Environment*, **140**, 309–313.

Keshwani DR, Cheng JJ (2009) Switchgrass for bioethanol and other value-added applications: a review. *Bioresource Technology*, **100**, 1515–1523.

Laird DA (2008) The charcoal vision: a win-win-win scenario for simultaneously producing bioenergy, permanently sequestering carbon, while improving soil and water quality. *Agronomy Journal*, **100**, 178–181.

Laird DA, Fleming P, Davis DD, Horton R, Wang B, Karlen DL (2010) Impact of biochar amendments on the quality of a typical Midwestern agricultural soil. *Geoderma*, **158**, 443–449.

Lawes RA, Oliver YM, Robertson MJ (2009) Integrating the effects of climate and plant available soil water holding capacity on wheat yield. *Field Crops Research*, **113**, 297–305.

Machlis L, Torrey JG (1956) *Plants in Action*. Freeman, San Francisco.

Meyer W, Green G (1980) Water use by wheat and plant indicators of available soil water. *Agronomy Journal*, **72**, 253–257.

Nguyen B, Koide RT, Dell C, Drohan P, Skinner H, Adler PR, Nord A (2014) Turnover of soil carbon following addition of switchgrass-derived biochar to four soils. *Soil Science Society of America Journal*. doi: 10.2136/sssaj2013.07.0258.

Niblick B, Monnell JD, Zhao X, Landis AE (2013) Using geographic information systems to assess potential biofuel crop production on urban marginal lands. *Applied Energy*, **103**, 234–242.

Novak JM, Watts DW (2013) Augmenting soil water storage using uncharred switchgrass and pyrolyzed biochars. *Soil Use and Management*, **29**, 98–104.

Novak J, Busscher W, Watts D *et al.* (2012) Biochars impact on soil-moisture storage in an Ultisol and two Aridisols. *Soil Science*, **177**, 310–320.

O'Neill M, Dobrowolski J (2011) Water and agriculture in a changing climate. *HortScience*, **46**, 155–157.

Parrish DJ, Fike JH (2005) The biology and agronomy of switchgrass for biofuels. *Critical Reviews in Plant Sciences*, **24**, 423–459.

Pereira RG, Heinemann AB, Madari BE (2012) Transpiration response of upland rice to water deficit changed by different levels of eucalyptus biochar. *Pesquisa Agropecuária Brasileira*, **47**, 716–721.

Sarkar D, Haldar A (2005) *Physical and Chemical Methods in Soil Analysis*. New Age International (P) Ltd., New Delhi, India.

Schlenker W, Hanemann WM, Fisher AC (2007) Water availability, degree days, and the potential impact of climate change on irrigated agriculture in California. *Climatic Change*, **81**, 19–38.

Seneviratne SI, Corti T, Davin EL *et al.* (2010) Investigating soil moisture–climate interactions in a changing climate: a review. *Earth-Science Reviews*, **99**, 125–161.

Skinner RH, Adler PR (2010) Carbon dioxide and water fluxes from switchgrass managed for bioenergy production. *Agriculture, Ecosystems & Environment*, **138**, 257–264.

Spokas KA, Cantrell KB, Novak JM *et al.* (2012) Biochar: a synthesis of its agronomic impact beyond carbon sequestration. *Journal of Environmental Quality*, **41**, 973–989.

STSC (1991) Statgraphics statistical graphics system. Version 5.1, Rockville, MD; STSC, Inc.

Ulyett J, Sakrabani R, Kibblewhite M, Hann M (2014) Impact of biochar addition on water retention, nitrification and carbon dioxide evolution from two sandy loam soils. *European Journal of Soil Science*, **65**, 96–104.

Vano JA, Scott MJ, Voisin N *et al.* (2010) Climate change impacts on water management and irrigated agriculture in the Yakima River Basin, Washington, USA. *Climatic Change*, **102**, 287–317.

Veihmeyer FJ, Hendrickson AH (1949) Methods of measuring field capacity and permanent wilting percentage of soils. *Soil Science*, **68**, 75–94.

Vuichard N, Ciais P, Wolf A (2009) Soil carbon sequestration or biofuel production: new land-use opportunities for mitigating climate over abandoned Soviet farmlands. *Environmental Science & Technology*, **43**, 8678–8683.

Wong M, Asseng S (2006) Determining the causes of spatial and temporal variability of wheat yields at sub-field scale using a new method of upscaling a crop model. *Plant and Soil*, **283**, 203–215.

6

Nitrogen rate and landscape impacts on life cycle energy use and emissions from switchgrass-derived ethanol

ERIC G. MBONIMPA[1], SANDEEP KUMAR[2], VANCE N. OWENS[3], RAJESH CHINTALA[2], HEIDI L. SIEVERDING[4] and JAMES J. STONE[4]

[1]Department of Systems Engineering and Management, Air Force Institute of Technology, WPAFB, OH, USA, [2]Plant Science Department, South Dakota State University, Brookings, SD, USA, [3]North Central Regional Sun Grant Center, South Dakota State University, Brookings, SD, USA, [4]Department of Civil and Environmental Engineering, South Dakota School of Mines and Technology, Rapid City, SD, USA

Abstract

Switchgrass-derived ethanol has been proposed as an alternative to fossil fuels to improve sustainability of the US energy sector. In this study, life cycle analysis (LCA) was used to estimate the environmental benefits of this fuel. To better define the LCA environmental impacts associated with fertilization rates and farm-landscape topography, results from a controlled experiment were analyzed. Data from switchgrass plots planted in 2008, consistently managed with three nitrogen rates (0, 56, and 112 kg N ha^{-1}), two landscape positions (shoulder and footslope), and harvested annually (starting in 2009, the year after planting) through 2014 were used as input into the Greenhouse gases, Regulated Emissions and Energy use in transportation (GREET) model. Simulations determined nitrogen (N) rate and landscape impacts on the life cycle energy and emissions from switchgrass ethanol used in a passenger car as ethanol–gasoline blends (10% ethanol:E10, 85% ethanol:E85s). Results indicated that E85s may lead to lower fossil fuels use (58 to 77%), greenhouse gas (GHG) emissions (33 to 82%), and particulate matter (PM2.5) emissions (15 to 54%) in comparison with gasoline. However, volatile organic compounds (VOCs) and other criteria pollutants such as nitrogen oxides (NOx), particulate matter (PM10), and sulfur dioxides (SO$_x$) were higher for E85s than those from gasoline. Nitrogen rate above 56 kg N ha^{-1} yielded no increased biomass production benefits; but did increase (up to twofold) GHG, VOCs, and criteria pollutants. Lower blend (E10) results were closely similar to those from gasoline. The landscape topography also influenced life cycle impacts. Biomass grown at the footslope of fertilized plots led to higher switchgrass biomass yield, lower GHG, VOCs, and criteria pollutants in comparison with those at the shoulder position. Results also showed that replacing switchgrass before maximum stand life (10–20 years.) can further reduce the energy and emissions reduction benefits.

Keywords: bioethanol, Emissions, energy use, greenhouse gases regulated emissions and energy use in transportation, life cycle analysis, switchgrass

Introduction

Switchgrass (*Panicum virgatum* L.), a tall grass native to USA, is considered as a promising feedstock to produce second generation biofuel (McLaughlin & Kszos, 2005). Second generation biofuels (e.g., lignocellulosic ethanol) were mandated by US Energy Independence and Security Act of 2007 (Congress, 2007). Fueling millions of North American light duty vehicles with domestic cellulose-derived fuels could improve environmental quality and sustainability of the energy sector (Spatari *et al.*, 2005). Currently, major bio-ethanol feedstocks include corn (*Zea mays* L.) and sugar cane (*Saccharum officinarum* L.). However, biofuels production from row crops (e.g., corn starch-derived ethanol) may negatively impact water quality because nutrients transport from fertilized cropland can contribute to eutrophication of water bodies (Simpson *et al.*, 2008). It may also impact food security if croplands that feed humans are transformed into fuel-feedstock production lands (e.g., Naylor *et al.*, 2007; Simpson *et al.*, 2008).

Switchgrass, like many other tall grasses such as miscanthus (*Miscanthus giganteus*) and prairie cordgrass (*Spartina pectinata*), may be preferable in certain areas because it is perennial, can be grown on land less suitable for row crops, and can be used alternatively for forage. Its benefits include carbon sequestration, high

Correspondence: Eric G. Mbonimpa
e-mail: eric.mbonimpa@afit.edu

biomass generation, and low fertilization and soil disturbance requirements (Wang et al., 2010; Hartman et al., 2011). Switchgrass production can be influenced by environmental conditions and agricultural management, especially in the first few years of establishment. Environmental conditions, such as landscape position, and fertilizer management can affect the biomass yield, greenhouse gas emissions, and eutrophication potential (Bai et al., 2010; Nikièma et al., 2011; Mbonimpa et al., 2015).

One of the challenges of switchgrass production is that the yield and nutrients requirement depend on location, weather, and agricultural management (Parrish & Fike, 2005). To increase economic and environmental competitiveness of switchgrass-based biofuel, research has been focused on efficient feedstock production and biomass-to-fuel conversion. But there is still a wide range of yield potentials reported by various studies. Parrish & Fike (2005) indicated that in US regions with sufficient rainfall (e.g., US Midwest down to Southeast), roughly 15 Mg ha^{-1} of biomass can be produced annually with approximately 50 kg N ha^{-1} of fertilizer application. Guretzky et al. (2011) indicated yields of up to 21 Mg ha^{-1} in US southern great plains with N fertilization rate up to 225 kg N ha^{-1}. Nikièma et al. (2011) in Michigan observed yield increases from 4.89 Mg ha^{-1} with increases in fertilization of 56 kg N ha^{-1} (1.5× yield) and 112 kg N ha^{-1} (2.5× yield). Mulkey et al. (2006) reported that switchgrass biomass yield ranged from 3.5 to 5.5 Mg ha^{-1} in South Dakota with 56 kg N ha^{-1} of fertilizer, and no increase in biomass yield beyond this fertilization rate. These yields are achieved after full stand establishment. Switchgrass can take 1 to 3 years after planting to reach maximum yield potential (Parrish & Fike, 2005) and has a stand life of 10–20 years (Monti et al., 2009; Sokhansanj et al., 2009). The relationship between yield and N fertilization rate was found to be location dependent; switchgrass response to nitrogen is likely to be less pronounced in northern US locations than southern locations (Guretzky et al., 2011). These yield–nitrogen relations may also vary depending on whether the management includes one or two harvests a year. One harvest could promote more carbon sequestration because, as suggested in Guretzky et al. (2011), it allows maximum translocation of nutrients and storage reserves in roots before harvest. In addition to N fertilizer, herbicides may be needed during establishment years; an established stand should outcompete weeds. Soil amendments such as phosphorus, lime and potassium may also be required (Bai et al., 2010).

The yield and nitrogen management may also be influenced by the landscape topography, especially on sloped landscapes. In previous studies, greater biomass yields at the deposition position (footslope) were observed and linked to soils with greater production potential compared to higher elevation position (shoulder) (Bachman, 1997; Harmoney et al., 2001). Higher carbon dioxide (CO_2) emissions and soil organic carbon were observed at the footslope compared to the shoulder position (Mbonimpa et al., 2015). Carbon storage also varies with depth; Liebig et al. (2008) reported a significant carbon sequestration with soil carbon increases of about 1.1 and 2.9 Mg-C ha^{-1} year^{-1} for the 0–30 cm and 0–120 cm depths, respectively, when growing switchgrass.

To understand switchgrass and derived fuels, life cycle analysis (LCA), a common process for environmental accounting, has been used. Past LCA studies on switchgrass-derived ethanol generated inventories of environmental impacts from various life cycle steps that include switchgrass production (based mostly on yield for one or few locations), transportation, conversion of biomass to ethanol, transportation of ethanol, blending with other fuels, and use in various types of vehicles (Wu et al., 2006; Bai et al., 2010). This pathway is referred as well-to-wheel or cradle-to-grave to indicate start and end points of analysis. Usually to show the environmental benefits of the cellulosic ethanol and other renewable energy used in vehicles, the pathway of comparison involves petroleum-derived fuels such as gasoline (Wang et al., 2012; Luk et al., 2013). However, results from these LCA studies vary considerably due to differences in assumptions, inputs, and system boundaries. In particular, studies that ignored electricity coproduced from ethanol production process wastes (lignin) indicated less environmental benefits compared to studies that considered coproducts (Wang et al., 2011; Luk et al., 2013). Nevertheless, substantial environmental benefits of switchgrass-derived ethanol in comparison with fossil fuels were reported by previous life cycle studies. Spatari et al. (2005) indicated that greenhouse gas (GHG) emissions are 57% lower for a switchgrass-derived E85-fueled (85% ethanol, 15% gasoline) automobile compared to petroleum gasoline-fueled automobile. Wu et al. (2006) reported reductions in petroleum and fossil fuels (66–93%), GHG emissions (82–87%), and SOx (39–43%) when they compared switchgrass-derived unblended ethanol to gasoline. Schmer et al. (2008) reported 94% lower GHG emissions and 540% more renewable energy than non-renewable energy consumed during switchgrass ethanol life cycle (with switchgrass yield between 5.2 and 11 ton ha^{-1}) in comparison with gasoline life cycle. Bai et al. (2010) indicated that driving with switchgrass ethanol (E85) leads to 65% less GHG emissions than gasoline. They also noticed adverse impacts in comparison with gasoline. With 100 kg N ha^{-1} of fertilizer application, Bai *et al.* (2010) calculated approximately 2.5 times more eutrophication potential (in terms of kilograms of PO$_4$ equivalent) in comparison with gasoline. Also, the addition of herbicides contributed to water ecotoxicity. However, the toxicity in Bai *et al.* (2010) may be overestimated because they applied herbicides every year, while in practice it is applied only in the first few (~2–3) years after planting switchgrass. They also did not clarify the impact management practices, such as fertilizer management, and location environmental aspects (e.g., soil, weather, and topography) have on their findings.

Although previous studies performed a complete LCA analysis for switchgrass-derived ethanol, most of them used the standard biomass yields, fertilizers rates, and included soil amendments which may not be needed at some locations. Most studies have not included field-scale agricultural conditions associated with nitrogen application rates and landscape positions which may influence the overall environmental impact of switchgrass-derived ethanol. To fill this gap, an LCA using the Greenhouse gases, Regulated Emissions, and Energy use in Transportation (GREET) model (Wang, 1999, 2008) was performed using data from a field-scale case study of switchgrass grown with different N application rates and landscape positions in US northern plains region. Data on specific energy used by machinery, soil GHG emissions under various treatment conditions, and biomass yield associated with N rates and landscape positions were used as inputs in the model. The primary objective of this study was to assess the impacts of field-scale agricultural management on the life cycle energy use and emissions of switchgrass-derived ethanol. Most vehicles currently on the road cannot operate on pure ethanol; therefore, this study focused on ethanol–gasoline blends as a realistic path forward for switchgrass-derived ethanol product development.

Materials and methods

System boundary and functional unit

The ethanol-blend life cycle was divided into three major steps: (i) the feedstock (switchgrass) production and delivery to the cellulosic ethanol production plant; (ii) the fuel production, mixing, and distribution; and (iii) vehicle operation. The production and disposal of machinery and vehicles are not included. The life cycle materials and processes are depicted in Fig. 1. Life cycle energy (Joules) and emissions (grams) were estimated per unit distance of vehicle operation (per kilometer). The feedstock production part of the study involved field experiments whereas fuel production and vehicle operation were accomplished using GREET model simulations (described later).

For the feedstock production, agricultural inputs including seed, nutrients, and herbicides were transported to the farm to grow switchgrass. The facilities that produce agricultural inputs acquired 75% of process energy from natural gas and 25% from electricity. The switchgrass production site was located near Bristol (45°16′24.55″N, 97°50′13.34″W), South Dakota, USA. The feedstock was produced on 12 plots (21.3 m wide and 366 m long each) that were historically seeded with soybeans (*Glycine max.* L.). Switchgrass was planted on May 17, 2008. The plots form a split plot factorial design comprised of three N treatments (0, 56, 112 kg N ha^{-1}) and two landscape positions (shoulder and footslope).

Fuel consumption of machinery for agricultural management activities was obtained from Grisso *et al.* (2004). The plots were not tilled before planting, and the switchgrass was planted using a 'Truax no-till drill' (Truax Company, New Hope, MN) pulled by a diesel-powered tractor. Herbicides were applied as needed from 2008-2011 using a diesel powered applicator.

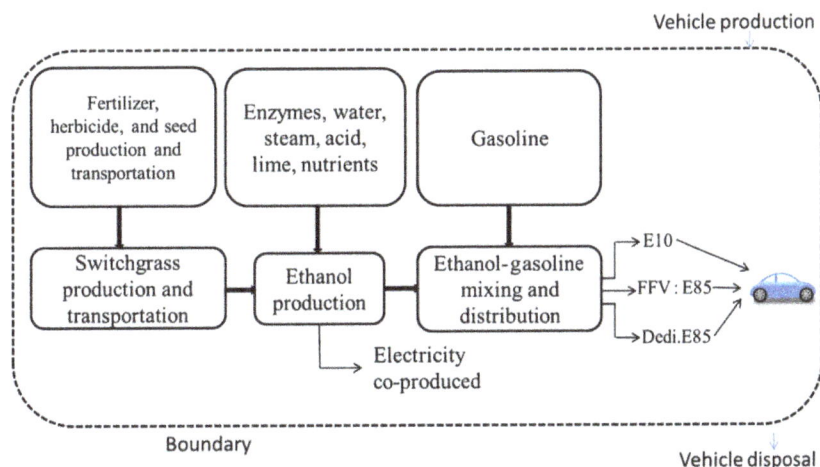

Fig. 1 Life cycle analysis system boundary for the study. Vehicle production and disposal were considered outside the scope of the study.

N was applied annually in late spring beginning in 2009 with the same type of equipment. Harvesting switchgrass started the year after planting (2009) and continued every year thereafter after a killling frost in autumn. Fertilizer continued to be applied each year in production years. Harvest was accomplished using a mower and baler pulled by a tractor. The farm activity of each machine was converted into energy inputs (Tables 1 and 2) using fuel consumed per unit area covered or per number of bales harvested. The switchgrass yield for each plot was calculated by weighing bales produced minus losses during transport and handling. The biomass yield at different landscapes was determined by sampling a square meter quadrat, mowing, and weighing for each landscape position. To account for moisture content, a subsample was collected, dried, and weighed. Soil GHG emissions from this switchgrass land were monitored as described in detail by Mbonimpa et al. (2015).

Life cycle inventory analysis and assessment using GREET model

The GREET (version1_2013) model developed by Argonne National Lab (ANL) was used for the LCA in this study (Wang, 1999). Fuels that were compared include gasoline, a mixture of 10% ethanol, and 90% gasoline by volume (E10) that can be used in most gasoline vehicles; and flex fuel a mixture of 85% ethanol and 15% gasoline by volume (E85) used for flex fuel vehicles (FFV). Total energy use, fossil energy use, petroleum use, GHG emissions, and emissions of criteria pollutants for a light duty passenger vehicle were obtained using GREET. This study compared the results from blended ethanol–gasoline technologies with those from gasoline. The technologies were assumed to be those available by the year 2015 (target year of simulations). Nitrous oxide (N_2O) emissions from switchgrass farming as a percentage of total nitrogen (N)-fertilizer were calculated from monitored N_2O emissions at various field experimental treatments, that is, fertilizer rate and landscape position (see Tables 1 and 2 for modeled scenarios). The simulations include potential CO_2 emissions reductions from land-use change that were estimated with Carbon Calculator for Land Use Change from Biofuel production (CCLUB) for the GREET model (Dunn et al., 2014). The CO_2 reduction due to land-use change was estimated to be approximately 129 g L^{-1} of ethanol produced. GREET estimates a soil carbon sequestration of about 48 800 g CO_2 per dry ton of switchgrass.

Ethanol generation. The bioprocessing of switchgrass into ethanol was simulated using the GREET model as described in Wu et al. (2008, 2006), and they describe the source of emissions during the conversion processes. This process is a net producer (coproduction) of electricity (Spatari et al., 2005). Ethanol yield used was 0.35 L kg^{-1} of dry biomass. This value (GREET (default value)) is based on assessment of recent conversion advances (saccharification/fermentation process).

Ethanol use. The GREET model vehicle fuel economy is adjusted for on-road performance using the US Environmental Protection Agency (EPA) kilometers-per-liter-based method and a split between city (43%) and highway (57%) vehicle kilometers traveled (VMT). In GREET (version1_2013), vehicle models are in 5-year increment; the target vehicle model in this study is 2010 with a fuel economy of 58 kilometers per liter. GREET also contains inventory of exhaust emissions (VOC, PM10, PM2.5, CH_4, and N_2O) in g km^{-1} traveled. Three types of alternative-fueled vehicles were compared with gasoline fueled. These include gasoline vehicles that consume low-percentage ethanol blends (E10), flex fuel vehicles (FFV, E85), and

Table 1 Farm-level information used in the GREET model for various N rates

Farm-level information	N levels		
	0 kg ha^{-1}	56 kg ha^{-1}	112 kg ha^{-1}
Moisture (average, %)	16.6	16.6	16.6
Yield (ton ha^{-1} yr^{-1})	3.74	5.11	5.12
Biomass (dry tons, d.ton ha^{-1} yr^{-1})	2.43	3.32	3.32
Fertilizer (g-N/d.ton)	0	13 126	26 226
Herbicides (g/d.ton)-15 year harvest	27.44	20.10	20.05
Farming energy (Btu/d.ton)-15 year harvest	174 401	151 679	151 531
Herbicides (g/d.ton)-1 year harvest	1365.6	998.8	997.8
Farming energy (Btu/d.ton)-1 year harvest	277 972	227 438	227 216
Seeds (kg ha^{-1})-sunburst variety	11.2	11.2	11.2
N content in biomass (%)	0.58	0.62	0.66
Harvest collection rate (%) (default-GREET)	90	90	90
N_2O* (% of N in fertilizer and biomass)	1.5	1.0	0.6

*Nitrous oxide.

Table 2 *Farm-level information used in GREET model for various landscape positions

Farm-level information	Landscape positions			
	Shoulder	Footslope	Shoulder	Footslope
N levels	0 kg ha^{-1}		112 kg ha^{-1}	
Biomass (Dry tons, d.ton ha^{-1} yr^{-1})	6.47	8.41	9.88	15.87
Fertilizer (g-N/d.ton)	0	0	8819	5498
Herbicides (g/d.ton)	10.26	7.89	6.72	4.18
Farming energy (Btu/d.ton)	65 496	59911	50 997	31 749

*These are the parameters that changed in reference to Table 1.

dedicated E85 (Dedi.E85) which are 6.5% more efficient than other vehicles (gasoline, E10, FFV).

Transportation and distribution. Switchgrass biomass was transported by heavy-duty trucks (25 metric tons of switchgrass payloads), and the ethanol produced is a distributed across country with barges, pipeline, rail road, and heavy-duty trucks. For the transportation, distribution distance values were obtained from GREET model default estimates. It was assumed that the switchgrass was transported to a conversion plant located at distance of 85 km one way (GREET model default).

Scenarios. A total of ten GREET models were setup and executed. Six models represented the three nitrogen rates (0, 56, 112 kg N ha^{-1}); each nitrogen rate had two models, one-year harvest or 15-year harvest. These two stand ages provided upper (full stand life, 15-year harvest.) and lower (abandonment, 1-year harvest) bounds. The resources used during stand establishment are distributed to years the switchgrass is harvested or abandoned. Four models represented shoulder

and footslope positions; each position had two models, with no N (0 kg·N ha^{-1}) or high N (112 kg N ha^{-1}). These model scenarios contain different yields, energy use, biomass nitrogen content, N$_2$O emissions, herbicides to yield, and fertilizer to yield ratios as shown in Tables 1 and 2.

Results

Impact of nitrogen fertilizer and landscape position on energy use

The GREET simulations compared the total energy consumed during the life cycle of switchgrass ethanol blended with gasoline (used in three types of passenger vehicles: Std: E10, FFV:E85, and Dedi.E85) to that of conventional gasoline. Results (shown in Fig. 2) indicate that switchgrass-based blended ethanol life cycles may consume more energy (8 to 120%) than that of gasoline life cycle. However, gasoline life cycle

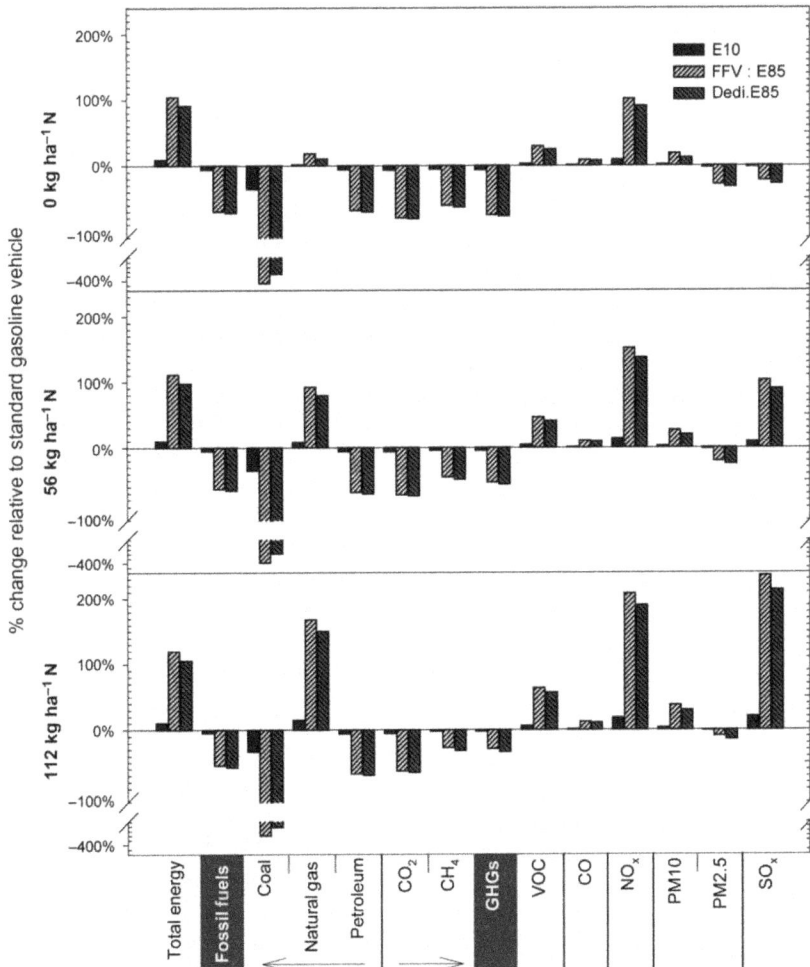

Fig. 2 Percent change of switchgrass ethanol life cycle impacts relative to gasoline for switchgrass grown with 0 kg ha^{-1} (top), 56 kg ha^{-1} (middle), and 112 kg ha^{-1} (bottom).

uses more fossil energy (total of coal, petroleum, and natural gas) than blended ethanol's (E10, E85, Dedi.E85). Use of E85s resulted in 58 to 77% less fossil fuel consumption. E10 use reduced fossil fuels consumption by 5 to 6.5% in comparison with gasoline. Life cycle energy use by blended ethanol fuels was impacted by nitrogen (N) fertilizer applied during feedstock production (Table 3). The addition of 56 and 112 kg N ha of fertilizer increased total energy and fossil fuel use by 8 and 16%, respectively, in comparison with no N addition. The natural gas use increased with N fertilization rate increase as well (Fig. 2). The results also demonstrated a large saving in coal use (up to 406%) for switchgrass-based ethanol life cycle compared to gasoline in the Northern Plains region of United States (Fig. 2).

Further, the simulations showed that the landscape topography slightly impacted life cycle energy use. As shown in Fig. 3, when comparing ethanol blends and gasoline for the shoulder and footslope, the difference is small and less noticeable. On plots that received 112 kg N ha^{-1}, the shoulder position was linked to 88 and 94 kJ km^{-1} (approximately 5.4% of total energy) more life cycle energy use for FFV: E85 and Dedi.E85, respectively (Table 4). The difference was very small (~3 kJ km^{-1}) for plots where no fertilizer was applied. At low ethanol blend (E10), the difference due to landscape was also small (8 kJ km^{-1} at 112 kg N ha^{-1}, 0.28 kJ km^{-1} for no fertilizer). The differences were largely attributed to fossil fuel as shown in Table 4.

Ethanol life cycle energy is distributed within three main steps: feedstock production, fuel production, and vehicle operation. For E85s, about 21, 38.3, and 40.7% of energy is used on feedstock production, fuel production, and vehicle operation, respectively. For low blend (E10), the distribution changes to 5.2, 20.4, and 74.4%. In comparison, the life cycle of gasoline is distributed among the three life cycle steps as 6, 13, and 81%. Application of fertilizer increased the share of the energy for feedstock production by 3 and 6% for 56 and 112 kg N ha^{-1}, respectively. The landscape change only led to 1% (higher at the shoulder) difference in the share of energy associated with feedstock production.

Impact of nitrogen fertilizer and landscape position on GHG emissions

The results indicated that the feedstock production lead to a net GHG sink as shown with negative numbers in Table 3. For E85s, 130–160 g km^{-1} of GHG was removed from the atmosphere in nonfertilized switchgrass. But, the addition of fertilizer increased GHG by

approximately 69 g km^{-1} and 139 g km^{-1} for 56 and 112 kg N ha^{-1}, respectively. The fuel production also removed a small amount of atmospheric GHG (~3–4 g km^{-1}). However, the GHG sink by the feedstock and fuel production was offset by vehicle operation and resulted into net GHG emissions of about 50–200 g km^{-1}. The landscape position also impacted GHG trends in fertilized plots. For E85s, an average of 18 g km^{-1} more GHGs was removed at the footslope in comparison with the shoulder position (Table 4). In comparison with gasoline, net GHG associated with E85s was between 33 and 82% lower for 56 and 112 kg N ha^{-1}, respectively. For E10, these were 2.8 to 7% lower.

Among individual GHGs, the CO_2 was reduced by the feedstock and fuel production steps for switchgrass ethanol used in E85s. Other GHGs, CH_4 and N_2O, were emitted by all steps of the fuel life cycle and increased with nitrogen rate and at the shoulder position (Fig. 4). N_2O was the GHG that increased the most with the increase in fertilizer rate. N_2O increased approximately 18- and 36-fold for 56 and 112 kg N ha^{-1} fertilizer application, respectively, when the feedstock was produced for E85s. But that increase was about 4.8- to 8.7-folds if the total N_2O from all life cycle steps were considered. For similar situations (adding 56 and 112 kg N ha^{-1}), CH_4 increased by about 73–148%. The N_2O emissions at the shoulder of fertilized plots were 35% higher in comparison with the footslope for E85s. In comparison with gasoline, N_2O emissions were between 540 and 2321% higher for E85s as shown in Fig. 5. For E10, the increase was about 50–200%.

Impact of nitrogen fertilizer and landscape position on volatile organic carbon (VOC) emissions

The results show that increase in fertilization rate to grow switchgrass is associated with increases in VOCs. Although the largest portion of VOCs originated from fuel production (37% for E10, 53% for E85) and vehicle operation (55% for E10, 33% for E85), the fertilizer increase led to increase in total VOCs (14% at 56 kg N ha^{-1}, 27% at 112 kg N ha^{-1}). This increase is most noticeable for E85 fuels, especially at the feedstock production stage where the highest fertilizer rate resulted in approximately 9 times higher VOCs in comparison with no fertilization (Table 3). In comparison with gasoline, the life cycle VOCs for E85 produced from switchgrass grown with no nitrogen were 23% higher than VOCs from gasoline. This percentage increased (by 39% at 56 kg N ha^{-1}, 55% at 112 kg N ha^{-1}) with increases in fertilization rate (Fig. 2). Life cycle VOC emissions from using blended ethanol produced from unfertilized switchgrass grown at the

Table 3 Energy and emissions from different fuels and sources as impacted by N rate; the number in middle brackets in italics is change due to increase of N rate to 56 kg ha^{-1} and the number in the following (last) bracket is increase due 112 kg per ha N rate

Fuel type	Feedstock	Fuel	Vehicle
Total E (KJ km^{-1})			
Gasoline	230	499	3070
E10	215	841 (28) (56)	3070
FFV-E85	1585 (314) (641)	2893	3070
Dedi E85	1481 (294) (599)	2704	2869
CO$_2$ (g km^{-1})			
Gasoline	19	38	223
E10	3	33 (3) (5)	223
FFV-E85	−164 (25) (51)	−12	219
Dedi E85	−153 (23) (47)	−11	205
Net GHG (g km^{-1}) 1 year-harvest			
Gasoline	25.8	39.8	224.9
E10	9.59	37.5 (5.6) (11.7)	224.5
FFV-E85	−144 (65) (134)	−3.85 (0.04) (0.08)	220.6
Dedi E85	−134 (60) (125)	−3.60 (0.04) (0.08)	206.2
VOC (g km^{-1})			
Gasoline	0.01	0.068	0.106
E10	0.01	0.074 (0.002) (0.005)	0.106
FFV-E85	0.008 (0.032) (0.064)	0.127	0.1
Dedi E85	0.008 (0.029) (0.060)	0.118	0.1
NOx (g km^{-1})			
Gasoline	0.077	0.055	0.075
E10	0.072	0.076 (0.009) (0.020)	0.075
FFV-E85	0.081 (0.113) (0.231)	0.224	0.075
Dedi E85	0.075 (0.107) (0.217)	0.209	0.075
PM2.5 (g km^{-1})			
Gasoline	0.005	0.045	0.007
E10	0.004	0.044	0.007
FFV-E85	0.005 (0.006) (0.012)	0.024	0.007
Dedi E85	0.005 (0.005) (0.011)	0.023	0.007

Fuel type	Feedstock	Fuel	Vehicle
Net GHG (g km^{-1})			
Gasoline	25.8	39.8	224.9
E10	9.59	36 (6) (12)	224.5
FFV-E85	−160 (69) (139)	−3.84 (0.04) (0.08)	220.6
Dedi E85	−150 (65) (130)	−3.57 (0.06) (0.10)	206.2
CH$_4$ (g km^{-1})			
Gasoline	0.276	0.068	0.007
E10	0.257	0.066 (0.005) (0.010)	0.007
FFV-E85	0.083 (0.061) (0.123)	0.026	0.007
Dedi E85	0.078 (0.057) (0.115)	0.024	0.007
Fossil E (KJ km^{-1})			
Gasoline	218	490	3070
E10	203	464 (27) (55)	2865
FFV-E85	223 (310) (633)	12	717
Dedi E85	208 (290) (592)	11	670
CO (g km^{-1})			
Gasoline	0.016	0.059	1.781
E10	0.015	0.072 (0.003) (0.007)	1.781
FFV-E85	0.027 (0.043) (0.076)	0.193	1.781
Dedi E85	0.025 (0.031) (0.071)	0.180	1.781
PM10 (g km^{-1})			
Gasoline	0.005	0.048	0.016
E10	0.005	0.049	0.016
FFV-E85	0.006 (0.007) (0.015)	0.054	0.016
Dedi E85	0.005 (0.007) (0.015)	0.050	0.016
SOx (g km^{-1})			
Gasoline	0.033	0.043	0.004
E10	0.031	0.042	0.003
FFV-E85	0.022 (0.101) (0.207)	0.020	0.002
Dedi E85	0.020 (0.095) (0.194)	0.019	0.002

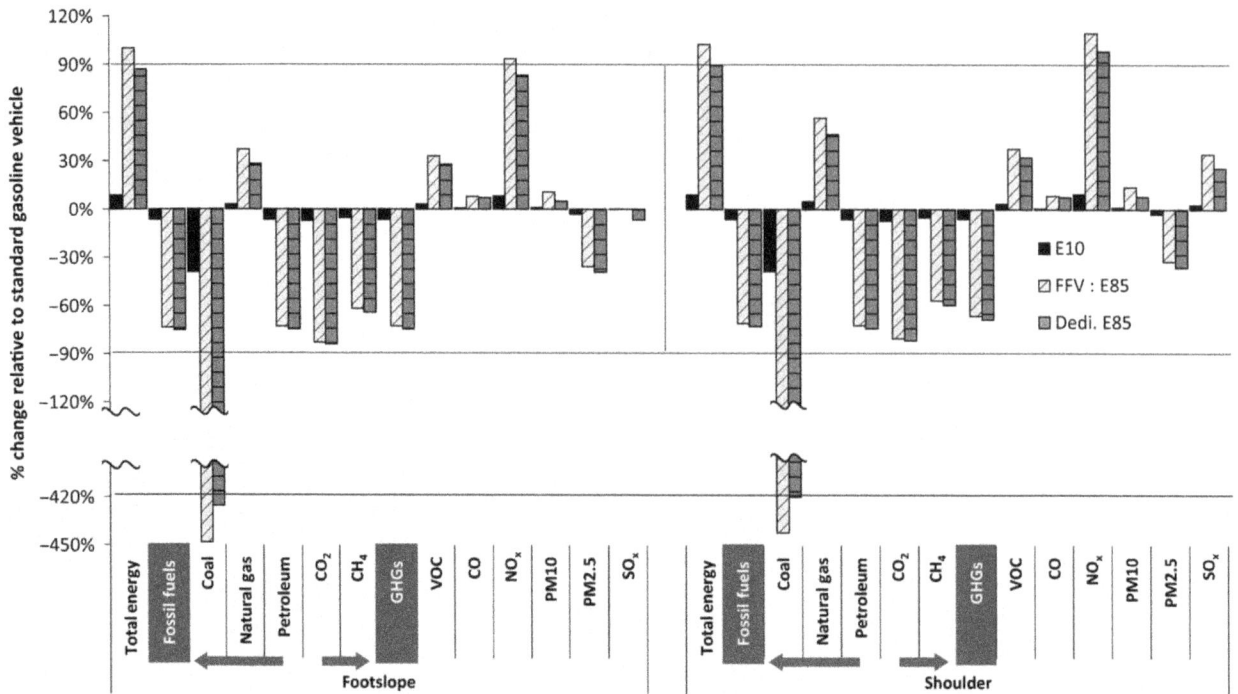

Fig. 3 Impact of landscape on percent change of switchgrass ethanol life cycle impacts relative to gasoline: switchgrass grown with 112 kg N ha^{-1}.

shoulder are similar to those grown at the footslope. In contrast, for plots where 112 kg N ha^{-1} of fertilizer was applied, the E85 life cycle's VOCs were higher (~by 3.5%) for shoulder biomass compared to the footslope biomass. If only VOCs from feedstock production are compared, the shoulder biomass accounts for 45% more VOCs than the footslope. For E10, there was little difference between shoulder and footslope (Table 4).

In comparison with gasoline, E85s (FFV, Dedi.E85, respectively) were linked to approximately 28–38% more VOCs those from gasoline. Life cycle VOCs were slightly higher (by ~3.6%) for E10 than gasoline.

Impact of nitrogen fertilizer and landscape position on criterial pollutants emissions

Carbon monoxide (CO). CO emissions are mainly attributable to vehicle operation (~97%) as shown in Table 3. Hence, the application of fertilizer had a small impact on the life cycle CO; it increased by about 1.7 and 3.6% for addition of 56 and 112 kg N ha^{-1}, respectively, for E85s. However, if only the feedstock production is considered, the increase in CO emissions due to nitrogen application was about 2.2 and 3.8 times higher for E85 with the application of 56 and 112 kg N ha^{-1}, respectively, in comparison with no nitrogen application. In general, the entire life cycle of ethanol–gasoline mix produced approximately 2 g of CO emissions per km.

In comparison with gasoline, these emissions were higher by about 7, 8.7, and 10.5% for 0, 56, and 112 kg-N ha^{-1}, respectively. The results also indicated that there were small differences in CO emissions based on landscape positioning. Vehicle operation is the primary source of CO.

NOx. The results showed that in plots where no fertilizer was applied, the majority of NO$_x$ originated from E85 fuel production (~59%) and the rest of NO$_x$ generated from feedstock production (~21%) and vehicle operation (~10%) (Table 3). The addition of fertilizer changed the distribution of NOx among life cycle steps. For high fertilizer rate, the majority (51%) of NO$_x$ were produced from feedstock production meant for E85s and the fuel production share reduced to ~37%. For gasoline and E10, NO$_x$ were approximately distributed equally on the three life cycle steps. Overall, the life cycle of ethanol–gasoline produced between 252 and 724 mg km^{-1} of NO$_x$ for E10 and E85s, respectively. In comparison with gasoline, these emissions were much higher by about 73.4, 124.5, and 177.5% for 0, 56, and 112 kg-N ha^{-1} fertilization, respectively (Fig. 2).

Results also indicated that the life cycle NO$_x$ emissions were approximately similar for blended fuels derived from the shoulder biomass and footslope biomass on plots with no fertilizer application. For fertilized plots, the footslope biomass (for E85 fuel) was

Table 4 Energy and emissions from different fuels as impacted by position: the numbers without brackets are for the shoulder and numbers in brackets is the difference between footslope and shoulder

N rate	Fuel type	Feedstock	Fuel	Vehicle
Total E (KJ km⁻¹)				
0 kg ha⁻¹	Gasoline	230	499	3070
	E10	215	836 (−0.28)	3070
	FFV-E85	1523 (−3.31)	2893	3070
	Dedi E85	1424 (−3.09)	2704	2869
112 kg ha⁻¹	Gasoline	230	499	3070
	E10	215	854 (−8)	3070
	FFV-E85	1735 (−94)	2893	3070
	Dedi E85	1622 (−88)	2704	2869
CO₂ (g km⁻¹)				
0 kg ha⁻¹	Gasoline	18.9	37.8	223.5
	E10	3.1	33.0	223.1
	FFV-E85	−169 (−0.25)	−11.8	219.2
	Dedi E85	−158 (−0.24)	−11.0	204.8
112 kg ha⁻¹	Gasoline	18.9	37.8	223.4
	E10	3.1	34.0 (−0.7)	223.1
	FFV-E85	−152 (−7.3)	−11.8	219.2
	Dedi E85	−142 (−6.9)	−11.0	204.8
N₂O (g km⁻¹)				
0 kg ha⁻¹	Gasoline	0.000	0.001	0.004
	E10	0.000	0.004	0.004
	FFV-E85	0.008	0.025	0.004
	Dedi E85	0.008	0.023	0.004
112 kg ha⁻¹	Gasoline	0.000	0.001	0.004
	E10	0.000	0.012 (−0.003)	0.004
	FFV-E85	0.102 (−0.036)	0.024	0.004
	Dedi E85	0.095 (−0.034)	0.023	0.004
VOC (g km⁻¹)				
0 kg ha⁻¹	Gasoline	0.010	0.068	0.106
	E10	0.010	0.074	0.106
	FFV-E85	0.006	0.127	0.100
	Dedi E85	0.006	0.118	0.100
112 kg ha⁻¹	Gasoline	0.010	0.068	0.106
	E10	0.010	0.075	0.106
	FFV-E85	0.027 (−0.008)	0.127	0.100
	Dedi E85	0.026 (−0.008)	0.118	0.100
NOx (g km⁻¹)				

N rate	Fuel type	Feedstock	Fuel	Vehicle
Net GHG (g km⁻¹)				
0 kg ha⁻¹	Gasoline	25.8	39.8	224.9
	E10	9.6	35.7	224.5
	FFV-E85	−164 (−0.25)	−3.8	220.6
	Dedi E85	−154 (−0.24)	−3.6	206.2
112 kg ha⁻¹	Gasoline	25.8	39.8	224.9
	E10	9.6	39.7 (−1.6)	224.5
	FFV-E85	−118.6 (−18.6)	−3.9	220.6
	Dedi E85	−110.9 (−17.4)	−3.6	206.2
CH₄ (g km⁻¹)				
0 kg ha⁻¹	Gasoline	0.276	0.068	0.007
	E10	0.257	0.065	0.007
	FFV-E85	0.077	0.026	0.007
	Dedi E85	0.072	0.024	0.007
112 kg ha⁻¹	Gasoline	0.276	0.068	0.007
	E10	0.257	0.069 (−0.002)	0.007
	FFV-E85	0.118 (−0.016)	0.026	0.007
	Dedi E85	0.111 (−0.016)	0.024	0.007
Fossil E (KJ km⁻¹)				
0 kg ha⁻¹	Gasoline	218	490	3070
	E10	203	457 (−0.28)	2865
	FFV-E85	163 (−3.21)	12	717
	Dedi E85	152 (−3.01)	11	670
112 kg ha⁻¹	Gasoline	218	490	3070
	E10	203	477 (−8)	2865
	FFV-E85	372 (−92)	12	717
	Dedi E85	348 (−87)	11	670
CO (g km⁻¹)				
0 kg ha⁻¹	Gasoline	0.016	0.059	0.106
	E10	0.015	0.071	0.106
	FFV-E85	0.017	0.193	0.100
	Dedi E85	0.016	0.180	0.100
112 kg ha⁻¹	Gasoline	0.016	0.059	0.106
	E10	0.015	0.073	0.106
	FFV-E85	0.040 (−0.011)	0.193	0.100
	Dedi E85	0.034 (−0.011)	0.180	0.100
PM10 (g km⁻¹)				

(continued)

Table 4 (continued)

N rate	Fuel type	Feedstock	Fuel	Vehicle	N rate	Fuel type	Feedstock	Fuel	Vehicle
0 kg ha⁻¹	Gasoline	0.077	0.055	0.075	0 kg ha⁻¹	Gasoline	0.005	0.048	0.016
	E10	0.072	0.074	0.075		E10	0.005	0.048	0.016
	FFV-E85	0.059	0.224	0.075		FFV-E85	0.004	0.054	0.016
	Dedi E85	0.056	0.209	0.075		Dedi E85	0.004	0.050	0.016
112 kg ha⁻¹	Gasoline	0.077	0.055	0.075	112 kg ha⁻¹	Gasoline	0.005	0.048	0.016
	E10	0.072	0.080 (−0.003)	0.075		E10	0.005	0.049	0.016
	FFV-E85	0.136 (−0.033)	0.224	0.075		FFV-E85	0.009 (−0.002)	0.054	0.016
	Dedi E85	0.127 (−0.031)	0.209	0.075		Dedi E85	0.008 (−0.002)	0.050	0.016
PM2.5 (g km⁻¹) 0 kg ha⁻¹	Gasoline	0.005	0.045	0.007	SOx (g km⁻¹) 0 kg ha⁻¹	Gasoline	0.033	0.043	0.004
	E10	0.004	0.044	0.007		E10	0.031	0.042	0.003
	FFV-E85	0.003	0.024	0.007		FFV-E85	0.014	0.020	0.002
	Dedi E85	0.003	0.023	0.007		Dedi E85	0.013	0.019	0.002
112 kg ha⁻¹	Gasoline	0.005	0.045	0.007	112 kg ha⁻¹	Gasoline	0.033	0.043	0.004
	E10	0.004	0.044	0.007		E10	0.031	0.050 (−0.003)	0.003
	FFV-E85	0.007 (−0.002)	0.024	0.007		FFV-E85	0.085 (−0.027)	0.020	0.002
	Dedi E85	0.007 (−0.002)	0.023	0.007		Dedi E85	0.080 (−0.026)	0.019	0.002

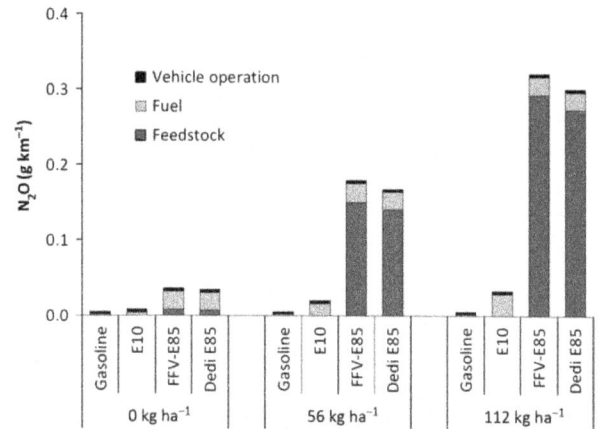

Fig. 4 Impact of nitrogen fertilization rate on life cycle nitrous oxide (N_2O) emissions for four types of fuels subdivided based on life cycle stage (i.e. vehicle operation, fuel or feedstock production).

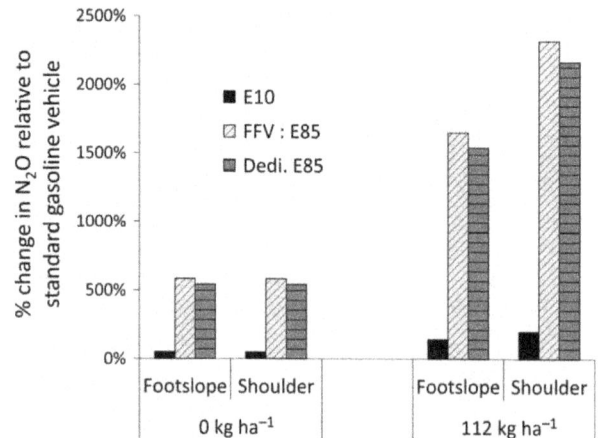

Fig. 5 Impact of landscape position and fertilization on percent change of switchgrass ethanol life cycle nitrous oxide (N_2O) relative to gasoline.

linked to lower (by approximately 8.4%) NO_x compared to the shoulder biomass (Table 4).

SO_x. The largest part of SO_x was contributed by the feedstock (~50%) and fuel production (~45%) for unfertilized plots. Addition of fertilizer increased the feedstock production share (85% for 56 kg N ha⁻¹; 91% for 112 kg N ha⁻¹) in SO_x emissions for E85s. The low ethanol blend (E10) and gasoline life cycle's SO_x emissions were almost similar as shown in Table 3. Vehicle operation contributed only between 0.8 and 4.5% for all fuels. In total, the life cycle of ethanol–gasoline mix produced between 23 and 251 mg of SO_x per km with the highest amount linked to highest N rate. In comparison with gasoline, SO_x emissions were lower (~54%) for

ethanol–gasoline mix produced from switchgrass that did not receive the fertilizer. However, there was a reversal when fertilizer was applied for E85, it led to 82 and 215% higher SO_x emissions, for 56 and 112 kg N ha^{-1}, respectively. Landscape positioning also impacted the life cycle SO_x emissions. Lower emissions (by approximately 35%) for E85 from footslope biomass in comparison with shoulder biomass were observed in plots that received 112 kg N ha^{-1}. Results indicated that life cycle SO_x emissions from shoulder biomass (used in E85) were higher than gasoline's emissions (Table 4), whereas SO_x emissions from the footslope biomass were approximately equal to emissions from gasoline (Fig. 3).

PM10 and PM2.5. The results indicated that the majority (50 to 79%) of the particulate matters are emitted during fuel production for both gasoline and ethanol–gasoline mix (Table 3). PM10 emissions from vehicle operation are more than three times higher than those from feedstock production, whereas PM2.5 emissions from vehicle operation are close to those from feedstock production. The application of 56 kg N ha^{-1} of fertilizer more than doubled feedstock production PMx emissions, and the application of 112 kg N ha^{-1} more than tripled feedstock PM emissions for E85 fuels. The results also showed that the life cycle of blended ethanol produced between 69 and 90 mg·PM10 km^{-1}, whereas it produced between 34 and 56 mg·PM2.5 km^{-1}. In comparison with gasoline (Fig. 2), PM10 from FFV:E85 was higher (by 4% for low N rate to 32% for high N rate), whereas those from PM2.5 were lower (by 15% for low N rate to 54% for high N rate). Further, the results indicated that field positioning on unfertilized plots did not impact life cycle PM. However, the addition of fertilizer increased PM10 by 3% for E85 from switchgrass grown at the shoulder in comparison with the footslope. Similarly, PM2.5 was 5% higher for E85 from switchgrass grown at the shoulder.

Impact of producing switchgrass for a period shorter than maximum stand life

In case the switchgrass is abandoned after only one harvest, this scenario would result in 2 to 3% more life cycle energy use, 20% more fossil fuel use in unfertilized plots and 8 to 12% more fossil fuels in fertilizer plots. It would also lead to 10 to 14% increase in life cycle PM from biomass grown in unfertilized plots and 4 to 9% for fertilized plots. No significant changes in N_2O and VOCs changes would be observed from shorting the stand life. The highest change would occur to life cycle SOx, with an increase of about 44% for unfertilized plots and 6 to 11% for fertilizer plots. The life

cycle environmental impacts due to shortening the stand life to one year were largely attributed to the feedstock production step.

Discussion

This discussion will explain the impact of nitrogen rate and landscape positions on emissions and energy use from ethanol-blend life cycle in the context of previous studies. However, it will not attempt to explain the source of all emissions and energy. This information can be obtained from the GREET tool database (downloadable from https://greet.es.anl.gov/), and it explains mass (including emissions) and energy flows from production of agricultural inputs, production of chemicals, and additives involved in ethanol (and other fuels involved such as petroleum, natural gas, coal) production, ethanol conversion and purification processes, and transportation.

Energy and emissions from switchgrass feedstock production constitute a key portion of the switchgrass-derived blended ethanol life cycle. These findings were consistent with previous studies: In Wang *et al.* (2012), a well-to-wheel analysis using GREET model showed that the GHG emissions from feedstock production were approximately half of the total GHG emissions for switchgrass-derived ethanol. Approximately 57–110% of GHG reductions with respect to gasoline were computed using GHGenius model (well to wheels) by Spatari *et al.* (2005), EBAMM model (well to pump) by Schmer *et al.* (2008), and GREET model (well to wheels) by Wu *et al.* (2006) and Wang *et al.* (2012). Fossil fuel replacement and other emissions in this study were also in line with findings in Wu *et al.* (2006) and Wang *et al.* (2012); Wu *et al.* (2006)'s GREET simulations. However, parameterization differences were discovered in some studies, for example, estimates of biomass yield and land-use change-related carbon sequestration in this study were conservative in comparison with Cherubini & Jungmeier (2010) who estimated 3 times higher yields and carbon sequestration. This variability was addressed by studies that conducted stochastic analysis of variables [e.g., Spatari & MacLean (2010)] and indicated that CO_2 emissions due land-use change and N_2O due to fertilizers are major source of uncertainty in LCA of cellulosic ethanol.

It is important to explore the impact of nitrogen fertilizer rate on switchgrass because it leads to increase in yield but adversely contributes to generation of emissions (N_2O, CO_2, NO_x) and increased energy use. Previous researchers doubted the environmental benefits of producing biofuel from plants that require fertilizers (Searchinger *et al.*, 2008). It is believed that N_2O emitted from soil due to nitrogen addition could offset the benefits of biofuels. But, land where no fertilizer was added also emitted N_2O due to natural cycling of nitrogen in soil, water, and the atmosphere (Vitousek *et al.*, 1997). Under anaerobic conditions, nitrogen in soil is transformed by microbes into N_2O (Smith *et al.*, 2008). This study demonstrated that increase in N_2O emissions is primarily offset by CO_2 taken up from the atmosphere to grow biomass and subsequently the biomass stored in the soil (roots) replenishes soil carbon. Croplands converted to grasslands (such as switchgrass) significantly improve soil carbon (Fazio & Monti, 2011).

The addition of regionally optimized N fertilization (56 kg N ha^{-1} for SD) during switchgrass production is likely economical; 37% more blended ethanol was generated with the same land than with unfertilized. Overfertilization (112 kg N ha^{-1} in this study) has no economic benefits and causes adverse environmental impacts. Field testing of soil nutrients (N, P, K, and S) and acidity can further reduce inputs, costs, and enhance yield during switchgrass production. Previous switchgrass LCA studies included addition of K, P, and lime and found these to add to the increase in energy use and emissions (Wu *et al.*, 2006; Bai *et al.*, 2010). In this study, K, P, S, and lime were not added because preliminary soil tests determined that it was not necessary. For high ratio of switchgrass ethanol fuel blends (E85), N optimization can improve GHGs, VOCs, NO_x, SO_x, and PM which are heavily generated from feedstock production. Overfertilization (112 kg N ha^{-1}) reduced life cycle environmental benefits almost by half or doubled the adverse impacts for switch-derived ethanol. Feedstock production involves more modeling uncertainty due to spatial variability in soils and climate, agricultural management, and the complex cycle of nitrogen and carbon at the farm level. The number of harvests before the switchgrass is replanted influences life cycle impacts, that is, the more years of productivity a stand provides, the greater the environmental benefits. Most previous studies indicate that switchgrass stands last between 10 and 20 years; however, stand duration depends on crop condition, marketability, and management decisions. Results demonstrated that, even with only one year of harvest, the switchgrass-derived ethanol would be better than gasoline in term of fossil fuel use, GHGs, and PM2.5. In other categories such as VOCs, CO, PM10, and NO_x, total energy gasoline is better than E85s for one year of harvest. This study showed that as a perennial crop, switchgrass has the benefit, once established (1–3 years), of avoiding energy use and pollution from planting and pesticides application, production of pesticides, and disruption of carbon sequestration.

Although using blended ethanol resulted in net GHG emissions, the primary goal is to develop renewable fuel sources better than existing nonrenewable fuels (e.g., gasoline). This study showed that GHGs were significantly reduced for switchgrass-derived E85 compared to gasoline. Agricultural management significantly influences switchgrass-based ethanol life cycle. Over the years, the emphasis was placed on making passenger vehicle more efficient to reduce emissions. However, increasing agricultural machinery efficiency can have a significant impact on emissions and fuel use reductions. For instance, Bochtis *et al.* (2010) showed that controlled traffic at the farm can itself significantly increase efficiency.

Landscape positioning where switchgrass is grown could have a significant impact on the life cycle of switchgrass-derived ethanol. Erosion and drainage from high elevation (shoulder) to low elevation (footslope) causes higher biomass growth at low elevation position due to higher moisture, nutrients, and organic matter (Mbonimpa *et al.*, 2015). The LCA in this study demonstrated that the footslope is linked to lower switchgrass-derived ethanol GHGs due to primarily higher biomass growth (CO_2 uptake) compared to the shoulder. Moreover, the analysis showed that the footslope was, in general, linked to better energy use and fewer emissions compared to the shoulder due to increased productivity. The increased productivity within the fertilized switchgrass appears likely due to downslope transport of fertilizer and soil rich in organic matter (erosion) from shoulder to the footslope.

Presently, corn ethanol dominates the US market for ethanol–gasoline blends and it is important to have a perspective on how switchgrass ethanol (cellulosic) compares to corn ethanol (starch). Using well-to-wheels GREET model simulations, Wang *et al.* (2012) and Wu *et al.* (2006) conducted simulations that compared life cycle impacts from corn ethanol, switchgrass ethanol, and gasoline (per unit joules contained in the fuel or on basis of using it in flex fuel vehicles-E85s). They indicated that when corn ethanol and switchgrass ethanol are compared to gasoline, switchgrass displaces about 57–66% of fossil fuels and 62–97% of GHGs whereas corn ethanol displaces about 32–57% of fossil fuels and 19–48% of GHGs. In comparison with gasoline, switchgrass ethanol produces 40% higher amount of NOx while corn ethanol produces 100% higher NOx. switchgrass ethanol showed benefits in reducing PM10 (by 30%) and SOx (by 40%) while corn ethanol contributed higher amount of PM 10 (>200%) and SOx (105%) in comparison with gasoline. Corn ethanol also had slightly higher amount of VOCs and CO. They attributed these differences to this fact: 'Instead of using coal or natural gas to fuel the ethanol plant, as in the corn ethanol production process, the conversion process in their study (similar to our study) relies on biomass residuals and methane gas from an on-site waste water treatment plant to generate heat and power, which decreases fossil energy consumption' (Wu *et al.*, 2006).

Although this study did not analyze other environmental impacts such as eutrophication and ecotoxicity potential, it is expected that these impacts can be reduced when a cropland is converted into a perennial grassland (Monti *et al.*, 2009). In addition, perennial grasses such as switchgrass improve the biodiversity of species in the area (Hartman *et al.*, 2011). While GREET is an efficient, effective model to analyze the life cycle of switchgrass fuel systems, it is a simplification of a complex system that would be strongly linked to policy and market effects (Stratton *et al.*, 2011). In the future, better understanding of switchgrass systems should be realized when mass commercialization of switchgrass-derived is further established.

In conclusion, generation and use of switchgrass ethanol–gasoline mixture can displace GHGs emissions. It can also lead to displacement of total fossil fuels because petroleum and coal reductions offset observed increase in natural gas. The increase in natural gas was mainly contributed by fertilizer application. Therefore, overfertilization should be avoided (e.g., rate above 56 kg N ha^{-1} in this case) because fertilizers-related energy and emissions are a significant portion of the entire switchgrass ethanol life cycle. There is also a difficulty to generalize the benefits or drawbacks of switchgrass ethanol due to the impact other environmental conditions such as landscape topography can have on switchgrass ethanol life cycle. Better soil conditions at the footslope of the hill, where eroded material deposited, led to greater life cycle environmental benefits. Selective topography-based planting of switchgrass in footslopes, such as riparian corridors, may be both economically and environmentally favorable. However, it is believed that in the long-term switchgrass could improve the soil quality at eroded higher grounds as well. For a future follow-up study, we recommend LCA for switchgrass from various geographical locations, ethanol produced using other process pathways, and exploration of other environmental impact categories.

Acknowledgements

This study was supported by the North Central Regional Sun Grant Center at South Dakota State University (SDSU) through a grant provided by the US Department of Energy Bioenergy Technologies Office under award number DE-FC36-05GO85041.

References

Bachman WJ (1997) 'Soil Survey of Day County, South Dakota,' Natural Resources Conservation Service.

Bai Y, Luo L, van der Voet E (2010) Life cycle assessment of switchgrass-derived ethanol as transport fuel. *The International Journal of Life Cycle Assessment*, **15**, 468–477.

Bochtis D, Sørensen C, Green O, Moshou D, Olesen J (2010) Effect of controlled traffic on field efficiency. *Biosystems Engineering*, **106**, 14–25.

Cherubini F, Jungmeier G (2010) LCA of a biorefinery concept producing bioethanol, bioenergy, and chemicals from switchgrass. *The International Journal of Life Cycle Assessment*, **15**, 53–66.

Congress (2007) Energy independence and security act of 2007. *Public Law* 110–140, 110th Congress.

Dunn JB, Qin Z, Mueller S, Kwon H, Wander M, Wang M (2014) Carbon Calculator for Land Use Change from Biofuels Production (CCLUB). User's Manual and Technical Documentation. Argonne National Laboratory.

Fazio S, Monti A (2011) Life cycle assessment of different bioenergy production systems including perennial and annual crops. *Biomass and Bioenergy*, **35**, 4868–4878.

Grisso RD, Kocher MF, Vaughan DH (2004) Predicting tractor fuel consumption. *Applied Engineering in Agriculture*, **20**, 553–561.

Guretzky J, Biermacher J, Cook B, Kering M, Mosali J (2011) Switchgrass for forage and bioenergy: harvest and nitrogen rate effects on biomass yields and nutrient composition. *Plant and Soil*, **339**, 69–81.

Harmoney KR, Moore KJ, Brummer EC, Burras CL, George JR (2001) Spatial legume composition and diversity across seeded landscapes. *Agronomy Journal*, **93**, 992–1000.

Hartman JC, Nippert JB, Orozco RA, Springer CJ (2011) Potential ecological impacts of switchgrass (*Panicum virgatum* L.) biofuel cultivation in the Central Great Plains, USA. *Biomass and Bioenergy*, **35**, 3415–3421.

Liebig MA, Schmer MR, Vogel KP, Mitchell RB (2008) Soil carbon storage by switchgrass grown for bioenergy. *Bioenergy Research*, **1**, 215–222.

Luk JM, Pourbafrani M, Saville BA, MacLean HL (2013) Ethanol or bioelectricity? Life cycle assessment of lignocellulosic bioenergy use in light-duty vehicles. *Environmental Science and Technology*, **47**, 10676–10684.

Mbonimpa EG, Hong CO, Owens VN et al. (2015) Nitrogen fertilizer and landscape position impacts on CO2 and CH4 fluxes from a landscape seeded to switchgrass. *GCB Bioenergy*, **7**, 836–849.

McLaughlin S, Kszos AL (2005) Development of switchgrass (*Panicum virgatum*) as a bioenergy feedstock in the United States. *Biomass and Bioenergy*, **28**, 515–535.

Monti A, Fazio S, Venturi G (2009) Cradle-to-farm gate life cycle assessment in perennial energy crops. *European Journal of Agronomy*, **31**, 77–84.

Mulkey V, Owens V, Lee D (2006) Management of switchgrass-dominated conservation reserve program lands for biomass production in South Dakota. *Crop Science*, **46**, 712–720.

Naylor RL, Liska AJ, Burke MB, Falcon WP, Gaskell JC, Rozelle SD, Cassman KG (2007) The Ripple effect: biofuels, food security, and the environment. *Environment: Science and Policy for Sustainable Development*, **49**, 30–43.

Nikièma P, Rothstein DE, Min D-H, Kapp CJ (2011) Nitrogen fertilization of switchgrass increases biomass yield and improves net greenhouse gas balance in northern Michigan, U.S.A. *Biomass and Bioenergy*, **35**, 4356–4367.

Parrish DJ, Fike JH (2005) The biology and agronomy of switchgrass for biofuels. *Critical Reviews in Plant Sciences*, **24**, 423–459.

Schmer MR, Vogel KP, Mitchell RB, Perrin RK (2008) Net energy of cellulosic ethanol from switchgrass. *Proceedings of the National Academy of Sciences*, **105**, 464–469.

Searchinger T, Heimlich R, Houghton RA et al. (2008) Use of US croplands for biofuels increases greenhouse gases through emissions from land-use change. *Science*, **319**, 1238–1240.

Simpson TW, Sharpley AN, Howarth RW, Paerl HW, Mankin KR (2008) The new gold rush: fueling ethanol production while protecting water quality. *Journal of Environmental Quality*, **37**, 318–324.

Smith P, Martino D, Cai Z et al. (2008) Greenhouse gas mitigation in agriculture. *Philosophical Transactions of the Royal Society B: Biological Sciences*, **363**, 789–813.

Sokhansanj S, Mani S, Turhollow A, Kumar A, Bransby D, Lynd L, Laser M (2009) Large-scale production, harvest and logistics of switchgrass (*Panicum virgatum* L.)–current technology and envisioning a mature technology. *Biofuels, Bioproducts and Biorefining*, **3**, 124–141.

Spatari S, MacLean HL (2010) Characterizing model uncertainties in the life cycle of lignocellulose-based ethanol fuels. *Environmental Science and Technology*, **44**, 8773–8780.

Spatari S, Zhang Y, MacLean HL (2005) Life cycle assessment of switchgrass-and corn stover-derived ethanol-fueled automobiles. *Environmental Science and Technology*, **39**, 9750–9758.

Stratton RW, Wong HM, Hileman JI (2011) Quantifying variability in life cycle greenhouse gas inventories of alternative middle distillate transportation fuels. *Environmental Science and Technology*, **45**, 4637–4644.

Vitousek PM, Aber JD, Howarth RW *et al.* (1997) Human alteration of the global nitrogen cycle: sources and consequences. *Ecological Applications*, **7**, 737–750.

Wang MQ (1999) 'GREET 1.5-transportation fuel-cycle model-Vol. 1: methodology, development, use, and results.' Argonne National Lab., IL (United States). Funding organisation: US Department of Energy (United States).

Wang M (2008) The Greenhouse Gases, Regulated Emissions, and Energy Use in Transportation (GREET) Model: Version 1.5. Center for Transportation Research, Argonne National Laboratory.

Wang DAN, Lebauer DS, Dietze MC (2010) A quantitative review comparing the yield of switchgrass in monocultures and mixtures in relation to climate and management factors. *GCB Bioenergy*, **2**, 16–25.

Wang M, Huo H, Arora S (2011) Methods of dealing with co-products of biofuels in life-cycle analysis and consequent results within the US context. *Energy Policy*, **39**, 5726–5736.

Wang M, Han J, Dunn JB, Cai H, Elgowainy A (2012) Well-to-wheels energy use and greenhouse gas emissions of ethanol from corn, sugarcane and cellulosic biomass for US use. *Environmental Research Letters*, **7**, 045905.

Wu M, Wu Y, Wang M (2006) Energy and emission benefits of alternative transportation liquid fuels derived from switchgrass: a fuel life cycle assessment. *Biotechnology Progress*, **22**, 1012–1024.

Wu M, Wu Y, Wang M (2008) 'Mobility Chains Analysis of Technologies for Passenger Cars and Light Duty Vehicles Fueled with Biofuels: Application of the Greet Model to Project the Role of Biomass in America's Energy Future (RBAEF) Project.' Argone National Lab.

7

On mitigating emissions leakage under biofuel policies

DEEPAK RAJAGOPAL[1,2]

[1]Institute of the Environment and Sustainability, University of California, Los Angeles, CA, USA, [2]School of Public and Environmental Affairs, Indiana University, Los Angeles, CA, USA

Abstract

A reason for much pessimism about the environmental benefits of today's biofuels, essentially corn and sugar-cane ethanol, is the so-called indirect land-use change (ILUC) emissions associated with expanding biofuel production. While there exist several simulation-based estimates of indirect emissions, the empirical basis underlying key input parameters to such simulations is not beyond doubt, while empirical verification of indirect emissions is hard. Regardless, regulators have adopted global warming intensity ratings for biofuels based on those simulations and in some case are holding regulated firms accountable for (some forms of) leakage. Suffice to say that both the estimates of and the approach to regulating leakage are controversial. The objective of this study is therefore to review a wider economic in order to identify a broader set of policy options for mitigating emissions leakage. We find that controlling leakage by affixing responsibility to regulated firms lacks support in the broader literature, which emphasizes alternative approaches.

Keywords: biofuels, climate policy, emissions, indirect emissions, indirect fuel market effects, indirect land-use change, leakage, price effects

Introduction

A basic insight of economics is that within a system of interconnected markets, a shock to one market would ripple throughout the system causing both intended and unintended effects. The latter are also referred to variously as indirect effects, spillovers, or leakage. In a pollution context, leakage is simply an increase in total pollution outside the policy jurisdiction in response to a decrease in total pollution within the jurisdiction. Leakage could be negative in that total emissions outside the jurisdiction could decline as well, but unless stated otherwise, leakage in our context refers to positive leakage. Leakage, therefore, undermines the effectiveness of policy intervention. Greenhouse gases (GHG) being global pollutants, and with global climate change policy expected to remain elusive (Diringer, 2013), it is therefore essential that leakage does not render subglobal efforts counterproductive.

Mitigating leakage presents two main challenges. The first is that quantifying leakage due to a policy, both *ex ante* and *ex post*, is shrouded in uncertainty. It is particularly so for GHGs given the ubiquitous use of fossil fuels, the global nature of commodity markets, and in the case of land uses, the heterogeneous and diffuse nature of emissions. The second is that as leakage emanates from unregulated activities, sources of leakage cannot be directly targeted. At the same time, leakage cannot

simply be ignored without careful consideration. In the case of biofuel policies, the policymakers's response to both these issues (discussed in detail later) is controversial. (For differing of views on indirect emissions, see RFA, 2008; UCS, 2008; Economist, 2009).

A sense of pessimism toward currently commercial biofuels is palpable in the current academic and policy literature, which is rife with support for the next-generation biofuels from ligno-cellulosic sources. These biofuels are predicted to provide substantially greater direct benefits and entail small negative spillovers (Schubert, 2006; Campbell *et al.*, 2008; Rubin, 2008; EPA, 2009; Somerville *et al.*, 2010; Mabee *et al.*, 2011). One might therefore conclude that today's biofuels are in decline and with them the concerns about their unintended effects as well. However, this might be an exaggeration. According to the US Energy Information Administration's projections for energy use in 2040, cellulosic biofuels account for less than 2% of total biofuel consumption with 98% still derived from current types of feedstock (Annual Energy Outlook, 2014). Today's biofuels appear to face greater challenges from recent trends in energy efficiency, infrastructure-related constraints, innovation in shale gas and shale oil, and electric vehicles relative to that from cellulosic biofuels. Regardless, it seems prudent to continue to innovate in the technically proven and commercially mature first-generation biofuels while implementing policies to manage their unintended consequences. This provides a context and motivation for this study.

Correspondence: Deepak Rajagopal
e-mail: rdeepak@ioes.ucla.edu

There has emerged a large literature on leakage under biofuel policies, and there have also been a few recent reviews of this literature (See Khanna & Crago, 2012; Rajagopal, 2013; Tokgoz & Laborde, 2014). There, however, has not been an attempt to relate either the severity of leakage from biofuels or the policies for controlling it to those in other contexts. This study is therefore motivated by the idea that a review of the broader economic literature on leakage could suggest alternative potentially less controversial and more effective approaches for controlling leakage. We focus on the following questions: (1) How do indirect emissions from biofuel-based policies compare with emission leakage under other types of policies, both in theory and the size of leakage relative to direct reduction in emissions? (2) How do current policy measures to limit indirect emissions compare with recommendations in the broader literature on leakage? Specifically, we compare leakage from biofuels to the literature on leakage under: (i) environmental regulation and international trade, (ii) agricultural and land-use policies, (iii) energy efficiency, and (iv) optimal depletion of exhaustible resources. We summarize the theoretical arguments, the quantitative estimates, and the policy insights from each area above that are applicable to biofuel policies and beyond.

Briefly, the following findings emerge. Rigorous econometric estimates of leakage on a global scale is hard and such estimates are either nonexistent or likely highly uncertain at best. Simulation-based approaches are currently the principal method for quantifying global leakage. Such estimates for biofuel policies, while they suggest a possibility of policy backfire, are also wide ranging and uncertain. Despite significant resources having been expended in quantifying leakage from biofuels, uncertainty is not declining. Policies employing a point estimate of leakage estimate are therefore controversial. Different from suggestions in the literature, some current regulations aim to control leakage from biofuels by directly penalizing regulated firm while ignoring some important indirect effects. We therefore explain why controlling leakage in this manner is better avoided and suggest alternative strategies, which merit further exploration.

The next section reviews a diverse literature on leakage, following which we discuss current approaches to control leakage from biofuel expansion. We conclude by discussing alternative leakage control policies that deserve further consideration.

Quantifying leakage

We focus on GHG leakage, which is, but one unintended consequence of biofuel expansion. Reduction in food supply, greater demand for farm chemicals, and biodiversity effects of agricultural expansion are a few other unintended consequences, which are beyond our scope. We also do not discuss the issue of 'problem shifting' from one type of burden to another such as a reduction in GHG emissions that inadvertently causes an increase in water pollution. Following Tinbergen (1952), in principle, each such unintended effect could be addressed by attaching at least one policy instrument targeting each such effect.

Indirect emissions of biofuel policies

Figure 1 provides a graphical intuition to price effects and leakage using biofuel shocks as an example. Leakage, more commonly indirect emissions in this literature, from biofuels began receiving serious attention following Searchinger et al.'s (2008) predictions of land-use change impacts of US ethanol mandates. The basic idea is that biofuel expansion increases demand (a shifting out of the demand curve) for cropland. This increases the price of agricultural land causing landowners to cultivate more land (a movement along and up the supply curve) by diverting land from its next best use. This in turn affects the returns to land in those uses and so on the effect ripples through until ultimately forested land is converted. These conversions lead to release of terrestrial carbon, referred to as indirect land-use change (ILUC) emissions. 'Indirect' here implies that emissions occur from land that may not directly be under cultivation for biofuel production. There is now a large literature predicting ILUC effects (Melillo et al., 2009; Hertel et al., 2010; Lapola et al., 2010; Tyner et al., 2010; Beach et al., 2012; Laborde & Valin, 2012). Whereas Searchinger et al. (Searchinger et al., 2008) concluded that average ILUC emissions per megajoule (MJ) of corn ethanol was 108 gram CO_2eq, Hertel et al. (Hertel et al., 2010) and Tyner et al. (Tyner et al., 2010) predict 27 and 12 gram CO_2eq per MJ, respectively. According to Searchinger et al., just the ILUC emissions are 10% greater the total life cycle GHG emissions per gasoline. If one includes the supply chain emissions of ethanol production, then it is almost twice the life cycle GHG emissions intensity of gasoline. Hertel et al. and Tyner et al.'s estimates suggest corn ethanol expansion is much less vulnerable to backfire in the long run. Expansion of sugarcane ethanol production in Brazil is also predicted to backfire on GHG mitigation efforts (Lapola et al., 2010). It should be pointed out ILUC associated with deforestation causes an instantaneous and one-time large increase in emissions. And so the average ILUC emission intensity is a metric derived by amortizing such emissions over an assumed project life span of 30 years or more. This is the idea of the

Table 1 Summary of literature predicting magnitude of leakage

Context	Leakage terminology	Typical sectoral coverage of reported estimates	Range of estimates	Prediction methodology	Policy approaches
Biofuels	Indirect or market-mediated emissions	Land, fuels	Wide ranging and uncertain	Simulation	Regulated firms accountable for indirect emissions*
Environmental regulation and international trade	Positive and negative leakage	Economy-wide but industrial only	5–25%, but at least one prediction >100%	Simulation	Import tariffs and subsidies for domestic production
Land set aside – agriculture and forestry	Slippage/spillover	Same sector	Typically low, <10% but at least one prediction > 90%	Simulation and econometric	Better targeting of land parcels, preemptive enrollment of sensitive parcels, adopt national/regional targets for land use as opposed to project level targets
Energy efficiency	Rebound effect	Own sector and in some cases economy-wide	Wide ranging but generally small (in a static context)	Simulation and econometric	Couple efficiency policy with emission pricing, target improvements in activities that are price inelastic
Optimal resource extraction	Leakage, supply-side leakage	Same sector	Potentially 100%	Simulation	Emission pricing, buy out rights to developing resources prone to leakage

*Unlike with other contexts, for biofuels, we report the actual approach adopted by two prominent regulations – the US RFS and California LCFS.

carbon debt, which means that certain biofuels provide positive net emission reduction only after a certain period, which may be in the order of decades (Fargione et al., 2008). It is a matter of another debate as to how to treat the benefits of emissions reduction which begin to accrue only after a few decades while the damages from ILUC emissions are immediate (Melillo et al., 2009; O'Hare et al., 2009).

Following ILUC, studies on leakage in the global market for transportation fuels, an indirect fuel use effect (IFUE) began to appear (de Gorter & Drabik, 2011; Hochman et al., 2011; Rajagopal et al., 2011; Rajagopal, 2013). Similar to ILUC, the IFUE effect arises because biofuel supply reduces the demand for oil depressing its world price. This leads to a partial rebound in the consumption of oil such that there is a less than 1 : 1 replacement of oil products with biofuel, which is a type of leakage. Rajagopal & Plevin (2013) show how different oil products are affected differently under an ethanol and biodiesel policies. Studies suggest that IFUE effect by itself could lead either to positive or to negative leakage. This was followed by studies analyzing the combined effect of ILUC and IFUE (Chen & Khanna, 2012; Bento et al., 2013; Huang et al., 2013; Rajagopal & Plevin, 2013). Another indirect effect termed as the indirect food effect (IFE), which refers to a net reduction in food consumption and hence avoids emissions associated with food production, has been argued to contribute to additional emissions reduction (Zilberman et al., 2013). Aggregate leakage, calculated by adding estimates of individual effects from different partial equilibrium analyses or estimated consistently in a multimarket or computable general equilibrium (CGE) framework, could, in theory, be net positive or negative.

The methodologies used for predicting indirect effects of biofuels are principally simulations of market equilibrium, which range from single-market single-region partial equilibrium to global CGE models, and mathematical programing (Khanna & Zilberman, 2012). There is a long history of application of such techniques to analyze trade policies, agricultural and energy market shocks, and climate policies. The weight of evidence from numerical models while it suggests leakage is likely positive in the short to medium term, and the predictions are simply too wide ranging, varying not only with the modeling framework but also across studies using a given computational model. See Dumortier et al. (2011) for a sensitivity of the FAPRI modeling system (Fabiosa et al., 2010), while Hertel et al. (2010) and Tyner et al. (2010) report widely varying point estimates using the GTAP model and database (Golub & Hertel, 2012). Rajagopal & Plevin (2013) do a Monte Carlo simulation of a simple partial model

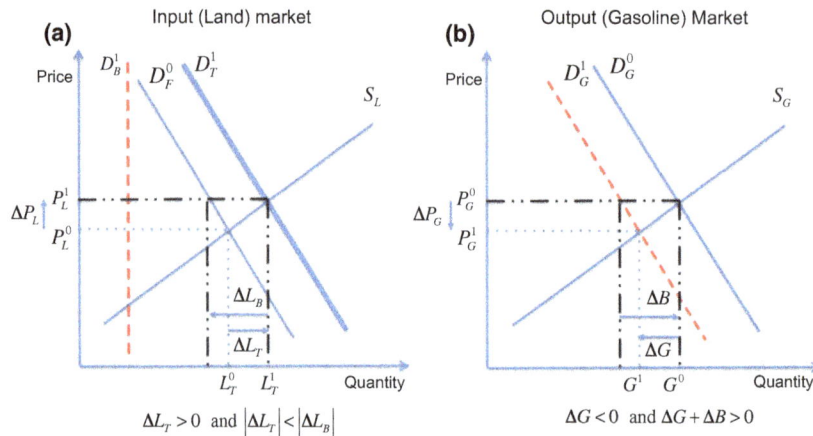

Fig. 1 Graphical intuition for leakage due to price effects in the short run for biofuels. Panels (a) and (b) depict the impact of an ethanol mandate on an input market (land) and an output (gasoline) market, respectively. The x and y axes denote quantity and price, respectively. Upward sloping lines represent the supply function of a commodity, and those sloping downward represent demand. An ethanol mandate is a positive demand shock to land, which shifts out. The vertical dotted line in the left panel denotes that a fixed quantity of land is demanded for producing crops that will be used to produce a mandated quantity of biofuel (not shown) and this is not a function of the price of land. The thicker downward sloping line is the aggregate demand for land for both biofuel and food. Before the biofuel mandate, the equilibrium, which is the intersection of supply (S) and demand (D_F^0), is (P_L^0, L_T^0). Postmandate, the new equilibrium is at (P_L^1, L_T^1). The equilibrium price of cropland increases and total cropland used increases from (L_T^0, L_T^1). Land allocated to food production declines more than the land allocated to biofuel production such that total expansion $\Delta L_T < \Delta L_B$. This is on account of the downward slope of demand for food. To the gasoline market, biofuel is a negative demand shock. The dotted line on the right panel denotes that demand for gasoline shifts inward, which means a decrease in demand, on account of biofuel supply (not shown). Before biofuel supply, the equilibrium in the gasoline market is at (P_G^0, G^0), and postbiofuel mandate, it is at (P_G^1, G^1). Both the equilibrium price of gasoline and quantity consumed decrease ($\Delta P_G < 0$, $\Delta G < 0$). However, consumption of gasoline decreases by less than the quantity of increase in biofuel, that is, $\Delta G + \Delta B > 0$. This is the basic mechanics of price effects, whose ultimate effect is that introducing (or eliminating) a fixed quantity of a new (or existing) product via a mandate (or a ban) does not reduce (or increase) an equal quantity of a substitute. Because of the price responsiveness of supply and demand, the total quantity of the basket of substitutes is not fixed. The pollution associated with the net change is the basis of emissions leakage. P, price; L, Quantity of Land; G, Quantity of Gasoline; B, Quantity of Biofuel; D, Demand; S, Supply. Superscripts 0 and 1 denote pre and post biofuel mandate respectively. Subscripts F, B and T denote Food, biofuel and total land respectively.

and report wide-ranging predictions of leakage for different plausible assumptions of behavioral and technical parameters, which any simulation model needs to assume. This literature also shows that the choice of policy instrument, the complementary policies that are employed to limit land-use change, and the role of technical change are each crucial in determining the magnitude of indirect emissions (Lemoine *et al.*, 2010; Bento *et al.*, 2013).

While there exist several simulation-based estimates of indirect emissions due to biofuel expansion, the empirical basis underlying key input parameters used in those simulations is not beyond doubt, while empirical verification of indirect emissions is hard. But there is some indirect evidence at hand. By econometrically estimating the elasticity of aggregate land supply with respect to land price for Brazil and the United States, Barr *et al.* (2011) conclude that current estimates of acreage expansion, and therefore, associated carbon emissions, from biofuels expansion may be biased upward.

Data also show that while the agricultural commodity price boom from 2006 to 2009 increased profitability per planted acre by about 64%, planted crop area increased only about 2% suggesting the landowners are more reluctant to expand crop area relative to the rate of conversion implied in simulation-based studies (Swinton *et al.*, 2011). Two recent papers also point out that existing simulation models in relying solely on yield response to higher crop prices have ignored other changes in land use at the intensive margin, such as double cropping, and a reduction in pre-existing farmland that is not under intensive cultivation (Babcock & Iqbal, 2014; Langeveld *et al.*, 2014).

Nevertheless, different biofuels have been assigned a global warming intensity rating by regulators based on available estimates. These estimates are being used for determining compliance with existing regulations. Suffice to say that both the estimates of and the approach to regulating indirect emissions are highly controversial (see RFA, 2008 and UCS, 2008).

Emissions leakage in the broader literature

We review the literature on the mechanisms of leakage and their implications in four different areas of the economic and policy literature.

International trade and environmental policy

Leakage here refers to an increase in emissions outside of the region undertaking domestic mitigation. Two distinct channels of leakage have been identified in this context (Elliott & Fullerton, 2014). One mechanism is that environmental policies that raise the domestic price of energy cause domestic producers to cede competitive advantage to producers abroad who then produce and emit more. This is termed as a terms-of-trade effect. When the policy-implementing region is large, then its policies could affect global demand for polluting resources (say, oil), which would lower the world price of those resources, causing an increase in their consumption abroad, a second mechanism of leakage. The second mechanism is believed to dominate the first (Zhang, 2012).

Quantitative estimates of leakage, which are almost entirely based on simulations of CGE models, are predicted to be 5% to 25% although particular economic sectors might be subject to large leakages (Harstad, 2012). An exception is Babiker (2005), which argues that if the structure of models estimating small leakage is extended to include features such as economies of scale, market power, and richer representations of international trade, leakage could exceed 100 percent in the worst case. On the contrary, Fullerton et al. (2011) derive conditions under which partial carbon regulation could lead to negative leakage, that is, reduction in emissions from the unregulated sectors as well and conditions exist to make this plausible in reality. Overall, this literature suggests that unilateral climate policies by industrialized countries will have small carbon leakage effects and unlikely to backfire (Mattoo et al., 2009; Burniaux et al., 2012). The focus in this literature is largely emissions from industrial activities.

The main policies suggested for mitigating competitive effects are border adjustment policies such as import tariffs and subsidies for home producers (Fischer & Salant, 2012; Zhang, 2012). While such policies may mitigate competitive effects and emissions leakage, they increase the social cost of achieving a given emissions target.

Slippage effects of agricultural and land-use-based policies

Leakage is a concern for land-use policies. For instance, the US Conservation Reserve Program (CRP) is a program that compensates landowners for setting aside agricultural lands, which lowers production and raises crop prices. This creates incentives to cultivate on land that is not enrolled in the CRP, which is termed slippage. It is estimated that less than 15% of intended benefits of the CRP is lost to slippage (Wu, 2000; Roberts & Bucholz, 2006). Similarly, forest protection projects have also been shown to shift deforestation to unprotected areas (Schwarze et al., 2002; Wunder, 2008). However, like with the CRP, an econometric study of slippage under a Mexican forest protection program also reports low levels of leakage (Alix-Garcia et al., 2012). Estimating leakage from different forest carbon sequestration activities using a combination of analytic, econometric, and sector-level optimization models, Murray et al. (2004) found leakage ranges from <10% to >90% depending on the activities affected and the project location. Urban and residential zoning has also been shown to increase demand for land in neighboring zones (Pollakowski & Wachter, 1990). Glaeser & Kahn (2010) econometrically predict that a hypothetical regulation that blocks one additional housing unit construction in San Francisco causes that activity to occur in a city whose energy mix is 50% more emission intensive. Overall, this literature implies that leakage effects, while important for various reasons, are not likely cause environmental impacts to worsen.

In terms of policies, this literature calls for superior targeting of locations for conservation to minimize both substitution and price effects (Wu et al., 2001). Preemptive enrollment of nearby lands that are prone to slippage is also suggested. A very different proposal is to track and control leakage at a national level as opposed to at a project level in order to minimize transaction cost, which refers to cost associated with the design, adoption, administration, monitoring, and enforcement of a program (Plantinga & Richards, 2008).

Rebound effects of energy efficiency and conservation

Leakage, described here as 'rebound' effects, refers to various mechanisms by which higher energy efficiency lowers the marginal cost of energy-consuming services, and the monetary savings get spent on activities that ultimately require energy (Greening et al., 2000; Sorrell, 2007). The total rebound effect is sometimes further disaggregated into a direct effect, which is an offsetting increase in energy use in the same service undergoing efficiency improvements, and an indirect effect which is an increase in energy use elsewhere. Given the ubiquitous use of energy, leakage from energy efficiency on a large scale manifests at multiple scales, within a single household, across different sectors within a region and across nations. As a result, the indirect rebound effect is hard to quantify. Empirical estimates of rebound effect are wide ranging varying with the application, whether direct or indirect effects are analyzed, short run vs. long run, developmental state of an economy, the region of study, etc. Although the estimates of rebound are wide ranging, there are claims that risk of backfire is overstated (Gillingham et al., 2013). One application with empirical evidence that the long-run rebound effect is well below 100% is personal transportation [see papers by Small & Van Dender (2007); Hughes et al. (2006)]. However, drawing general conclusions about the magnitude of the rebound effect is to be avoided (Van den Bergh, 2011).

A simple response to the rebound effect is to couple it with emission pricing, either through a fee or a cap. That said, a justification for energy efficiency policies is itself the political infeasibility of this approach. Another suggestion is to target efficiency improvements in those services whose demand is inelastic, that is, not price responsive, specifically, price reductions (Davis et al., 2014). However, this does not ensure that the savings do not lead to a preference for bigger or more powerful products that are more energy intensive and it also does not preclude indirect rebound effects.

Dynamics of resource use and intertemporal leakage

A different type of leakage to the above is intertemporal leakage, which occurs when the flow of emissions over time is altered by the policy. Cumulative emissions, however, might or might not be affected. Two hypotheses have been put forward here. Jevons (1906) wrote that more efficient utilization of a resource such as coal would create new sources of demand for that resource over time, which ultimately accelerates its depletion. Termed as the Jevons' paradox, it implies 100% leakage in that the resource is nevertheless exhausted and perhaps sooner too. A second hypothesis is that a reduction in demand for polluting resources due to policies, such as carbon pricing, could lead to faster extraction of those resources if owners of such resources are pessimistic about future demand. This is termed the Green Paradox (Sinn, 2008). This type of leakage is also referred to as 'supply-side' leakage, one that is distinct from 'demand-side' leakage, which refers to greater consumption of polluting resources in unregulated markets. In other words, even though cumulative emissions are unaffected, environmental degradation accelerates under pure demand-side policies such as carbon tax or tradable permits that ignore supplier behavior. When supply-side leakage occurs outside the policy jurisdiction, the more effective approach is to procure property rights to vulnerable resources to prevent their exploitation. A type of intertemporal leakage is also discussed

in the land-use literature, where it is argued that conservation achieved by paying landowners are contingent on such payments being maintained (Murray et al., 2004). Biofuels are also argued to entail carbon debt on account of initial high emissions from land-use change, which they pay back through reduction in fossil carbon emissions over several years or decades (Fargione et al., 2008; O'Hare et al., 2009).

Empirical evidence for intertemporal leakage is hard to find although there are some simulation-based estimates. While the simplest of models assuming homogenous resource deposits predict 100% leakage under policies such as emission taxes and clean energy mandates, more realistic assumptions, such as rising extraction costs, suggest leakage would be lower (Fischer & Salant, 2012). Leakage is also predicted to decline with increase in stringency for demand-side policies.

Summary of leakage modeling literature

Analyzing and quantifying leakage from energy and environmental policies has a rich history in the economic literature (See Table 1 for a tabular summary of the literature discussed). Emissions leakage is a concern that is not specific to any specific technology or policy instrument but arises because not all the relevant sources of pollution are affected uniformly by a technology or policy shock. The basic starting point for leakage is price effects. On account of limited past experience, our ability to econometrically predict leakage, and to verify estimates derived from numerical approaches, is limited, and biofuel policies are no exception. However, the range of numerical estimates does suggest greater vulnerability to leakage for biofuel policies relative to the other contexts. Furthermore, biofuels are a heterogeneous resource, in that there are multiple types of biofuels and any given type could be produced with varying amount of emissions depending on the agro-climatic, technical, economic, and policy parameters affecting their production. The variability in the direct supply chain emissions is higher relative to other technologies.

Controlling leakage

Current approaches to controlling GHG leakage from biofuel policies

We discuss only approaches that go beyond a mere statement of intent to control leakage and adopt specific regulatory measures. We describe two such regulations today, both within the United States, namely the US Federal Renewable Fuel Standard (RFS) and the California low carbon fuel standard (LCFS). The RFS requires fuel retailers to ensure that either biofuels sales (or an equivalent amount of purchased biofuel credits) comprise a certain minimum share of their total sales. Under this regulation, only biofuels whose total life cycle footprint taking into indirect emissions is below an upper limit are permitted for compliance. For a more detailed discussion of the various categories of biofuels, their target shares and their specific emission thresholds relative to gasoline refer EPA (2009). The RFS relies on a combination of FASOM (Adams et al., 1996) and FAPRI (Fabiosa et al., 2010) computational models to determine the indirect emissions intensity of each type of biofuel. The LCFS requires each regulated fuel supplier to ensure that the sales-weighted life cycle emissions intensity of all fuels sold be below a common upper limit. Under this regulation, the life cycle emission intensity of each batch of fuel is the simple sum of a direct supply chain emission intensity that is specific to a fuel producer and an average global warming intensity rating for the indirect emissions intensity from each type of fuel. The latter is determined by the regulator based on the biofuel pathway used by the firm and is not specific to the regulated firm. The LCFS relies on the GTAP modeling framework (Golub & Hertel, 2012) to compute indirect emissions intensity of each biofuel pathway. Neither of these regulations regulates indirect emissions other than ILUC emissions. We categorize the approach of both the RFS and LCFS to indirect emissions a micro- or 'firm-level' regulation of leakage.

Arguments against firm-level regulation of leakage

A key argument here is the range of uncertainty in numerical estimates of leakage coupled with a lack of an empirical basis for any single estimate, which is necessary under this approach. If there is reason to believe that empirical evidence could become available *ex post*, then one could true-up emissions and a firm could be held accountable for it's actual level of indirect emissions, forgetting for a moment the ethical and legal implications of this approach discussed later. However, there is little rigorous evidence in the literature to suggest that there exist proven and reliable techniques to predict, ex ante or ex post, the indirect emissions of a policy, let alone at a firm-specific level. We thus also lack the tools to verify whether the firm-level approach was successful in controlling leakage and if so to what extent, not to mention the cost of monitoring emissions to collect data for empirical investigation.

A second related argument is the inherent instability in any estimate because of dynamic economic processes underlying the changes we observe. For instance, a historical view of agricultural development in the large economies suggests extensification as a primary mode of expanding production during initial stages of development followed by intensification during the later stages. For instance, the peak in US agricultural acreage occurred during early 1900s and was the result of land settlement policies after agricultural intensification started to occur. As a result, total agricultural acreage in the United States and other industrial countries has continued to decline and is now largely stable (Cochrane, 1993). Sustained high food prices have also spurred innovation whose long-run impact was a surfeit of supply, which then led to farm subsidies for reducing food production and land retirement through programs such as CRP (Gardner, 1992). In response, it has also been out that in the absence of biofuels, these same dynamic processes would have returned land to nature leading to greater carbon sequestration. Resolving this requires predicting how much extra innovation would occur because of biofuels than otherwise and what is its implications for land use, which is itself an area of debate (Keeney & Hertel, 2009; Berry, 2011. http://www.arb.ca.gov/fuels/lcfs/workgroups/ewg/010511-berry-rpt.pdf; Nassar et al., 2011).

A third argument is the harshness and inequity in holding a few firms who exercise little control over leakage accountable for what is a consequence of larger forces at play. The fact that leakage estimates are uncertain accentuates the problem (Eco, 2009; Zilberman et al., 2011). At best, one could derive an estimate of the average indirect emissions per unit of a biofuel, but a firm-specific value is unrealistic. However, any average value that is chosen could force some firms to shut down for no fault of their own because despite adopting best practices, compliance could be infeasible for some firms.

A fourth argument is that there is little support in the literature for penalizing individuals firms on the basis of predicted leakage effects. In the international trade literature, the recommended approach is border tariffs and subsidies to domestic firms to reduce competitive effects. The slippage literature calls for superior targeting of land retirements to minimize and preemptively enrolling susceptible land parcels. The energy efficiency literature suggests subsidies target efficiency in services with inelastic demand. Finally, intertemporal leakage is addressed by choosing the right types of policies and the right stringency of policies.

Last but not least, firm-level approaches do not guarantee that the policy will not backfire. One reason for this is that all forms of leakage may not have been considered in estimating leakage. For instance, Rajagopal & Plevin (2013) show that the LCFS regulation might backfire despite considering ILUC emissions because it ignores IFUE. In any case, as mentioned before, we also lack the tools to verify the actual effect on emissions.

Looking ahead: alternative strategies for limiting leakage

We now suggest approaches that obviate the need for a point estimate of leakage. The literature on their applicability to biofuel policies is scarce and therefore is an area for further research. A targeted and effective solution to leakage is to expand the scope of the regulation to include unregulated emitters. For leakage crossing political boundaries, this requires intergovernmental environmental agreements, on which a rich literature exists and we have little to add here. Instead, we focus on adjustments that the jurisdiction in question could pursue unilaterally. Given a prediction of leakage under a proposed policy instrument, we discuss (i) altering the stringency of the instrument, (ii) adding complementary instruments, and (iii) adopting an altogether different instrument. Mixing the first two strategies is of course an option as well.

In choosing from the above approaches, we argue that three factors merit consideration – (i) the level of confidence in the estimates of leakage, (ii) the severity of the problem implied by those estimates, and (iii) whether the type of pollution in question is the primary objective of the policy. If leakage is estimated to a high level of confidence, and the estimates suggest a substantial risk of backfire, and if emissions reduction is the primary objective of the policy, then the policy merits a complete reconsideration. If, however, any one of the above conditions is not true, then elimination of the policy altogether appears extreme.

Biofuel (and renewable energy) policies are multiobjective. For this reason, we emphasize approaches that do not *de jure* or *de facto* ban any given technology on account of leakage. Typically, one or more among consumption mandates, subsidies, and performance standards have been employed as policy instruments to support the adoption of new technologies to reduce pollution. We emphasize mandates, which have been the main driver of renewable energy expansion worldwide. Much of the discussion that follows is also applicable to leakage under a broad range of policies.

The size of the mandate has a strong effect on the size of leakage. Their relationship is, however, not necessarily monotonic. Let us take the case of ILUC. A larger biofuel mandate, all else equal, implies bigger increase in land value, which implies larger land-use change and emissions. Higher prices, however, also induce innovation that shrinks land use in the long run, a phenomenon with empirical support (see Discussion in Section 4.2). With regard to IFUE emissions, bigger mandates would lead to higher domestic prices and lower world oil prices. This means lower domestic fuel consumption (and emissions) and higher consumption (and emissions) abroad, respectively. Under certain conditions, the combined effect could be smaller leakage under a bigger mandate (Rajagopal *et al.*, 2011). To determine the optimal size of the mandate given leakage, we, however, need to again rely on those same computational models, whose calculations, we argued earlier, depend on uncertain model parameters.

A second modification to a renewable energy mandate is to limit the most risky compliance pathways. The firm-level approaches under the RFS and LCFS discussed above are of this type. An alternative would be to simply cap the total quantity or the share of the risky pathways and distribute permits to regulated parties up to the cap. It is worth pointing out that current RFS mandate of 15 billion gallons of first-generation ethanol is a floor and not a cap. The adoption of explicit volumetric targets for second-generation fuels under the RFS II regulations partially remedies this situation. To go further, policymakers could cap each specific type of biofuel as well. A complementary modification would be to ramp up the stringency of the mandate ever more slowly over time in order to mitigate price effects and, therefore, mitigate leakage. Another response is to index mandates to the size of grain inventory to limit adverse price shocks. The benefit of both these approaches is that they simultaneously help address leakage globally.

One suggestion from the slippage literature is to track and control leakage at a regional or national level as opposed to a project level (Plantinga & Richards, 2008). In the case of biofuels, they are but one driver of land-use change in addition to demand for food, feed, timber, and forestry products and supply-side shocks such as weather and energy shocks. Therefore, a more direct approach would be to adopt national targets for land-use patterns and pursue international agreements to limit adverse land-use change abroad. As there exist multinational programs such as the United Nations' Reducing Emissions from Deforestation and Forest Degradation (REDD), one option is to further strengthen such programs in the face of additional burden from specific policies. Countries adopting biofuel mandates could commit additional funds in lieu of the additional stress they cause to existing international policies and programs. These funds could be used to purchase additional set-aside land to compensate for slippage. Within the United States, additional funds could be used to ensure landowners continue to enroll in the Conservation Reserve Program. This is consistent with the supply-side approach to limiting leakage that emerged from studies on the green paradox and intertemporal leakage (Harstad, 2012).

Finally, with regard to mandates, the combination of facts that biofuel mandates are not a pollution control policy, estimates of leakage are highly uncertain, and that large investments have already been undertaken (despite the sunk cost argument being a noneconomic one), the option of altogether eliminating the mandate is likely therefore a politically infeasible option.

The implications for limiting leakage under a performance-based standard such as the LCFS are similar to that we derive for renewable fuel mandates. Performance standards are in theory technology neutral. But if there is concern that a risky technology might comprise the principal compliance mechanism then, one response is to establish an upper bound on the quantity or share of the risky technology as opposed to tampering with that technology's performance rating itself, which is the approach under the LCFS. Another strategy is to regulate only direct emissions, but make either the performance standard more stringent or impose an upper bound on the direct emission intensity of the risky technologies and make the upper bound more stringent over time. The intuition for this approach is that as the direct benefits of a technology increase, a lesser quantity of that technology is required to reach a given standard, reducing its vulnerability to leakage. For leakage attributable to a subsidy policy, lowering it, eliminating it altogether, or indexing it to some measure of a performance are direct and feasible responses. For instance, the excise tax exemption on domestic ethanol consumption and import tariffs on ethanol were both eliminated by the US federal government in 2012.

To summarize, the alternatives to firm-level regulation do not eliminate the risk of backfire but such is the case with firm-level approaches as well. Nevertheless, by not requiring precise estimates of a highly uncertain variable, the alternative approaches are simpler and may therefore engender less controversy and transaction costs. Furthermore, a small amount of leakage might be desirable if it improves socioeconomic outcomes. For instance, availability of additional to cropland allows a supply response that mitigates food price inflation.

In trying to limit the supply of the most risky technologies, care should be taken so that it does not result in an effective ban on a technology, whose performance could be improved through innovation. Such innovation can lead to the generation of new technology that could have positive spillover effects for other markets. The impact of alternative leakage control policies on innovation is an important topic for future research.

Significant time and resources have been expended in not just computational modeling but also in debating, lobbying, and litigating around various issues related to precise quantification of leakage for the purpose of regulating leakage at the firm level. In economic terms, current approaches involve high transaction costs. There are several potential alternatives to firm-level regulation of leakage that appear to be simpler, more transparent,

and impose low transaction costs. Further research on mitigating leakage should therefore focus on analyzing the costs and benefits of such alternatives.

References

Adams DM, Alig RJ, Callaway JM, McCarl BA, Winnett SM (1996) The forest and agricultural sector model (fasom): Model structure and policy applications. Forest Service, United States Department of Agriculture, Research Paper PNW-RP-495.

Alix-Garcia JM, Shapiro EN, Sims KRE (2012) Forest conservation and slippage: Evidence from Mexico's national payments for ecosystem services program. *Land Economics*, **88**, 613–638.

Annual Energy Outlook (2014) *United States Energy Information Administration*, 2014. Available at: http://www.eia.gov/forecasts/aeo/er/early_fuel.cfm. (accessed Mar 26 2015).

Babcock BA, Iqbal Z (2014) Using recent land use changes to validate land use change models. Technical Report Staff Report 14-SR 109, Center for Agricultural and Rural Development (CARD) at Iowa State University.

Babiker MH (2005) Climate change policy, market structure, and carbon leakage. *Journal of International Economics*, **65**, 421–445.

Barr KJ, Babcock BA, Carriquiry MA, Nassar AM, Harfuch L. (2011) Agricultural land elasticities in the United States and Brazil. *Applied Economic Perspectives and Policy*, **33**, 449–462.

Beach RH, Zhang YW, McCarl BA (2012) Modeling bioenergy, land use, and ghg emissions with fasomghg: model overview and analysis of storage cost implications. *Climate Change Economics*, **3**, 1250012. Available at: http://www.worldscientific.com/action/doSearch?pubType=specific&AllField=beach+zhang&publication=40000029 (accessed on 26 March 2015).

Bento A, Landry J, Klotz R (2013) Are there carbon savings from us biofuel policies? The critical importance of accounting for leakage in land and fuel markets. *Energy Journal*, **36**, 2015.

Berry ST (2011) Biofuels policy and the empirical inputs to gtap models. Available at: http://www.arb.ca.gov/fuels/lcfs/workgroups/ewg/010511-berry-rpt.pdf (accessed 26 March 2105).

Burniaux JM, Martins JO (2012) Carbon leakages: a general equilibrium view. *Economic Theory*, **49**, 473–495.

Campbell JE, Lobell DB, Genova RC, Field CB (2008) The global potential of bioenergy on abandoned agriculture lands. *Environmental Science & Technology*, **42**, 5791–5794.

Chen X, Khanna M (2012) The market-mediated effects of low carbon fuel policies. *AgBioForum*, **15**, 1–17.

Cochrane WW (1993) *Development of American Agriculture: A Historical Analysis*. University of Minnesota Press, Minneapolis, MN.

Davis LW, Fuchs A, Gertler P (2014) Cash for coolers: evaluating a large-scale appliance replacement program in Mexico. *American Economic Journal Economic Policy*, **6**, 207–238.

Diringer E (2013) Climate change: A patchwork of emissions cuts. *Nature*, **501**, 307–309.

Dumortier J, Hayes DJ, Carriquiry M, Dong F, Elobeid A, Fabiosa JF, Tokgoz S (2011) Sensitivity of carbon emission estimates from indirect land-use change. *Applied Economic Perspectives and Policy*, **33**, 428–448.

ECO (2003) Maized and confused. *The Economist*, Available at: http://www.economist.com/node/14205727/ (accessed on 26 March 2015).

Elliott J, Fullerton D (2014) Can a unilateral carbon tax reduce emissions elsewhere? *Resource and Energy Economics*, **36**, 6–21.

EPA (2009) Regulation of fuels and fuel additives: Changes to renewable fuel standard program. Notice of proposed rulemaking 40 CFR Part 80; EPA-HQ-OAR-2005-0161, US Environmental Protection Agency.

Fabiosa JF, Beghin JC, Dong F, Elobeid A, Tokgoz S, Yu T-H (2010) Land allocation effects of the global ethanol surge: predictions from the international fapri model. *Land Economics*, **86**, 687–706.

Fargione J, Hill J, Tilman D, Polasky S, Hawthorne P (2008) Land clearing and the biofuel carbon debt. *Science*, **319**, 1235–1238.

Fischer C, Salant SW (2012) Alternative climate policies and intertemporal emissions leakage: quantifying the green paradox. *Resources for the Future Discussion Paper*, Report # RFF DP 12-16.

Fullerton D, Karney D, Baylis K (2011) *Negative leakage*. Technical report, National Bureau of Economic Research.

Gardner BL (1992) Changing economic perspectives on the farm problem. *Journal of Economic Literature*, **30**, 62–101. ISSN 0022-0515.

Gillingham K, Kotchen MJ, Rapson DS, Wagner G (2013) Energy policy: The rebound effect is overplayed. *Nature*, **493**, 475–476.

Glaeser EL, Kahn ME (2010) The greenness of cities: carbon dioxide emissions and urban development. *Journal of Urban Economics*, **67**, 404–418.

Golub AA, Hertel TW (2012) Modeling land-use change impacts of biofuels in the gtap-bio framework. *Climate Change Economics*, **3**, 1250015–1.

de Gorter H, Drabik D (2011) Components of carbon leakage in the fuel market due to biofuel policies. *Biofuels*, **2**, 119–121.

Greening LA, Greene DL, Difiglio C (2000) Energy efficiency and consumption – the rebound effect – a survey. *Energy Policy*, **28**, 389–401.

Harstad B°. (2012) Buy coal! a case for supply-side environmental policy. *Journal of Political Economy*, **120**, 77–115.

Hertel TW, Golub A, Jones AD, O'Hare M, Plevin RJ, Kammen DM (2010) Effects of us maize ethanol on global land use and greenhouse gas emissions: estimating market-mediated responses. *BioScience*, **60**, 223–231.

Hochman G, Rajagopal D, Zilberman D (2011) The effect of biofuels on the international oil market. *Applied Economic Perspectives and Policy*, **33**, 402–427.

Huang H, Khanna M, Oʻnal H, Chen X (2013) Stacking low carbon policies on the renewable fuels standard: economic and greenhouse gas implications. *Energy Policy*, **56**, 5–15.

Hughes JE, Knittel CR, Sperling D (2006) *Evidence of a shift in the short-run price elasticity of gasoline demand*. Technical report, National Bureau of Economic Research.

Jevons WS (1906) *The coal question: an inquiry concerning the progress of the nation, and the probable exhaustion of our coal-mines*. The Macmillan Company, London, UK.

Keeney R, Hertel TW (2009) The indirect land use impacts of United States biofuel policies: the importance of acreage, yield, and bilateral trade responses. *American Journal of Agricultural Economics*, **91**, 895–909.

Khanna M, Crago CL (2012) Measuring indirect land use change with biofuels: Implications for policy. *Annual Review of Resource Economics*, **4**, 161–184.

Khanna M, Zilberman D (2012) Modeling the land-use and greenhouse-gas implications of biofuels. *Climate Change Economics*, **3**, 1250016–1.

Laborde D, Valin H (2012) Modeling land-use changes in a global cge: assessing the eu biofuel mandates with the mirage-biof model. *Climate Change Economics*, **3**, 1250017–1.

Langeveld JWA, Dixon J, vanKeulen H, Quist-Wessel PM (2014) Analyzing the effect of biofuel expansion on land use in major producing countries: evidence of increased multiple cropping. *Biofuels, Bioproducts and Biorefining*, **8**, 49–58.

Lapola DM, Schaldach R, Alcamo J, Bondeau A, Koch J, Koelking C, Priess JA (2010) Indirect land-use changes can overcome carbon savings from biofuels in Brazil. *Proceedings of the National Academy of Sciences of the United States of America*, **107**, 3388–3393.

Lemoine DM, Plevin RJ, Cohn AS, Jones AD, Brandt AR, Vergara SE, Kammen DM (2010) The climate impacts of bioenergy systems depend on market and regulatory policy contexts. *Environmental Science & Technology*, **44**, 7347–7350.

Mabee WE, McFarlane PN, Saddler JN (2011) Biomass availability for lignocellulosic ethanol production. *Biomass and Bioenergy*, **35**, 4519–4529.

Mattoo A, Subramanian A, van der Mensbrugghe D, He J (2009) Reconciling climate change and trade policy. *Research Working Papers*, **1**, 1–46.

Melillo JM, Reilly JM, Kicklighter DW *et al.* (2009) Indirect emissions from biofuels: how important? *Science*, **326**, 1397–1399.

Murray BC, McCarl BA, Lee H-C (2004) Estimating leakage from forest carbon sequestration programs. *Land Economics*, **80**, 109–124.

Nassar AM, Harfuch L, Bachion LC, Moreira MR (2011) Biofuels and land-use changes: searching for the top model. *Interface Focus*, **1**, 224–232.

O'Hare M, Plevin RJ, Martin JI, Jones AD, Kendall A, Hopson E (2009) Proper accounting for time increases crop-based biofuels' greenhouse gas deficit versus petroleum. *Environmental Research Letters*, **4**, 024001.

Plantinga AJ, Richards KR (2008) International forest carbon sequestration in a post-kyoto agreement. *Harvard Project on International Climate Agreements Discussion Paper*, 08-11.

Pollakowski HO, Wachter SM (1990) The effects of land-use constraints on housing prices. *Land Economics*, **66**, 315–324.

Rajagopal D (2013) The fuel market effects of biofuel policies and implications for regulations based on lifecycle emissions. *Environmental Research Letters*, **8**, 024013.

Rajagopal D, Plevin RJ (2013) Implications of market-mediated emissions and uncertainty for biofuel policies. *Energy Policy*, **56**, 75–82.

Rajagopal D, Hochman G, Zilberman D (2011) Indirect fuel use change and the environmental impact of biofuel policies. *Energy Policy*, **39**, 228–233.

RFA (2008) Understanding Land Use Change and U.S. Ethanol Expansion. Renewable Fuels Association. Available at: http://www.ethanolrfa.org/page/-/objects/documents/2042/understanding_iluc_-_exec_summary.pdf. (accessed 26 March 2015).

Roberts MJ, Bucholz S (2006) Slippage in the conservation reserve program or spurious correlation? A rejoinder. *American Journal of Agricultural Economics*, **88**, 512–514.

Rubin EM (2008) Genomics of cellulosic biofuels. *Nature*, **454**, 841–845.

Schubert C (2006) Can biofuels finally take center stage? *Nature Biotechnology*, **24**, 777–784.

Schwarze R, Niles JO, Olander J (2002) Understanding and managing leakage in forest-based greenhouse-gas-mitigation projects. *Philosophical Transactions of the Royal Society of London*, **360**, 1685–1703.

Searchinger T, Heimlich R, Houghton RA *et al.* (2008) Use of US croplands for biofuels increases greenhouse gases through emissions from land-use change. *Science*, **319**, 1238.

Sinn H-W (2008) Public policies against global warming: a supply side approach. *International Tax and Public Finance*, **15**, 360–394.

Small KA, Van Dender K (2007) Fuel efficiency and motor vehicle travel: the declining rebound effect. *The Energy Journal*, **28**, 25–51.

Somerville C, Youngs H, Taylor C, Davis SC, Long SP (2010) Feedstocks for lignocellulosic biofuels. *Science*, **329**, 790–792.

Sorrell S (2007) The rebound effect: an assessment of the evidence for economy-wide energy savings from improved energy efficiency. Technical Report ISBN 1-903144-0-35, Sussex Energy Group for the UK Energy Research Centre.

Swinton SM, Babcock BA, James LK, Bandaru V (2011) Higher us crop prices trigger little area expansion so marginal land for biofuel crops is limited. *Energy Policy*, **39**, 5254–5258.

The Economist. Maized and confused, August 10th 2009.

Tinbergen J (1952) *On the theory of economic policy*. North Holland, Amsterdam.

Tokgoz S, Laborde D (2014) Indirect land use change debate: What did we learn? *Current Sustainable and Renewable Energy Reports*, **1**, 104–110.

Tyner WE, Taheripour F, Zhuang Q, Birur DK, Baldos U (2010) *Land use changes and consequent co2 emissions due to us corn ethanol production: A comprehensive analysis*. Technical report, Department of Agricultural Economics, Purdue University.

UCS (2008) Letter to the California Air Resources Board. *Union of Concerned Scientists*. Available at: http://www.ucsusa.org/assets/documents/clean_vehicles/call_to_action_biofuels_and_land_use_change.pdf (accessed 26 March 2015).

Van den Bergh J (2011) Energy conservation more effective with rebound policy. *Environmental and Resource Economics*, **48**, 43–58.

Wu JJ (2000) Slippage effects of the conservation reserve program. *American Journal of Agricultural Economics*, **82**, 979–992. ISSN 1467-8276.

Wu JJ, Zilberman D, Babcock BA (2001) Environmental and distributional impacts of conservation targeting strategies. *Journal of Environmental Economics and Management*, **41**, 333–350.

Wunder S (2008) *Moving ahead with REDD: issues, options and implications*. chapter 7. Center for International Forestry Research, Bogor Barat, Indonesia.

Zhang Z (2012) Competitiveness and leakage concerns and border carbon adjustments. *International Review of Environmental and Resource Economics*, **6**, 225–287.

Zilberman D, Hochman G, Rajagopal D (2011) On the inclusion of indirect land use in biofuel. *University of Illinois Law Review* 413–434.

Zilberman D, Barrows G, Hochman G, Rajagopal D (2013) On the indirect effect of biofuel. *American Journal of Agricultural Economics*, **95**, 1332–1337.

8

Life cycle analysis of biochemical cellulosic ethanol under multiple scenarios

COLIN W. MURPHY[1] and ALISSA KENDALL[2]

[1]Institute of Transportation Studies and Energy Institute, University of California, 1605 Tilia St. Suite 100, Davis, CA 95616, USA, [2]Department of Civil and Environmental Engineering, University of California, 1 Shields Ave., Davis, CA 95616, USA

Abstract

Cellulosic ethanol is widely believed to offer substantial environmental advantages over petroleum fuels and grain-based ethanol, particularly in reducing greenhouse gas emissions from transportation. The environmental impacts of biofuels are largely caused by precombustion activities, feedstock production and conversion facility operations. Life cycle analysis (LCA) is required to understand these impacts. This article describes a field-to-blending terminal LCA of cellulosic ethanol produced by biochemical conversion (hydrolysis and fermentation) using corn stover or switchgrass as feedstock. This LCA develops unique models for most elements of the biofuel production process and assigns environmental impact to different phases of production. More than 30 scenarios are evaluated, reflecting a range of feedstock, technology and scale options for near-term and future facilities. Cellulosic ethanol, as modeled here, has the potential to significantly reduce greenhouse gas (GHG) emissions compared to petroleum-based liquid transportation fuels, though substantial uncertainty exists. Most of the conservative scenarios estimate GHG emissions of approximately 45–60 g carbon dioxide equivalent per MJ of delivered fuel (g CO_2e MJ^{-1}) without credit for coproducts, and 20–30 g CO_2e MJ^{-1} when coproducts are considered. Under most scenarios, feedstock production, grinding and transport dominate the total GHG footprint. The most optimistic scenarios include sequestration of carbon in soil and have GHG emissions below zero g CO_2e MJ^{-1}, while the most pessimistic have life-cycle GHG emissions higher than petroleum gasoline. Soil carbon changes are the greatest source of uncertainty, dominating all other sources of GHG emissions at the upper bound of their uncertainty. Many LCAs of biofuels are narrowly constrained to GHG emissions and energy; however, these narrow assessments may miss important environmental impacts. To ensure a more holistic assessment of environmental performance, a complete life cycle inventory, with over 1100 tracked material and energy flows for each scenario is provided in the online supplementary material for this article.

Keywords: biofuels, carbon intensity, cellulosic ethanol, climate change, corn stover, greenhouse gas, LCA, life cycle assessment, switchgrass

Introduction

The environmental attributes of lignocellulosic biofuels have been a subject of interest to researchers for decades (Fu et al., 2003; Spatari et al., 2005; Wu et al., 2006; Pimentel & Patzek, 2008; Delucchi, 2010; Hsu et al., 2010). They may have the potential to displace a meaningful fraction of petroleum-based transportation fuels and reduce transport-related greenhouse gas (GHG) emissions, but technological and economic hurdles remain (Searchinger et al., 2008; Campbell et al., 2009; Bonin & Lal, 2012). The life-cycle GHG impacts from cellulosic biofuels are still quite uncertain. Lab and

Correspondence: Alissa Kendall
e-mail: amkendall@ucdavis.edu

pilot-scale studies of several lignocellulosic biofuel technologies indicate that many of the proposed conversion pathways have the potential to produce low-carbon fuels, but some have higher life-cycle GHG emissions than the petroleum fuels they seek to displace (Kendall & Yuan, 2013). Awareness of the importance of feedstock production on biofuels' life-cycle GHG footprint is increasing, as studies have revealed feedstock production practices can create a massive 'carbon debt' that offsets years of potential benefits from biofuel use (Fargione et al., 2008).

Over a dozen commercial-scale cellulosic ethanol facilities are currently planned or under construction, though at present only a few small commercial-scale facilities and a handful of pilot plants have been completed in the US (Advanced Ethanol Council, 2013;

Brown & Brown, 2013). Understanding the environmental impacts of each pathway is important to policymakers. Agreement on a common set of modeling assumptions, system boundaries, coproduct allocation practices and environmental accounting standards would help facilitate comparisons between pathways (Bonin & Lal, 2012; Wardenaar *et al.*, 2012).

Life cycle analysis (LCA) can serve a critical role in the development of advanced biofuels by determining the carbon intensity of a fuel and by linking particular environmental impacts to certain elements in the production process. LCAs also highlight areas of uncertainty within the production cycle and inform design decisions. For biofuel production, direct and indirect land-use change (LUC) and associated soil carbon fluxes, nitrous oxide (N_2O) emissions from soils, biorefinery conversion efficiencies and the material inputs to bio-refineries have all been identified as areas of significant uncertainty (Sheehan *et al.*, 2003; Blanco-Canqui & Lal, 2007; Hoskinson *et al.*, 2007; Anderson-Teixeira *et al.*, 2009; Kendall & Chang, 2009; Cherubini & Stromman, 2010; Hoben *et al.*, 2011; Khanna *et al.*, 2011; Sanchez *et al.*, 2012).

This study advances understanding of this subject in a number of ways. Most published LCAs of cellulosic ethanol production pathways omit substantial detail regarding the full production cycle; most focus their attention on one element such as feedstock production (Spatari *et al.*, 2005; Kim *et al.*, 2009) or conversion processes (e.g., Fu *et al.*, 2003; Borrion *et al.*, 2012; Iribarren *et al.*, 2012). This often leads to combinations of data and models that may not reflect consistent assumptions in terms of region or scope. For example, the Argonne National Laboratory's GREET model is often used for part or all of the biofuel production chain not modeled by the authors (Burnham *et al.*, 2006; Kendall & Chang, 2009). Independent analyses are needed to ensure that limitations of existing models do not constrain insights for improving biofuel pathways, that uncertainties in life-cycle stages are identified and understood, and to insure that unanticipated consequences have the greatest chance of being identified.

The LCA presented in this article is predominantly based on original modeling of each phase of cellulosic ethanol production, which maximizes transparency and facilitates thorough sensitivity analysis. In particular, this model allows for hot-spot analysis of the elements of production which have the greatest contribution to life-cycle environmental impacts.

In addition, this study provides a comprehensive LCI tracking over 1100 mass and energy flows for each scenario, which is available in the supporting information for this paper (Data S1). Providing these full LCIs to fellow researchers allows for future comparisons and

greater transparency. By focusing on the material and energy flows from biofuel production, rather than the combustion phase in a vehicle or exposure to pollutants, the effects of modeling assumptions are more transparent. Thus, the provided LCIs are based on a functional unit of 1 MJ of ethanol, rather than other potential functional units such as distance traveled in a vehicle.

One particular challenge for modeling cellulosic ethanol production is that few facilities have been constructed, so their performance under mature, commercial-scale conditions is unknown. As cellulosic conversion technology matures, future plants may have higher yields, beneficial coproducts or lower input requirements as compared to current technology (Lau & Dale, 2009; Unruh *et al.*, 2010; ZeaChem Inc., 2012; Ishola *et al.*, 2013). This article describes an LCA model of a technology process pathway currently under development by Edeniq, Inc., and projects a variety of possible scenarios for near-term (pilot plant) and intermediate-term (5–10 years) development. The Edeniq technology utilizes thermal exposure coupled with mechanical size reduction, and microorganisms which are engineered to produce the hydrolysis enzymes necessary to convert cellulose into fermentable sugars. These design choices help the process approach the minimum level of inputs required for the biochemical production of cellulosic ethanol. Variations on the conversion process, such as additional chemical additives, are tested in scenario analysis. While no set of scenarios can hope to exhaustively characterize all possible development paths, these reflect several likely variations on the process. In addition, the role of coproducts is explored through scenario analysis, because they prove to be critical to determining the environmental performance of the cellulosic ethanol.

Materials and methods

Goal and scope definition

This model evaluates the environmental flows consumed and generated during the production of biochemical cellulosic ethanol on a field-to-blending-terminal basis, using a mixture of attributional and consequential LCA. The system boundaries (Fig. 1) include the agronomic processes for feedstock production (equipment use, agrochemical production and application, and biogeochemical field emissions), transport of feedstock to the conversion facility, the conversion process and transport of denatured ethanol to the blending terminal where ethanol is mixed into retail motor vehicle fuels. The LCA presents results in terms of anhydrous ethanol because the denaturant is transported into, then out of the system with minimal modification, which means that its only effect on the system is the additional transportation activity required, which is accounted for in the modeling. This system boundary (field-to-blending terminal)

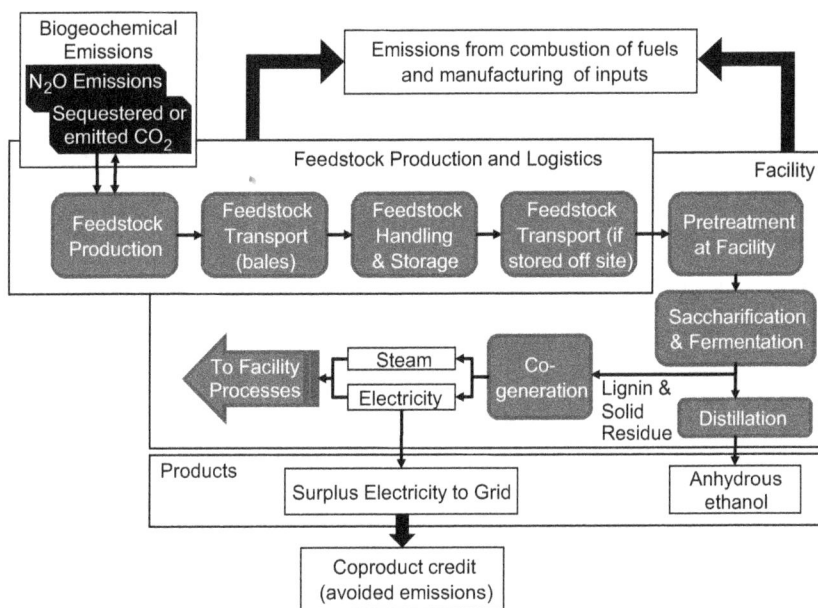

Fig. 1 Schematic of Life cycle analysis (LCA) system boundaries.

allows for fuels to be compared on an energy-equivalent basis, independent of any effects from vehicle fuel economy.

The system boundaries represent a typical or generic system assumed to be located in the Midwestern or Central US While site-specific characteristics, such as regional feedstock yield and access to freight transportation, could exert a significant effect on the net life-cycle environmental impacts of any given project, this study intends to evaluate effects of an average future conversion pathway, to allow for assessment of the potential outcomes from large-scale adoption of these, or similar, technologies. Thus, average or typical values for data that vary spatially, such as feedstock yield or average freight transport distance, were used.

Other system boundary considerations in this model include simplified modeling of the ethanol production facility's construction, which assesses major construction materials only (concrete, steel, and aggregate) and does not consider facility decommissioning. Previous studies have shown that the construction and demolition of buildings has negligible life-cycle impacts compared to materials production and building operation (Scheuer et al., 2003). Production of mobile equipment, including farm equipment, is also excluded on similar grounds.

Waste treatment at the conversion facility is likely to have minimal impact. Most of the solid waste from the conversion process is combusted to generate energy. In the base case facility, wastewater is condensed to a syrup and combusted, the future facility has an on-site wastewater treatment plant with anaerobic digester. Both facilities recycle water through the process where possible; water losses are primarily through evaporation, rather than discharge. Residual ash from biomass combustion and other un-recycled materials are produced in small quantities and disposed of in a landfill, where they are largely inert. Ash from combustion may have value as a soil additive (Risse & Gaskin, 2013), however, its value for this use

is uncertain and thus excluded from this analysis (Pandey & Singh, 2010). Any other waste products are assigned neither environmental burdens for disposal, nor credits as coproducts.

Life cycle inventory

The various LCIs reported by this model are developed in Microsoft Excel, with VBA scripting used to automate some processes (Microsoft Inc., 2006). LCI data for material and energy inputs are primarily derived from LCI databases such as Ecoinvent and GaBi Professional (PE International, 2008; Ecoinvent Ecoinvent Centre, 2011). A detailed table describing the LCIs used in the model, and their source, is provided in Data S2. Inventory datasets may not be geographically or temporally appropriate, however, which presents a challenge to researchers. Shortcomings in LCI data are well-documented and widely discussed in literature, (e.g., Huijbregts et al., 2001; Björklund, 2002). Recent comprehensive LCIs of chemical inputs to the biofuel system, including pesticides, fertilizers, antibiotics and propagation nutrients are notably scarce. Wherever possible, recent US data were used in this LCA to maintain consistent geographic and temporal context, though in some cases European data or data more than 10 years old were the only available option. Though uncertainty is introduced due to the paucity of regionally and temporally relevant LCIs, in general the flows which relied on the most uncertain LCIs are relatively small inputs to the process.

Stover and switchgrass feedstock production

Two feedstocks are modeled in this LCA: corn stover and switchgrass.

Corn stover is the above-ground biomass left after corn grain harvest and has traditionally been considered a residue. It is typically reincorporated into the soil to reduce erosion, return nutrients and maintain soil condition. When harvested for another use, stover becomes a coproduct of corn production and presents several allocation challenges.

In the base case of this LCA stover is treated using a consequential approach, where only changes to *business-as-usual* (BAU) practices are allocated to stover. Alternative attribution methods are discussed in depth in Murphy & Kendall (2013). BAU is assumed to be repeated no-till corn cultivation with grain harvested and stover left on the field. When stover is removed from the field, embodied nutrients [nitrogen (N), phosphorous (P) and potassium (K)], along with carbon, are removed as well. This requires the replacement of nutrients removed in stover by additional mineral or petrochemical fertilizer. The mass of N, P and K replaced is calculated using stover composition values from Wortmann & Klein (2008). Agricultural equipment activity, for collecting, baling and stacking baled stover is also attributed to the resulting biomass, since it would not have occurred in the BAU case. Average residue removal rate is assumed to be 38% and the net annual yield of corn stover was assumed to be 3.95 dry tons per hectare (Hess *et al.*, 2009).

Switchgrass production, in contrast to corn, has not been done at commercial scale, so fewer data exist on cultivation and harvest practices. The switchgrass cultivation data used in this study were provided by Ceres Inc., a company that has extensively researched grassy crop cultivation (Ceres Inc., 2013). The data rely on the University of Nebraska crop budgets for agricultural equipment work rates and energy consumption (Klein & Wilson, 2013), as well as proprietary research into fertilization parameters. Fuel consumption values used here approximately agree with values elsewhere in literature (e.g., Hess *et al.*, 2009). Switchgrass differs from corn in that there may be predictable and significant differences in activity on a year-to-year basis due to its growing cycle. After a 2 year establishment phase, in which harvests are limited and agronomic practices are tailored to maximize the growth of young plants, there is a 6–10 year period of maturity, with full harvests each year. After this maturity period, the plot is reseeded to maintain high yields. These variable emission and materials use rates are averaged across the lifespan of the plot. Average annual yield, including the low-yielding postestablishment years, was set at 5.7 dry tons per hectare, which conservatively reflects the near-term expectations of growers, and is somewhat lower than other literature (Hess *et al.*, 2009; Jung & Lal, 2011).

Both corn and switchgrass require fertilizer to achieve high yields. Total N, P and K application is determined by soil conditions and plant characteristics, but the choice of which formulation to apply may be determined largely by local supply, past experience and economic conditions (Nutrient Stewardship, 2011). The compounds used for corn fertilization are based on those most commonly used on corn in the United States according to USDA National Agricultural Statistics Service data (Economic Research Service, 2012). Ammonium phosphate nitrate is applied in sufficient quantities to meet P requirements. N requirements beyond what is provided by ammonium phosphate nitrate are met with liquid ammonia, and

K requirements are met by potassium chloride. For switchgrass, diammonium phosphate (DAP) is used to meet P requirements. Excess N demand is met with liquid ammonia, and K demand with potassium chloride. Total applications of N, P and K assigned to corn stover are 32, 8 and 57 kg ha^{-1}-year (28, 7, and 52 lb acre^{-1}-year), respectively, which approximately agree with the values published in Sindelar *et al.* (2013).

The N fertilizer needs of switchgrass, compared to those for corn stover, have been the subject of disagreement in existing literature. The modeling work presented here finds that switchgrass requires more N than stover on a per-dry-ton basis when high biomass yields are achieved, though some of this difference is explained by the allocation procedure for assigning fertilizer between corn grain and corn stover; most of the N applied to corn fields is assigned to the grain. Other studies find smaller differences between stover and switchgrass N fertilizer requirements (Sanderson *et al.*, 1999; Vogel *et al.*, 2002; Bonin & Lal, 2012). In contrast, Adler *et al.* (2007) find switchgrass needing substantially less N per ton of produced biomass than corn stover, based on modeling. The data used to model switchgrass production in this study represent current practices from a commercial producer growing switchgrass on marginal land. In addition, switchgrass has one or two establishment years in every planting cycle, in which there is little or no harvest, which reduces the effective yield without a commensurate drop in fertilizer demand. Finally, while switchgrass may not demonstrate a significant benefit when only on-field processes and N are modeled, switchgrass is likely to have a substantially more beneficial impact on soil organic carbon (SOC) than corn stover, due to its extensive root system, which is not included in the base case scenario.

Nitrogen fertilization can result in volatilization of N as N$_2$, NH$_4$, N$_2$O or nitrogen oxides (NO$_x$). Only NO$_x$ and N$_2$O are tracked because N$_2$ is inert and NH$_4$ is quickly removed from the atmosphere (Seinfeld & Pandis, 2006). N$_2$O volatilization was estimated as 1.06% of applied N (Linquist *et al.*, 2012) and NO$_x$ volatilization at rate of 238 mg m^{-2} (Akiyama *et al.*, 2000).

Since most US corn production uses tile drainage, nutrient leaching (both N and P) into groundwater was not considered in this model, but run-off to surface water was estimated as 31.6% of applied N and 4.66% of applied P (Powers, 2005). Powers concluded that potassium loss to water was minimal due to soil microbial activity and the relatively low solubility of potassium salts under conditions found in soil. All impacts from lime application to soils are allocated to grain production and so are not considered (Murphy & Kendall, 2013).

Field equipment activity generally follows the feedstock harvest and processing system described by the *Pioneer Scenario* from the Idaho National Laboratories (INL) Uniform feedstock model (Hess *et al.*, 2009), using the LCIs developed in Murphy & Kendall (2013).

Equipment operations result in emissions of criteria pollutants and GHGs from fuel combustion. Criteria pollutant emissions other than SO$_2$ are calculated using the US EPA Tier III nonroad emission factors based on the power output of the engine (EPA, 1994). Emissions of sulfur (as SO$_2$) and CO$_2$ are calculated by determining the mass of fuel consumed and typical carbon and sulfur content of diesel fuel. The life-cycle environmental impacts of diesel production, prior to combustion, are based on the GaBi PE database (PE International, 2008). Data S2 provides additional detail on model inputs for corn stover production and switchgrass production.

Soil carbon changes

Removing large amounts of biomass from a field can directly impact soil characteristics, including carbon and N content. These are highly dependent on local biogeochemical conditions, and a significant amount of uncertainty exists regarding their magnitude (Lal, 2006; Spatari & MacLean, 2010). Soil carbon effects are subdivided into two categories, direct and indirect. The base case of this model assumes no significant direct long-term changes in SOC as a result of feedstock collection. These assumptions are then tested in low and high SOC change scenarios for switchgrass and corn stover, based on the models presented in a meta-analysis by Anderson-Teixeira et al. (2009) as well as a corn stover analysis based on DAYCENT modeling reported by Kim et al. (2009). SOC changes are amortized over a 20-year period in the modeling.

In addition to direct changes to soil conditions, biomass production activity can cause indirect changes to land use and cover, through market or policy mechanisms. For example, when corn acreage which was previously used to supply human consumption is used for biofuel, the un-met demand will be satisfied with corn or substitute grains grown elsewhere. Since neither corn stover nor switchgrass have been collected on a large scale at present, no relevant data exist regarding these indirect changes. Indirect LUC (iLUC) and cultivation intensity changes from stover and grassy crop utilization may be relatively small, since stover is currently a residue and grassy crops can be grown on marginal agricultural land, but it is too early in the development of large-scale biofuel production capacity to reach a firm conclusion on this subject. The United States has enacted policies designed to minimize iLUC emissions (Schnepf & Yaccobucci, 2013), which supports the inference that such emissions for corn stover and switchgrass are likely to be minimal. Accordingly, they are omitted in this study, but future research should evaluate this assumption directly.

Feedstock pretreatment

Pretreatment operations include all processes from the point where bales of stover are stacked in road-side storage to where they begin the conversion process, excluding transportation. These are modeled based on pretreatment processes described in the INL Advanced Uniform Feedstock *Pioneer Scenario* (Hess et al., 2009). The required energy consumption for feedstock handling and grinding is estimated using mass-based energy consumption factors for each step in the pretreatment process. The base case assumes that stationary equipment (e.g., grinders and conveyor systems) are powered by grid electricity. Diesel combustion emissions for mobile equipment, such as forklifts and bale handlers, are estimated using the same methodology as used for agricultural equipment and, similarly, assume EPA Nonroad Tier III compliant engines. Data S2 provides additional detail on pretreatment parameters.

Ethanol conversion process

The conversion process refers to all of the activities which occur at the biofuel conversion facility. Two production scales are modeled representing near-term (smaller) and future (larger) facilities. These two production scales are important factors in the modeling scenarios explored in this study.

The near-term conversion facility (used in the base case) is sized to produce 62.7 million liters (16.6 million gallons) of anhydrous ethanol per year at a conversion efficiency of 243 l per dry ton (64 gallons per dry ton) of biomass. This yield rate is taken from Lau & Dale (2009) and represents a conservative estimate of potential ethanol production, as compared to theoretical yields from similar pathways, which can be more than twice as high (Matsushika et al., 2009).

In future facility scenarios, a 231 million liter per year (61 million gallon per year) facility with a 265 liter per dry ton (70 gallon per dry ton) yield is assumed. The future facility attempts to represent a reasonable estimate of facility parameters that have benefited from several years of technological learning and process refinement and a scale that is similar to existing commercial ethanol facilities (Humbird et al., 2011).

A primary obstacle to using cellulosic feedstocks for liquid fuel production is the recalcitrance of cellulose to break down into its constituent sugars, which can then be fermented into ethanol. Most processes require significant additions of acids, or enzymes such as cellulase, to liberate sugars from cellulosic materials (e.g., Humbird et al., 2011). The model described in this paper reflects a production process under development and currently in pilot-scale deployment. This process utilizes a genetically engineered fermentation microorganism that produces part of the cellulase enzymes which would otherwise have to be externally added; the balance of enzyme demand is produced at the same facility in the manner described by the NREL process design (Humbird et al., 2011). Accordingly, the base case scenario presented in this article assumes that no enzymes need be purchased from a third party; several additional scenarios are reported here which add external enzymes to the process, based on data from MacLean & Spatari (2009).

Several chemicals are required to assist in hydrolysis, pH control, microorganism growth, enzyme production, water absorption and cleaning. These are listed, along with the source of their LCI data in Data S2 and they, or similar chemicals, would likely be common to many biochemical cellulosic ethanol production processes. Caustic (sodium hydroxide) used for the clean-in-place system is assumed to be recycled at 90% efficiency (Koch Membrane Systems, 2012); all other chemicals are assumed to be completely consumed in the process or removed by the wastewater treatment plant, with minimal environmental impact.

Production of steam and electricity dominates environmental flows associated with the conversion facility. Most cellulosic ethanol process designs assume that energy needs are met by combusting process byproducts, in this case, lignin cake, spent cell mass and unconsumed carbohydrates from the conversion process as well as condensed syrup from the liquid waste stream for the base case facility. The future facility burns biogas from the wastewater treatment digester as well; methane potential of this digester was estimated based on Tian et al. (2013).

The combustion of byproducts is assumed to occur in a fluid-ized bed boiler, driving an extraction turbine. Overall thermal efficiency of electricity production is assumed to be 22.9% for the base case (Grass & Jenkins, 1994) and 25% for the future facility (Jenkins, 2012). Under these conditions, the base case facility could produce sufficient electricity and high-pressure steam for internal needs and have a surplus of approximately 5.3 MW of electrical generation capacity. This excess energy may be used to provide electricity to a local grid, or be consumed by a colocated industrial demand. Due to the developmental nature of cellulosic ethanol conversion technology and the lack of similar commer-cial-scale plants, the base case scenario conservatively assumes that the plant meets its own energy needs but does not export electricity, so no credit is assigned to the surplus energy.

Many cellulosic biofuel production processes are expected to produce electricity, heat or steam in addition to their fuel prod-uct. This presents an allocation problem for modelers; some fraction of environmental impacts should be assigned to each coproduct, but there are multiple allocation methodologies which have been proposed (Guinee et al., 2002; Luo et al., 2009; Cherubini et al., 2011; Murphy & Kendall, 2013). In addition to methodological debates, coproduct allocation is affected by real-world differences in how a coproduct is used based on market or policy conditions affected by geographic, economic and his-torical factors. To the greatest extent possible, the modeling effort presented in this article tries to focus on inherent charac-teristics of the feedstock production and conversion process. Doing so improves transparency and reduces uncertainty by minimizing the effect of assumptions regarding how allocation is done, and what design choices a conversion facility operator might make with regard to producing and using coproducts. Accordingly, the base case scenario focuses only on the neces-sary core elements of the process, and assumes that surplus heat and power are not utilized. Then, in scenario analyses that build on the base case, a suite of alternative utilization possibilities are examined. Where electricity coproduct allocation is handled through displacement, it is assigned an LCI that reflects US grid average electricity, as opposed to marginal generation, because generation occurs on a continuous basis.

Air pollutant emissions from the boiler are assumed to be controlled with a multi-stage cyclone and baghouse (for partic-ulate matter) and by good combustion practices and exhaust gas recirculation for reducing NO_x emissions. Emissions from the boiler are modeled after a similar biomass boiler from another proposed cellulosic ethanol facility (AMEC Earth & Environmental, 2009). VOC emissions from hydrolysis and fer-mentation tanks are controlled with a wet scrubber and leakage of VOCs from the process or during transfer between vessels is assumed to be negligible. All CO_2 emitted from combusting process byproducts is biogenic and assumed to not to contrib-ute to changing atmospheric CO_2 and, in accordance with widely accepted carbon accounting methods, is not included in calculations (BSI, 2011).

Facility construction

A simplified treatment of facility construction impacts is included in this LCI. The two largest elements of environmental impact from construction are the production of concrete and steel, which represent the greatest mass of material within the proposed bio-refinery. Facility lifespan was, upon recommen-dation from industry partners, conservatively assumed to be 20 years and no building decommissioning was considered. Impacts were distributed over the facility lifetime by straight-line amortization. This analysis indicates that construction materials accounted for less than 0.5% of total GHG impact of each unit of resultant fuel, in the base case.

Transportation

Previous geospatial modeling of biofuel systems has indicated that average transport distances for feedstocks between field and processing facility can exceed 50 miles (80.4 km, Parker et al., 2010). The *Pioneer scenario* envisions processing and stor-age facilities located relatively near to fields to minimize the bale transport distance, which is less efficient than bulk trans-port. This study assumes an average bale transport distance of 25 miles (40.2 km) and an average bulk transport distance of 50 miles (80.4 km) for the future scenario, which models a 61 mil-lion gallon per year conversion facility, and is based on the conclusions of Parker et al. (2010). Feedstock logistics for smal-ler, near-term facilities have not been as well described in liter-ature, so the values for the larger facility were scaled down assuming an approximately homogenous distribution of feed-stock in a circular area around the facility, which yielded 17 mile (25.7 km) bale and 34 (51.4 km) mile bulk transport distances.

Feedstock transport was modeled as described in the uni-form feedstock model (26 bales per 53 foot trailer, or 21.8 met-ric tons of minimally compressed bulk feedstock). Agricultural transport trucks are typically older than the fleet average, since they travel shorter distances and haul lower value goods. Feed-stock transport trucks are assumed to have California 2002 model year compliant engines with 120 000 engine hours per truck; this assumption was picked because older California trucks are often sold into agricultural use on the secondary market and the type of engine technology from this period approximately represents what would be commonly in usage for this activity today. Estimates of diesel consumption were generated using representative values for fuel economy and assuming empty back haul at 1.4 times the fuel economy of loaded trucks (Frey et al., 2008). Emission factors for the criteria pollutants were derived from EMFAC (CARB, 2007), CO_2 and sulfur emissions were estimated based on the mass of fuel combusted. Data S2 provides additional detail on transporta-tion parameters.

Scenario analysis

In addition to the base case scenarios for corn stover and switchgrass-based ethanol, multiple scenarios were evaluated as well (Tables 2 and 3 respectively) These scenarios are not intended to be an exhaustive set of possibilities, but rather reflect potentially important conditions for future cellulosic eth-anol production systems. The modeling presented in this paper does not attempt to predict which scenario, or combination of

Table 1 Base case results for multiple flows and indicators. Indicators are reported in units of g CO_2-equivalent (IPCC, 2007), g SO_2-equivalent, (Guinee et al., 2002), g PO_4-equivalent (Guinee et al., 2002) and units of MJ fossil energy per liter delivered ethanol

	Base case		Future facility	
	corn stover	switchgrass	corn stover	switchgrass
Air pollutants				
CO (kg l^{-1})	4.70×10^{-2}	4.68×10^{-2}	1.33×10^{-1}	1.33×10^{-1}
NO_x (kg l^{-1})	4.23×10^{-2}	4.10×10^{-2}	1.11×10^{-1}	1.09×10^{-1}
SO_x (kg l^{-1})	3.43×10^{-3}	3.83×10^{-3}	2.10×10^{-3}	2.47×10^{-3}
$PM_{2.5}$ (kg l^{-1})	3.78×10^{-3}	3.74×10^{-3}	7.91×10^{-3}	7.88×10^{-3}
PM_{10} (kg l^{-1})	4.16×10^{-3}	4.11×10^{-3}	8.46×10^{-3}	8.42×10^{-3}
Pb (air) (kg l^{-1})	1.33×10^{-7}	1.09×10^{-7}	1.11×10^{-7}	8.96×10^{-8}
Ozone (kg l^{-1})	1.78×10^{-6}	1.76×10^{-6}	1.55×10^{-6}	1.53×10^{-6}
NMVOC (kg l^{-1})	3.87×10^{-3}	4.14×10^{-3}	4.01×10^{-3}	4.26×10^{-3}
GHGs				
CO_2 (kg l^{-1})	8.20×10^{-1}	8.57×10^{-1}	6.77×10^{-1}	7.11×10^{-1}
CH_4 (kg l^{-1})	2.50×10^{-3}	2.78×10^{-3}	1.54×10^{-3}	1.80×10^{-3}
N_2O (kg l^{-1})	6.46×10^{-4}	9.07×10^{-4}	5.66×10^{-4}	8.06×10^{-4}
SF_6 (kg l^{-1})	5.83×10^{-9}	5.66×10^{-9}	4.35×10^{-9}	4.19×10^{-9}
Indicators				
GWP_{100} (kg CO_2e l^{-1}, 100 year IPCC basis)	1.06	1.19	8.74×10^{-1}	9.85×10^{-1}
GWP_{20} (kg CO_2e l^{-1}, 20 year IPCC basis)	1.18	1.31	9.43×10^{-1}	1.06
Acidification potential (kg SO_2-equivalent l^{-1})	3.33×10^{-2}	3.27×10^{-2}	7.97×10^{-2}	7.92×10^{-2}
Eutrophication potential (kg PO_4-equivalent l^{-1})	4.93×10^{-3}	7.71×10^{-3}	$4.44 \times 10^{--3}$	6.99×10^{-3}
Fossil energy (MJ fossil fuel l^{-1})	1.30×10	1.38×10	9.92	1.06×10

scenarios is most likely. Each scenario describes a plausible real-world condition.

The base case is close to a theoretically minimal process for producing cellulosic ethanol by a biochemical pathway; it uses only the chemicals necessary for microorganism propagation, enzyme production, moisture removal and system cleaning. The system is closed from an energy standpoint, the only energy which crosses the system boundary is the embodied energy of the feedstock, which includes diesel used for trucks and agricultural equipment, electricity used for particle size reduction before final transport to the facility. In the base case, no coproducts are created, obviating the need for allocation procedures. The scenarios build upon this and reflect plausible changes that would be observed in commercial facilities, including the generation of electricity as a coproduct, additional chemical inputs to the process, SOC changes caused by feedstock production, etc.

Corn stover scenarios

SOC impacts may play a major role in determining the net GHG intensity of stover-based fuels. The effects of stover collection on soil condition and soil chemistry are still being researched and debated (Karlen et al., 2011a,b; Anderson-Teixeira et al., 2013; Johnson et al., 2013). Accordingly, scenarios C1 and C2 evaluate the effects of SOC depletion based on the meta-analysis of field tests reported in Anderson-Teixeira et al. (2009), and scenarios C3 and C4 evaluate N_2O and SOC parameters based on DAYCENT modeling reported in Kim et al. (2009).

Coproduct credits, conversion efficiency and the need for additional chemical inputs are also a source of uncertainty in the life cycle of biofuels. The cellulosic biofuel described in this article has only one important potential coproduct; electricity. Scenarios C5 and C6 evaluate the impacts of the facility producing a net surplus or net demand for grid electricity. In scenario C6, where excess electricity is produced, a fraction of it is consumed by the satellite feedstock processing facilities. When all energy demands are considered, the base case facility can deliver approximately 3 MW of electricity to the grid. Scenario C7 further explores the use of surplus energy produced at the facility by evaluating the expected displacement of natural gas if excess heat and electricity are provided to a colocated industrial consumer by a shared CHP system. Scenarios C8 and C9 evaluate higher or lower conversion efficiency.

Scenarios C10 evaluates the effect of adding additional cellulose enzymes, based on the estimates provided by MacLean & Spatari (2009); C11 evaluates the effect of adding a dilute acid pretreatment step based on the loading rates of Humbird et al. (2011).

Switchgrass scenarios

The set of scenarios examined for switchgrass-based fuels is smaller, due to the characteristics of switchgrass and the absence of data for commercial-scale production and processing of switchgrass. Only one SOC change scenario is considered (S1), based on the results of Anderson-Teixeira et al. (2009). In contrast to corn stover production

Table 2 List of corn stover feedstock scenarios

Scenario number	Scenario name	Changes from base case
C1	Corn stover high SOC change	Corn stover SOC changes per Anderson-Teixeira *et al.* (2009) Full-Form model, 30 cm sampling depth, 25% residue removal, 20 year sample
C2	Corn stover low-SOC change	Corn stover SOC changes per Anderson-Teixeira *et al.* (2009) Reduced-Form model, 30 cm sampling depth, 25% residue removal, 20 year sample
C3	Corn stover w/N_2O reduction	Reduces N_2O emissions by 0.5 kg N_2O-N per hectare-year, based on average from Table 3 Kim *et al.* (2009)
C4	Corn stover w/DAYCENT soil change parameters	Applies N_2O, NO_x and SOC changes based on average of all sites reported in Table 3, Kim *et al.* (2009)
C5	Corn stover 5 MW electricity deficit	As base case + facility requires 5 MW of additional electricity from grid, using Ecoinvent US average grid LCI. Electricity is assigned to facility processes based on proportion of total energy demand in base case.
C6	Corn stover 3 MW electricity surplus	As base case, except facility exports 3 MW of electricity to grid, displacing US average grid LCI generation (Ecoinvent LCI)
C7	Corn stover, shared CHP displacing natural gas	As base case, except HP steam above plant requirements assumed to displace equivalent natural gas generation (GaBI PE LCI)
C8	Corn stover, high conversion efficiency	As base, except facility produces 10% more ethanol per unit of feedstock (70.6 gallon per dry ton) input, with no other changes to system
C9	Corn stover, low conversion efficiency	As base, except facility produces 10% less ethanol per unit of feedstock (57.9 gallon per dry ton) input, with no other changes to system
C10	Corn stover, external enzymes	As base, but requires 10 g kg^{-1} dry matter of cellulase enzymes instead of on-site production, LCI as reported by (MacLean & Spatari 2009)
C11	Corn stover, dilute acid pretreatment	As base, but adds additional acid and neutralization chemicals based on (Humbird *et al.*, 2011)
CF1	Corn stover, future facility, base case	60 MGY Facility, 70 gallon per ton yield, same chemical demands as C1 facility.
CF2	Corn stover, future facility, high SOC change	As CF1, with SOC changes per Anderson-Teixeira *et al.* (2009) Full-Form model, 30 cm sampling depth, 25% residue removal, 10 year sample
CF3	Corn stover, future facility, low-SOC change	As CF1, with Corn Stover SOC changes per Anderson-Teixeira *et al.* (2009) Reduced-Form model, 30 cm sampling depth, 25% residue removal, 10 year sample
CF4	Corn stover, future facility, 6.3 MW electricity surplus	As CF1, with 6.3 MW exported electricity, displacing US average grid mix (Econinvent LCI)
CF5	Corn stover, future facility, shared CHP displacing natural gas	As base case, except HP steam above plant requirements assumed to displace equivalent natural gas generation (GaBI PE LCI)
CF6	Corn stover, future facility, high conversion efficiency	As CF1, with 77 gallon per ton yield
CF7	Corn stover, future facility, with dilute acid pretreatment	As CF1, but adds additional acid and neutralization chemicals per (Humbird *et al.*, 2011)

systems, switchgrass systems appear to have the potential to increase SOC, due largely to extensive root systems. Scenario S2 evaluates the effect of switchgrass yields at the upper end of their uncertainty range, 9.1 dry ton ha^{-1}. Scenarios S3–S5 evaluate the importance of electricity surplus or deficit, as well as shared CHP, high and low conversion efficiency, in a manner similar to corn stover scenarios C5–C9. S8 and S9 examine the effects of additional cellulase enzymes and a dilute acid pretreatment phase, similar to scenarios C10 and C11. The model assumes that the additional pretreatment steps for switchgrass would be identical to corn stover. This assumption should be tested in future research.

Future facility scenarios

The effects of improvements and efficiency gains in facilities over time are estimated in future facility scenarios, CF1–CF7 and SF1–SF7. These scenarios model a facility with a 231 million liter per year (40 million gallon per year) capacity and 265 l per dry ton (70 gallon per dry ton) baseline conversion efficiency. Due to the larger facility output, the average transportation distances for feedstock are greater than for the baseline scenarios. Several additional scenarios build off the future facility scenario to evaluate circumstances equivalent to those in some of the baseline scenarios. When all demands are considered, in a manner equivalent to the surplus electricity

Table 3 List of switchgrass feedstock scenarios

Scenario number	Scenario name	Changes from base case
S1	Switchgrass with SOC change	SOC changes per Anderson-Teixeira *et al.* (2009) Reduced-Form model, 30 cm sampling depth, 10 years harvest
S2	Switchgrass high biomass yield	As Base, but 7.4 Mg ha^{-1} average annual switchgrass yield.
S3	Switchgrass 5 MW electricity deficit	As base case, except facility requires 5 MW of additional electricity from grid, using Ecoinvent US average grid LCI
S4	Switchgrass 3 MW electricity surplus	As base case, except facility exports 3 MW of additional electricity to grid, displacing using Ecoinvent US average grid LCI generation
S5	Switchgrass, shared CHP displacing natural gas	As base case, except HP steam above plant requirements assumed to displace equivalent natural gas generation (GaBI PE LCI)
S6	Switchgrass, high conversion process efficiency	As base, except facility produces 10% more ethanol per unit of feedstock (70.6 gallon per dry ton) input, with no other changes to system
S7	Switchgrass, low conversion process efficiency	As base, except facility produces 10% less ethanol per unit of feedstock (57.9 gallon per dry ton) input, with no other changes to system
S8	Switchgrass, External enzymes	As base, but requires 10 g kg^{-1} dry matter of cellulase enzymes instead of on-site production, LCI as reported by (MacLean & Spatari 2009)
S9	Switchgrass, dilute acid pretreatment	As base, but adds additional acid and neutralization chemicals per (Humbird *et al.*, 2011)
SF1	Switchgrass, future facility, base case	60 MGY Facility, 70 gallon per ton yield, same chemical demands as Base scenario facility.
SF2	Switchgrass, future facility, with SOC change	As SF1, with SOC changes per Anderson-Teixeira *et al.* (2009) Full-Form model, 30 cm sampling depth, 25% residue removal, 10 year sample
SF3	Switchgrass, future facility, high biomass yield	As SF1, but 7.4 Mg ha^{-1} average annual switchgrass yield.
SF4	Switchgrass, future facility, 6.3 MW electricity surplus	As SF1, with 6.3 MW exported electricity, displacing US average grid mix (Ecoinvent LCI)
SF5	Switchgrass, future facility, shared CHP displacing natural gas	As base case, except HP steam above plant requirements assumed to displace equivalent natural gas generation (GaBI PE LCI)
SF6	Switchgrass, future facility, high conversion efficiency	As SF1, with 77 gallon per ton yield
SF7	Switchgrass, future facility, with dilute acid pretreatment	As SF1, but adds additional acid and neutralization chemicals per (Humbird *et al.*, 2011)

scenarios for the base case facility, the future facility can return 6.3 MW of power to the grid. While the facility is over three times as large, the excess electricity does not scale linearly; the higher conversion rate and lower nutrient consumption of the future facility leaves less matter in the waste stream to combust. Tables 2 and 3 give more details about the scenarios under evaluation.

It is important to note that the scenarios presented here are not mutually exclusive. To maximize transparency, only one key effect, such as soil carbon loss or conversion facility yield, was examined in each scenario. The conditions described in two or more scenarios could cooccur, however. As an exhaustive set of all permutations of scenario combinations is not practical, comprehensive LCIs for each scenario are presented in Data S1. Combinations of multiple scenario effects can be created through appropriate sums and differences of these LCIs.

Results

Under the base case conditions, producing one MJ of ethanol emits 50.3 g CO_2e over its life cycle for stover-based fuel, and 56.1 g CO_2e for switchgrass-based fuel using 100-year GWPs from the IPCC's *Fourth Assessment Report* (IPCC, 2007). Table 1 shows key results for both base cases, as well as for the future facility. The base case does not include credit for coproducts (electricity or displaced natural gas); this credit reduces the life-cycle GHG intensity by 25–30 g CO_2e MJ^{-1} for the base case facilities. For comparison, California Reformulated Gasoline has life-cycle emissions of approximately 99 g CO_2e MJ^{-1} when calculated according to the California Low Carbon Fuel Standard (LCFS) methodology (CARB, 2012). By the same methods and including indirect land-use change, corn ethanol has emissions typically between 80 and 100 g CO_2e MJ^{-1} (CARB, 2012). This study considers national average or typical conditions; actual evaluations of biofuel GHG impacts should be made on a geographically and technologically explicit basis.

Under base case conditions, both feedstock pathways have the potential to yield life-cycle GHG reductions compared to petroleum gasoline of 40–50%, or by more than 70% when coproducts are considered. Corn stover has a slight advantage in feedstock production impacts, due to lower nitrogen requirements. However, the amount of N fertilizer allocated to corn stover depends on the allocation methods used during LCI development for corn stover; different allocation methods can assign more or less of the total fertilizer impact from corn cultivation to the stover (Murphy & Kendall, 2013).

The life-cycle GHG and energy flows for the base case scenario generally agree with other LCAs of cellulosic biofuels (Wang et al., 2007; Spatari & MacLean, 2010), though they are somewhat higher than those of Aden & Heath (2009). The variability between estimates reported in different studies is well within the anticipated range of uncertainty, given data limitations and differing modeling assumptions. Both stover and switchgrass-derived ethanol yield similar results for non-GHG emissions flows and indicators; the notable exception being significantly higher switchgrass VOC emissions due to the different fertilizer compounds used. Because the selected fertilizer compounds are based on test plots rather than commercial production, they might change in the future.

Production and preprocessing [grinding to a 10 mm (0.4 in) maximum particle size] of feedstocks dominate the life-cycle GHG impacts of this cellulosic ethanol production process; over 60% of base case GHG emissions are attributed to these phases. This observation has been noted by other studies in this field (e.g., Hsu et al., 2010). The potential excess electricity available for export to the grid is reduced by around half due to the demands of feedstock handling and grinding. Previous studies reported much greater surpluses of electricity (Humbird et al., 2011), however, they omitted the energy required to grind their biomass prior to conversion, which may explain the difference.

Production of N fertilizer is the dominant contributor to agricultural impacts, responsible for approximately two-thirds of the impact from feedstock. This illustrates a common problem for all energy crops; nitrogen is a limiting factor for yields, absent nitrogen fixing crops or cocrops. While P and K demands may be met to some extent by returning the ash from biomass combustion to the field, ash supplies only a limited amount of N (Mozaffari et al., 2000; Schiemenz & Eichler-Löbermann, 2010), which must be supplied from other sources.

The ratio of ethanol energy output to fossil energy input was around 2 : 1 for the base case facilities (Table 1), and approximately 4 : 1 for most scenarios in which surplus electricity was exported to the grid and negative (indicating absolute reductions in fossil fuel consumption) where surplus CHP energy was used to displace natural gas combustion. This indicates that even under very conservative assumptions, cellulosic ethanol systems as described here can reduce fossil fuel consumption.

Figure 2 shows the life-cycle GHG impacts of the corn stover scenarios differentiated by life-cycle phase. Scenarios C1, C2 and C4 demonstrate that soil carbon effects are an element of substantial uncertainty and that there are large discrepancies between SOC effects modeled by DAYCENT as reported by Kim et al., and the meta-analysis of field studies as reported by Anderson-Teixeira et al. leading to a range of 49–164 g CO_2e MJ^{-1} ethanol. Scenarios where coproduct credits occur result in the lowest net GWP (scenarios C6, C7, CF4 and CF5), and demonstrate the potential importance of energy coproducts in determining the net life-cycle GHG intensity of cellulosic ethanol. Only when the higher range of SOC losses from stover harvest are considered does the fuel become worse than a petroleum alternative.

Switchgrass-based scenarios where no SOC sequestration is modeled have slightly higher life-cycle GHG footprints than comparable corn ones, due to the higher consumption of fertilizer. However, Fig. 3 illustrates that when SOC increases during switchgrass cultivation are included (scenarios S1 and SF2) switchgrass ethanol performs extremely well (−9 to −4.5 g CO_2e MJ^{-1}). These scenarios result in sufficient soil carbon sequestration that a net reduction in GHG concentration in the atmosphere occurs due to switchgrass ethanol production; however, this conclusion depends on several assumptions about soil carbon dynamics and fossil fuel substitution. Further research is required before carbon negative pathways can be fully justified.

Coproduct credits (scenarios S4, S5, SF4 and SF5) also lead to significant reductions in life-cycle GHG intensity, resulting in intensities of 21–31 g CO_2e MJ^{-1} ethanol, which are much lower than the GHG intensity of petroleum gasoline. Even in cases where the facility requires 5 MW of additional grid electricity (scenario S3) the resulting fuels are lower in GHG intensity than petroleum gasoline.

Discussion

Uncertainties in LCA results

There is still uncertainty in switchgrass cultivation parameters because it is not yet commercially produced for ethanol. Yield, stand age at first harvest, stand age at removal and replanting, and fertilizer demand may change as agronomic practices are improved. The higher GHG footprint of switchgrass compared to corn stover should be considered in the context of potential SOC changes; corn stover has a significant risk of

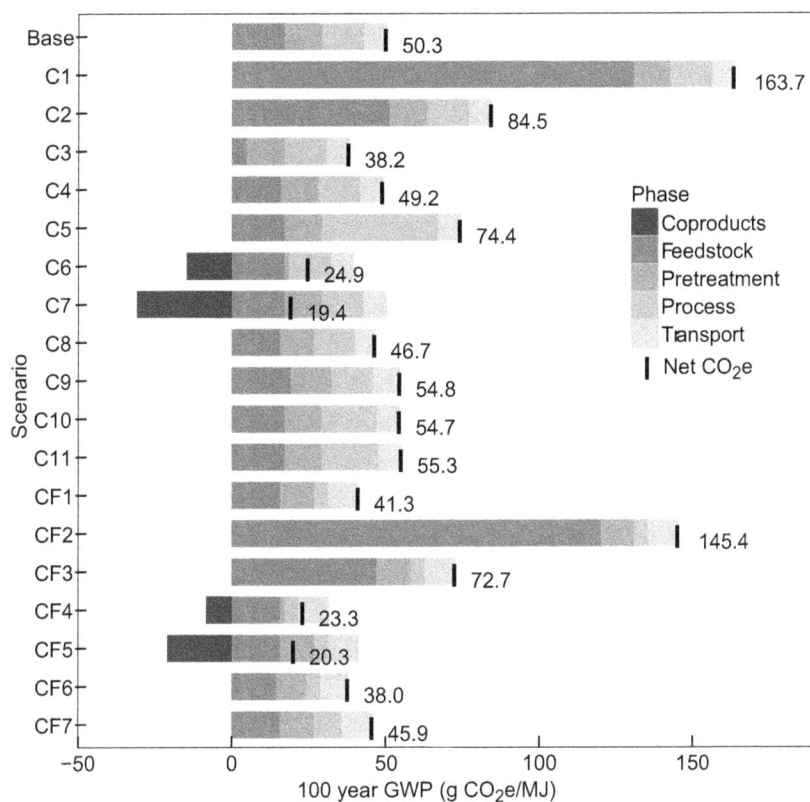

Fig. 2 Green house gas (GHG) footprint of all corn stover scenarios, with contributions from each phase of production identified. See Table 2 for description of scenarios.

decreasing SOC as a result of repeated harvest while switchgrass would likely *maintain or increase* SOC, and soil carbon changes have a greater effect on GHG intensity than any other parameters in this study.

SOC change is particularly important for corn stover. The base case assumes some portion of stover can be removed without reducing SOC levels, as has been claimed in a number of studies (Hess *et al.*, 2009; Follett *et al.*, 2012). In contrast, the soil carbon model from Anderson-Teixeira *et al.* (with SOC changes amortized over 20 years and a 30% removal rate) demonstrates SOC losses of a greater magnitude than all other GHG impacts in the stover ethanol production system combined, even under their more optimistic reduced-form model (the low-SOC change values used in this research). Their full form model predicts SOC losses so high that stover-based ethanol is more carbon-intensive than petroleum gasoline. Unfortunately, there are a limited number of studies on the SOC effects of stover removal, and the meta-analysis results from Anderson-Teixeira *et al.* may not reflect real-world SOC changes due to small-sample error.

An alternative source for estimating SOC changes, DAYCENT modeling by Kim *et al.* (2009), also predicts a loss of SOC under sustained stover harvest, though

the magnitude is significantly smaller than that predicted by Anderson-Teixeira *et al.* In addition, DAYCENT accounts for reductions in N_2O emissions due to stover removal, which ultimately outweigh SOC losses on a CO_2-equivalent basis. These N_2O reductions are highly uncertain and depend largely on local conditions. For example, Németh (2012) determined that stover removal could actually increase N_2O emissions by altering soil microbial communities and exacerbating temperature cycles. A detailed, region-specific analysis of soil effects due to stover removal may be necessary to accurately assess the GHG impacts of any stover-based biofuel. For example, it is very likely that particular farms may be able to recover stover without SOC loss (Vogt, 2012), and in these cases stover-based cellulosic ethanol may provide significant GHG benefits.

Based on this LCA, the most important difference between corn stover and switchgrass as biofuel feedstock is the *potential* difference between SOC effects, assuming switchgrass cultivation does not induce iLUC. Some authors, notably Follett *et al.* (2012), find that SOC gains can occur under sustained corn stover harvest, however, these studies typically assume a conversion from conventional-tillage to no-tillage, or production on low-SOC land. In some situations, these assumptions

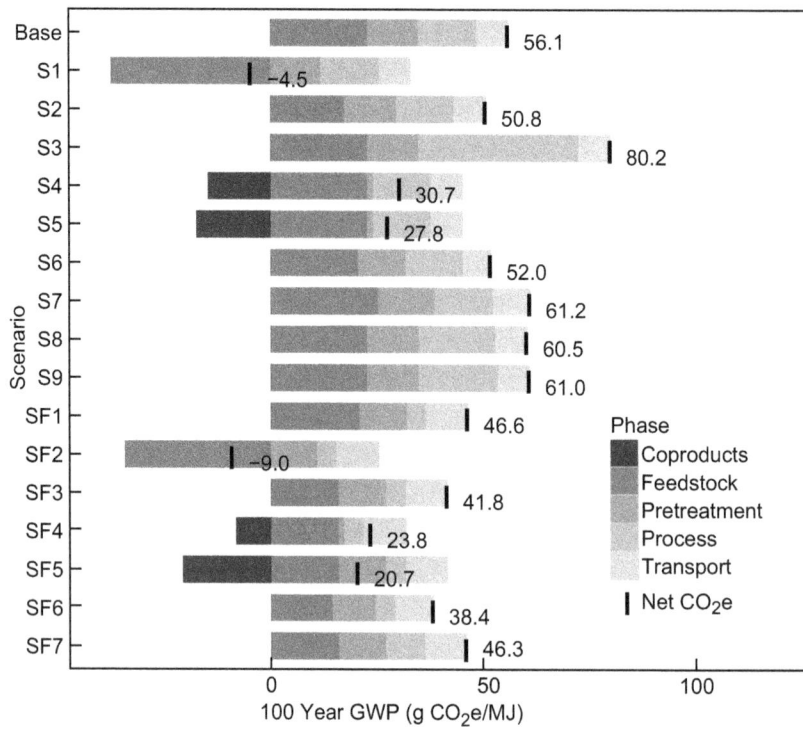

Fig. 3 Green house gas (GHG) footprint of all switchgrass scenarios, with contributions from each phase of production identified. See Table 3 for description of scenarios.

may hold true, but it is likely that the bulk of corn sto-ver for future biofuel systems would be produced from existing corn acreage, so it is more appropriate to model likely changes from BAU practices, which would not necessarily include a change in tillage. Switchgrass, on the other hand, is seldom grown at present, so any culti-vation of switchgrass would likely represent a change from BAU and the SOC changes, whether gains or losses, could be assigned to the products of the cultiva-tion system.

The second greatest area of uncertainty in LCA results is the electricity deficit or surplus generated by the conversion facility. The baseline scenario predicts a small (3 MW) surplus of electricity. The baseline sce-nario does not credit the produced ethanol for produc-ing grid electricity due to the substantial uncertainty surrounding what form this excess energy might take, or whether it is even economically feasible to recover.

Implications for cellulosic fuels and the renewable fuels standard (RFS)

When SOC changes are omitted, and assuming no iLUC occurs, both corn stover and switchgrass biofuels, under the assumptions described in the base case sce-nario, have the potential to produce biofuels with substantial life-cycle GHG reductions compared to petroleum-based fuels. Under base case conditions, corn

stover-based fuels approach the 50% life-cycle GHG reduction target required to achieve the 'Advanced' bio-fuel category of the RFS. The 60% reduction target for the 'Cellulosic' designation may be achievable for both corn stover and switchgrass fuels when coproduct cred-its are considered. Achieving even the Advanced cate-gory at commercial scale would represent a substantial milestone in sustainable transportation, if produced at commercial scale.

The GHG intensity of the produced ethanol is largely dependent on the ability of the facility to meet its energy needs using the byproducts of the conversion process, including lignin cake and biogas. Even in cases where significant additional chemicals are required in the conversion process, such as those reported in sce-nario C11 (the dilute acid hydrolysis scenarios) and C10 (additional enzyme scenarios), the lifecycle GHG impacts are still dominated by feedstock production and processing. Under these scenarios, the resultant fuels mostly meet the RFS 'Advanced' biofuel target, but would not meet the 'Cellulosic' target without addi-tional coproduct credits or improvements to conversion efficiency.

With conversion process GHG impacts relatively small, SOC becomes the most important area of uncer-tainty in the cellulosic ethanol life cycle. It is critical that SOC impacts of feedstock production be better under-stood, particularly from corn stover since it risks SOC loss when harvested. There may be certain soil conditions, such as those with high clay content, very high biomass production or very low initial SOC in which the base case assumption of SOC neutrality for partial corn stover harvest hold true. Accurately estimating the SOC effects of stover collection on a site-specific basis may be required to quantify the GHG impacts of stover-based biofuels. Under typical conditions, however, the consensus in literature is that routine stover collection would likely lead to some decrease in SOC, which imposes a substantial GHG penalty on the resultant fuels. If stover-based fuels are to meet the RFS targets, this penalty would have to be offset in some fashion; for example, by GHG credits from renewable electricity or reductions in N_2O from field emissions.

Switchgrass, on the other hand, appears to present an opportunity for reducing life-cycle GHG emissions by increasing SOC during cultivation. While the magnitude of this effect is uncertain, it is likely that it will offset the marginally higher GHG emissions from N fertilization than those of corn stover. It is important to note that SOC sequestration by switchgrass may decline over time as SOC equilibrates at a higher level, and may be lost when switchgrass is replanted and soil is disturbed. Further research is required to determine the duration of SOC sequestration as well as the persistence of SOC in soils used for bioenergy crops after repeated harvest.

The importance of SOC in the biofuel life cycle highlights an inherent limitation in using large quantities of biomass for energy purposes: the embodied nutrients in the feedstock must, at some point, be replaced to avoid soil depletion. At present, the default method for replacing these nutrients is petrochemical fertilizers. More environmentally friendly methods have been proposed to address nutrient replacement, such as compost or residual ash from biomass combustion, but it is unclear what impacts these methods might have (Pandey & Singh, 2010; Brown & Cotton, 2011; Risse & Gaskin, 2013). Future research is required to better evaluate alternatives to petrochemical fertilization.

Acknowledgements

This research was supported by the Department of Energy under Award Number DE-EE0002881. The content is solely the responsibility of the authors and does not necessarily represent the official view of the Department of Energy. The authors would like to thank Daniel Derr for his assistance throughout this project, and the cooperation of Edeniq, Inc. in providing facility data.

References

Aden A, Heath GA (2009) Critical review environmental and sustainability factors associated with next-generation biofuels in the US: what do we really know? *Environmental Science & Technology*, **43**, 4763–4775.

Adler PR, Del Grosso SJ, Parton WJ (2007) Life-cycle assessment of net greenhouse-gas flux for bioenergy cropping systems. *Ecological Applications*: A Publication of the Ecological Society Of America, **17**, 675–691.

Advanced Ethanol Council (2013) *Cellulosic Biofuels: Industry Progress Report 2012–2013*. Renewable Fuels Association, Washington, DC, USA.

Akiyama H, Tsuruta H, Watanabe T (2000) N_2O and NO emissions from soils after the application of different chemical fertilizers. *Chemosphere - Global Change Science*, **2**, 313–320.

AMEC Earth & Environmental (2009) Appendix B - emissions calculations. In: *Verenium Application for PSD Air Construction Permit, Highlands Ethanol, Highlands County, Florida*, pp. 1–52. Highlands Ethanol, LLC, Tampa, FL, USA.

Anderson-Teixeira KJ, Davis SC, Masters MD, Delucia EH (2009) Changes in soil organic carbon under biofuel crops. *GCB Bioenergy*, **1**, 75–96.

Anderson-Teixeira KJ, Masters MD, Black CK, Zeri M, Hussain MZ, Bernacchi CJ, DeLucia EH (2013) Altered belowground carbon cycling following land-use change to perennial bioenergy crops. *Ecosystems*, **16**, 508–520.

Björklund A (2002) Survey of approaches to improve reliability in LCA. *The International Journal of Life Cycle Assessment*, **7**, 64–72.

Blanco-Canqui H, Lal R (2007) Soil and crop response to harvesting corn residues for biofuel production. *Geoderma*, **141**, 355–362.

Bonin C, Lal R (2012) Bioethanol potentials and life-cycle assessments of biofuel feedstocks. *Critical Reviews in Plant Sciences*, **31**, 271–289.

Borrion AL, McManus MC, Hammond GP (2012) Environmental life cycle assessment of lignocellulosic conversion to ethanol: a review. *Renewable and Sustainable Energy Reviews*, **16**, 4638–4650.

Brown TR, Brown RC (2013) A review of cellulosic biofuel commercial-scale projects in the United States. *Biofuels, Bioproducts and Biorefining*, **7**, 235–245.

Brown S, Cotton M (2011) Changes in soil properties as a result of compost or mulch application: results of on-farm Sampling. *Compost Science & Utilization*, **19**, 88–97.

BSI (2011) *PAS2050: 2011 Specification for the Assessment of the Life Cycle Greenhouse Gas Emissions of Goods and Services*. British Standards Institute, London, UK.

Burnham A, Wang M, Wu Y (2006) *Development and Applications of GREET 2.7–The Transportation Vehicle-Cycle Model*. Argonne National Laboratory, Argonne, IL, USA.

Campbell JE, Lobell DB, Field CB (2009) Greater transportation energy and GHG offsets from bioelectricity than ethanol. *Science*, **324**, 1055–1057.

CARB (2007) *EMFAC 2007*. Mobile Source Emissions Inventory, California Air Resources Board, Sacramento, CA, USA.

CARB (2012) *Carbon Intensity Lookup Table for Gasoline and Fuels that Substitute for Gasoline*. California Air Resources Board, Sacramento, CA, USA. Available at: http://www.arb.ca.gov/fuels/lcfs/lu_tables_11282012.pdf.

Ceres Inc. (2013) Personal Communication to Colin Murphy.

Cherubini F, Strømman AH (2010) Production of biofuels and biochemicals from lignocellulosic biomass: estimation of maximum theoretical yields and efficiencies using matrix algebra. *Energy & Fuels*, **24**, 2657–2666.

Cherubini F, Strømman AH, Ulgiati S (2011) Influence of allocation methods on the environmental performance of biorefinery products—A case study. *Resources, Conservation and Recycling*, **55**, 1070–1077.

Delucchi MA (2010) Impacts of biofuels on climate change, water use, and land use. *Annals of the New York Academy of Sciences*, **1195**, 28–45.

Ecoinvent Centre (2011) *Ecoinvent v3*. Swiss Center for Life Cycle Inventories, Zürich, Switzerland.

Economic Research Service (2012) *USDA Economic Research Service - Fertilizer Use and Price*. Retrieved March 17, 2013, from http://www.ers.usda.gov/data-products/fertilizer-use-and-price.aspx#.UUZFexx9B8E. US Department of Agriculture, Washington, DC, USA.

EPA (1994) Part 89 - control of emissions from new and in-use nonroad compression-ignition engines. In *Title 40, Chapter 1, Subchapter C of the Code of Federal Regulations*. US Government Printing Office, Washington, DC, USA.

Fargione J, Hill J, Tilman D, Polasky S, Hawthorne P (2008) Land clearing and the biofuel carbon debt. *Science*, **319**, 1235–1238.

Follett RF, Vogel KP, Varvel GE, Mitchell RB, Kimble J (2012) Soil carbon sequestration by switchgrass and no-till maize grown for bioenergy. *BioEnergy Research*, **5**, 866–875.

Frey H, Rouphail NN, Zhai H (2008) Link-based emission factors for heavy-duty diesel trucks based on real-world data. *Transportation Research Record*, **2058**, 23–32.

Fu GZ, Chan AW, Minns DE (2003) Life cycle assessment of bio-ethanol derived from cellulose. *The International Journal of Life Cycle Assessment*, **8**, 137–141.

Grass SW, Jenkins BM (1994) Biomass fueled fluidized bed combustion: atmospheric emissions, emission control devices and environmental regulations. *Biomass and Bioenergy*, **6**, 243–260.

Guinee J, Gorrée M, Heijungs R et al. (2002) Handbook on life cycle assessment operational guide to the ISO standards. *The International Journal of Life Cycle Assessment*, **7**, 311–313.

Hess JR, Kenney KL, Ovard LP, Searcy EM, Wright CT (2009) *Commodity-Scale Production of an Infrastructure-Compatible Bulk Solid from Herbaceous Lignocellulosic Biomass*. Idaho National Lab, Idaho Falls, ID, USA.

Hoben JP, Gehl RJ, Millar N, Grace PR, Robertson GP (2011) Nonlinear nitrous oxide (N2O) response to nitrogen fertilizer in on-farm corn crops of the US Midwest. *Global Change Biology*, **17**, 1140–1152.

Hoskinson RL, Karlen DL, Birrell SJ, Radtke CW, Wilhelm WW (2007) Engineering, nutrient removal, and feedstock conversion evaluations of four corn stover harvest scenarios. *Biomass and Bioenergy*, **31**, 126–136.

Hsu DD, Inman D, Heath GA, Wolfrum EJ, Mann MK, Aden A (2010) Life cycle environmental impacts of selected US ethanol production and use pathways in 2022. *Environmental Science & Technology*, **44**, 5289–5297.

Huijbregts MAJ, Norris G, Bretz R et al. (2001) Framework for modelling data uncertainty in life cycle inventories. *The International Journal of Life Cycle Assessment*, **6**, 127–132.

Humbird D, Davis R, Tao L et al. (2011) *Process Design and Economics for Biochemical Conversion of Lignocellulosic Biomass to Ethanol - Dilute-Acid Pretreatment and Enzymatic Hydrolysis of Corn Stover*. National Renewable Energy Laboratory, Golden, CO, USA.

IPCC (2007) Climate change 2007 - the physical science basis. In: *Contribution of Working Group I to the Fourth Assessment Report of the Intergovernmental Panel on Climate Change* (eds Solomon S, Qin D, Manning M, Chen Z, Marquis M, Averyt KB, Tignor M, Miller HL). Cambridge University Press, Cambridge, UK.

Iribarren D, Peters JF, Dufour J (2012) Life cycle assessment of transportation fuels from biomass pyrolysis. *Fuel*, **97**, 812–821.

Ishola MM, Jahandideh A, Haidarian B, Brandberg T, Taherzadeh MJ (2013) Simultaneous saccharification, filtration and fermentation (SSFF): a novel method for bioethanol production from lignocellulosic biomass. *Bioresource Technology*, **133**, 68–73.

Jenkins BM (2012) Personal Communication to Colin Murphy.

Johnson J, Acosta-Martinez V, Cambardella C, Barbour N (2013) Crop and soil responses to using corn stover as a bioenergy feedstock: observations from the Northern US corn belt. *Agriculture*, **3**, 72–89.

Jung JY, Lal R (2011) Impacts of nitrogen fertilization on biomass production of switchgrass (Panicum Virgatum L.) and changes in soil organic carbon in Ohio. *Geoderma*, **166**, 145–152.

Karlen DL, Birell SJ, Hess JR (2011a) A five-year assessment of corn stover harvest in central Iowa, USA. *Soil and Tillage Research*, **115–116**, 47–55.

Karlen DL, Varvel GE, Johnson JMF et al. (2011b) Monitoring soil quality to assess the sustainability of harvesting corn stover. *Agronomy Journal*, **103**, 288–295.

Kendall A, Chang B (2009) Estimating life cycle greenhouse gas emissions from corn–ethanol: a critical review of current US practices. *Journal of Cleaner Production*, **17**, 1175–1182.

Kendall A, Yuan J (2013) Comparing life cycle assessments of different biofuel options. *Current Opinion in Chemical Biology*, **17**, 439–443.

Khanna M, Crago CL, Black M (2011) Can biofuels be a solution to climate change? The implications of land use change-related emissions for policy. *Interface Focus*, **1**, 233–247.

Kim S, Dale B, Jenkins R (2009) Life cycle assessment of corn grain and corn stover in the United States. *The International Journal of Life Cycle Assessment*, **14**, 160–174.

Klein RN, Wilson RK (2013) *Nebraska Crop Budgets, 2013*. University of Nebraska-Lincoln, Lincoln, NE, USA.

Koch Membrane Systems (2012) *Recovery of Caustic Acids in the Dairy Industry*. Retrieved February 3, 2014. http://www.kochmembrane.com/resources/application-bulletins/recovery-of-caustic-and-acids-in-the-dairy-industr.aspx.

Lal R (2006) Soil and environmental implications of using crop residues as biofuel feedstock. *International Sugar Journal*, **108**, 161–167.

Lau MW, Dale BE (2009) Cellulosic ethanol production from AFEX-treated corn stover using Saccharomyces cerevisiae 424A(LNH-ST). *Proceedings of the National Academy of Sciences of the United States of America*, **106**, 1368–1373.

Linquist B, Groenigen KJ, Adviento-Borbe MA, Pittelkow C, Kessel C (2012) An agronomic assessment of greenhouse gas emissions from major cereal crops. *Global Change Biology*, **18**, 194–209.

Luo L, van der Voet E, Huppes G, Udo de Haes H (2009) Allocation issues in LCA methodology: a case study of corn stover-based fuel ethanol. *The International Journal of Life Cycle Assessment*, **14**, 529–539.

MacLean HL, Spatari S (2009) The contribution of enzymes and process chemicals to the life cycle of ethanol. *Environmental Research Letters*, **4**, 014001.

Matsushika A, Inoue H, Kodaki T, Sawayama S (2009) Ethanol production from xylose in engineered Saccharomyces cerevisiae strains: current state and perspectives. *Applied Microbiology and Biotechnology*, **84**, 37–53.

Microsoft Inc. (2006) *Microsoft Excel 2007*. Redmond, Washington, DC, USA.

Mozaffari M, Rosen CJ, Russelle MP, Nater EA (2000) Corn and soil response to application of ash generated from gasified alfalfa stems. *Soil Science*, **165**, 896–907.

Murphy CW, Kendall A (2013) Life cycle inventory development for corn and stover production systems under different allocation methods. *Biomass and Bioenergy*, **58**, 67–75.

Németh D (2012) Nitrous oxide emission and abundance of N-cycling microorganisms in corn-based biofuel cropping systems. MS Thesis. University of Guelph, Canada.

Pandey VC, Singh N (2010) Impact of fly ash incorporation in soil systems. *Agriculture, Ecosystems & Environment*, **136**, 16–27.

Parker N, Hart Q, Tittmann P et al. (2010) *National Biorefinery Siting Model: Spatial Analysis and Supply Curve Development*. University of California, Davis, CA, USA.

PE International (2008) *GaBi 4 System - Software and Databases for Life Cycle Engineering*. Leinfelden-Echterdingen, Germany.

Pimentel D, Patzek T (2008) Ethanol production using corn, switchgrass and wood; biodiesel production using soybean. In: *Biofuels, Solar and Wind as Renewable Energy Systems* (ed. Pimentel D), pp. 373–394. Springer, Netherlands, Dordrecht, The Netherlands.

Powers SE (2005) *Quantifying Cradle-to-Farm Gate Life-Cycle Impacts Associated with Fertilizer Used for Corn, Soybean, and Stover Production*. NREL/TP-510-37500. National Renewable Energy Laboratory, Golden, CO, USA.

Risse LM, Gaskin JW (2013) Best management practices for wood ash as agricultural soil amendment. *University of Georgia Cooperative Extension Bulletin*, **1142**, 1–4.

Sanchez ST, Woods J, Akhurst M et al. (2012) Accounting for indirect land-use change in the life cycle assessment of biofuel supply chains. *Journal of the Royal Society Interface*, **9**, 1105–1119.

Sanderson MA, Read JC, Reed RL (1999) Pasture management & forage utilization - harvest management of switchgrass for biomass feedstock and forage production. *Agronomy Journal*, **91**, 5–10.

Scheuer C, Keoleian GA, Reppe P (2003) Life cycle energy and environmental performance of a new university building: modeling challenges and design implications. *Energy and Buildings*, **35**, 1049–1064.

Schiemenz K, Eichler-Löbermann B (2010) Biomass ashes and their phosphorus fertilizing effect on different crops. *Nutrient Cycling in Agroecosystems*, **87**, 471–482.

Schnepf R, Yaccobucci BD (2013) Renewable fuel standard (RFS): overview and issues. Report R40155. Congressional Research Service, Washington, DC, USA.

Searchinger T, Heimlich R, Houghton RA et al. (2008) Use of US croplands for biofuels increases greenhouse gases through emissions from land-use change. *Science*, **319**, 1238–1240.

Seinfeld JH, Pandis S (2006) *Atmospheric Chemistry and Physics* (2nd edn). John Wiley & Sons, Hoboken, NJ, USA.

Sheehan J, Aden A, Paustian K, Killian K, Brenner J, Walsh M, Nelson R (2003) Energy and environmental aspects of using corn stover for fuel ethanol. *Journal of Industrial Ecology*, **7**, 117–146.

Sindelar AJ, Lamb JA, Sheaffer CC, Rosen CJ, Jung HG (2013) Fertilizer nitrogen rate effects on nutrient removal by corn stover and cobs. *Agronomy Journal*, **105**, 437–445.

Spatari S, MacLean HL (2010) Characterizing model uncertainties in the life cycle of lignocellulose-based ethanol fuels. *Environmental Science & Technology*, **44**, 8773–8780.

Spatari S, Zhang Y, MacLean HL (2005) Life cycle assessment of switchgrass-and corn stover-derived ethanol-fueled automobiles. *Environmental Science & Technology*, **39**, 9750–9758.

Stewardship Nutrient (2011) *Soil Sampling Enhances Crop, Maximizes Fertilizer Use*. The Fertilizer Institute, Washington, DC, USA.

Tian Z, Mohan GR, Ingram L, Pullammanappallil P (2013) Anaerobic digestion for treatment of stillage from cellulosic bioethanol production. *Bioresource Technology*, **144**, 387–395.

Unruh D, Pabst K, Schaub G (2010) Fischer–tropsch synfuels from biomass: maximizing carbon efficiency and hydrocarbon yield. *Energy & Fuels*, **24**, 2634–2641.

Vogel K, Brejda J, Walters D, Buxton D (2002) Switchgrass biomass production in the Midwest USA. *Agronomy Journal*, **94**, 413–420.

Vogt W (2012) Managing corn stover residues offers new feed opportunity. *Southern Farmer*, **11**, 10.

Wang M, Wu M, Huo H (2007) Life-cycle energy and greenhouse gas emission impacts of different corn ethanol plant types. *Environmental Research Letters*, **2**, 024001.

Wardenaar T, van Ruijven T, Beltran AM, Vad K, Guinée J, Heijungs R, Ruijven T (2012) Differences between LCA for analysis and LCA for policy: a case study on the consequences of allocation choices in bio-energy policies. *The International Journal of Life Cycle Assessment*, **17**, 1059–1067.

Wortmann C, Klein R (2008) *Considerations for the Harvest of Corn Stover*. Natural Resources Conservation Service, United States Department of Agriculture, Washington, DC, USA.

Wu M, Wang M, Huo H (2006) *Fuel-Cycle Assessment of Selected Bioethanol Production Pathways in the United States*. Argonne National Laboratory, Argonne, IL, USA.

ZeaChem Inc. (2012) Advantages of the ZeaChem process. Retrieved June 22, 2013, from http://www.zeachem.com/technology/advantages.php.

9

Direct N_2O emission factors for synthetic N-fertilizer and organic residues applied on sugarcane for bioethanol production in Central-Southern Brazil

MARCOS SIQUEIRA NETO[1], MARCELO V. GALDOS[2], BRIGITTE J. FEIGL[1], CARLOS E. P. CERRI[3] and CARLOS C. CERRI[1]

[1]Centro de Energia Nuclear na Agricultura, Universidade de São Paulo (CENA/USP), Av. Centenário, 303, P.O. Box. 96, 13400-970 Piracicaba, SP, Brazil, [2]Laboratório Nacional de Ciência e Tecnologia do Bioetanol (CTBE), R. Giuseppe Máximo Scalfaro, 10.000, P.O. Box. 6170, 13083-970 Campinas, SP, Brazil, [3]Escola Superior de Agricultura Luiz de Queiroz, Universidade de São Paulo (ESALQ/USP), Av. Pádua Dias, 11, 13400-970 Piracicaba, SP, Brazil

Abstract

The production and use of biofuels have increased rapidly in recent decades. Bioethanol derived from sugarcane has become a promising alternative to fossil fuel for use in automotive vehicles. The 'savings' calculated from the carbon footprint of this energy source still generates many questions related to nitrous oxide (N_2O) emissions from sugarcane cultivation. We quantified N_2O emissions from soil covered with different amounts of sugarcane straw and determined the direct N_2O emission factors of nitrogen fertilizers (applied at the planting furrows and in the topdressing) and the by-products of sugarcane processing (filter cake and vinasse) applied to sugarcane fields. The results showed that the presence of different amounts of sugarcane straw did not change N_2O emissions relative to bare soil (control). N-fertilizer increased N_2O emissions from the soil, especially when urea was used, both at the planting furrow (plant cane) and during the regrowth process (ratoon cane) in relation to ammonium nitrate. The emission factor for N-fertilizer was $0.46 \pm 0.33\%$. The field application of filter cake and vinasse favored N_2O emissions from the soil, the emission factor for vinasse was $0.65 \pm 0.29\%$, while filter cake had a lower emission factor of $0.13 \pm 0.04\%$. The experimentally obtained N_2O emission factors associated with sugarcane cultivation, specific to the major sugarcane production region of the Brazil, were lower than those considered by the IPCC. Thus, the results of this study should contribute to bioethanol carbon footprint calculations.

Keywords: bioethanol, carbon footprint, filter cake, N-fertilizer, Saccharum officinarum L., vinasse

Introduction

N_2O is a long-lived trace gas that is naturally present in the atmosphere. This gas is able to absorb infrared radiation and relay it in the form of thermic energy. The heating potential of this gas is 298 times higher than CO_2 (IPCC, 2001), and it participates directly in stratospheric ozone depletion (Ravishankara et al., 2009). The atmospheric N_2O concentration has steadily remained at 270 ppbv since the last glacial period (Flückiger et al., 1999). However, by 2013, the concentration increased to 325.9 ppbv, with an average absolute increase during the last 10 years of 0.82 ppbv yr^{-1} (WMO, 2014). The major N_2O producers are the soil microorganisms that are responsible for N transformation (Bouwman, 1998). Agricultural activity with large N inputs from the Haber–Bosch synthesis is the most significant N_2O source to the atmosphere (Butterbach-Bahl et al., 2013).

Brazil's 2nd National Communication to the Framework Convention of the United Nations on Climate Change estimated that 84% (456.8 Gg) of the total N_2O emissions were from agricultural soils, of which 17% (77.8 Gg) were from the use of synthetic fertilizers and crop residues (Brazil, 2010).

Currently, Brazilian biofuel production is based on sugarcane. In 2013, bioethanol production was 21 million m^3 (UNICA, 2014) from a planted area of 9.1 million ha, of which 2.1 million ha were from new areas and reform/planting (CANASAT, 2014). The success of bioethanol production is due to the hardiness of the sugarcane crop, enabling regrowth after harvesting, high tillering production and positive energy balance (Macedo et al., 2008). The expansion of sugarcane production in the central-southern region of Brazil is due to the increase in the domestic fleet of flex fuel vehicles

Correspondence: Dr Marcos Siqueira Neto
e-mail: msiqueir@usp.br

and the export demand for bioethanol (Rudorff *et al.*, 2010).

The uncertainty associated with the emissions of non-CO_2 gases affects the carbon footprint calculation of the bioethanol derived from sugarcane. N_2O emissions from the use of different N-sources (fertilizer, by-products and crop residues) are considered to be major negative contributions to the 'savings' of this biofuel compared with fossil fuels (Smeets *et al.*, 2009). This consideration is because the estimate of N_2O emissions uses a single default emission factor based only on the amount of applied N, which ignores the complex interactions between the microorganisms responsible for N_2O production and environmental factors (IPCC, 2006).

The most common fertilization management practices on sugarcane plantations in the central-southern region of Brazil are N applied to planting furrows for reform and topdressing for sugarcane ratoons. Synthetic N-sources, urea and ammonium nitrate, account for over 85% of the fertilizer used in sugarcane cultivation. In addition to synthetic fertilizers, by-products of the sugarcane agribusiness production phase (filter cake and vinasse) can cause considerable N input. The application of these byproducts to the crop is common and has been extensively studied as a means of nutrient cycling for plant growth, potentially reducing input costs (Prado *et al.*, 2013).

The filter cake is obtained during the sugar manufacture, with every ton of sugarcane processed generating close to 30 kg (Veiga *et al.*, 2006) from the clarification of the juice obtained during milling. This composite is rich in phosphorus, in addition to calcium, magnesium, sulfur and micronutrients (Fravet *et al.*, 2010). The filter cake is applied to the planting furrow in doses between 15 and 40 ton ha^{-1}. The vinasse is considered to be the major residue of the ethanol production; every liter of ethanol produced generates between 10 and 18 L of vinasse (Freire & Cortez, 2000). Vinasse is rich in organic matter and stands out as a potassium source, in addition to other minerals (Silva *et al.*, 2014). The vinasse is commonly applied as fertirrigation directly onto sugarcane straws in the ratoons at a dose of 150 m^3 ha^{-1}.

The different sources (synthetic or organic), amounts applied, the application forms (furrow or topdressing, single or combined) and edaphoclimatic factors influence the resulting N_2O emissions and thereby necessitate different emission factors for each N-fertilizer management process on a sugarcane plantation (Choudhard *et al.*, 2001; Bouwman *et al.*, 2002; Khalil *et al.*, 2004; Stehfest & Bouwman, 2006; Denmead *et al.*, 2010; Aguilera *et al.*, 2013; Gu *et al.*, 2013).

To increase the knowledge of the contribution of N_2O emissions of sugarcane production to bioethanol carbon footprint calculations, we quantified N_2O emissions from soil covered with different amounts of sugarcane straw and determined the direct N_2O emission factors of N-fertilizers (applied at planting furrows and top-dressing) and by-products of sugarcane agribusiness (filter cake and vinasse) applied to sugarcane fields in the south central region of Brazil.

Materials and methods

Site description

The experiments were conducted in an area of 4.83 ha (22°36'35.7"S 47°36'06.5"W) that has been cultivated continuously with sugarcane since 1971. This area is located in the southwest region of Brazil, municipality of Piracicaba, São Paulo State. The regional climate is classified as Köppen's Cwa – mesothermal humid subtropical, with a dry winter and a hot and wet summer. The mean annual precipitation is 1400 mm yr^{-1}, and the mean annual temperature is 22.5°C. The soil is classified as clayey Oxisol, a Typic Acrustox (Soil Survey Staff, 1999). Soil samples were taken to characterize the physical and chemical properties, shown in Table 1.

The experimental area was cultivated with sugarcane variety RB 86-7515. At the time of the experiments, the crop was in the third growth cycle, or second ratoon. Sugarcane has been harvested mechanically without straw burning in this area since 2009.

Three experiments were sequentially installed: the first evaluated N_2O emissions from different quantities of straw deposited on the ground, the second experiment evaluated emissions due to the application of N-fertilizer and vinasse to ratoon sugarcane, and the third trial measured N_2O emissions from the application of N-fertilizer and filter cake in the planting furrow during the reform of the sugarcane plantation.

Table 1 Soil physic-chemical characteristics in the experimental area with sugarcane cultivation in the Central-southern part of the Brazil

Soil parameter	Soil layer (m)	
	0.0–0.1	0.1–0.2
Texture (g kg^{-1})		
Clay	676	684
Silt	101	93
Sand	223	223
Bulk density (g dm^{-3})	1.18	1.17
pH (CaCl2)	4.2	4.2
Total C (g kg^{-1})	15.1	14.5
Total N (g kg^{-1})	1.2	1.1
Avail. P (mg dm^{-3})	8	16
K$^+$ (mmol$_c$ dm^{-3})	2.7	2.1
Ca^{++} (mmol$_c$ dm^{-3})	12	12
Mg^{++} (mmol$_c$ dm^{-3})	4	4
CEC (mmol$_c$ dm^{-3})	70.3	76.0
BS (%)	26	24

CEC, cation exchanged capacity; BS, base saturation.

Experimental design and treatments

The first experiment started with the quantification of the newly deposited plant residues on the soil surface by the sugarcane harvested on July 6, 2012. To that end, we collected all the straw in ten 1 m^2 quadrants randomly distributed in the area. All of the collected materials were dried at 60°C until a constant weight was reached. Total C and N were determined using an elemental analyzer (LECO© CN 2000®, St. Joseph, Michigan). The amount of straw produced was 15 Mg ha^{-1}, with a total C content of 41.2% and 0.82% N. The field experiment was installed on August 8 and 9, 2012. We used a completely randomized design with four treatments (quantities of straw) and five replications. The quantities of straw were equivalent to (i) a full dose, 15 Mg ha^{-1} dry mass straw; (ii) a 66% dose, 10 Mg ha^{-1} dry mass straw; (iii) a 33% dose, 5 Mg ha^{-1} dry mass straw; and (iv) no dose (control). The sample chambers were installed in the inter-row, in an area over 450 m^2 (an useful area of twelve 32-m-long rows of sugarcane in 1.2-m intervals). Gas sampling to determine N_2O emissions began the following day and was performed weekly between August 2012 and March 2013, totaling 176 days.

The second experiment, designed to determine N_2O emissions from N-fertilizer and vinasse application to sugarcane ratoons, was installed on November 12 and 13, 2012. We delimited a plot of 0.25 ha of the total area (an useful area of twenty five 32-m-long rows of sugarcane in 1.2-m intervals), and the experiment was allocated using a completely randomized design consisting of eight situations – six treatment (mineral and organic N-sources in two doses), an interaction between sources (mineral and organic) and a control, each with five replicates. N_2O emissions from fertilizer were evaluated from two N-sources (ammonium nitrate at 35% of N and urea at 46% N) in doses of 80 and 120 kg N ha^{-1}. The vinasse treatments were equivalent to doses of 150 and 300 m^3 ha^{-1} of vinasse (6.45 g C L^{-1} and 0.57 g N L^{-1}) applied directly on the straw, and there was additionally a control treatment (without N-fertilizer or vinasse). In addition, to verify the interaction between sources (N-fertilizer plus vinasse) that usually occurs along the regrowth process, we evaluated a treatment of an 80 kg N ha^{-1} dose of urea followed by 150 m^3 ha^{-1} of vinasse. To apply the exact N-fertilizer quantity of each source, the doses were weighed on a precision balance and surface-applied in a band about 0.1 m from the plant row directly inside the chambers on the straw. The vinasse application was made using a garden watering can to an area of 1 m^2 bordering the row where N-fertilizer was applied. Gas sampling to determine N_2O emissions began the day after the experiment installation and was performed from November to December 2012, first daily for 15 days, and then every 2–3 days following this period, for a total of 30 days.

The third experiment, designed to determine the N_2O emissions from N-fertilizer and filter cake application in the sugarcane planting furrow, began with the conventional preparation of the soil for sugarcane planting on a 0.25-ha plot (an useful area of ten 40-m-long planting furrows of 1.2-m intervals). This preparation consisted of manually harvesting the sugarcane without burning the straw followed by soil tillage operations for the physical destruction of the stumps, plowing to 30 cm

and making the planting furrows. These three operations were performed three days prior to the experiment installation, which occurred on April 29, 2013. The experimental design was completely randomized consisting of six situations with five replications. To evaluate N_2O emissions from N-fertilizer, two N-sources (ammonium nitrate and urea) were tested at a single dose of 60 kg N ha^{-1}. To determine N_2O emissions from the application of filter cake, doses equivalent to 25 and 40 Mg ha^{-1} (wet mass) of filter cake (29.9% C and 1.67% N in dry weight) were applied to the planting furrow, in addition to a control without N-fertilizer or filter cake. In addition to these treatment conditions, another treatment was performed to evaluate the interaction between the mineral and organic sources of nitrogen that usually occurs in the planting furrow. Therefore, in the bottom furrow were applied an equivalent to 25 Mg ha^{-1} of filter cake plus an equivalent to 60 kg N ha^{-1} urea. To apply the exact N-fertilizer or filter cake (wet mass) quantity, the doses were weighed on a precision balance and applied directly to 1 m of a planting furrow. Sugarcane stalks were placed on top of the different N-sources and the furrows were then covered with soil, as is usually done during the planting process. After the furrow was covered, the chambers were installed on the row. Gas sampling to determine N_2O emissions began the day after experiment installation and was performed from April to June 2013, daily for 15 days, every 2–3 days for the next 30 days, and every 5–10 days after this period, for a total of 60 days.

N_2O sampling and analysis

The static chambers used to collect N_2O emitted by the soil and added materials consisted of two parts, a base and a lid. The dimensions of the base were 45 (width) × 70 (length) × 30 (height) cm; it was buried in the soil between 5 and 7 cm deep. Air samples of the chamber's headspace were collected with a nylon syringe of 20 mL (Becton Dickinson Ind. Surgical Inc.) at four fixed time intervals (0, 10, 20 and 30 minutes) to determine the N_2O concentration. The measurement of N_2O concentration in the syringes was performed using a Shimadzu© GC-2014® (Kyoto, Japan) gas chromatograph with a packed Porapak™ Q column (80–100 mesh) maintained at 82°C to separate molecular gases. N_2O was quantified using an electron capture detector (ECD) operating at 325°C. The N_2O fluxes were calculated by the linear change in the amount of N_2O in the chambers as a function of sampled time. During the sampling period, we also monitored environmental temperature and precipitation as well as ambient N_2O concentration to check the order of magnitude of the N_2O concentration in the chambers.

Result analysis

In the N_2O concentrations obtained for each sampling time, critical limits (lower and upper) were calculated by the box chart type (quartile). Thus, the 'outlier' values were discarded (missing subplots). We performed classical statistical analyses of the soil N_2O emission results for each treatment to verify the frequency and distribution of the data. The cumulative

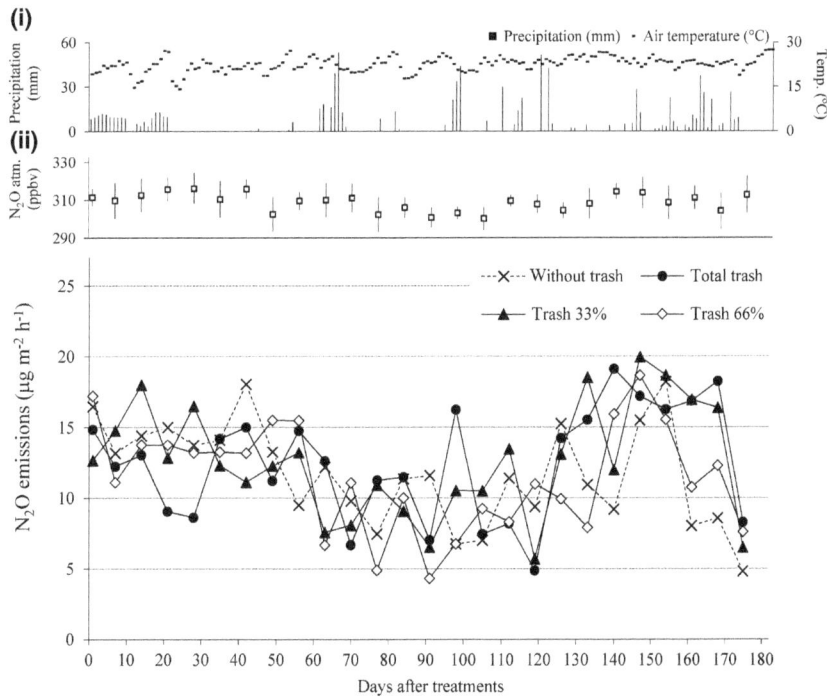

Fig. 1 N$_2$O emissions (μg m^{-2} h^{-1}) from different levels of sugarcane trash added to the soil surface. In the top figures (i and ii) represented environmental information during the experiment: (i) Precipitation (mm) and air temperature (°C); (ii) N$_2$O atmospheric concentration (ppmv).

(integral) N$_2$O emissions were calculated considering the period (days after application of nitrogen source) at which N$_2$O emission of N treatments no longer presented a difference significantly higher ($P < 0.05$) than the control treatment. Analysis of variance was performed using the Kruskal–Wallis test for the values of the cumulative N-N$_2$O emissions for each treatment compared with their respective controls. Means of the cumulative N$_2$O and emissions factors in different doses (same source – mineral or organic) were statistically separated using the Student–Newman–Keuls test ($P < 0.05$).

Calculating the N$_2$O emission factor

The emission factor for N$_2$O was calculated only when the treatments with N-sources showed significant differences from their respective controls. In other words, the difference in the integral ($\Sigma_{N_2O} - \Sigma_{Co}$) must always be positive.

To calculate the N$_2$O emission factor (*EF*) due to the application of N-fertilizer or by-products, we used the methodology described in Guidelines for National Inventories of Greenhouse Gases (IPCC, 2006), according to Eqn (1):

$$EF = \left(\frac{\Sigma_{N_2O} - \Sigma_{Co}}{Napll.}\right) \qquad (1)$$

Where Σ_{N_2O} is the total emission of N$_2$O in the chamber from the treatment when a source containing N was applied, Σ_{Co} is the total N$_2$O emission in the chamber from the control area, and Napll. is the amount of nitrogen applied as fertilizer or by-product in the chamber.

Results

N$_2$O emissions from soil amended with different amounts of sugarcane straw

During the sampling period, the average environmental temperature was 22.8 ± 2.5°C, and 86 days with precipitation events were recorded, totaling 1025.7 mm (Fig. 1i). The atmospheric concentration of N$_2$O was 309.8 ± 5.1 ppbv (Fig. 1ii).

The presence of different amounts of straw and even the absence of crop residues did not alter N$_2$O emissions (Fig. 1). On average, treatments emitted ~ 12 μg N$_2$O m^{-2} h^{-1} (minimum = 0.0 to maximum = 40.6 μg N$_2$O m^{-2} h^{-1}) with an accumulated value of ~ 30 mg N-N$_2$O m^{-2} in 176 days. No significant differences were found among treatments (Table 2).

In this experiment, emission factors were not determined because the results showed that straw should not be considered a N$_2$O source, in that treatments (quantities of straw) showed no significant cumulative emissions compared with the control (without straw).

Soil N$_2$O emissions derived from the sugarcane planting process

During the period of the experiment, the average temperature was 20.5 ± 2.0°C and 29 days with precipitation

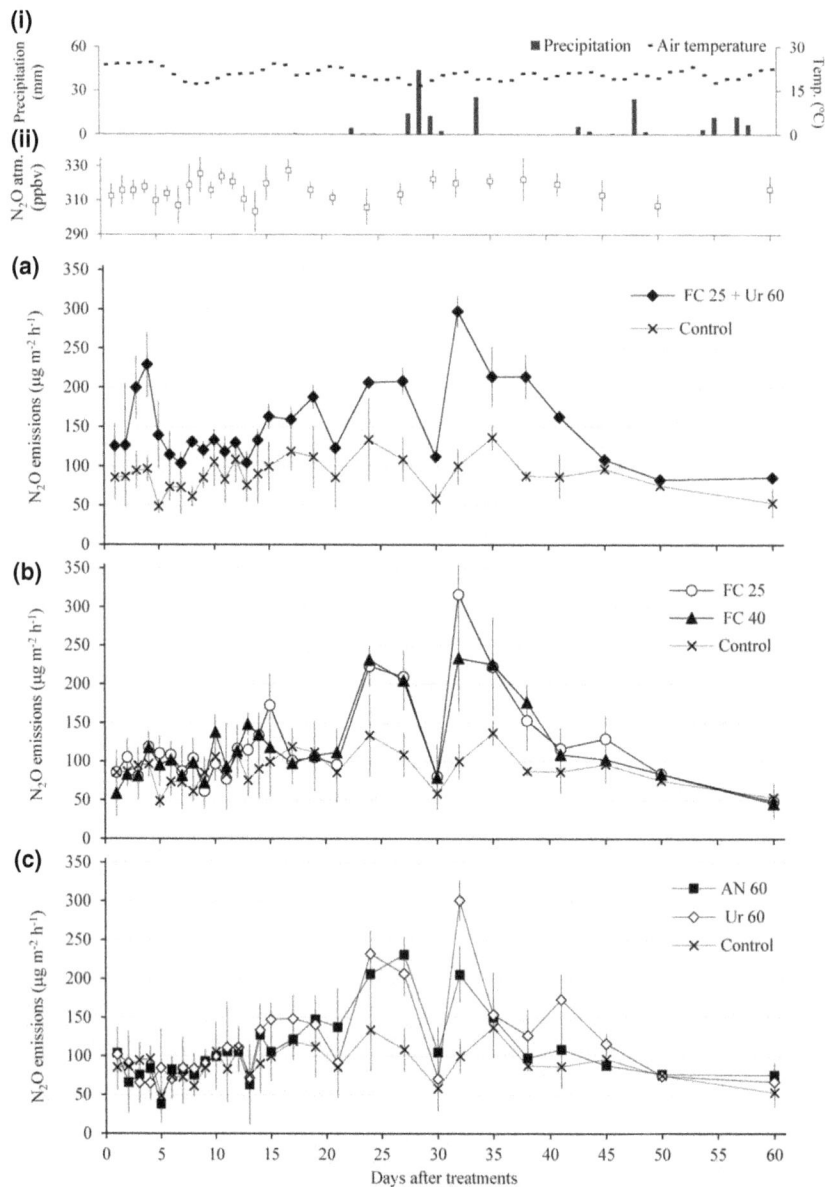

Fig. 2 N_2O emissions (μg m^{-2} h^{-1}) from filter cake (FC 25 = 25 Mg ha^{-1} $\approx$ 195 kg N ha^{-1}) and N-urea fertilizer (Ur 60 = 130 kg ha^{-1} $\approx$ 60 kg N ha^{-1}) applied in the furrow to the sugarcane plantation (a). From different doses of the Filter cake: FC 25 and FC 40 (40 Mg ha^{-1} $\approx$ 310 kg N ha^{-1}) (b). From different synthetic N-sources Ammonium nitrate – AN 60 (170 kg ha^{-1} $\approx$ 60 kg N ha^{-1}) and Urea – Ur 60 (c). In the top figures (i and ii) represented environmental information during the experiment: (i) Precipitation (mm) and air temperature ($^\circ$C); (ii) N_2O atmospheric concentration (ppmv).

events were recorded, for a total of 173.6 mm (Fig. 2i). The mean atmospheric N_2O concentration was 316.0 $\pm$ 16.1 ppbv (Fig. 2ii).

Figure 2a shows the daily average N_2O emissions resulting from the treatment combining filter cake (FC 25 Mg ha^{-1}) with N-fertilizer urea (Ur 60 kg N ha^{-1}), compared with the control. N_2O emissions from the organic and synthetic nitrogen sources (FC + Ur) were higher than the control from the start and throughout the sampling period, maintaining an average emission

level of 130 μg N_2O m^{-2} h^{-1}, higher than the 90 μg N_2O m^{-2} h^{-1} observed for the control.

On the 3rd day after treatment (DAT), N_2O emissions increased in treatment FC + Ur and reached an emission three times greater than the control by the 4th DAT. This increase in N_2O emissions was observed exclusively in the treatment with the combined of organic and synthetic sources. Other treatments in which synthetic (urea and ammonium nitrate) or organic (filter cake) sources were applied alone did not increase N_2O emissions before the

Table 2 N_2O emissions (mean $\pm$ standard deviation and range), cumulative N-N_2O emissions from different levels to the sugarcane trash deposited on the soil surface

Treatments	Mean $\pm$ SD*	Range†	Cumulative N-N_2O‡
	$\mu g\ N_2O\ m^{-2}\ h^{-1}$		mg N-$N_2O\ m^{-2}$
Without trash	11.7 ± 7.0	1.1–34.1	28.7 ± 4.3 [ns]
Trash 33%	12.6 ± 7.5	1.2–33.8	30.6 ± 3.0
Trash 66%	11.6 ± 10.0	0.6–39.7	28.0 ± 3.0
Total trash	12.5 ± 11.1	0.5–40.6	29.9 ± 3.7

*Mean $\pm$ SD = standard deviation.

†Range = Minimum and maximum

‡Cumulative N-N_2O = sampled time = 175 days

ns = Cumulative means without significant differ in the Student–Newman–Keuls test ($P < 0.05$).

20th DAT, when the first precipitation occurred since the beginning of the experiment (Fig. 2b and c).

At the 20th DAT, the first rainfall of the period occurred (0.5 mm – 18th DAT) and emissions began to respond to rainfall. The lower N_2O emissions at the 30th DAT can be attributed to consecutive high rainfall events (~70 mm) that occurred the day before sampling. After the effect of the heavy rainfall ceased, the highest average emission of N_2O was recorded at the 32th DAT (~300 $\mu g\ N_2O\ m^{-2}\ h^{-1}$), approximately three times the emission observed for the control. After the 40th DAT, emission values of the treatment with N application decreased until they were similar to the control.

Cumulative N_2O emissions from the filter cake application were significantly higher than the control (40%), but the two doses (25 and 40 Mg ha^{-1}) did not show a significant difference in N_2O emissions (Fig. 2b).

In general, the pattern of N_2O emissions from soils treated with ammonium nitrate and urea fertilizers were similar (Table 3). However, more pronounced peaks were observed in the treatment with the urea application (Fig. 2c).

The accumulated values showed no significant difference between the N-sources, but they were 36 and 40% (ammonium nitrate and urea, respectively) higher than the control (Table 3).

Table 3 shows that not only were the average N_2O emissions similar, but the ranges of presented values were also similar. The accumulation for the period from the different doses of filter cake was on average 40% higher than the control.

The use of synthetic N-fertilizer during sugarcane planting generated an emission factor of 0.48%, while the filter cake application showed an average value three times lower (0.13%). The combination of organic and synthetic sources (FC + Ur), which is the most common condition, showed an emission factor of 0.21% (Table 3).

N_2O emissions from soil for ratoon sugarcane

During this experimental period, the average temperature was $24.7 \pm 1.2°C$, and 12 days with precipitation events were recorded, totaling 172.3 mm (Fig. 3i). The

Table 3 N_2O emissions (mean $\pm$ standard deviation and range), cumulative N-N_2O emissions and N_2O emissions factor from N-sources (mineral and organic) applied in the sugarcane planting

Treatments (sources/levels)	N_2O emissions			N_2O emission factor	
	Mean $\pm$ SD*	Range†	Cumulative N-N_2O‡	N-source/level§	N-source¶
	$\mu g\ N_2O\ m^{-2}\ h^{-1}$		mg N-$N_2O\ m^{-2}$	% N-applied	
Control	89.3 ± 31.0	12.9–190.8	80.3 ± 9.7^B		
Mineral source					
Ammonium nitrate (60 kg N ha^{-1})	110.9 ± 53.7	10.8–268.6	108.1 ± 4.9^A	0.44 ± 0.08[ns]	0.48 ± 0.12^a
Urea (60 kg N ha^{-1})	119.3 ± 60.8	15.5–335.4	113.0 ± 9.5^A	0.52 ± 0.15	
Organic source					
Filter cake (25 Mg ha^{-1})	123.3 ± 60.3	12.2–332.6	112.0 ± 8.8^A	0.17 ± 0.02[ns]	0.13 ± 0.04^b
Filter cake (40 Mg ha^{-1})	123.5 ± 57.8	14.0–326.5	114.6 ± 4.7^A	0.10 ± 0.03	
Filter cake (25 Mg ha^{-1}) + urea (60 kg N ha^{-1})	156.7 ± 75.3	10.4–463.7	132.5 ± 14.7^A	0.21 ± 0.05	

*Mean $\pm$ SD = standard deviation.

†Range = Minimum and maximum.

‡Cumulative N_2O = sampled time in the sugarcane planting = 60 days.

§N_2O emissions factor from each N-sources and level applied in the sugarcane planting.

¶N_2O emissions factor from N-source applied in the sugarcane planting. Different capital letter to the Cumulative N-N_2O (N-source and dose × control) and small letter to the Emission Factor (N-source/level and N-source) with significant differ in Student–Newman–Keuls test ($P < 0.05$); ns, not significant.

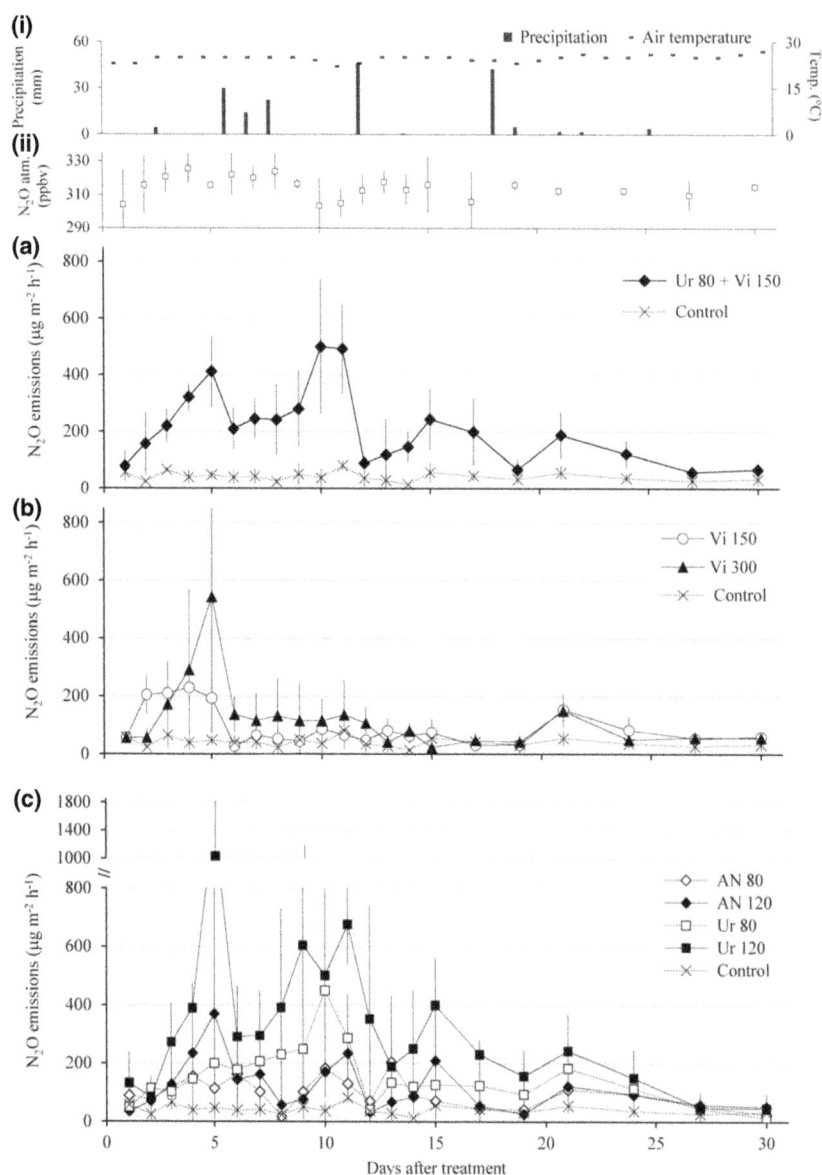

Fig. 3 N_2O emissions ($\mu g \ m^{-2} \ h^{-1}$) from Urea – Ur 80 (174 kg $ha^{-1} \approx 80$ kg N ha^{-1}) and vinasse – Vi 150 (150 $m^3 \ ha^{-1} \approx 195$ kg N ha^{-1}) applied in the sugarcane ratoon (a). From different doses of the vinasse – Vi 150 and Vi 300 (300 $m^3 \ ha^{-1} \approx 390$ kg N ha^{-1}) (b). From different synthetic N-sources and doses: Ammonium nitrate – AN 80 (225 kg $ha^{-1} \approx 80$ kg N ha^{-1}) and AN 120 (340 kg $ha^{-1} \approx 120$ kg N ha^{-1}); and Urea – Ur 80 and Ur 120 (260 kg $ha^{-1} \approx 120$ kg N ha^{-1}) (c). In the top figures (i and ii) represented environmental information during the experiment: (i) Precipitation (mm) and air temperature ($^\circ$C); (ii) N_2O atmospheric concentration (ppmv).

atmospheric N_2O concentration was 314.5 ± 18.1 ppbv (Fig. 3ii).

Figure 3a presents the daily N_2O emissions from the treatment with the application of N-fertilizer urea (Ur 80 kg N ha^{-1}) with vinasse (Vi 150 $m^3 \ ha^{-1}$) compared with the control. N_2O emissions from the combination of synthetic and organic N-sources (Ur + Vi) were higher than the control starting from the 2nd DAT and maintained a median level of 188 $\mu g \ N_2O \ m^{-2} \ h^{-1}$, while the control emission was 41 $\mu g \ N_2O \ m^{-2} \ h^{-1}$.

This means that the combination of organic and synthetic sources was 4.5 times higher than the control.

On the 5th DAT, N_2O emissions from the Ur + Vi treatment were nine times higher than the control value. This high value of N_2O emissions early in the experimental period was observed in all treatments in which synthetic (urea and ammonium nitrate) and organic (vinasse) sources were applied alone (Fig. 3b and c). The maximum values in the range of results clearly expresses the magnitude of N_2O emissions using differ-

Table 4 N_2O emissions (mean ± standard deviation and range), cumulative $N-N_2O$ emissions and N_2O emissions factor from N-sources (mineral and organic) applied in the sugarcane ratoon

Treatments (sources/levels)	N_2O emissions			N_2O emission factor	
	Mean ± SD*	Range†	Cumulative $N-N_2O$‡	N-source/level§	N-source¶
	$\mu g\ N_2O\ m^{-2}\ h^{-1}$		$mg\ N-N_2O\ m^{-2}$	% N-applied	
Control	41.9 ± 24.6	5.2–103.2	17.9 ± 1.0[B]		
Mineral source					
Ammonium nitrate (80 kg N ha^{-1})	97.5 ± 77.5	5.0–478.3	39.1 ± 5.8[A]	0.25 ± 0.07[ns]	0.24 ± 0.06[b]
Ammonium nitrate (120 kg N ha^{-1})	119.2 ± 112.8	4.9–530.2	47.3 ± 8.6[A]	0.23 ± 0.07	
Urea (80 kg ha^{-1})	153.4 ± 145.0	2.5–998.4	61.8 ± 12.9[A]	0.52 ± 0.15[ns]	0.68 ± 0.24[a]
Urea (120 kg ha^{-1})	323.9 ± 329.7	3.8–2188.4	122.0 ± 28.1[A]	0.83 ± 0.22	
Organic source					
Vinasse (150 m^3 ha^{-1})	97.3 ± 77.7	3.8–312.1	38.6 ± 4.4[A]	0.77 ± 0.16[ns]	0.65 ± 0.29[a]
Vinasse (300 m^3 ha^{-1})	125.5 ± 170.8	3.5–1117.9	46.9 ± 19.9[A]	0.54 ± 0.37	
Urea (80 kg N ha^{-1}) + vinasse (150 m^3 ha^{-1})	212.6 ± 156.5	3.6–831.3	83.6 ± 21.3[A]	0.59 ± 0.19	

*Mean ± SD = standard deviation.

†Range = Minimum and maximum.

‡Cumulative N_2O = sampled time in the sugarcane ratoon = 30 days.

§N_2O emissions factor from each N-sources and level applied in the sugarcane ratoon.

¶N_2O emissions factor from N-source applied in the sugarcane ratoon. Different capital letter to the Cumulative $N-N_2O$ (N-source and dose × control) and small letter to the Emission Factor (N-source/level and N-source) with significant differ in Student–Newman–Keuls test ($P < 0.05$); ns, not significant.

ent N-sources (synthetic and organic) applied on the surface (Table 4). The reduction in N_2O emissions between the 6th and 9th DAT can be related to consecutive high rainfall events (~ 66 mm) that occurred during the sampling period. The same occurred at 12 (46 mm) and 19 DAT (42 mm), when precipitation events occurred the day before sampling. At the 10th DAT, a second increase in N_2O emissions was observed in the treatment with N application at a magnitude thirteen times higher than the control. Two other increases in emissions were observed: one at the 15th DAT, four times higher than the control, and another at the 21th DAT, three times higher than the control. After the 24th DAT, N_2O emissions from the treatment with N application dropped to levels similar to the control.

The vinasse applied to the ratoon produced cumulative emissions 110% and 160% (for 150 and 300 m^3 ha^{-1}, respectively) higher than the control. Although N_2O emissions were higher as a result of the applied dose, there was no significant difference between them (Fig. 3b, Table 4).

The patterns of N_2O emissions from N-fertilizers (ammonium nitrate and urea) were quite different (Fig. 3c, Table 4). Ammonium nitrate had lower emissions than urea. The cumulative emissions for the different doses of ammonium nitrate showed differences of more than 115 and 160% (80 and 120 kg N ha^{-1}, respectively) compared with the control. However, the increase in the applied dose did not cause a proportional increase in N_2O emissions, and no showed significant difference.

N_2O emissions derived from urea were high in both applied doses. The highest dose had an accumulated value about twice the lowest dose. Thus, the source urea emitted 240 and 570% (80 and 120 kg N ha^{-1}, respectively) more than the control.

By the sugarcane ratoon, the use of ammonium nitrate presented an emission factor of 0.24%, while the source urea generated an emission factor almost three times higher (0.68%). The vinasse application presented an emission factor close to that observed for urea (0.65%). The combination of synthetic and organic sources (Ur + Vi), which is the most normal field condition, presented an emission factor of 0.59% (Table 4).

Discussion

N_2O emissions from soil and sugarcane straws

N_2O emissions from different quantities of sugarcane straw did not cause measurable changes. For the conditions of this study, our goal was to determine whether during the course of decomposition of sugarcane straw, it presents itself as a source for soil N, which would contribute significantly to N_2O emissions, because from the first cut, the straw covering the soil remains throughout the conduction period after the ratoon re-growth, in a cycle of deposition and decomposition.

Carmo *et al.* (2012) found increased N_2O emissions with an increasing amount of sugarcane residues deposited on the ground; however, this study included application of N-fertilizer and vinasse. Malhi & Lemke (2007) found no significant differences in N_2O emissions due to the maintenance or removal of different crop residues. The authors also commented that due to the high variability in the results, it is difficult to determine the interaction between factors (e.g., soil water content and temperature) that may influence the N_2O emissions, even with the application of the N-fertilizer. Wang et al. (2008) reported a reduction between 24 and 30% in N_2O emissions due to removal of the sugarcane straw and attributed this finding to the importance of organic compounds in the regulation of N_2O fluxes.

Our study evaluated only sugarcane straw as a variable, without introducing other N-sources. Sugarcane straw is a plant residue with a high C:N ratio (approximately 50–80). Moreover, sugarcane straw has a large amount of lignin and polyphenols that reduce the decomposition rate (Abiven et al., 2005) and the availability of the minimal N (0.8%) is slow (Fortes et al., 2012; Leal et al., 2013). Huang et al. (2004) evaluated the influence of crop residue application with different C:N ratios on N_2O emissions, including sugarcane straw. This study showed that increasing the C:N ratio of the plant residues incorporated into the soil promoted reduction of N_2O emissions.

Currently, it is known that the amount of straw added by sugarcane is approximately 14.1 (7.4 to 24.3) Mg ha^{-1} yr^{-1}; however, there are different methods of managing the straw in the field (Leal et al., 2013). In the central-southern region of Brazil, it is common to bunch the excess straw between the rows of crops because the producers claim that the high amount deposited along the cycles of regrowth promotes a shallower root system, which reduces the length of the crop cycle. The growth in the production of second generation ethanol can increase an industrial demand for excess straw remaining in field (Goldemberg, 2008).

It is worth mentioning that the N_2O emission was approximately 12 μg N_2O m^{-2} h^{-1}, a low value for agricultural soils. However, one must consider that the experiment was conducted at the beginning of a regrowth cycle, after exportation of nutrients by the crop (stalks) with no contribution of N-sources. In an experiment to assess N_2O emissions from elephant grass cultivation for biomass production, Morais et al. (2013) found N_2O emissions ranging from 0.1 to 10 μg m^{-2} h^{-1}, which increased due to soil management and N-fertilizer application.

Throughout the experimental period (~180 days), interventions related to sugarcane management in the area were made only to meet the goals of the experiment (i.e., herbicide application), without the normal crop management practices in the region (i.e., topdressing N-fertilizer or vinasse application), so as to priori-

tize only the N_2O emissions from sugarcane straw. Additionally, one must note that the determinations were made between crop rows (Cai et al., 2012). The ground in this area has a high clay content (> 60%), and the third regrowth is sugarcane, which means that it has high soil bulk density (maximum 1.52 g cm^{-3}) as a result of machinery traffic and consequently lower pore space for N_2O production and gas exchange.

N_2O emissions from organic and synthetic N-sources in sugarcane cultivation

The N_2O emissions in the control plots at sugarcane planting (~ 90 μg N_2O m^{-2} h^{-1}) were considered to be high for agricultural soils. This result is most likely due to soil disruption in the reform area cultivation due to tillage operations (physical destruction of stumps, plowing, and opening and closing of the furrows) causing the incorporation of crop residues and aeration into the soil surface layer, which favors the priming effect and soil organic matter mineralization (La Scala et al., 2006; Silva-Olaya et al., 2013). These processes accelerate the labile-C availability in the soil, and thus, the set of microbiological processes enabled by the formation of anaerobic sites with high CO_2 concentrations together with the mineralization-N increase and N_2O emissions from the soil (Rochette, 2008; Morais et al., 2013). The same effect was observed by Morais et al. (2013), who observed higher N_2O emissions after soil plowing of between 30 and 100 μg m^{-2} h^{-1} (maximum 450 μg m^{-2} h^{-1}) for elephant grass management. In the same period, the authors also found higher CO_2 and inorganic N concentrations.

The increase in N_2O emissions at the beginning of the measurements in the experiment involving sugarcane planting that occurred for the trial with a combination of organic and synthetic sources (filter cake + urea) shows the effect of the interaction between the source; however, it has not been possible to establish with certainty whether this synergistic effect was due to a reaction between the filter cake and urea, or, more simplistically, as soil moisture was low, the filter cake moisture would have initiated the urea transformation, favoring N_2O emissions.

In the regrowth crop, the N_2O emissions control had a value (~ 40 μg N_2O m^{-2} h^{-1}) that is considered normal for cultivated agricultural soils in tropical regions (Gomes et al., 2009; Kachenchart et al., 2012; Morais et al., 2013). The increase in N_2O emissions during the initial period of sampling is most likely associated with transformations in soil N because the application of the fertilizer+vinasse (Vi + Ur) promoted the same effect on the soil in sugarcane planting as filter cake plus urea. The urea and vinasse provided N-reactive and a faster increase in soil labile-C and humidity by vinasse.

Ball *et al.* (1999) showed that high doses of N-fertilizer promoted changes in N_2O emissions for a period of up to six weeks. Morais *et al.* (2013) showed that 80% of the N_2O emissions due to N-fertilizer occurred in the 60 days after the N-fertilizer was applied.

N_2O emissions from vinasse were high when compared to the control or to other sources containing nitrogen (e.g. ammonium nitrate). This was most likely due to the available N form in this by-product in addition to the high organic content readily available to microorganisms in the soil, in which the higher microbial activity produced anaerobic sites with high N_2O production.

During the experiments, the N_2O emissions were largely influenced by precipitation, both the lack and excess thereof. There was no irrigation during planting; thus, the lack of moisture hindered the availability of N-fertilizer. In low soil moisture conditions, there was little development of the microbiota responsible for the N transformation (immobilization, mineralization and denitrification). The onset of precipitation and the resulting increased soil moisture promoted the development of microorganisms, increasing the availability of the reactive forms of soil N.

Regardless of the situation (planting or regrowth), the entry of water into the porous soil system promoted the 'expulsion' of gases contained in the pore spaces (physical movement) generated by soil disturbances at planting and in the macro- and mesopores formed between the particles and aggregates in the soil of regrowth.

Excessive rainfall on consecutive days, even in tropical soils with good drainage, causes most of the pore space to temporarily fill with water (~100% saturation of WFPS), thereby decreasing microbial activity and reducing greenhouse gas emissions (Bouwman, 1998). Hence, at higher WFPS (> 80%), the major product of the denitrifying community is N_2 (complete denitrification) (Davidson *et al.*, 2000). Leaching may also occur because N_2O has a high solubility in water (Heincke & Kaupenjohann, 1999).

Emissions factor from N applied in sugarcane cultivation

For the synthetic N-fertilizer applied in the sugarcane, the overall EF was 0.46 ± 0.33%. The ammonium nitrate has a lower emission factor than the urea. This behavior was most evident during the sugarcane regrowth and surface N application. This difference in N_2O emissions between sources is related to changes occurring in the soil related to the transformation of each source especially changes in soil pH. When applied, the breakdown of urea by the enzyme urease leads to a rapid pH increase (Lara-Cabeza & Souza, 2008); this brief increase in the soil pH is sufficient to promote the conversion of ammonium to nitrate (Wickramasinghe *et al.*, 1985) and,

consequently, start the denitrification. However, the conversion of ammonium to nitrate promotes soil acidification, which hinders nitrification (Page *et al.*, 2002), also affecting the denitrifying community (Daum & Schenk, 1998).

Carmo *et al.* (2012) calculated EF for N-fertilizer in sugarcane as 1.1% for N applied at planting as urea at the same dose as in this study (60 kg N ha^{-1}) and, for the N applied to the topdressing, the EF was 0.76% for a dose of 120 kg N ha^{-1}, a factor very similar to that determined in our experiment (0.83%). Lisboa *et al.* (2011) presented an emission factor of 3.8% for N-fertilizer in calculating the carbon footprint of Brazilian ethanol, based mostly on literature review. Morais *et al.* (2013) evaluated urea application in elephant grass, which has a similar management as sugarcane involving harvest and regrowth cycles. In this situation, the authors found an EF of 0.51% after three cycles.

The EF for filter cake and vinasse did not showed the expected behavior, or, the increasing the dose did not increase N_2O emissions linearly. Thus, the highest dose showed a smaller EF than the lowest dose, this occurs because the EF calculation is based on the amount of N applied.

The filter cake applied at sugarcane planting had a lower EF relative to other synthetic and organic sources, even with similar cumulative emissions to synthetic sources used in planting (0.13%). This occurred because even with a low N content (~ 2%), the amount of filter cake applied at planting (25 and 40 Mg ha-1) was high. However, the amount of N applied was not available to denitrifying organisms in the soil, and when the EFs were calculated, the values were low.

The vinasse applied generated a high emission factor (0.65%) analogous to that observed for urea application. N contained in the vinasse, together with the application form of fertirrigation, favored N_2O production in the soil. In a study of short duration with application of 200 m^{-3} ha^{-1} of vinasse, Oliveira *et al.* (2013) calculated an EF of 0.44%.

The EF calculated from the sources combination filter cake + urea at planting and urea + vinasse in covering regrowth did not produce values representing a summation of sources when individually evaluated, but rather demonstrated an interaction between sources leading to a specific emissions factor.

Carmo *et al.* (2012) evaluated emissions from mixing organic and synthetic sources for the supply of N to the crop of sugarcane planting and regrowth. The specific EF were much higher for the combination planting (filter cake + urea); EF was 1.1% vs. the 0.21% observed in our study. In the regrowth crop period (urea + vinasse), the difference was even greater, with an EF of the 3.0%, vs. 0.59% in our study.

Schils *et al.* (2008) studied the effect of the combination of fertilizer with an organic source (liquid slurry) in ryegrass in the Netherlands and found an EF between 0.13 and 0.17% in the application of calcium ammonium nitrate, an EF of 0.12% in the application of liquid slurry, and values of 0.35 to 0.39% when a mix of fertilizer and slurry was applied. Dambreville *et al.* (2008), in a maize crop in the Brittany region (France), found an EF between 0.01 and 0.23% from liquid pig slurry application alone and of 0.01% for ammonium nitrate.

In the literature, N_2O emission factors from N-fertilizer application in sugarcane cultivation mostly come from studies conducted in Australia, with values ranging from 1.0 to 21% (Weier, 1999; Galbally *et al.*, 2005; Wang *et al.*, 2008; Allen *et al.*, 2010; Denmead *et al.*, 2010). In terms of commercial sugarcane cultivation in Australia, the nitrogen cycle peculiarities, which reflect on the direct and indirect N_2O emissions and consequently on the EFs, are related especially to the use of irrigation, drainage and presence of the seasonal fluctuations in the groundwater levels that directly influence soil oxygenation (and gas diffusion) against water-filled pore space (WFPS%). Abbasi & Adams (2000) found that for the same amount of N-fertilizer applied, the EF responded differently to increased WFPS%, with EFs of 0.15, 0.40 and 2.8%, respectively, for 63, 71 and 84% WFPS% values.

In the central-southern region of Brazil, where more than 85% of the Brazilian sugarcane is cultivated (UNICA, 2014), the management does not include the irrigation, or the soil does not report the occurrence of this specific phenomenon of poor drainage or seasonal fluctuations in the groundwater.

As shown by Smeets *et al.* (2009), our results indicate that the use of the default value of IPCC (2006) for the conditions of sugarcane cultivation in the central-southern region of Brazil overestimates the contribution of direct N_2O emissions from N-fertilizer (synthetic or organic). Our results not only supply measured values for EFs from N-applied but also help to reduce the variance of the values. Combining these results with other values for this region (Carmo *et al.*, 2012; Morais *et al.*, 2013; Oliveira *et al.*, 2013) should facilitate eliminating the uncertainties associated with these emission sources.

Other studies showed that sugarcane usually does not respond to increasing rates of N (Reis Jr *et al.*, 2000; Franco *et al.*, 2011). This is due to the contribution of biological nitrogen fixation (BNF) by groups of diazotrophic endophytics responsible for 25–60% of the N supply to the crop (Reis Junior *et al.*, 2000; Boddey *et al.*, 2001; Resende *et al.*, 2003). Thus, the N management in sugarcane varies considerably with respect to sources and doses. Evaluating different variety behavior to reduce the external N supply (synthetic and/or organic) and

the possibility of finding the best economic level and not 'standardized' quantities of fertilizers based on expected yields can effectively reduce N_2O emissions.

In the case of greenhouse gas inventories from biofuels, we suggest a top-down approach using EFs evaluated for specific situations (Tier 2 – IPCC, 2006) in addition to extrapolating N_2O emission values to regional scales to include more sophisticated mathematical models with a bottom-up approach (Del Grosso *et al.*, 2008) that seek a better understanding of the N cycle, including the influence of other factors (e.g., water, temperature, aeration, labile-C availability and plant N demand), rather than only the amount of N-applied and amount of N_2O emitted.

Acknowledgements

We would like to thank Usina Capuava, Piracicaba, SP, for allowing us to conduct our experiments in their sugarcane areas. We also thank Admilson Margato and Ralf Araújo for assistance in the field and laboratory analysis. Support for this research was provided by the São Paulo Research Foundation (FAPESP) through the Regular Research Program (Process Number 2011/07276-6) and, to the first author, through the Post-doctoral Program (Process Number 2010/20065-1).

References

Abbasi MK, Adams WA (2000) Gaseous N emission during simultaneous nitrification-denitrification associated with mineral N fertilization to a grassland soil under field conditions. *Soil Biology & Biochemistry*, **32**, 1251–1259.

Abiven S, Recous S, Reyes V, Oliver R (2005) Mineralisation of C and N from root, stem and leaf residues in soil and role of their biochemical quality. *Biology and Fertility of Soils*, **42**, 119–128.

Aguilera E, Lassaletta L, Sanz-Cobena A, Garnniere J, Vallejo A (2013) The potential of organic fertilizers and water management to reduce N_2O emissions in Mediterranean climate cropping systems. A review. *Agriculture, Ecosystems & Environment*, **164**, 32–52.

Allen DE, Kingston G, Rennenberg H, Dalal RC, Schmidt S (2010) Effect of nitrogen fertilizer management and waterlogging on nitrous oxide emission from subtropical sugarcane soil. *Agriculture, Ecosystems & Environment*, **136**, 209–217.

Ball BC, Parker JP, Scott A (1999) Soil and residue management effects on cropping conditions and nitrous oxide fluxes under controlled traffic in Scotland 2. Nitrous oxide, soil N status and weather. *Soil & Tillage Research*, **52**, 191–201.

Boddey RM, Polidoro JC, Resende AS, Alves BJR, Urquiaga S (2001) Use of the ^{15}N natural abundance technique for the quantification of the contribution of N_2 fixation to sugarcane and other grasses. *Australian Journal of Plant Physiology*, **28**, 889–895.

Bouwman AF (1998) Nitrogen oxides and tropical agriculture. *Nature*, **392**, 866–867.

Bouwman AF, Bouwman LJM, Batjes NH (2002) Modeling global annual N_2O and NO emissions from fertilized fields. *Global Biogeochemistry Cycles*, **16**, 1080–1088.

Brazil. (2010) *Brazil's 2nd National Communication to the Framework Convention of the United Nations on Climate Changes*. Ministério de Ciência & Tecnologia, Brasília, DF.

Butterbach-Bahl K, Baggs EM, Dannenmann M, Kiese R, Zechmeister-Boltenstern S (2013) Nitrous oxide emissions from soils: how well do we understand the processes and their controls? *Philosophical Transaction of The Royal Society B*, **368**, 20130122.

Cai Y, Ding W, Luo J (2012) Spatial variation of nitrous oxide emission between interrow soil and interrow plus row soil in a long-term maize cultivated sandy loam soil. *Geoderma*, **181–182**, 2–10.

CANASAT (2014) Sugarcane crop monitoring in Brazil by Earth observing satellite images. Available online: http://www.dsr.inpe.br/laf/canasat/ (accessed on 14 May 2014).

Carmo JB, Filoso S, Zotelli LC *et al.* (2012) Infield greenhouse gas emissions from sugarcane soils in Brazil: effects from synthetic and organic fertilizer application and crop trash accumulation. *Global Change Biology Bioenergy*, **5**, 267–280.

Choudhard MA, Akramkhanov A, Saggar S (2001) Nitrous oxide emissions in soils cropped with maize under long-term tillage and under permanent pasture in New Zeland. *Soil & Tillage Research*, **62**, 61–71.

Dambreville C, Morvan T, Germon JC (2008) N₂O emission in maize-crops fertilized with pig slurry, matured pig manure or ammonium nitrate in Brittany. *Agriculture, Ecosystems & Environment*, **123**, 201–210.

Daum D, Schenk MK (1998) Influence of nutrient solution pH on N₂O and N₂ emissions from a soil less culture system. *Plant and Soil*, **209**, 279–287.

Davidson EA, Keller M, Erickson HE, Verchot LV, Veldkamp E (2000) Testing a conceptual model of soil emissions of nitrous and nitric oxides. *BioScience*, **50**, 667–680.

Del Grosso SJ, Wirth T, Ogle SM, Parton WJ (2008) Estimating agricultural nitrous oxide emissions. *Eos*, **89**, 259–540.

Denmead OT, Macdonald BCT, Bryant G *et al.* (2010) Emissions of methane and nitrous oxide from Australian sugarcane soils. *Agricultural and Forest Meteorology*, **150**, 748–756.

Flückiger J, Dällenbach A, Blunier T, Stauffer B, Stocker TF, Raynaud D, Barnola J-M (1999) Variations in atmospheric N₂O concentration during abrupt climatic changes. *Science*, **285**, 227–230.

Fortes C, Trivelin PCO, Vitti AC (2012) Long-term decomposition of sugarcane harvest residues in Sao Paulo state, Brazil. *Biomass and Bioenergy*, **42**, 189–198.

Franco HCJ, Otto R, Faroni CE, Vitti AC, Oliveira ECA, Trivelin PCO (2011) Nitrogen in sugarcane derived from fertilizer under Brazilian field conditions. *Field Crops Research*, **121**, 29–41.

Fravet PRF, Soares RAB, Lana RMQ, Lana AMQ, Korndörfer GH (2010) Effect of filter cake doses and methol of application on yield and technologycal quality of sugar cane ratoon. *Ciência e Agrotecnologia*, **34**, 618–624.

Freire WJ, Cortez LAB (2000) *Vinhaça de Cana-de-Açúcar.* Agropecuária, Guaíba.

Galbally I, Meyer M, Bentley S *et al.* (2005) A study of environmental and management drivers of nitrous oxide emissions in Australian agro-ecosystems. *Environmental Sciences*, **123**, 225–237.

Goldemberg J (2008) The Brazilian biofuels industry. *Biotechnology for Biofuels*, **1**, 6.

Gomes J, Bayer C, Costa FS, Piccolo MC, Zanatta JA, Vieira FCB, Six J (2009) Soil nitrous oxide emissions in long-term cover crops-based rotations under subtropical climate. *Soil & Tillage Research*, **106**, 36–44.

Gu J, Nicoullaud B, Rochette P, Grossel A, Hénault C, Cellier P, Richard G (2013) A regional experiment suggests that soil texture is a major control of N₂O emissions from tile-drained winter wheat fields during the fertilization period. *Soil Biology & Biochemistry*, **60**, 134–141.

Heincke M, Kaupenjohann M (1999) Effects of soil solution on the dynamic of N₂O emissions: a review. *Nutrient Cycling in Agroecosystems*, **55**, 133–157.

Huang Y, Zou J, Zheng X, Wang Y, Xu Y (2004) Nitrous oxide emissions as influenced by amendment of plant residues with different C: N ratios. *Soil Biology & Biochemistry*, **36**, 973–981.

IPCC (2001) Technical summary. In: *Climate Change 2001. The Scientific Basis. Contributions of Working Group I of the Third Assessment Report of the Intergovernmental Panel on Climate Change* (eds Houghton JT, Ding Y, Griggs DJ, Noguer M, van der Linden PJ, Dai X, Maskell K, Johnson CA), pp. 239–288. Cambridge University Press, Cambridge.

IPCC (2006) Guidelines for National Greenhouse Gas Inventories. Vol. 4. *Agriculture, Forestry and Other Land Use.* Prepared by the National Greenhouse Gas Inventories Program (eds Eggleston HS, Buendia L, Miwa K, Ngara T, Tanabe K), pp. 5.1–5.50. IGES, Hayama, Kanagawa.

Kachenchart B, Jones DL, Gajaseni N, Edwards-Jones G, Limsakul A (2012) Seasonal nitrous oxide emissions from different land uses and their controlling factors in a tropical riparian ecosystem. *Agriculture, Ecosystems & Environment*, **158**, 15–30.

Khalil K, Mary B, Renault P (2004) Nitrous oxide production by nitrification and denitrification in soil aggregates as affected by O₂ concentration. *Soil Biology & Biochemistry*, **36**, 687–699.

La Scala NJr, Bolonhezi D, Pereira GT (2006) Short-term soil CO₂ emission after conventional and reduced tillage of a no-till sugar cane area in southern Brazil. *Soil & Tillage Research*, **91**, 244–248.

Lara-Cabeza WAR, Souza MA (2008) Ammonia volatilization, leaching of nitrogen and corn yield in response to the application of mix of urea and ammonium sulphate or gypsum. *Brazilian Journal of Soil Science*, **32**, 2331–2342.

Leal MRLV, Galdos MV, Scarpare FV, Seabra JEA, Walter A, Oliveira COF (2013) Sugarcane straw availability, quality, recovery and energy use: a literature review. *Biomass and Bioenergy*, **53**, 11–19.

Lisboa CC, Butterbach-Bahl K, Mauder M, Kiese R (2011) Bioethanol production from sugarcane and emissions of greenhouse gases. Know and unknowns. *Global Change Biology Bioenergy*, **3**, 1–16.

Macedo IC, Seabra JEA, Silva JEAR (2008) Green house gases emissions in the production and use of ethanol from sugarcane in Brazil: The 2005/2006 averages and a prediction for 2020. *Biomass and Bioenergy*, **32**, 582–595.

Malhi SS, Lemke R (2007) Tillage, crop residue and N fertilizer effects on crop yield, nutrient uptake, soil quality and nitrous oxide gas emissions in a second 4-yr rotation cycle. *Soil & Tillage Research*, **96**, 269–283.

Morais RF, Boddey RM, Urquiaga S, Jantalia CP, Alves BJR (2013) Ammonia volatilization and nitrous oxide emissions during soil preparation and N fertilization of elephant grass (*Pennisetum purpureum* Schum.). *Soil Biology & Biochemistry*, **64**, 80–88.

Oliveira BG, Carvalho JLN, Cerri CEP, Cerri CC, Feigl BJ (2013) Soil greenhouse gas fluxes from vinasse application in Brazilian sugarcane areas. *Geoderma*, **200**, 77–84.

Page KL, Dalal RC, Menzies NW, Strong WM (2002) Nitrification in a Vertisol subsoil and its relationship to the accumulation of ammonium-nitrogen at depth. *Australian Journal of Soil Research*, **40**, 727–735.

Prado RM, Caione G, Campos CNS (2013) Filter cake and vinasse as fertilizers contributing to conservation agriculture. *Applied and Environmental Soil Science*, **1**, 1–9.

Ravishankara AR, John SD, Robert WP (2009) Nitrous oxide (N₂O): the dominant ozone-depleting substance emitted in 21st century. *Science*, **326**, 123–125.

Reis Junior FB, Reis VM, Urquiaga S, Döbereiner J (2000) Influence of nitrogen fertilisation on the population of diazotrophic bacteria *Herbaspirillum* spp. and *Acetobacter diazotrophicus* in sugar cane (*Saccharum* spp.). *Plant and Soil*, **219**, 153–159.

Resende AS, Xavier RP, Quesada DM, Urquiaga S, Alves BJR, Boddey RM (2003) Use of green manures in increasing inputs of biologically fixed nitrogen to sugarcane. *Biology and Fertility of Soils*, **37**, 215–220.

Rochette P (2008) No-till only increases N₂O emissions in poorly-aerated soils. *Soil & Tillage Research*, **101**, 97–100.

Rudorff BFT, Aguiar DA, Silva WF, Sugawara LM, Adami M, Moreira MA (2010) Studies on the rapid expansion of sugarcane for ethanol production in São Paulo State (Brazil) Using landsat data. *Remote Sensing*, **2**, 1057–1076.

Schils RLM, Groenigen JW, Velthof GL, Kuikman PJ (2008) Nitrous oxide emissions from multiple combined applications of fertilizer and cattle slurry to grassland. *Plant and Soil*, **310**, 89–101.

Silva APM, Bono JAM, Pereira FAR (2014) Fertigation with vinasse in sugarcane crop: effect on the soil and on productivity. *Revista Brasileira de Engenharia Agrícola e Ambiental*, **18**, 38–43.

Silva-Olaya AM, Cerri CEP, La Scala Jr N, Dias CTS, Cerri CC (2013) Carbon dioxide emissions under different soil tillage systems in mechanically harvested sugarcane. *Environmental Research Letters*, **8**, 1–9.

Smeets EMW, Bouwman LF, Stehfest E, van Vuuren DP, Posthuma A (2009) Contribution of N₂O to the greenhouse gas balance of first-generation biofuels. *Global Change Biology*, **15**, 1–23.

Soil Survey Staff (1999) *Soil Taxonomy: A Basic System of Soil Classification for Making and Interpreting Soil Surveys*, 2nd edn. Natural Resources Conservation Service. U.S. Department of Agriculture Handbook, Washington, DC, USA.

Stehfest E, Bouwman L (2006) N₂O and NO emission from agricultural fields and soils under natural vegetation, summarizing available measurement data and modeling of global annual emissions. *Nutrient Cycling in Agroecosystems*, **74**, 207–228.

UNICA (2014) Union of the Sugarcane industries. Available online: http://www.unicadata.com.br/. (Accessed on 14 May 2014).

Veiga CFM, Vieira JR, Morgado IF (2006) *Diagnóstico da Cadeia Produtiva da Cana-de-Açúcar do Estado do Rio de Janeiro: Relatório de Pesquisa.* FAERJ:SEBRAE, Rio de Janeiro.

Wang WJ, Moody PW, Reeves SH, Salter B, Dalal RC (2008) Nitrous oxide emissions from sugarcane soils: effects of urea forms and application rate. *Proceedings of the Australian Society of Sugar Cane Technologists*, **30**, 87–94.

Weier KL (1999) N₂O and CH₄ consumption in a sugarcane soil after variation in nitrogen and water application. *Soil Biology & Biochemistry*, **31**, 1931–1941.

Wickramasinghe KN, Rodgers GA, Jenkinson DS (1985) Transformations of nitrogen fertilizers in soil. *Soil Biology & Biochemistry*, **17**, 625–630.

World Meteorological Organization. 2014WMO Greenhouse gas bulletin: the state of greenhouse gases in the atmosphere based on observations through 2013. Available online: http://www.wmo.int/pages/prog/arep/gaw/gaw_home_en.html. (Accessed on 21 Oct 2014).

10

Nitrous oxide emissions during establishment of eight alternative cellulosic bioenergy cropping systems in the North Central United States

LAWRENCE G. OATES[1], DAVID S. DUNCAN[1], ILYA GELFAND[2,3], NEVILLE MILLAR[2,3], G. PHILIP ROBERTSON[2,3] and RANDALL D. JACKSON[1]

[1]DOE-Great Lakes Bioenergy Research Center & Department of Agronomy, University of Wisconsin–Madison, Madison, WI 53706, USA, [2]DOE-Great Lakes Bioenergy Research Center & W.K. Kellogg Biological Station, Michigan State University, Hickory Corners, MI 49060, USA, [3]Department of Plant, Soil and Microbial Sciences, Michigan State University, East Lansing, MI 48824, USA

Abstract

Greenhouse gas (GHG) emissions from soils are a key sustainability metric of cropping systems. During crop establishment, disruptive land-use change is known to be a critical, but under reported period, for determining GHG emissions. We measured soil N_2O emissions and potential environmental drivers of these fluxes from a three-year establishment-phase bioenergy cropping systems experiment replicated in southcentral Wisconsin (ARL) and southwestern Michigan (KBS). Cropping systems treatments were annual monocultures (continuous corn, corn–soybean–canola rotation), perennial monocultures (switchgrass, miscanthus, and poplar), and perennial polycultures (native grass mixture, early successional community, and restored prairie) all grown using best management practices specific to the system. Cumulative three-year N_2O emissions from annuals were 142% higher than from perennials, with fertilized perennials 190% higher than unfertilized perennials. Emissions ranged from 3.1 to 19.1 kg N_2O-N ha^{-1} yr^{-1} for the annuals with continuous corn > corn–soybean–canola rotation and 1.1 to 6.3 kg N_2O-N ha^{-1} yr^{-1} for perennials. Nitrous oxide peak fluxes typically were associated with precipitation events that closely followed fertilization. Bayesian modeling of N_2O fluxes based on measured environmental factors explained 33% of variability across all systems. Models trained on single systems performed well in most monocultures (e.g., $R^2 = 0.52$ for poplar) but notably worse in polycultures (e.g., $R^2 = 0.17$ for early successional, $R^2 = 0.06$ for restored prairie), indicating that simulation models that include N_2O emissions should be parameterized specific to particular plant communities. Our results indicate that perennial bioenergy crops in their establishment phase emit less N_2O than annual crops, especially when not fertilized. These findings should be considered further alongside yield and other metrics contributing to important ecosystem services.

Keywords: Bayesian model averaging, cellulosic biofuels, corn, greenhouse gas, miscanthus, poplar, restored prairie, switchgrass

Introduction

Nitrous oxide (N_2O) is a potent greenhouse gas (GHG) and the main contribution to radiative forcing in the atmosphere by agriculture (Robertson *et al.*, 2000). For first generation biofuels derived from edible oils and starches, N_2O emissions from feedstock production comprise a substantial proportion of their total carbon footprint (Gelfand *et al.*, 2011). The development of second-generation biofuels derived from cellulosic materials offers the potential to substantially reduce N_2O

emissions associated with feedstock production (Sanderson & Adler, 2008; Smith *et al.*, 2013). As bioenergy cropping system viability is considered, the greenhouse gas emissions of these systems will be a key component of sustainability evaluation (Reay *et al.*, 2012).

Nitrous oxide emitted from soils is primarily the product of microbially driven nitrification and denitrification. These processes are influenced by a broad range of environmental factors including temperature, oxygen availability, rates of microbial activity, and the availability of nitrogen substrates (Robertson & Groffman, 2015). The effect of these factors can depend on soil-specific properties (Henault *et al.*, 2005) including the composition of the microbial community (Cavigelli & Robertson, 2001),

Correspondence: Lawrence G. Oates
e-mail: goates@glbrc.wisc.edu

while their relative importance may differ among cropping systems (Dechow & Freibauer, 2011). Numerous process-based models have been developed in an attempt to account for these contextual effects within a generalized framework, but they are rarely calibrated to the actively managed and harvested perennial cropping systems that have emerged as leading candidates for second-generation biofuel feedstocks (Chen *et al.*, 2008). More broadly, the dynamics governing N_2O emissions in perennial cropping systems managed for biofuel feedstock production are poorly represented in the literature (but see Nikièma *et al.*, 2011, 2012; Palmer *et al.*, 2014).

While systems based on polycultures and perennial species are anticipated to emit less N_2O than conventional agricultural cropping systems, this reduction is likely to be contingent on previous land-use and conversion methodology, phase and length of establishment, soil type, and management of inputs and production processes. Land conversion to cropping systems either through extensification or intensification of open land systems such as pasture is known to significantly increase soil organic carbon (SOC) loss (Adler *et al.*, 2007; Zenone *et al.*, 2011; Sanford *et al.*, 2012). Vegetation removal and cultivation may also affect the N cycle, especially during conversion and establishment (Bouwman *et al.*, 2010; Gelfand *et al.*, 2011; Nikièma *et al.*, 2012; Ruan & Robertson, 2013) and lead to significant nitrogen loss through leaching and gaseous emissions (Robertson *et al.*, 2012; Smith *et al.*, 2013). We must improve our understanding of N_2O emissions of likely biofuel feedstock systems to help ensure that expansion of bioenergy production generates expected societal benefits (Robertson *et al.*, 2008; Dale *et al.*, 2011, 2014).

We compared the establishment-phase N_2O emissions of annual monocultures of continuous corn and corn–soybean–canola rotations; perennial monocultures of switchgrass, miscanthus, and hybrid poplar; and perennial polycultures of early successional species, native grasses, and native prairie species. Our results cover the 2- to 4-year period following planting over which many perennial crops attain 'full capacity' biomass production (McLaughlin & Adams Kszos, 2005; Anderson-Teixeira *et al.*, 2013). Our aims were to (i) provide a direct comparison of the aggregate N_2O emissions from a broad range of feedstock production systems; (ii) characterize the effects of location and year on N_2O emissions from these cropping systems; and (iii) evaluate cropping system impacts on relationships among N_2O fluxes and environmental factors.

Materials and methods

One study site was located in southwestern Michigan at Michigan State University's W.K. Kellogg Biological Station (KBS) in Hickory Corners, MI (42°23′47″ N, 85°22′26″ W and 288 m asl), and another site was located in southcentral Wisconsin, USA, at the University of Wisconsin's Arlington Agricultural Research Station (ARL) in Arlington, WI (43°17′45″ N, 89°22′48″ W and 315 m asl). Mean annual air temperature at KBS is 9.9 °C, and annual precipitation is 1027 mm (MSCO, 2013). Soils are well-drained Kalamazoo loam (fine-loamy, mixed, semi-active, mesic Typic Hapludalfs with soil C < 15 g kg^{-1}, N ≤ 0.13 g kg^{-1}) developed over glacial outwash (Crum & Collins, 1995). At ARL, mean annual temperature and precipitation are 6.9 °C and 869 mm, respectively (NWS, 2013). Soils at the site are classified as Plano silt loam (fine-silty, mixed, superactive, mesic Typic Argiudolls with soil C > 20 g kg^{-1}, N ≥ 0.19 g kg^{-1}) developed over glacial till (Jokela *et al.*, 2011). Experiments at both sites were established in spring 2008 in a randomized complete block design. Ten treatments (eight cropping systems including each phase of the three-phase corn–soybean–canola rotation) were represented in five blocks of 30 × 40 m plots at each location, for a total of 100 plots. The systems under study were (i) continuous no-till corn (*Zea mays* L.); (ii) corn–soybean (*Glycine max* [L.] Merr.)–canola (*Brassica napus* L.) rotation with all 3 phases represented; (iii) monoculture switchgrass (*Panicum virgatum* L.); (iv) monoculture miscanthus (*Miscanthus x giganteus*); (v) hybrid poplar (*Populus nigra x P. maximowiczii* 'NM6') on a 6-year coppicing rotation; (vi) a mixture of five native grass species; (vii) a mixture of 18 native prairie species; and (viii) an early successional community defined by the pre-existing seed bank and novel recruitment with no management other than fertilizer application and harvest. Of note, miscanthus at ARL suffered >95% mortality over the 2008–2009 winter and was subsequently replanted in May 2010. With the exception of miscanthus at KBS and the early successional community at both sites, perennial systems did not receive N fertilizer in 2009, while poplar at both sites received N fertilizer in 2010 only (full crop and management details are given in Table S1).

Estimating nitrous oxide emissions

When soils were consistently >0 °C, N_2O fluxes were measured biweekly with additional sampling to characterize episodic events (i.e., fertilizer application and precipitation events) using vented static chambers. All measurements were made between 1000 and 1600 h local time. Cylindrical chamber bases of 28.5 cm diameter were inserted ~5 cm below the soil surface. With the chamber lid installed, the chamber had an effective headspace volume of ~10 l (~17 cm height). Lids had a septum for gas extraction and a 2-mm diameter vent and vent tube to allow for chamber pressure equilibration. Headspace gas from within the chambers was extracted immediately following lid placement with a 30-ml nylon syringe and a 23-gauge needle. Three subsequent extractions were made at 20-min intervals over a 60-min period. Glass 5.9-ml Exetainer vials (Labco Limited, Buckinghamshire, UK) were flushed with 20 ml of extracted sample and then overpressurized with 10 ml of sample to avoid contamination and facilitate analysis. At each sampling event, field standards (1 ppm N_2O, 1 ppm CH_4, and 400 ppm CO_2) and ambient air were loaded into vials to assess potential sample loss prior to analysis. Sample CO_2, N_2O, and CH_4 concentrations were determined by gas chromatography using an infrared gas analyzer (IRGA, LiCor 820, Lincoln, NE, USA) for CO_2, an electron capture detector (micro-ECD, Agilent 7890A GC System, Santa Clara, CA, USA) for N_2O, and a flame ionization detector (FID, Agilent 7890A) for CH_4.

Visual inspection of CO_2 accumulation curves identified samples with lost pressure or other measurement problems, for which fluxes were discarded (~2% of total measured fluxes). Remaining fluxes were analyzed with the HMR package (v0.3.1, Pedersen, 2011) in the R statistical environment (v3.0.3, R Core Team, 2014) to fit gas concentrations against time with a nonlinear model (Hutchinson & Mosier, 1981), a linear regression, and a null flux based on root mean squared error minimization. Of the 4139 flux estimates, a nonlinear model was used for the 691 (16.7%) fluxes where the 95% confidence interval for the nonlinear estimate excluded the corresponding linear estimate. In all other cases, including the case of a null flux, the linear flux estimate was used. This estimate was used to calculate the aggregate flux for the day it was sampled (daily flux) by assuming the estimate was the average flux during that day. Annual fluxes were calculated by integrating the linear interpolation of daily fluxes over one calendar year (Smith & Dobbie, 2001). Cumulative three-year emissions were then calculated for each experimental plot by summing the aggregated annual emissions from the three study years within a plot.

Assessing soil environmental variables

Concurrent with trace gas sampling, soil volumetric water content (VWC) (m^3 m^{-3}) and soil temperature (°C) were measured within 1 m of the chamber with a time domain reflectometer using 20-cm rods (FieldScout 300, Spectrum Technologies, Inc., Plainfield, IL, USA) and a 15-cm soil temperature probe (Checktemp 1C, Hanna Instruments, Smithfield, RI, USA), respectively. For analysis, VWC was converted to proportion of water-filled pore space (WFPS) using the equation:

$$\text{WFPS} = \frac{\theta}{TP} \quad (1)$$

where θ is equal to VWC, and TP is total porosity (m^3 m^{-3}) calculated using the equation:

$$TP = \left(1 - \frac{Bd}{Pd}\right) \times 100\% \quad (2)$$

where Bd is the bulk density at each site (g cm^{-3}) and Pd is particle density, assumed to be 2.65 g cm^{-3} for both sites. Castellano et al. (2010) found WFPS was of limited utility as a predictor of N_2O flux across soils with differing textures. Accordingly, we scaled and centered WFPS at each site to create a new variable (WFPS$_C$) for use in analyses.

In 2009, inorganic soil nitrogen was estimated using ion resin strips (General Electric, Watertown, MA, USA) placed in the field for 1-month periods from March to November. Matched pairs of anion and cation strips were placed at each of three persistent sampling stations per plot. Each month, all six strips in each plot were collected from the field, cleaned with deionized water to remove visible soil and then extracted in

2 M KCl. Extracted strips were regenerated with 0.5 M HCl and 0.5 M NaHCO$_3$ prior to next use. For the years 2010 and 2011, soil cores to 15 cm depth were taken concurrently with N_2O measurements to estimate soil inorganic nitrogen (N) pools. A 10 g wet-weight subsample was weighed out for immediate inorganic N extraction in 2 M KCl following Robertson et al. (1999). Potassium chloride extracts were stored in 20-mL polyethylene scintillation vials frozen at −20 °C prior to analysis. Colorimetric determination of extracts for ammonium (NH_4^+) (USEPA method-Pub# 27200110) and nitrate (NO_3^-) (USEPA method-Pub# 27190110) was performed on a Flow Solution 3100 segmented flow injection analyzer (OI Analytical, College Station, TX, USA).

Data analysis

Emissions were analyzed with linear mixed-effect models using the NLME package (v3.1, Pinheiro et al., 2013) in the R statistical environment (v3.1.1, R Core Team, 2014). Nitrous oxide emissions were summed over three calendar years; sums were log-transformed prior to analysis to approximate normally distributed data. Models were constructed to analyze response variables as a function of the fixed effects of treatment (cropping system) and site, accounting for the random effect of block nested within site (site/block). Models were improved by allowing for distinct variances among cropping systems, sites, or both, and evaluated with likelihood ratio tests. With the variance structure in place, significant fixed effects and interactions were determined by sequentially collapsing treatment levels and comparing subsequent models with likelihood ratio tests. This process continued iteratively until none of the remaining groups could be collapsed. Annual emissions considered year as an additional potential component of the variance structure and used crop rather than cropping system as a factor and each combination of year and site was analyzed separately. Variance structure optimization and assessment of treatment differences were conducted as described above.

Bayesian model averaging (BMA) was used to evaluate relationships between measured environmental variables and daily N_2O fluxes for cropping system level (Hoeting et al., 1999; Maroja et al., 2009). We conducted the model averaging process using the bic.glm function of the R package BMA (v 3.16.2.2, Raftery et al., 2013). Because N_2O fluxes are seldom normally distributed, and negative fluxes are biologically relevant (Schlesinger, 2013), we used a hyperbolic arcsine transformation (Burbidge et al., 2013) on daily flux data prior to analysis. The maximal model was defined as soil temperature, WFPS$_C$, NO_3^-–NH_4^+, year, and site, as well as all second-order interactions among these terms. Because we used a different method for estimating inorganic N in 2009, we only used 2010 and 2011 data. Of the 2657 data points in these two years, 2176 (82%) included all environmental measures. Annual emissions recalculated from this subset were highly correlated to those obtained from the full dataset ($R^2 = 0.90$), although the range of values was slightly greater. We trained the model on the full dataset and also subsets of the data by cropping system to generate system-specific models. We then used environmental data from the full dataset to evaluate the capacity of models trained from a given system to predict emissions from other systems or from the full dataset.

Statistical analyses are discussed in greater detail in Appendix S1.

Results

Cumulative nitrous oxide emissions over the 3-year establishment phase

Cropping systems at the Wisconsin research station (ARL) emitted 23% more N_2O than their Michigan (KBS) counterparts. The only systems for which this pattern did not hold were continuous corn and miscanthus, where emissions were not significantly different between the sites, and the native grasses, where KBS had higher cumulative emissions (Fig. 1). Note that while we did not have miscanthus data from ARL in 2009 because of winter kill, cumulative 2010–2011 emissions were similar at both sites. Across both sites, N_2O emissions relative to aboveground yield of continuous corn were slightly greater than the rotation (0.88), switchgrass (0.67), and early successional community (0.79). The emissions relative to yield in native grasses were just under half that of continuous corn (0.43), while the restored prairie (0.18), miscanthus (0.15), and poplar (0.14) were the lowest.

Annual nitrous oxide emissions by year and site

We analyzed annual N_2O emissions separately by *year* and *site* (Fig. 2) due to a significant *site* × *year* × *cropping system* interaction ($P < 0.001$). For 2009 treatment comparisons, model selection indicated that continuous corn and miscanthus at KBS were not significantly different and had the highest emissions, followed by the rotational phase of corn and poplar; all other systems had emissions that were not significantly different from each other (Fig. 2a). The treatments responded differently at ARL where the corn phase of the rotation had more than twice the emissions of continuous corn. After continuous corn and the corn phase of the rotation, there were no differences among other systems with the exception of restored prairie, which had the lowest emissions (Fig. 2b).

The general patterns of emissions from treatments in 2010 were similar at both sites. Continuous corn and the corn phase of the rotation had the highest emissions at both KBS and ARL (Fig. 2c,d). Most systems at KBS had relatively low emissions (Fig. 2c), similar to emissions from native grass mix and restored prairie at ARL (Fig. 2d). With the exception of continuous corn, which was 21% higher at KBS, average emissions at ARL were 12% higher from annual systems, and 68% higher from perennial systems than the respective treatments at KBS.

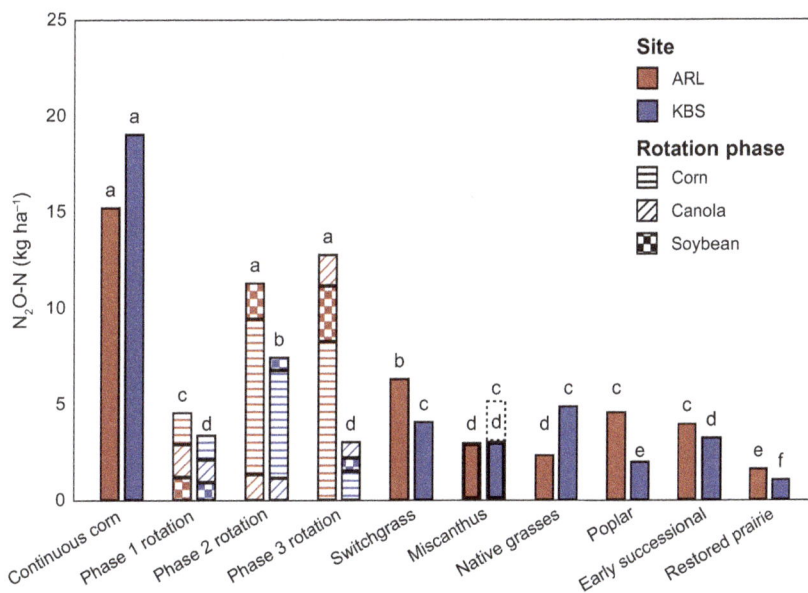

Fig. 1 Cumulative 2009–2011 N_2O fluxes from 8 bioenergy cropping systems grown at Arlington Agricultural Research Station, WI (ARL) and Kellogg Biological Station, MI (KBS). Values presented are geometric means with $n = 5$ at ARL and $n = 4$ at KBS. Phases of a corn–soybean–canola rotation are separated by the contribution of each specific rotation phase in ascending chronological order; analysis was conducted on the summed fluxes. Bars sharing a letter can be grouped during stepwise factor level collapse without significantly reducing the explanatory power of the model ($P > 0.05$). Miscanthus bars and letters correspond to 2010–2011 fluxes; for KBS, the dotted line and corresponding letter show the contribution of 2009 flux.

Emissions in 2011 differed from the patterns observed in previous years. At KBS, the highest emissions were from the native grasses, followed by continuous corn, which grouped with the switchgrass, miscanthus, and early successional community systems. Poplar, restored prairie, and the soybean phase of the rotation responded with the lowest emissions (Fig. 2e). At ARL, the switchgrass monoculture deviated from prior patterns and together with continuous corn averaged 34% higher emissions than the average across all phases of the rotation, 54% higher than miscanthus, poplar, and early successional community, and 82% higher than the average of the native grass mix and restored prairie systems (Fig. 2f).

Environmental predictors of daily nitrous oxide flux

ARL and KBS differed substantially in their soil moisture; water-filled pore space (WFPS) values at ARL were almost universally higher than at KBS (Table 1). Nevertheless, seasonal patterns within sites were largely similar, with reduced summer WFPS in 2009 and 2011, but sustained WFPS during the wet summer in 2010 (Fig. S1). Median NO_3^- and NH_4^+, as measured by resin

strips, were similar at both sites although the range of values for both species observed at ARL was greater. Extractable values of NH_4^+ were very similar at KBS and ARL for both 2010 and 2011, but the range of values for extractable NO_3^- was again greater at ARL (Table 1).

The timings of precipitation events, fertilizer applications, and N_2O flux measurements varied among sites and years, with sharp increases in daily N_2O fluxes tending to occur when these events synchronized (Fig. 3). In 2011 at KBS, for instance, fertilization of most perennial crops was followed in rapid succession by a 10-mm precipitation event and a very large N_2O flux (Fig. 3f); aggregate annual emissions from perennial systems were very high at KBS that year (Fig. 2e). During the late spring and early summer, systems receiving no N fertilization tended to show limited changes in their emissions. By contrast, fluxes from corn systems varied over several orders of magnitude. N fertilization in corn (blue arrows, Fig. 3) preceded a sharp increase in N_2O emission; the sole exception occurred in 2011 at ARL, when there were no precipitation events between fertilization and the next flux measurement (Fig. 3e). The canola rotational phase and most perennial systems

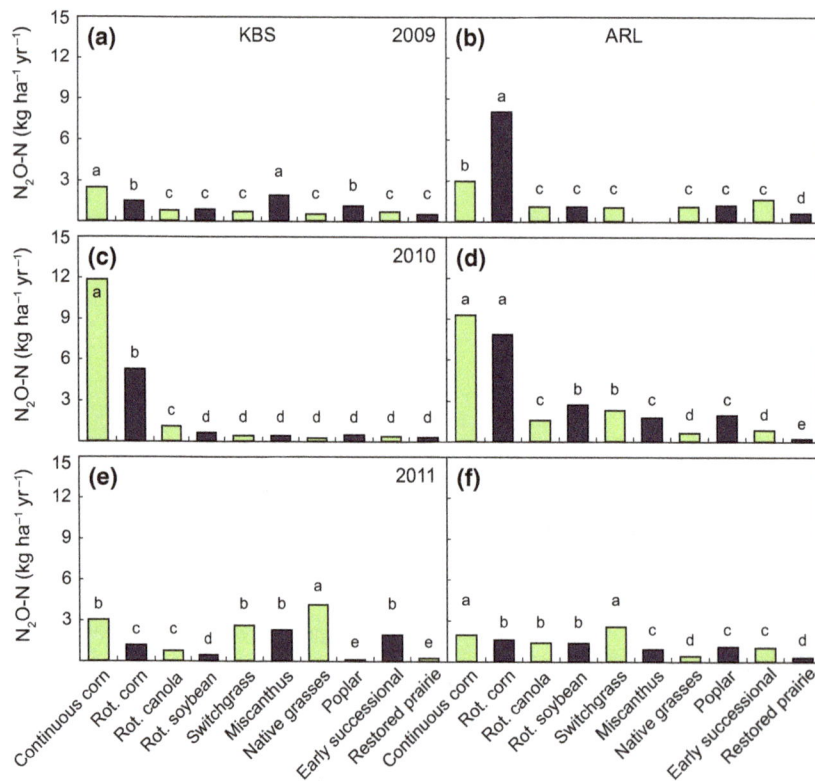

Fig. 2 Geometric mean of annual nitrous oxide emissions from the Biofuels Cropping System Experiment for the periods of 2009, 2010, and 2011. Left panels (a, c, and e) show results from KBS, $n = 4$, and right panels (b, d and f) show results from ARL, $n = 5$. See Fig. 1 legend for further information.

Table 1 Environmental factors observed at the GLBRC Bioenergy Cropping System Experiment colocated in Michigan (KBS) and Wisconsin (ARL)

Year	Site	Soil temp (°C)	WFPS (%)	NH_4^+ Resin strip ($\mu g\ N\ cm^{-2}\ day^{-1}$)	NH_4^+ Soil pool ($\mu g\ N\ g^{-1}$ soil)	NO_3^- Resin strip ($\mu g\ N\ cm^{-2}\ day^{-1}$)	NO_3^- Soil pool ($\mu g\ N\ g^{-1}$ soil)
2009	KBS	12.5 (3.0–22.0)	20 (7–29)	0.010 (0.002–0.071)		0.38 (0.03–3.68)	
	ARL	13.7 (1.1–23.3)	71 (40–89)	0.008 (0.001–0.149)		0.39 (0.04–5.59)	
2010	KBS	14.0 (5.4–21.0)	20 (13–27)		2.9 (0.7–12.6)		2.6 (1.5–5.3)
	ARL	15.5 (2.9–22.8)	72 (53–93)		3.7 (1.9–17.4)		5.3 (0.9–56.3)
2011	KBS	15.0 (4.0–23.5)	22 (9–31)		1.7 (0.7–7.7)		2.3 (1.3–4.4)
	ARL	15.0 (−0.4–25.0)	67 (37–83)		1.7 (0.9–7.7)		2.7 (0.4–25.2)

Median values are presented, with 5th and 95th percentile values in parentheses. Sites were Arlington Agricultural Research Station, WI (ARL) and Kellogg Biological Research Station, MI (KBS). WFPS is water-filled pore space.

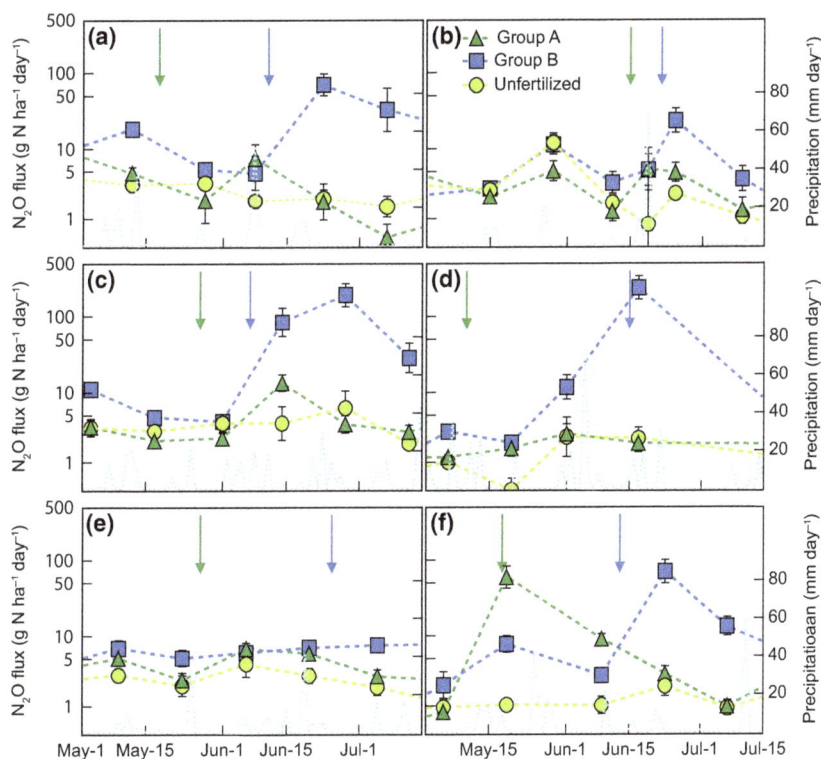

Fig. 3 Average daily precipitation and nitrous oxide flux during late spring/early summer. Data presented separately by site and year: (a) 2009 ARL, (b) 2009 KBS, (c) 2010 ARL, (d) 2010 KBS, (e) 2011 ARL, and (f) 2011 KBS. All treatments within a group were fertilized at the same time, denoted by arrows. Group A (gray arrow) consisted of rotational canola, early successional community, switchgrass (post-2009), native grasses (post-2009), poplar (2010 only), and miscanthus (2010 at KBS, 2011 both sites). Group B (black arrow) consisted of continuous and rotational corn and miscanthus (2009 at KBS). Unfertilized treatments included rotational soybeans, restored prairie, poplar (except 2010), switchgrass (2009 only), native grasses (2009 only), and miscanthus (2010 at ARL). Data are presented with inverse hyperbolic sine scaling, with error bars ± 1 SE.

were fertilized earlier in the year and at a lower rate than corn (Table S1; gray arrows, Fig. 3). The range of emissions from these systems tended to be lower than for corn, as were the emission increases following fertilization events. Our data are too limited to

broadly infer a generalized relationship between precipitation amounts and subsequent N_2O fluxes, but we note that in all cases where N_2O flux increased sharply after fertilization there was at least one precipitation event of 10 mm or more between fertilization and flux measurement, whereas among the cases where fertilization was not followed by a large flux increase, there was only one instance where a precipitation event of more than 10 mm was observed prior to flux measurement (Fig. 3a, Group A).

Bayesian model averaging of environmental predictors

Models trained on specific systems varied greatly in their capacity to explain variation in both their training data and the full data (Table 2). Specifically, poplar, miscanthus, and systems containing corn were relatively well modeled while the polycultures (native grasses, early successional community, and restored prairie) were not. There were no environmental factors that substantially contributed to model fits across all systems (Table S2). There were also substantial differences in how well models trained on a given system predicted fluxes from other systems (Table

S3). This relationship was not reciprocal. For example, the model based on data from native grass system predicted fluxes from poplar system better than the model based on poplar data predicted fluxes from native grasses.

We tested how effectively models trained on specific systems captured the temporal dynamics of fluxes from their own and other systems (Fig. 4). Flux dynamics at ARL were better modeled than those from KBS, where the models failed to predict major emission events (Fig. 4d–f). Despite the high correlation between predictions from the corn and native grass-based models ($R^2 = 0.58$), actual values from the native grass model were systematically lower than those from the corn model. The switchgrass-based model was inconsistent in its relationship to the other two models, sometimes tracking the corn model (Fig. 4b,d) and at other times the native grass model (Fig. 4f). Overall, each model's performance was poor on systems other than the one on which it had been trained.

Table 2 Bayesian model averaged posterior probabilities of inclusion for environmental factors used to predict N_2O fluxes

| Factor | Training dataset | | | | | | | |
	Full data	Corn	Miscanthus	Native grasses	Early successional community	Poplar	Restored Prairie	Switchgrass
Site	1.00	1.00	0.53	0.60	0.25	1.00	0.03	0.17
Year	0.61	0.23	0.01	0.05	0.03	0.02	0.19	0.12
NH_4^+	0.81	0.06	0.13	0.01	0.04	0.09	0.05	0.90
NO_3^-	0.95	0.05	0.12	0.54	0.53	0.04	0.12	0.04
Soil temperature (ST)	0.00	0.27	0.23	0.24	0.04	0.04	0.01	0.14
$WFPS_C$	0.03	0.02	0.04	0.12	0.07	0.02	0.03	0.07
Site × Year	0.00	0.02	0.05	0.82	0.03	0.01	0.04	0.02
Site × NH_4^+	1.00	0.29	0.05	0.01	0.02	0.04	0.18	1.00
Site × NO_3^-	1.00	0.02	0.04	0.32	0.78	0.01	0.06	0.96
Site × ST	1.00	1.00	0.55	0.61	1.00	0.01	0.03	0.20
Site × $WFPS_C$	0.00	0.57	0.02	0.01	0.38	0.21	0.07	0.05
Year × NH_4^+	0.68	0.01	0.01	0.08	0.03	0.05	0.02	0.02
Year × NO_3^-	1.00	0.03	0.02	0.01	0.75	0.10	0.04	0.89
Year × ST	0.00	0.19	0.02	0.17	0.03	0.02	0.02	0.07
Year × $WFPS_C$	1.00	0.01	0.50	0.05	0.09	0.69	0.02	0.02
NH_4^+ × NO_3^-	0.00	0.09	0.08	0.02	0.16	0.02	0.04	0.02
NH_4^+ × ST	1.00	0.03	0.07	0.01	0.12	0.23	0.03	0.90
NH_4^+ × $WFPS_C$	0.06	1.00	0.01	0.01	0.02	0.03	0.02	0.02
NO_3^- × ST	0.08	0.76	0.86	0.42	0.43	0.87	0.03	0.02
NO_3^- × $WFPS_C$	1.00	0.01	0.75	0.06	0.04	0.01	0.04	0.02
ST × $WFPS_C$	1.00	0.03	0.99	0.95	0.09	1.00	0.03	0.96
Training R^2	0.33	0.46	0.39	0.21	0.17	0.52	0.06	0.27
Full data R^2	0.33	0.22	0.27	0.24	0.07	0.22	0.02	0.20

Models were trained on N_2O flux data obtained from individual cropping systems. NH_4^+ and NO_3^- concentrations were log-transformed for analysis. The two sites were Arlington Agricultural Research Station (ARL) and Kellogg Biological Research Station (KBS). $WFPS_C$ is water-filled pore space scaled and centered separately for each site. See Table S2 for factor coefficients and probabilities for all cropping systems.

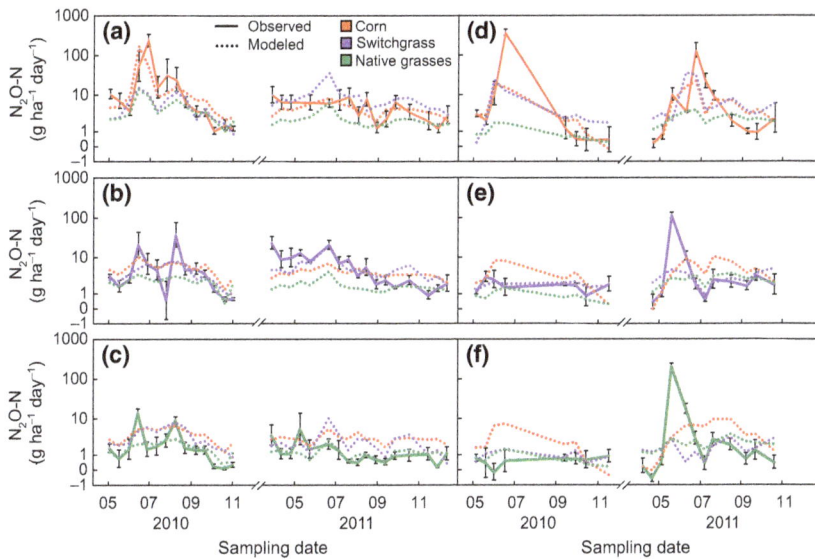

Fig. 4 Observed and modeled daily N_2O fluxes at ARL and KBS. Solid lines indicate mean observed N_2O fluxes of: (a) continuous corn ARL, (b) switchgrass ARL, (c) native grasses ARL, (d) continuous corn KBS, (e) switchgrass KBS, and (f) native grasses KBS. Dotted lines indicate mean daily predictions from models trained on data collected from the three cropping systems. The vertical axis has inverse hyperbolic sine scaling, which asymptotically approaches logarithmic scaling. Error bars indicate ±1 SE. A 90-day period during which no fluxes were measured was removed from the horizontal axis to conserve space.

Discussion

Cropping systems that included corn were consistently higher in N_2O emissions for any given site year, notwithstanding a high degree of interannual variability. With few exceptions, N_2O emissions from fertilized perennial systems were much lower than annual systems, while those from unfertilized restored prairie were lowest overall.

Previous work has demonstrated that rate of N fertilizer application has a strong influence on N_2O emissions from both annual and perennial systems (Dobbie et al., 1999; Millar et al., 2010). The percentage of applied fertilizer emitted as N_2O varies with site conditions and management, but, in general, N_2O emissions are highest where inorganic N is readily available. This is reflected in emissions from the continuous corn system which received ~ $3x$ the rate of N (160 kg ha^{-1}) as the fertilized perennial systems (56 kg ha^{-1}). Other factors such as crop rotation (i.e., crop diversity) can influence N_2O emission rates. Crop rotation is often associated with reduced N demand which results in reduced N inputs and N_2O emissions (Drury et al., 2008; Osterholz et al., 2014). The effect of crop rotation in our study was less clear with cumulative emissions being highly dependent on climatic conditions during the corn phase, indicating a strong interaction between crop phase and year.

Cumulative N_2O emissions can be driven by short-duration, high-intensity flux events (Molodovskaya et al., 2012). These events typically occur when coincidence of fertilization and precipitation results in limited soil oxygen and readily available reactive N (Dobbie & Smith, 2003; Castellano et al., 2010; Matthews et al., 2010). In general, we observed this pattern, where nearly concurrent fertilization and precipitation events resulted in substantial emissions spikes; of all the fertilized systems, this effect was most clearly observed in corn while emissions for crops grown in the absence of fertilizer were almost invariant throughout this period at ARL and were only slightly variable at KBS.

The difference in cumulative N_2O emissions between ARL and KBS is likely driven by soil properties. Soils at the two sites differed in both their order and texture, with ARL soils consisting of fine textured mollisols and KBS a coarser-textured alfisol. Of the two soil orders, mollisols are typically associated with higher carbon and nutrient contents, consistent with the higher carbon and inorganic N values recorded at ARL. Given that both sites received comparable precipitation, greater WFPS at ARL appears to stem from greater water-holding capacity emerging from high soil organic matter and finer soil texture. We thus attribute the consistently higher emissions at ARL to soil moisture-driven increases in anaerobic microsite abundance and longevity (Bouwman, 1996) coupled with greater N and C availability. Our findings are consistent with studies in both agricultural and wildland ecosystems that have linked finer soil texture and greater soil carbon availability to increased N_2O emissions (Bouwman et al., 2002; Stehfest & Bouwman, 2006). While the restored prairie system was not fertilized at either site, greater soil C at ARL suggests potential for rapidly mineralizing N, and this coupled with higher WFPS, may have facilitated the slightly higher fluxes observed at ARL. In contrast, between-site differences in the poplar system were at least partially attributable to an infestation of the ARL plots with the fungal leaf pathogen marssonina (*Marssonina populi* (Lib.) Magnus). The infestation peaked in mid-August 2010 leading to complete defoliation by 15 September; this likely reduced plant N uptake in the only year that N fertilizer was applied to poplar, leaving more N available for microbial conversion and loss as N_2O.

Cropping systems based on perennial species require multiple years to become fully established (Parrish & Fike, 2005). This development is most evident in the delay in attaining maximum yields, but N-cycling processes may also change during this period (Smith et al., 2013; Lesur et al., 2014). During the period of this study, N_2O emissions per unit aboveground yield were much lower in the perennial systems. It is likely this ratio will improve as the systems come into full production phase and as farmers become more efficient, both in harvest timing and mechanical efficiency, at harvesting perennial biomass. Perennial systems also produce greater biomass belowground which over time will likely improve soil organic matter and site fertility. While our study was not structured to explicitly explore the effects of the establishment period, it is still a potentially relevant contextual element for interpreting results from the perennial cropping systems. Our results were largely comparable to studies of perennial systems similarly conducted over the establishment phase (Hernandez-Ramirez et al., 2009; Smith et al., 2013). However, given the high influence of interannual variability on our results, it is clear that long-term studies will be required to bound this variability and determine whether establishment-phase N_2O emissions are representative of established perennial cropping systems.

A consideration for our results is that our study lacked measurements during winter (December–February), which could have resulted in underestimation of N_2O emissions. Substantial N loss can occur during this fallow period, especially in conventionally managed annual row crops when bare soil is subjected to freeze–thaw events (Johnson et al., 2010). Soil N is susceptible to denitrification during this period, especially for systems where manure has been applied after the primary crop has been harvested (Parkin et al., 2006), or where vegetative cover is not present during winter (McSwiney et al., 2010). However, only our annual cropping systems had significant bare soil during winter, and as

discussed above, N_2O emissions in these systems were dominated by relatively brief spikes following fertilizer events. Nitrogen was not applied in the fall to any of our cropping systems so we expect that winter measurements would have negligible influence on the magnitude and comparison of N_2O emissions.

The Bayesian averaged models we analyzed in detail suggested alternative cropping systems would produce different N_2O fluxes under a given set of environmental conditions. Our statistical models based on emissions measured from our various cropping systems had very distinct parameterizations, even when the systems were as similar as continuous and rotational corn. Previous studies have similarly found that key environmental predictors of N_2O emissions vary among systems (Dechow & Freibauer, 2011; Imer et al., 2013). The BMA based on switchgrass, for instance, frequently predicted substantially higher emissions than the model based on native grasses, implying that at a given soil moisture, soil temperature, and inorganic N concentration, switchgrass and native grasses would have different N_2O emissions. Cropping systems may differ not only in their effect on environmental parameters (e.g., through crop species differences in N uptake and water use), but also in their response to these parameters. The role of plant community composition and diversity in determining trace gas fluxes from soil has received little attention (but see Hoeft et al., 2012).

Given our results, there may be a significant gap in our ability to account for how plant community composition influences the response of N_2O fluxes to environmental drivers. Single-species monocultures have typically been used to model the broader category of herbaceous biomass crops (Surendran Nair et al., 2012), with switchgrass used as a model for exploring the properties and environmental responses of bioenergy crops (Lewandowski et al., 2003; Tulbure et al., 2012). The cropping system specificity we observed in the response of N_2O fluxes to environmental parameters suggests it may be risky to rely on model systems to predict the behaviors of perennial and polycultural biomass cropping systems, particularly with the potential for high variability during the establishment phase.

In summary, across years with highly variable climate, N_2O emissions were consistently higher from annual than perennial cropping systems. Under particular conditions, namely rainfall following fertilizer, emissions from corn dwarfed all other systems. N_2O emissions were consistently low for unfertilized restored prairie harvested for biomass. Perennial cropping systems on highly productive mollisols had higher N_2O emissions than the same systems growing on moderately productive alfisols. Finally, N_2O flux responses to environmental conditions during establishment were

not generalizable across cropping systems, indicating that use of model systems, especially for perennials and polycultures, should be performed cautiously.

Acknowledgements

We thank S. Hamilton, A. Dean, J. Tesmer, J. Sustachek, N. Tautges, A. Miller, Z. Andersen, B. Faust, K. Kahmark, S. VanderWulp, and many others for assistance in the field and laboratory. We also thank S. Bohm for providing the FluxQC framework, S. Sippel for database management, and anonymous reviewers for comments that improved the manuscript. Funding was provided by the DOE Great Lakes Bioenergy Research Center (DOE BER Office of Science DE-FC02-07ER64494) and the DOE OBP Office of Energy Efficiency and Renewable Energy (DE-AC05-76RL01830), and at KBS the NSF Long-term Ecological Research Program and Michigan State University AgBioResearch.

References

Adler PR, Del Grosso SJ, Parton WJ (2007) Life-cycle assessment of net greenhouse-gas flux for bioenergy cropping systems. *Ecological Applications*, **17**, 675–691.

Anderson-Teixeira KJ, Masters MD, Black CK, Zeri M, Hussain MZ, Bernacchi CJ, DeLucia EH (2013) Altered belowground carbon cycling following land-use change to perennial bioenergy crops. *Ecosystems*, **16**, 508–520.

Bouwman AF (1996) Direct emission of nitrous oxide from agricultural soils. *Nutrient Cycling in Agroecosystems*, **46**, 53–70.

Bouwman AF, Boumans LJM, Batjes NH (2002) Emissions of N₂O and NO from fertilized fields: summary of available measurement data. *Global Biogeochemical Cycles*, **16**, 6-1-6-13.

Bouwman AF, Van Grinsven JJM, Eickhout B (2010) Consequences of the cultivation of energy crops for the global nitrogen cycle. *Ecological Applications*, **20**, 101–109.

Burbidge JB, Magee L, Robb AL (2013) Alternative transformations to handle extreme values of the dependent variable. *Journal of the American Statistical Association*, **83**, 123–127.

Castellano MJ, Schmidt JP, Kaye JP, Walker C, Graham CB, Lin H, Dell CJ (2010) Hydrological and biogeochemical controls on the timing and magnitude of nitrous oxide flux across an agricultural landscape. *Global Change Biology*, **16**, 2711–2720.

Cavigelli MA, Robertson GP (2001) Role of denitrifier diversity in rates of nitrous oxide consumption in a terrestrial ecosystem. *Soil Biology and Biochemistry*, **33**, 297–310.

Chen D, Li Y, Grace P, Mosier AR (2008) N₂O emissions from agricultural lands: a synthesis of simulation approaches. *Plant and Soil*, **309**, 169–189.

Crum JR, Collins HP (1995) *KBS Soils*. Kellogg Biological Station Long-Term Ecological Research, Michigan State University, Hickory Corners, MI. Available at: http://lter.kbs.msu.edu/research/site-description-and-maps/soil-description (accessed 19 September 2014).

Dale VH, Kline KL, Wright LL, Perlack RD, Downing M, Graham RL (2011) Interactions among bioenergy feedstock choices, landscape dynamics, and land use. *Ecological Applications*, **21**, 1039–1054.

Dale BE, Anderson JE, Brown RC et al. (2014) Take a closer look: biofuels can support environmental, economic and social goals. *Environmental science and Technology*, **48**, 7200–7203.

Dechow R, Freibauer A (2011) Assessment of German nitrous oxide emissions using empirical modelling approaches. *Nutrient Cycling in Agroecosystems*, **91**, 235–254.

Dobbie K, Smith K (2003) Nitrous oxide emission factors for agricultural soils in Great Britain: the impact of soil water-filled pore space and other controlling variables. *Global Change Biology*, **9**, 204–218.

Dobbie KE, McTaggart IP, Smith KA (1999) Nitrous oxide emissions from intensive agricultural systems: variations between crops and seasons, key driving variables, and mean emission factors. *Journal of Geophysical Research*, **104**, 26891.

Drury CF, Yang XM, Reynolds WD, McLaughlin NB (2008) Nitrous oxide and carbon dioxide emissions from monoculture and rotational cropping of corn, soybean and winter wheat. *Canadian Journal of Soil Science*, **88**, 163–174.

Gelfand I, Zenone T, Jasrotia P, Chen J, Hamilton SK (2011) Carbon debt of Conservation Reserve Program (CRP) grasslands converted to bioenergy production. *Proceedings of the National Academy of Sciences*, **108**, 13864–13869.

Henault C, Bizouard F, Laville P, Gabrielle B, Nicoullaud B, Germon JC, Cellier P (2005) Predicting *in situ* soil N₂O emission using NOE algorithm and soil database. *Global Change Biology*, **11**, 115–127.

Hernandez-Ramirez G, Brouder SM, Smith DR, Van Scoyoc GE (2009) Greenhouse gas fluxes in an eastern Corn Belt soil: weather, nitrogen source, and rotation. *Journal of Environmental Quality*, **38**, 841–854.

Hoeft I, Steude K, Wrage N, Veldkamp E (2012) Response of nitrogen oxide emissions to grazer species and plant species composition in temperate agricultural grassland. *Agriculture, Ecosystems and Environment*, **151**, 34–43.

Hoeting JA, Madigan D, Raftery AE, Volinsky CT (1999) Bayesian model averaging: a tutorial. *Statistical Science*, **14**, 382–417.

Hutchinson GL, Mosier AR (1981) Improved soil cover method for field measurement of nitrous oxide fluxes. *Soil Science Society of America Journal*, **45**, 311.

Imer D, Merbold L, Eugster W, Buchmann N (2013) Temporal and spatial variations of soil CO₂, CH₄ and N₂O fluxes at three differently managed grasslands. *Biogeosciences*, **10**, 5931–5945.

Johnson JMF, Archer D, Barbour N (2010) Greenhouse gas emission from contrasting management scenarios in the northern Corn Belt. *Soil Science Society of America Journal*, **74**, 396–406.

Jokela W, Posner J, Hedtcke J, Balser T, Read H (2011) Midwest cropping system effects on soil properties and on a soil quality index. *Agronomy Journal*, **103**, 1552–1562.

Lesur C, Bazot M, Bio-Beri F, Mary B, Jeuffroy M-H, Loyce C (2014) Assessing nitrate leaching during the three-first years of *Miscanthus* × *giganteus* from on-farm measurements and modeling. *GCB Bioenergy*, **6**, 439–449.

Lewandowski I, Scurlock JMO, Lindvall E, Christou M (2003) The development and current status of perennial rhizomatous grasses as energy crops in the US and Europe. *Biomass and Bioenergy*, **25**, 335–361.

Maroja LS, Andrés JA, Walters JR, Harrison RG (2009) Multiple barriers to gene exchange in a field cricket hybrid zone. *Biological Journal of the Linnean Society*, **97**, 390–402.

Matthews RA, Chadwick DR, Retter AL, Blackwell MSA, Yamulki S (2010) Nitrous oxide emissions from small-scale farmland features of UK livestock farming systems. *Agriculture, Ecosystems and Environment*, **136**, 192–198.

McLaughlin SB, Adams Kszos L (2005) Development of switchgrass (*Panicum virgatum*) as a bioenergy feedstock in the United States. *Biomass and Bioenergy*, **28**, 515–535.

McSwiney CP, Snapp SS, Gentry LE (2010) Use of N immobilization to tighten the N cycle in conventional agroecosystems. *Ecological Applications*, **20**, 648–662.

Millar N, Robertson GP, Grace PR, Gehl RJ, Hoben JP (2010) Nitrogen fertilizer management for nitrous oxide (N₂O) mitigation in intensive corn (Maize) production: an emissions reduction protocol for US Midwest agriculture. *Mitigation and Adaptation Strategies for Global Change*, **15**, 185–204.

Molodovskaya M, Singurindy O, Richards BK, Warland J, Johnson MS, Steenhuis TS (2012) Temporal variability of nitrous oxide from fertilized croplands: hot moment analysis. *Soil Science Society of America Journal*, **76**, 1728.

MSCO (2013) Michigan State Climatologist's Office: 27 year summary of annual values for Gull Lake (3504) 1981–2010, Available at: http://climate.geo.msu.edu/climate_mi/stations/3504/1981-2010%20annual%20summary.pdf (accessed 2 April 2014).

Nikièma P, Rothstein DE, Min D-H, Kapp CJ (2011) Nitrogen fertilization of switchgrass increases biomass yield and improves net greenhouse gas balance in northern Michigan, USA. *Biomass and Bioenergy*, **35**, 4356–4367.

Nikièma P, Rothstein DE, Miller RO (2012) Initial greenhouse gas emissions and nitrogen leaching losses associated with converting pastureland to short-rotation woody bioenergy crops in northern Michigan, USA. *Biomass and Bioenergy*, **39**, 413–426.

NWS (2013) National Weather Service: Wisconsin 30 year avereage temperature and precipitation 1981–2010, Available at: www.crh.noaa.gov/images/mkx/climate/avg_30_year_precip.png and www.crh.noaa.gov/images/mkx/climate/avg_30_year_temp.png (accessed 2 April 2014).

Osterholz WR, Kucharik CJ, Hedtcke JL, Posner JL (2014) Seasonal nitrous oxide and methane fluxes from grain- and forage-based production systems in Wisconsin, USA. *Journal of Environment Quality*, **43**, 1833–1843.

Palmer MM, Forrester JA, Rothstein DE, Mladenoff DJ (2014) Conversion of open lands to short-rotation woody biomass crops: site variability affects nitrogen cycling and N₂O fluxes in the US Northern Lake States. *GCB Bioenergy*, **6**, 450–464.

Parkin TB, Kaspar TC, Singer JW (2006) Cover crop effects on the fate of N following soil application of swine manure. *Plant and Soil*, **289**, 141–152.

Parrish DJ, Fike JH (2005) The biology and agronomy of switchgrass for biofuels. *Critical Reviews in Plant Sciences*, **24**, 423–459.

Pedersen AR (2011) 'HMR': Flux estimation with static chamber data. R package version 0.3.1.

Pinheiro J, Bates D, DebRoy S, Sarkar D, The R Core Team (2013) nlme: Linear and Nonlinear Mixed Effects Models. R package version 3.1.

R Core Team (2014) R: A language and environment for statistical computing. R Foundation for Statistical Computing, Vienna, Austria. URL http://www.R-project.org/.

Raftery AE, Hoeting JA, Volinsky CT, Painter I, Yeung KY, (2013) *BMA: Bayesian Model Averaging*. R package version 3.16.2.2. Available: http://cran.r-project.org/package=BMA (accessed 25 September 2013).

Reay DS, Davidson EA, Smith KA, Smith P, Melillo JM, Dentener F, Crutzen PJ (2012) Global agriculture and nitrous oxide emissions. *Nature Climate Change*, **2**, 410–416.

Robertson GP, Groffman PM (2015) Nitrogen transformations. In: *Soil Soil Microbiology, Ecology and Biochemistry* (ed. Paul EA), pp. 421–446. Academic Press, Burlington, MA.

Robertson GP, Coleman DC, Bledsoe CS, Sollins P (eds.) (1999) *Standard Soil Methods for Long-Term Ecological Research*, pp. 1–462. Oxford University Press, New York.

Robertson GP, Paul EA, Harwood RR, (2000) Greenhouse gases in intensive agriculture: contributions of individual gases to the radiative forcing of the atmosphere. *Science*, **289**, 1922–1925.

Robertson GP, Dale VH, Doering OC *et al.* (2008) Sustainable biofuels redux. *Science*, **322**, 49–50.

Robertson GP, Bruulsema TW, Gehl RJ, Kanter D, Mauzerall DL, Rotz CA, Williams CO (2012) Nitrogen–climate interactions in US agriculture. *Biogeochemistry*, **114**, 41–70.

Ruan L, Robertson GP (2013) Initial nitrous oxide, carbon dioxide, and methane costs of converting conservation reserve program grassland to row crops under no-till vs. conventional tillage. *Global Change Biology*, **19**, 2478–2489.

Sanderson MA, Adler PR (2008) Perennial forages as second generation bioenergy crops. *International Journal of Molecular Sciences*, **9**, 768–788.

Sanford GR, Posner JL, Jackson RD, Kucharik CJ, Hedtcke JL, Lin T-L (2012) Soil carbon lost from Mollisols of the North Central USA with 20 years of agricultural best management practices. *Agriculture, Ecosystems and Environment*, **162**, 68–76.

Schlesinger WH (2013) An estimate of the global sink for nitrous oxide in soils. *Global Change Biology*, **19**, 2929–2931.

Smith KA, Dobbie KE (2001) The impact of sampling frequency and sampling times on chamber-based measurements of N_2O emissions from fertilized soils. *Global Change Biology*, **7**, 933–945.

Smith CM, David MB, Mitchell CA, Masters MD, Anderson-Teixeira KJ, Bernacchi CJ, DeLucia EH (2013) Reduced nitrogen losses after conversion of row crop agriculture to perennial biofuel crops. *Journal of Environment Quality*, **42**, 219.

Stehfest E, Bouwman L (2006) N_2O and NO emission from agricultural fields and soils under natural vegetation: summarizing available measurement data and modeling of global annual emissions. *Nutrient Cycling in Agroecosystems*, **74**, 207–228.

Surendran Nair S, Kang S, Zhang X *et al.* (2012) Bioenergy crop models: descriptions, data requirements, and future challenges. *GCB Bioenergy*, **4**, 620–633.

Tulbure MG, Wimberly MC, Boe A, Owens VN (2012) Climatic and genetic controls of yields of switchgrass, a model bioenergy species. *Agriculture, Ecosystems and Environment*, **146**, 121–129.

Zenone T, Chen J, Deal MW *et al.* (2011) CO_2 fluxes of transitional bioenergy crops: effect of land conversion during the first year of cultivation. *GCB Bioenergy*, **3**, 401–412.

11

Offsetting high water demands with high productivity: Sorghum as a biofuel crop in a high irradiance arid ecosystem

PATRICIA Y. OIKAWA[1], G. DARREL JENERETTE[2] and DAVID A. GRANTZ[2]

[1]*Department of Environmental Science, Policy and Management, University of California, Berkeley, CA 94702, USA,*
[2]*Department of Botany and Plant Sciences, University of California, Riverside, CA 92521, USA*

Abstract

High irradiance arid environments are promising, yet understudied, areas for biofuel production. We investigated the productivity and environmental trade-offs of growing sorghum (*Sorghum bicolor*) as a biofuel feedstock in the low deserts of California (CA). Using a 5.3 ha experimental field in the Imperial Valley, CA, we measured aboveground biomass production and net ecosystem exchange of CO_2 and H_2O via eddy covariance over three growing periods between February and November 2012. Environmental conditions were extreme, with high irradiance, vapor pressure deficit (VPD), and air temperature throughout the growing season. Air temperature peaked in August with a maximum of 45.7 °C. Sorghum produced an annual aboveground biomass yield of 43.7 Mg per hectare. Net ecosystem exchange (NEE) was highest during the summer growth period and reached a maximum of -68 μmol CO_2 m^{-2} s^{-1}. Water use efficiency, or biomass water ratio (BWR), was high (4.0 g dry biomass kg^{-1} H_2O) despite high seasonal evapotranspiration (1094 kg H_2O m^{-2}). The BWR of sorghum surpassed that of many C4 biofuel candidate crops in the United States, as well as that of alfalfa which is currently widely grown in the Imperial Valley. Sorghum also outperformed many US biofuel crops in terms of radiation use efficiency (RUE), achieving 1.5 g dry biomass MJ^{-1}. We found no evidence of saturation of NEE at high levels of photosynthetically active radiation (PAR) (up to 2250 μmol m^{-2} s^{-1}). In addition, we found no evidence that NEE was inhibited by either high VPD or air temperature during peak photosynthetic phases. The combination of high productivity, high BWR, and high RUE suggests that sorghum is well adapted to this extreme environment. The biomass production rates and efficiency metrics spanning three growing periods provide fundamental data for future Life Cycle Assessments (LCA), which are needed to assess the sustainability of this sorghum biofuel feedstock system.

Keywords: biofuel, biomass water ratio, eddy covariance, radiation use efficiency, *Sorghum bicolor*, water use efficiency

Introduction

With the passage of the Energy Independence and Security Act of 2007, US biofuel production is expected to increase to 36 billion gallons of biofuel by 2022 (RFA, 2010). Currently, the dominant first generation biofuel feedstock crops are US maize and Brazilian sugarcane (Mussatto *et al.*, 2010). However, extensive research is being conducted to diversify this market and identify the best candidates and environments for second generation biofuel production that can be used to produce lignocellulosic ethanol. For example, recent work has focused on crops that can replace maize in the Midwestern United States such as Miscanthus and switchgrass

(Schmer *et al.*, 2008; Dohleman & Long, 2009; Hickman *et al.*, 2010; Davis *et al.*, 2011; Zeri *et al.*, 2011; Vanloocke *et al.*, 2012; Gelfand *et al.*, 2013). States such as California (CA) are actively seeking ways to lower the GHG footprint of transportation fuels (AB32, the Global Warming Solutions Act, CARB, 2009), including local production in place of liquid fuel shipped from the Midwestern United States or Brazil.

High irradiance arid environments such as the Imperial Valley, CA may be promising areas for biofuel feedstock production (Bazdarich & Sebasta, 2001; Kaffka, 2009). These environments typically have long growing seasons and abundant light, providing an environment that could maximize biofuel feedstock production. The Imperial Valley is characterized by high irradiance [>3000 MJ photosynthetically active radiation (PAR) m^{-2} yr^{-1}], however, this extreme environment

Correspondence: Patricia Y. Oikawa
e-mail: patty.oikawa@gmail.com

also has high vapor pressure deficit (VPD) (>7 kPa) and high temperature (>40 °C in the summer). Many potential biofuel feedstocks are inefficient in converting high irradiance to biomass under high VPD and temperature (Shurpali *et al.*, 2009; Archontoulis *et al.*, 2012). Therefore, an important first step in evaluating biofuel feedstock production in the Imperial Valley is to measure sensitivity to extreme environmental conditions. This includes radiation use efficiency (RUE), a metric of a crop's ability to convert light to biomass (Monteith & Moss, 1977) and water use efficiency (WUE), a metric of the trade-off between biomass and water use (Jackson *et al.*, 2005; Gerbens-Leenes *et al.*, 2009; Zeri *et al.*, 2013). Finally, the evaluation of biofuel feedstock production in high irradiance arid environments requires assessment of resource use efficiency variability throughout the long growing season, including periods of high VPD and temperature.

Despite the trade-offs associated with high irradiance arid environments, certain biofuel crops have the capability to thrive under extreme conditions. For example, sorghum, a C4 annual grass originally from arid northeastern Africa, is considered water use efficient and heat and drought tolerant (Rooney *et al.*, 2007; Shoemaker & Bransby, 2010; de Vries *et al.*, 2010; Xie & Su, 2012). However, water use of sorghum can vary substantially depending on climate and other production conditions (Gerbens-Leenes *et al.*, 2009). Quantitative evaluation of resource use efficiency is a necessary first step in assessing the potential of sorghum as a biofuel feedstock in high irradiance arid environments including the Imperial Valley (Shoemaker & Bransby, 2010; Zegada-Lizarazu & Monti, 2012).

To address the lack of data concerning biomass production potential and resource use efficiency of sorghum in high irradiance arid environments, we addressed the following questions: (i) Are high evapotranspiration (ET) rates in high irradiance arid environments offset by high rates of biomass production? (ii) Is sorghum capable of efficiently exploiting the high irradiance environment? (iii) Is CO_2 uptake inhibited due to high VPD and/or air temperature? (iv) How does sorghum performance vary as environmental conditions change across the growing season? To address these questions, we measured aboveground biomass production and net ecosystem exchange of CO_2 (NEE) and evapotranspiration (ET) with eddy covariance in the Imperial Valley. We used these measurements to compute water use and energy conversion metrics, and to evaluate the responses of sorghum to light, VPD, and air temperature. Measurements were collected over three growing periods to evaluate the influence of environmental change occurring during the long (10 months) growing season on sorghum performance.

In addition, we compared sorghum yield, water use, and energy conversion metrics with other potential biofuel crops, sorghum grown outside CA, and biomass crops currently grown in the Imperial Valley. The presented biomass production data and resource use efficiency analyses are an important step toward determining the profitability, sustainability, and relative benefit of sorghum cultivation in high irradiance arid environments.

Materials and methods

Site description

All measurements were conducted in 2012 at a 5.3 ha experimental field in the low elevation (−18 m ASL) University of California Desert Research and Extension Center (DREC), El Centro, CA (32°N 48′42.6″, 115°W 26′37.5″) characterized by deep alluvial soil (42% clay, 41% silt, 16% sand). While some parts of the United States experienced a heat wave in 2012, our field site experienced historically typical air temperatures. Weather station data collected at DREC [operated by the California Irrigation Management Information System (CIMIS) since December 1989] report the annual minimum and maximum monthly air temperatures in 2012 to be 4.2 and 41.0 °C, respectively, compared to 21-year means of 3.9 and 41.6 °C across 1990–2011. The field was left fallow for 8 months prior to planting. All sensor systems were deployed in the summer of 2011. The field was tilled prior to planting seed of *Sorghum bicolor* (cv. Photoperiod LS; Scott Seed Inc., Hereford, TX, USA) on 16 February 2012 at 90 000 plants ha^{-1} with beds separated by 1.5 m and 20 cm deep furrows. This variety of Sorghum is typically grown for hay production and is a fast-growing annual plant. The field was gravity-fed flood irrigated as needed, usually every 10 days or when soil surface volumetric water content fell below 0.10 cm^3 cm^{-3}. The primary water source in the Imperial Valley is the Colorado River. Fertilizer treatments of 90 kg N ha^{-1} were applied on 10 February, 18 June, and 16 August 2012 (270 kg N ha^{-1} yr^{-1}). Pesticides were applied at 2.1 l ha^{-1} on 30 April 2012 (Lorsban® insecticide, Dow AgroSciences, Indianapolis, IN, USA) and herbicides were applied at 0.84 kg ha^{-1} on 27 March 2012 (Maestro®, Nufarm Americas Inc., Alsip, IL, USA). Data presented here were collected during sorghum cultivation spanning 28 February to 12 November 2012 which included three harvests. Harvests were conducted when the plants started to flower. The first harvest occurred on 4 June (DOY 156) after which the biomass was left to dry on the soil surface for an additional 10 days before removal. The second harvest occurred on 14 August (DOY 227) and was conducted using large tractors that removed the biomass the day of cutting. The final harvest occurred on 12 November (DOY 318) and was conducted similarly to the first harvest. The second and third growth periods are referred to as ratoon crops, defined as a crop that is harvested two or more times from a single planting during the growing season (Rao *et al.*, 2013).

Micrometeorological and eddy covariance measurements

We used eddy covariance to estimate evapotranspiration (ET) from the experimental field and the net ecosystem exchange of CO_2 (NEE). This approach requires a suite of micrometeorological sensors and uses the covariance between vertical wind speed and a scalar gas to estimate trace gas flux (Baldocchi et al., 1988). Micrometeorological and eddy covariance systems were mounted on a tower in the center of the field. The height of the eddy covariance system was maintained above the roughness sublayer and was moved up the tower as the canopy grew (Raupach, 1994). Prevailing winds came from the south-southwest. Air temperature and relative humidity were measured with a shielded Platinum Resistance Temperature (PRT) and capacitive sensor probe (HMP45C, Vaisala, Helsinki, Finland). In addition, we measured precipitation with a tipping bucket rain gage (TE525, Campbell Scientific, Logan, UT, USA) and photosynthetically active radiation with a quantum sensor (PAR, PQS1, Kipp & Zonen, Delft, The Netherlands). A four-component net radiation sensor was used to measure down-welling and upwelling shortwave and longwave radiation (NR01, Hukseflux Thermal Sensors, Delft, The Netherlands). Ground heat storage and surface heat flux were calculated using soil heat flux plates (HFP01, Hukseflux Thermal Sensors), soil temperature probes (108, Campbell Scientific), and a soil moisture probe (CS616, Campbell Scientific) buried at the tower location. Two heat flux plates were buried at 8 cm depth, each with two temperature probes buried directly above them at 2 and 8 cm depth. A soil moisture probe was also buried above the flux plates at 2.5 cm depth. All of these sensors were sampled at 1 Hz, with 30 min averages stored. A 3D sonic anemometer measured wind speed at 10 Hz in three directions (CSAT3, Campbell Scientific). CO_2 and H_2O concentrations were measured at 10 Hz with an open path IRGA analyzer (LI-7500, LI-COR Biosciences, Lincoln, NE, USA). All micrometeorological and belowground data were stored on a datalogger (CR5000, Campbell Scientific).

Data processing

Postprocessing of raw data for eddy covariance flux estimates was conducted with EddyPro 4.0 (LI-COR). Preliminary processing included filtering data for diagnostic instrument flags. Following Vickers & Mahrt (1997), raw data were screened for artificial spikes defined as values ± 5 SD for wind and ± 3.5 SD for scalars based on a 5 min averaging window. Data were also tested for amplitude resolution, dropouts, absolute limits, skewness and kurtosis, discontinuities, time lags, and steadiness of horizontal wind following Vickers & Mahrt (1997). We employed a double rotation method for aligning the coordinate system to the main wind direction and making the average cross-stream and vertical wind components equal to zero. Measurements were corrected for air density fluctuations using the WPL correction (Webb et al., 1980). Temperature measured by the sonic anemometer was corrected for humidity (Schotanus et al., 1983). In addition, fluxes were quality filtered following Mauder & Foken (2006). Following these

initial steps, fluxes were computed over 30 min intervals and expressed in $\mu mol \ m^{-2} \ s^{-1}$ for NEE and $mm \ day^{-1}$ for ET where positive fluxes are net emissions and negative fluxes are net uptake.

Computed fluxes were then filtered for high and low friction velocities (velocities never exceeded 1.1 m s^{-1}). Friction velocity (u^*) thresholds were computed for each season following Reichstein et al. (2005) and were 0.035, 0.040, 0.032, and 0.028 m s^{-1} for February–March, April–June, July–September, October–November, respectively. We used a default threshold of 0.1 m s^{-1} as all computed u^* threshold values were below 0.1 m s^{-1}. The footprint of the tower was analyzed using the Hsieh footprint model (Hsieh et al., 2000). The distance from the tower that encompassed 70% of the footprint ranged between 15 and 155 m, with an average of 49.2 m. All fluxes where the 70% footprint exceeded the edge of the field were excluded from the analysis. Excluded or missing data were gapfilled following Reichstein et al. (2005). If NEE or ET data were missing, they would first be filled with the average of available values within ± 7 days under similar meteorological conditions defined as net radiation $\pm 50 \ W \ m^{-2}$, air temperature $\pm 2.5 \ °C$, and VPD ± 5.0 hPa. If necessary, this process was followed by expanding the window to ± 14 days. The last gap-filling level used averaged values from ± 7 days where similar meteorological conditions were based only on net radiation. Instrument and power failure resulted in 13% loss of 30 min flux data (34 days) over the course of the growing season. The remaining gaps were due to footprint, u^*, and quality filtering, where 30% of gaps occurred during the day and 62% occurred at night.

Energy balance was analyzed using data from the net radiometer and soil heat flux measurements with eddy covariance measurements of latent (LE) and sensible (H) heat flux. In addition, we computed the energy stored as photosynthesis (S_{ph}) according to Nobel (1974) where 1 W m^{-2} is stored for every 2.03 $\mu mol \ m^{-2} \ s^{-1}$. Ground heat flux (G) was estimated following Foken (2008):

$$G = G_z + \frac{\Delta\Theta(\theta_w m_{sw} c_w + p_s c_s)\Delta z}{\Delta t}$$

where G_z is the ground heat flux measurements averaged across heat flux plates at 8 cm depth, $\Delta\Theta$ is the change in temperature averaged between soil temperatures measured at 2 and 8 cm depth over the period Δt, θ_w is the volumetric water content measured at 2.5 cm, m_{sw} is the density of water, c_w is the specific heat capacity of water, p_s is the soil bulk density, and c_s is the specific heat capacity of soil.

Dry biomass and theoretical ethanol yields

Sorghum dry yield estimates are based on total aboveground plant biomass hand harvested in nine evenly spaced sections of the field (1.5 m² each). Dry to wet biomass weight ratio was on average 0.16 and applied consistently across growth periods. Theoretical ethanol yield (l ha^{-1}) is calculated according to lignocellulosic conversion rates where sorghum produces 0.26 l ethanol kg^{-1} dry biomass (Chandel et al., 2010; Singh et al., 2012).

Water and energy use efficiency metrics

Water use efficiency is calculated as a biomass water ratio (BWR), the ratio of dry harvested biomass to seasonal water requirements derived from ET (units: g dry biomass kg H_2O^{-1}) (Stanhill, 1986; Monteith, 1993). BWR was computed both seasonally and by growth period.

Radiation use efficiency (RUE) is the ratio of cumulative dry biomass (Wb; g m^{-2}) to cumulative PAR (MJ m^{-2}) expressed in g dry biomass MJ^{-1} following Monteith & Moss (1977). To convert PAR from photon units to radiometric units, we assumed a constant 218.6 KJ mol^{-1} photons (Beale & Long, 1995; Curt et al., 1998). RUE was computed both seasonally and by growth period.

NEE response to environmental variables

To further evaluate the RUE of sorghum and how it changed over the growing season, we compared NEE light response across growth periods. ANCOVA analyses were conducted on daytime data (PAR>10 µmol m^{-2} s^{-1}) for each growth period (period 1 = DOY 59–156; period 2 = DOY 157–227; period 3 = DOY 228–317) as well as during peak photosynthetic phases for each growth period (period 1 = DOY 140–145; period 2 = DOY 210–215; period 3 = DOY 260–265). Analyzing peak photosynthetic phases gave insight into light response during periods of growth that were important to biomass production at the site. NEE residuals from the linear relationships between NEE and PAR for each growth period were regressed against VPD and air temperature. Linear functions were used because hyperbolic functions did not increase model fit between NEE and PAR. The residual analyses gave insight into the influence of high temperature and aridity on NEE while controlling for changes due to light (similar to Huxman et al., 2003). Statistical analyses were performed in R (v. 2.15.1, Vienna, Austria).

Results

Sorghum produced three harvests in 2012, spanning February–November. Annual yield of sorghum in the experimental field was high, 43.7 Mg ha^{-1} (SD = 2.7; Table 1). The second growth period was the most productive (16.2 Mg ha^{-1}, SD = 1.9), followed by the first (15.8 Mg ha^{-1}, SD = 1.3) and third (11.7 Mg ha^{-1}, SD = 2.3; Table 1). Annual theoretical lignocellulosic ethanol yield was 11.4×10^3 l ha^{-1} (Table 1).

Annual precipitation measured at the experimental field was 20 mm. Vapor pressure deficit (VPD) hourly averages ranged between 1.1 and 4.6 kPa throughout the growing season and was highest in summer months during the second growth period, followed by the third and first (Fig. 1a). Average air temperature peaked in August during the second growth period (maximum recorded temperature = 45.7 °C) and was on average the lowest during the first growth period (Fig. 1b), with a minimum recorded air temperature of 2.8 °C. Irradiance was high throughout the growing season, however, the first growth period had higher amounts of shortwave downwelling radiation, followed by the second and third growth periods (average daily maximum was 962, 880, 777 W m^{-2} for the first, second, and third growth periods, respectively; Fig. 1c).

A good quality control assessment of eddy covariance measurements is evaluating energy balance closure. We used micrometeorological measurements (G and R_{net}) and tower data (LE, H, S_{ph}) to evaluate energy balance and found strong closure at our site (slope = 0.93, r^2 = 0.86; Fig. 2a). Similar closure was found using daily averages of variables R_{net}, H, LE, G, and S_{ph}

Table 1 Sorghum biomass yields (Mg dry biomass ha^{-1}), theoretical ethanol yield (l ha^{-1}), cumulative photosynthetically active radiation (PAR), cumulative water requirements (measured as evapotranspiration), biomass water ratios (BWR), and radiation conversion efficiencies are shown by growth period and annually. Theoretical ethanol yield is calculated according to lignocellulosic conversion rates (Chandel et al., 2010; Singh et al., 2012). BWR (g dry biomass kg^{-1} water) is computed by the ratio of cumulative dry biomass (kg m^{-2}) to cumulative evapotranspiration (kg m^{-2} $season^{-1}$). Conversion efficiency (g dry biomass MJ^{-1} PAR) is computed by the ratio of cumulative dry biomass (kg m^{-2}) to cumulative PAR (MJ m^{-2})

Sorghum growth period	Total biomass dry yield (Mg ha^{-1})	Theoretical ethanol yield (l ha^{-1})	PAR (MJ m^{-2})	Water requirements (kg m^{-2})	BWR (g dry biomass kg^{-1} H_2O)	Radiation conversion efficiency (g dry biomass MJ^{-1} PAR)
Growth period 1 (28 February – 4 June)	15.8	4108	1189	352	4.5	1.3
Growth period 2 (5 June – 14 August)	16.2	4212	960	458	3.5	1.7
Growth period 3 (15 August – 12 November)	11.7	3042	849	284	4.1	1.4
Annual production	43.7	11362	2997	1094	4.0	1.5

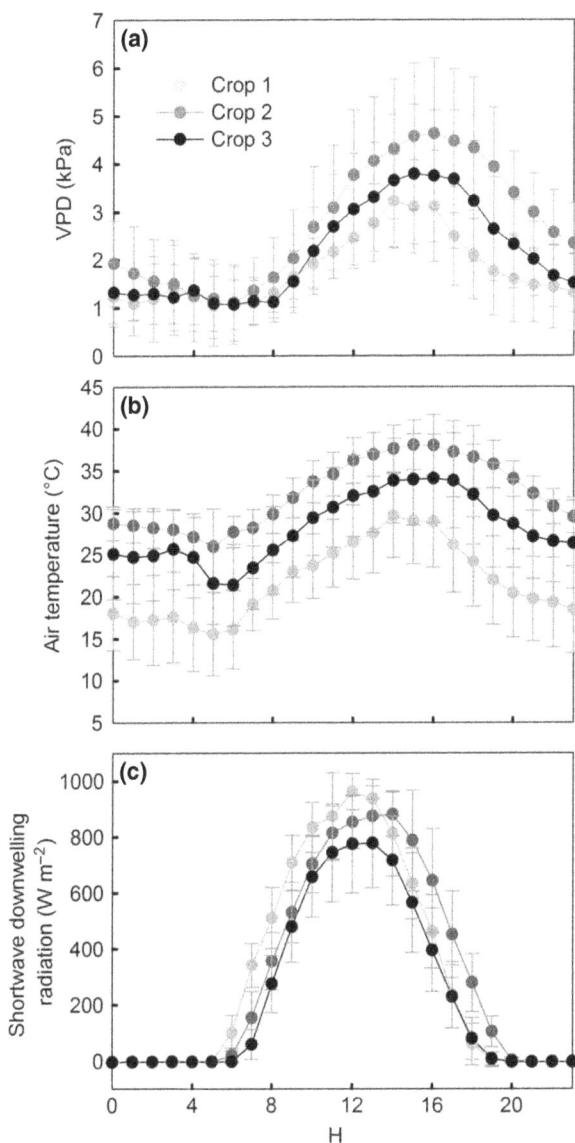

Fig. 1 Hourly averages ± SD for meteorological variables are shown including (a) vapor pressure deficit (VPD; kPa), (b) air temperature (°C), and (c) shortwave downwelling radiation (W m^{-2}). Averages were computed independently for each growth period (growth period 1 = February–June, growth period 2 = June–August, growth period 3 = August–November, 2012).

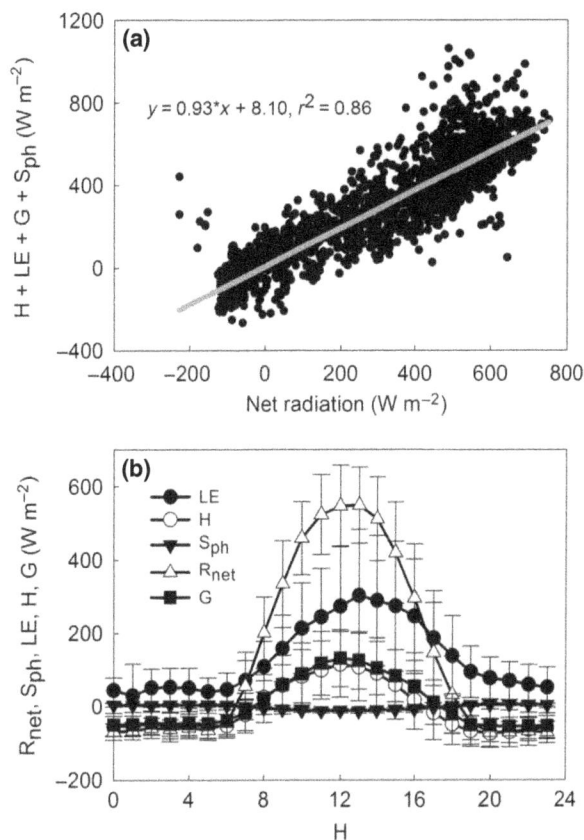

Fig. 2 Evaluation of annual energy balance. (a) Sensible heat (H), latent heat (LE), ground heat flux (G), and energy stored as photosynthesis (S_{ph}) are regressed against net radiation. (b) Hourly annual averages of energy balance components (R_{net}, net radiation; S_{ph}, energy stored in photosynthesis; LE, latent heat flux; H, sensible heat flux; G, ground heat flux; SD bars shown).

(slope = 0.80, r^2 = 0.70; data not shown). These components are also shown in hourly averages across the growing season (Fig. 2b).

Cumulative NEE over the entire growing season was −8.8 Mg C ha^{-1} (−3.1, −3.1, and −2.6 for growth periods 1, 2, and 3, respectively; Fig. 3a). The seasonal BWR of sorghum was high (4.0 g kg^{-1}) due to high biomass offsetting high water demands (1094 kg H$_2$O m^{-2} season^{-1}; Table 1). Radiation use efficiency (RUE) of sorghum was also high (1.5 g MJ^{-1}; Table 1).

Although the second growth period was only slightly more productive than the first (3% increase in dry yield), cumulative ET for the second growth period was 30% higher than the first (cumulative ET for the first and second growth periods was 352 and 458 mm, respectively; Fig. 3b). The third growth period had the lowest cumulative ET at 284 mm accounting for 26% of total cumulative ET (1.1 m; Fig. 3b) and 27% of total dry yield. The first and second growth periods exhibited strong uptake of CO$_2$ (Fig. 3c) with a maximum recorded NEE of −68 μmol CO$_2$ m^{-2} s^{-1} on DOY 213. ET peaked during the second growth period at a rate of 35 mm day^{-1} on DOY 221, which corresponded with the highest temperatures of the growing season (43.5 °C at midday on DOY 221; Fig. 3d).

To further evaluate the water use efficiency of each growth period, we calculated BWR by growth period and found that the first growth period of the season was more water use efficient (4.5 g kg^{-1}) compared to growth periods 2 (3.5 g kg^{-1}) and 3 (4.1 g kg^{-1}).

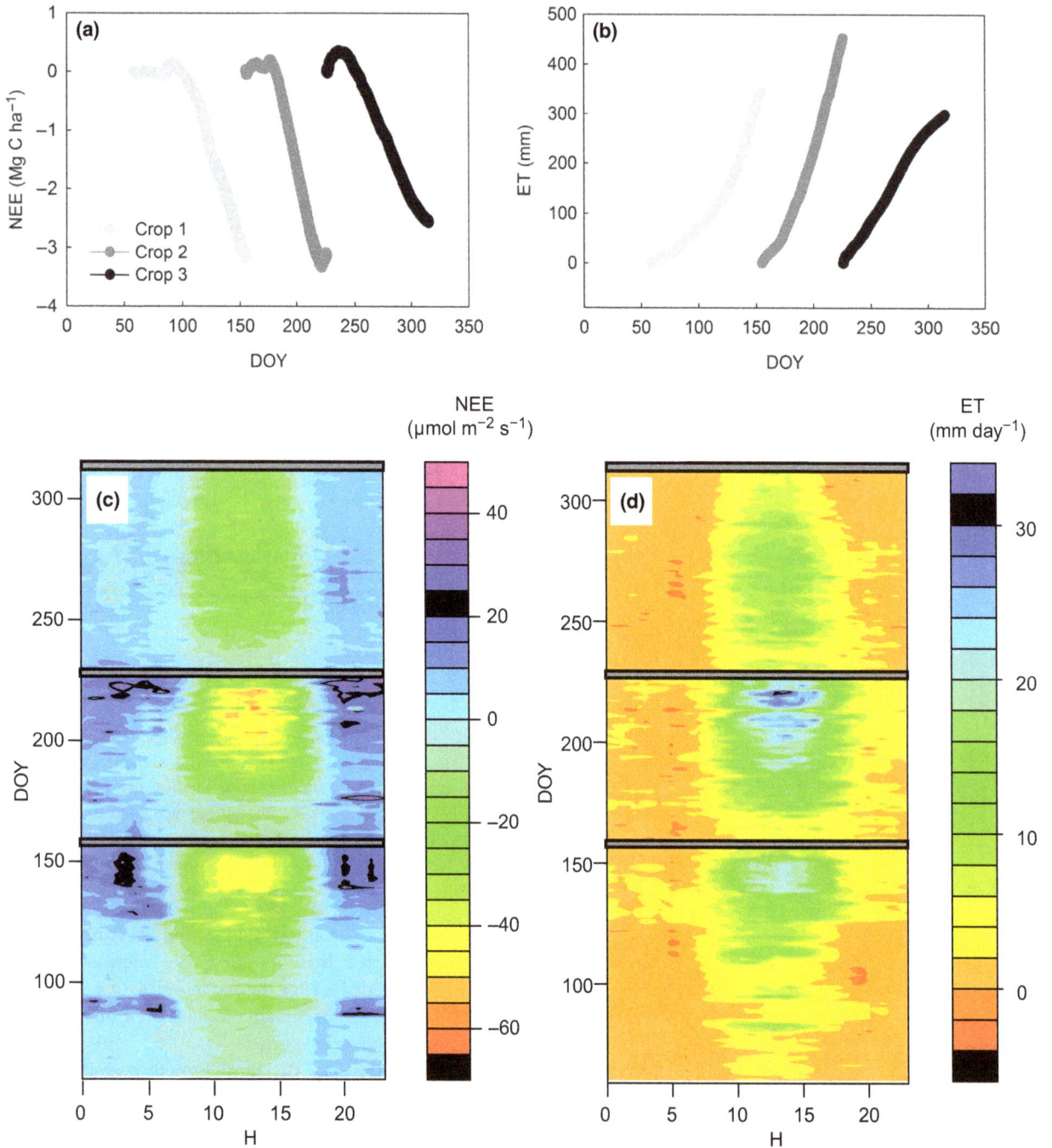

Fig. 3 Cumulative (a) NEE (μmol m^{-2} s^{-1}) and (b) ET (mm day^{-1}) from February to November 2012 are shown by growth period including periods during harvest (growth period 1 = February–June, growth period 2 = June–August, growth period 3 = August–November). Hourly (c) NEE (Mg C ha^{-1}) and (d) ET (mm) are also shown for the period of February to November 2012 (black boxes indicate harvests).

The first growth period started from seed and took approximately 45 days before entering a fast-growth stage (or five-leaf growth stage when typical midday NEE < −20 μmol m^{-2} s^{-1}; Fig. 3c), however, the second growth period started from ratoon and more quickly entered a fast-growth stage (<20 days). The second growth period was harvested with heavier machinery than the first harvest and was more damaging to the field. The third growth period therefore also started from ratoon, but had the smallest yield (Fig. 3c).

We evaluated light response of sorghum by growth period (growth period 1 = DOY 59–156; growth period 2 = DOY 157–227; growth period 3 = DOY 228–317) and during peak photosynthetic phases for each growth period (growth period 1 = DOY 140–145; growth period 2 = DOY 210–215; growth period 3 = DOY 260–265). We found no evidence of NEE saturation at high levels of PAR (up to 2250 μmol m^{-2} s^{-1}) in either analysis (only peak photosynthetic phases are presented as the trends were similar for both analyses; Fig. 4). During peak photosynthetic phases, the relationship between NEE and PAR significantly differed between growth periods (ANCOVA; $F = 127.01$, $P < 0.0001$; Fig. 4a), where growth periods 1 and 2 had higher radiation use efficiency (slope = -0.028) compared to growth period 3 (slope = -0.015). However, when accounting for biomass production, the second growth period had the highest RUE at 1.7 g dry biomass MJ^{-1}, while growth periods 1 and 3 had RUEs of 1.3 and 1.4 g dry biomass MJ^{-1}, respectively (Table 1). Residuals of NEE and PAR relationships during the peak photosynthetic phases were not related to air temperature (ANCOVA; $F = 0.814$, $P = 0.37$; Fig. 4b) or VPD (ANCOVA; $F = 1.19$, $P = 0.28$; Fig. 4c). In addition, no differences in NEE residual response to air temperature (ANCOVA; $F = 0.43$, $P = 0.51$) and VPD (ANCOVA; $F = 0.37$, $P = 0.54$) were detected between growth periods. However, when a similar analysis was conducted on the complete data sets (not just peak photosynthetic phases), both VPD (ANCOVA; $F = 688.3$, $P < 0.0001$) and air temperature (ANCOVA; $F = 836.4$, $P < 0.0001$) were related to NEE residuals, where NEE was inhibited with increasing VPD and air temperature. While growth periods responded similarly to changes in VPD, seasonal NEE residuals of growth period 2 responded more strongly to changes in air temperature (slope = -1.2, $r^2 = 0.21$) compared to growth periods 1 (slope = -0.5, $r^2 = 0.18$) and 3 (slope = -0.5, $r^2 = 0.17$).

Discussion

High growth rates and resource use efficiency of sorghum suggest that the high irradiance arid environment of the Imperial Valley, CA could be an advantageous location for sorghum production. The seasonal production rate was 43.7 Mg of dry biomass per hectare, which surpasses seasonal production rates of many C4 biofuel crops in the Midwestern United States, including maize, Miscanthus, and switchgrass (19, 30, and 7 Mg ha^{-1} growing season^{-1}, respectively) (Dohleman & Long, 2009; Hickman et al., 2010) and alfalfa grown in the Imperial Valley (24 Mg ha^{-1} growing season^{-1}) (Sanden et al., 2011). Overall, NEE was inhibited by high VPD and temperature, however, sorghum biomass

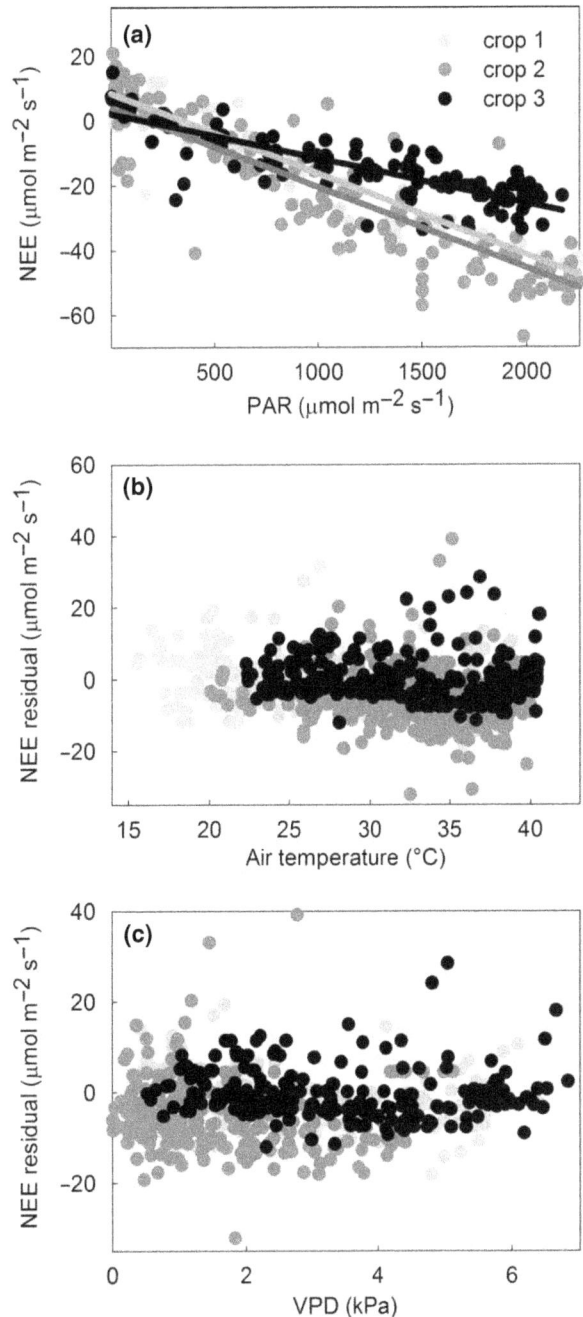

Fig. 4 Relationships between half hourly averages of net ecosystem exchange of CO_2 (NEE) and meteorological variables during peak photosynthetic phases for each growth period (growth period 1 = DOY 140–145; growth period 2 = DOY 210–215; growth period 3 = DOY 260–265). The relationship between NEE and (a) photosynthetically active radiation (PAR) significantly differed between growth periods (ANCOVA, $F = 127.01$, $P < 0.0001$). NEE residuals from the relationships between NEE and PAR were plotted against (b) air temperature and (c) vapor pressure deficit (VPD) during the same three growth periods (growth periods 1–3). Residuals of NEE were not related to air temperature or VPD.

production rates still outperformed previously reported rates for sorghum in the semiarid Mediterranean (31.7 Mg ha^{-1} season^{-1}) (Mastrorilli *et al.*, 1999) and Southeastern United States (19.4 Mg ha^{-1} season^{-1}) (Singh *et al.*, 2012). Net ecosystem exchange of CO_2 reached −68 µmol m^{-2} s^{-1} during peak season, exceeding rates observed in other biofuel cropping systems in the United States (Hollinger *et al.*, 2005; Zeri *et al.*, 2011) and tropical forests (Goulden *et al.*, 2004). No light saturation was detected and water and radiation use efficiency was high. Overall, our results suggest that sorghum is well suited to the low deserts of CA, allowing productivity levels that compare favorably with previously reported sorghum production rates and are high relative to other biofuel and forage crops grown in the United States.

As water becomes a limiting global resource, the viability of biofuel feedstock production will depend on water requirements and water use efficiency. Seasonal water evaporation for sorghum was 1094 mm H_2O (Table 1), similar to alfalfa in the Imperial Valley (1270 mm season^{-1}) (Sanden *et al.*, 2011). However, the productivity of sorghum was almost twice that of alfalfa (Sanden *et al.*, 2011), therefore the growing season BWR of sorghum was relatively high (4.0 and 1.9 g kg^{-1} for sorghum and alfalfa, respectively; Table 1). Sorghum also had a high BWR compared to biofuel crops grown in the Midwestern United States where seasonal water use and biomass production rates are lower (maize: 612 mm season^{-1}, BWR 3.1 g kg^{-1}; Miscanthus: 955 mm season^{-1}, BWR 3.1 g kg^{-1}; switchgrass: 764 mm season^{-1}, BWR 1.0 g kg^{-1}) (Dohleman & Long, 2009; Hickman *et al.*, 2010). Tropical sugarcane in São Paulo, Brazil typically has high yields (75.1 Mg ha^{-1} season^{-1}) with water needs similar to sorghum (1227 mm season^{-1}), resulting in a high BWR of 6.1 g kg^{-1} (Gouvêa *et al.*, 2009; I.B.G.E., 2013). Overall, sorghum's seasonal water use was high compared to biofuels in the Midwest, but similar to those of other crops grown in the Imperial Valley and sugarcane in Brazil. However, the high biomass production rates of sorghum offset ET rates resulting in a relatively high BWR compared to crops in the United States. Irrigation is required for production of all crops in high irradiance arid environments and will be a key focus of any life cycle assessment (LCA) of sorghum production in this region. Water is currently allocated to agricultural crops in the Imperial Valley, so that replacing crops such as alfalfa with more efficient sorghum may not increase water demands. However, current irrigation practices in the Imperial Valley may not be sustainable as the Colorado River, the water source of irrigated crops in the Imperial Valley, has an uncertain future (Gleick, 2010). Therefore, while the BWR of sorghum was high

compared to other biofuel crops in the United States, water inputs for irrigated crops in the Imperial Valley must be considered in future assessments of sustainability.

Sorghum exhibited high RUE and no photosynthetic saturation at high light levels. Sorghum's RUE (1.5 g MJ^{-1}) exceeded alfalfa grown in the Imperial Valley (0.8 g MJ^{-1}) (Sanden *et al.*, 2011), as well as biofuels in the Midwestern United States including maize (0.7 g MJ^{-1}), Miscanthus (1.1 g MJ^{-1}), and switchgrass (0.3 g MJ^{-1}) (Dohleman & Long, 2009; Hickman *et al.*, 2010). Sorghum RUE was lower than the typical RUE of sugarcane in São Paulo, Brazil (2.2 g MJ^{-1}; where yield data (I.B.G.E., 2013) were paired with cumulative PAR derived from daily average solar irradiance (18.6 MJ m^{-2} day^{-1}) (Marin *et al.*, 2008), assuming PAR is 50% of total irradiance). Residuals of the relationships between NEE and PAR indicated that NEE was inhibited by high VPD and temperature, however, these relationships were not detectable during peak photosynthetic phases. Overall, during high growth phases, VPD and air temperature did not appear to inhibit photosynthesis allowing the radiation use efficient crop to take advantage of consistent high light conditions in this extreme environment.

We observed significant variation in crop performance across the three growth periods. Despite experiencing the highest air temperatures of the season (Fig. 1) and exhibiting the greatest sensitivity to air temperature, growth period 2 produced the greatest amount of biomass. Therefore, although air temperature was inhibitory to NEE, the highest temperatures of the season corresponded with the highest biomass production rates. Biomass production rates also varied across growth periods. Most strikingly, the first and second growth periods had higher biomass production rates compared to the third growth period (Table 1). The decrease in production rates in the third growth period did not appear to be caused by VPD or temperature, as NEE response to these environmental variables was similar to growth period 1 and less sensitive compared to growth period 2. The third growth period tended to experience lower light levels compared to periods 1 and 2 (Fig. 1c) with shorter day lengths associated with autumn and winter (on average 2–4 less daylight hours in autumn–winter compared to spring–summer). It is also possible that damage during the second harvest could have contributed to the relatively low biomass production during the third growth period. Lower productivity in ratoon crops has been previously shown to be caused by unavoidable damage during harvest (e.g. in sugarcane (Bull, 2000; Cabral *et al.*, 2012) and in sorghum (Han *et al.*, 2012; Rao *et al.*, 2013). It has also been reported that certain sorghum genotypes exhibit lower biomass production in the second and third ratoon crop of the growing season compared to the crop grown from seed (Rooney et al., 2007; Rao et al., 2013), however, these genetic drivers are not well resolved.

Sorghum's biomass production rates in a high irradiance arid environment were comparable with those of the most successful biofuels worldwide. However, a LCA is necessary to make a full accounting of costs and benefits (Lund & Biswas, 2008) and thereby evaluate the sustainability of sorghum in the low desert. For example, a LCA would consider the costs associated with fertilizer inputs. The fertilizer inputs for sorghum were 270 kg N ha^{-1}, higher than those typically required for sugarcane systems in Brazil (Otto et al., 2013) and Miscanthus and switchgrass systems in the Midwest (Dohleman & Long, 2009; Hickman et al., 2010). Nitrogen fertilizers are energetically costly and lead to higher emissions of nitrogen trace gases such as N_2O and NO which have implications for climate change and air quality. Like all other energy sources, there are complex trade-offs that need to be considered to achieve sustainable energy production.

We have demonstrated that a sorghum-based biofuel production system in a high irradiance arid environment can have high productivity despite the challenging environment. The high water and radiation use efficiency of sorghum suggest that it may be a viable biofuel crop in the low desert. Our study highlights the potential of these environments for future biofuel feedstock production and provides fundamental data for future LCA studies of biofuel sustainability.

Acknowledgements

The authors are grateful to F. Miramontes and F. Maciel and the staff of the University of California Desert Research and Extension Center for much skillful assistance, to K. Kitajima for technical advice, and to Lindy Allsman and many dedicated undergraduate field assistants. We also thank R. Scott, C. Fertitta, and anonymous reviewers for helpful suggestions on early versions of this manuscript. We thank Coby Kriegshauser at Scott Seed Co. for providing seed for these experiments. This work was supported by the USDA-NIFA Award No. 2011-67009-30045, and by U.C. Riverside.

References

Archontoulis S, Yin X, Vos J, Danalatos N, Struik P (2012) Leaf photosynthesis and respiration of three bioenergy crops in relation to temperature and leaf nitrogen: how conserved are biochemical model parameters among crop species? *Journal of Experimental Botany*, **63**, 895–911.

Baldocchi DD, Hincks BB, Meyers TP (1988) Measuring biosphere-atmosphere exchanges of biologically related gases with micrometeorological methods. *Ecology*, **69**, 1331–1340.

Bazdarich M, Sebasta P (2001) *On the Economic Feasibility of Sugar Cane-to-Ethanol Operations in the Imperial Valley*. UCR Forecasting Center, A. Gary Anderson Graduate School of Management and UC Desert Research and Extension Center, University of California, Riverside, UCR Forecasting Center.

Beale CV, Long SP (1995) Can perennial C4 grasses attain high efficiencies of radiant energy conversion in cool climates? *Plant, Cell & Environment*, **18**, 641–650.

Bull T (2000) The sugar-cane plant. In: *Manual of Cane Growing* (eds Hogarth M, Allsop P), pp. 71–83. Australian Bureau of Sugar Experimental Stations, Indooroopilly.

Cabral OMR, Rocha HR, Gash JH, MaV Ligo, Tatsch JD, Freitas HC, Brasilio E (2012) Water use in a sugarcane plantation. *GCB Bioenergy*, **4**, 555–565.

California Air Resources Board (CARB) (2009) Assembly Bill 32: Global Warming Solutions Act. Available at: http://www.arb.ca.gov/cc/ab32/ab32.htm (accessed 10 April 2014).

Chandel AK, Singh OV, Chandrasekhar G, Rao LV, Narasu ML (2010) Key drivers influencing the commercialization of ethanol-based biorefineries. *Journal of Commercial Biotechnology*, **16**, 239–257.

Curt MD, Fernandez J, Martinez M (1998) Productivity and radiation use efficiency of sweet sorghum [Sorghum bicolor (L.) Moench] cv. Keller in central Spain. *Biomass and Bioenergy*, **14**, 169–178.

Davis SC, Parton WJ, Grosso SJD, Keough C, Marx E, Adler PR, Delucia EH (2011) Impact of second-generation biofuel agriculture on greenhouse-gas emissions in the corn-growing regions of the US. *Frontiers in Ecology and the Environment*, **10**, 69–74.

Dohleman FG, Long SP (2009) More productive than maize in the Midwest: how does miscanthus do it? *Plant Physiology*, **150**, 2104–2115.

Foken T (ed.) (2008) *Micrometeorology*. Springer-Verlag, Berlin Heidelberg.

Gelfand I, Sahajpal R, Zhang X, Izaurralde RC, Gross KL, Robertson GP (2013) Sustainable bioenergy production from marginal lands in the US Midwest. *Nature*, **493**, 514–517.

Gerbens-Leenes W, Hoekstra AY, Van Der Meer TH (2009) The water footprint of bioenergy. *Proceedings of the National Academy of Sciences*, **106**, 10219–10223.

Gleick PH (2010) Roadmap for sustainable water resources in southwestern North America. *Proceedings of the National Academy of Sciences*, **107**, 21300–21305.

Goulden ML, Miller SD, Da Rocha HR, Menton MC, De Freitas HC, Silva Figueira E, AM, De Sousa CaD (2004) Diel and seasonal patterns of tropical forest CO2 exchange. *Ecological Applications*, **14**, 42–54.

Gouvêa JRF, Sentelhas PC, Gazzola ST, Santos MC (2009) Climate changes and technological advances: impacts on sugarcane productivity in tropical southern Brazil. *Scientia Agricola*, **66**, 593–605.

Han KJ, Alison MW, Pitman WD, Day DF, Kim M, Madsen L (2012) Planting date and harvest maturity impact on biofuel feedstock productivity and quality of sweet Sorghum grown under temperate Louisiana conditions. *Agronomy Journal*, **104**, 1618–1624.

Hickman GC, Vanloocke A, Dohleman FG, Bernacchi CJ (2010) A comparison of canopy evapotranspiration for maize and two perennial grasses identified as potential bioenergy crops. *GCB Bioenergy*, **2**, 157–168.

Hollinger SE, Bernacchi CJ, Meyers TP (2005) Carbon budget of mature no-till ecosystem in North Central Region of the United States. *Agricultural and Forest Meteorology*, **130**, 59–69.

Hsieh C-I, Katul G, Chi T-W (2000) An approximate analytical model for footprint estimation of scalar fluxes in thermally stratified atmospheric flows. *Advances in Water Resources*, **23**, 765–772.

Huxman TE, Turnipseed AA, Sparks JP, Harley PC, Monson RK (2003) Temperature as a control over ecosystem CO2 fluxes in a high-elevation, subalpine forest. *Oecologia*, **134**, 537–546.

I.B.G.E. (2013) *Systematic Survey of Agricultural Production (LASP)*. Sistema IBGE de Recuperação Automática. Available at: http://www.ibge.gov.br/home/ (accessed 10 April 2014).

Jackson RB, Jobbágy EG, Avissar R et al. (2005) Trading water for carbon with biological carbon sequestration. *Science*, **310**, 1944–1947.

Kaffka S (2009) Can feedstock production for biofuels be sustainable in California? *California Agriculture*, **63**, 202–207.

Lund C, Biswas W (2008) A review of the application of lifecycle analysis to renewable energy systems. *Bulletin of Science, Technology & Society*, **28**, 200–209.

Marin FR, Lopes-Assad ML, Assad ED, Vian CE, Santos MC (2008) Sugarcane crop efficiency in two growing seasons in Sao Paulo State, Brazil. *Pesquisa Agropecuaria Brasileira*, **43**, 1449–1455.

Mastrorilli M, Katerji N, Rana G (1999) Productivity and water use efficiency of sweet sorghum as affected by soil water deficit occurring at different vegetative growth stages. *European Journal of Agronomy*, **11**, 207–215.

Mauder M, Foken T (2006) Impact of post-field data processing on eddy covariance flux estimates and energy balance closure. *Meteorologische Zeitschrift*, **15**, 597–609.

Monteith J (1993) The exchange of water and carbon by crops in a Mediterranean climate. *Irrigation Science*, **14**, 85–91.

Monteith J, Moss C (1977) Climate and the efficiency of crop production in Britain [and discussion]. *Philosophical Transactions of the Royal Society of London B, Biological Sciences*, **281**, 277–294.

Mussatto SI, Dragone G, Guimarães PMR *et al.* (2010) Technological trends, global market, and challenges of bio-ethanol production. *Biotechnology Advances*, **28**, 817–830.

Nobel P (ed.) (1974) *Introduction to Biophysical Plant Physiology*. Freeman, WH, San Francisco.

Otto R, Mulvaney RL, Khan SA, Trivelin PCO (2013) Quantifying soil nitrogen mineralization to improve fertilizer nitrogen management of sugarcane. *Biology and Fertility of Soils*, **49**, 893–904.

Rao PS, Rathore A, Reddy BVS (2013) Interrelationship among biomass related traits and their role in sweet Sorghum cultivar productivity in main and ratoon crops. *Sugar Tech*, **15**, 278–284.

Raupach MR (1994) Simplified expressions for vegetation roughness length and zero-plane displacement as functions of canopy height and area index. *Boundary-Layer Meteorology*, **71**, 211–216.

Reichstein M, Falge E, Baldocchi D *et al.* (2005) On the separation of net ecosystem exchange into assimilation and ecosystem respiration: review and improved algorithm. *Global Change Biology*, **11**, 1424–1439.

RFA (Renewable Fuel Association) (2010) Renewable Fuel Standard. Available at: http://www.ethanolrfa.org/pages/renewable-fuel-standard (accessed 10 April 2014).

Rooney WL, Blumenthal J, Bean B, Mullet JE (2007) Designing sorghum as a dedicated bioenergy feedstock. *Biofuels Bioproducts and Biorefining*, **1**, 147–157.

Sanden B, Klonsky K, Putnam D, Schwankl LKB (2011) Comparing costs and efficiencies of different alfalfa irrigation systems. In: *Western Alfalfa and Forage Conference*. Las Vegas, NV, USA, UC Cooperative Extension, Plant Sciences Department, University of California, Davis.

Schmer MR, Vogel KP, Mitchell RB, Perrin RK (2008) Net energy of cellulosic ethanol from switchgrass. *Proceedings of the National Academy of Sciences*, **105**, 464–469.

Schotanus P, Nieuwstadt FTM, HaR Bruin (1983) Temperature measurement with a sonic anemometer and its application to heat and moisture fluxes. *Boundary-Layer Meteorology*, **26**, 81–93.

Shoemaker CE, Bransby DI (2010) The role of sorghum as a bioenergy feedstock. In: *Sustainable Alternative Fuel Feedstock Opportunities, Challenges and Roadmaps for Six US Regions* (eds Braun R, Karlen D, Johnson D), pp. 149–159. Proceeedings of the Sustainable Feedstocks for Advance Biofuels Workshop, Atlanta, GA.

Shurpali NJ, Hyvönen NP, Huttunen JT *et al.* (2009) Cultivation of a perennial grass for bioenergy on a boreal organic soil–carbon sink or source? *GCB Bioenergy*, **1**, 35–50.

Singh MP, Erickson JE, Sollenberger LE, Woodard KR, Vendramini JMB, Fedenko JR (2012) Mineral composition and biomass partitioning of sweet sorghum grown for bioenergy in the southeastern USA. *Biomass and Bioenergy*, **47**, 1–8.

Stanhill G (1986) Water use efficiency. *Advances in Agronomy*, **39**, 53–85.

Vanloocke A, Twine TE, Zeri M, Bernacchi CJ (2012) A regional comparison of water use efficiency for miscanthus, switchgrass and maize. *Agricultural and Forest Meteorology*, **164**, 82–95.

Vickers D, Mahrt L (1997) Quality control and flux sampling problems for tower and aircraft data. *Journal of Atmospheric and Oceanic Technology*, **14**, 512–526.

de Vries SC, van de Ven GW, van Ittersum MK, Giller KE (2010) Resource use efficiency and environmental performance of nine major biofuel crops, processed by first-generation conversion techniques. *Biomass and Bioenergy*, **34**, 588–601.

Webb EK, Pearman GI, Leuning R (1980) Correction of flux measurements for density effects due to heat and water vapour transfer. *Quarterly Journal of the Royal Meteorological Society*, **106**, 85–100.

Xie T, Su P (2012) Canopy and leaf photosynthetic characteristics and water use efficiency of sweet sorghum under drought stress. *Russian Journal of Plant Physiology*, **59**, 224–234.

Zegada-Lizarazu W, Monti A (2012) Are we ready to cultivate sweet sorghum as a bioenergy feedstock? A review on field management practices. *Biomass and Bioenergy*, **40**, 1–12.

Zeri M, Anderson-Teixeira K, Hickman G, Masters M, Delucia E, Bernacchi CJ (2011) Carbon exchange by establishing biofuel crops in Central Illinois. *Agriculture, Ecosystems & Environment*, **144**, 319–329.

Zeri M, Hussain MZ, Anderson-Teixeira KJ, Delucia E, Bernacchi CJ (2013) Water use efficiency of perennial and annual bioenergy crops in central Illinois. *Journal of Geophysical Research: Biogeosciences*, **118**, 581–589.

12

Utilizing biofuels for sustainable development in the panel of 17 developed and developing countries

ILHAN OZTURK

Faculty of Economics and Administrative Sciences, Cag University, 33800 Mersin, Turkey

Abstract

This study investigates the dynamic linkages between biofuels production and sustainable indicators in the panel of 17 developed and developing countries, over the period of 2000–2012. The study emphasized the role of biofuels production in the sustainable development of the region. For this purpose, the study utilized four main sustainable indicators including carbon dioxide emissions, energy intensity, renewable energy generation, and total population that have a significant impact on the biofuels production. The study used dynamic heterogeneous panel econometric technique – Generalized Method of Moments and found that carbon dioxide emissions increase along with the increase in biofuels production. Therefore, the caution should be applied when burning the biofuels during the production process. In addition, renewable electricity generation also increases the biofuels production in the region. The results of robust least square regression confirmed that all of the sustainable indicators have a significant association with the biofuels production, as total primary energy consumption increases the biofuels production, while total population significantly decreases the biofuels production in the region. The results derived to the conclusion that for sustainable development in the region, the policymakers should have to formulate carbon free policies that coupled with the renewable energy sources for emphasizing the life cycle of bioenergy during the production process.

Keywords: biofuels production, carbon dioxide emissions, total population, total primary energy consumption, total renewable electricity generation

Introduction

Biofuels production merely based on living organisms that contain more than 80% of renewable materials, as it is derived from the process of photosynthesis, therefore, mostly referred as solar energy source. The term biofuels is hard to explain, as we used biofuel as like fossil fuel. However, biofuel has a unique correspondence from other energy sources, as it considered being carbon dioxide emissions neutral (AENews, 2014). There are different counter arguments poses the biofuels production. One argued in the favor of biofuels production that addresses the greenhouse gas emissions and removes carbon dioxide emissions from the atmosphere. However, the second stream of argument against the biofuels production poses a major threat to global food security and for biodiversity (Christopherson, 2008). According to UNCTAD (2008, P. vii), *'There is growing interest in biofuels in many developing countries as a means of 'modernizing' biomass use and providing greater access to clean liquid fuels while helping to address energy costs, energy security and global warming concerns associated with petroleum fuels'.* There is undeniable fact that fossil fuel is one of the main concerns in the circle of environmentalists and academicians toward environmental degradation. The quest for the alternative energy sources always is the need of the globe to sustain and formulates green policies. The biofuels somehow a convenient solution to preserve environment and substitute of high petroleum; however, biofuel energy should be evaluated with some cautions, as it is so far highly environmental friendly energy sources across the globe (Chauhan & Shukla, 2011). Biodiesel fuel is the only option to serve healthy nation and fulfill the requirements of Clean Air Act of 1990, as the particulate emissions associated with the biodiesel fuels is lowering around more than 50% with fossil source diesel.

Bioenergy is roughly estimated around more than 10%, that is, 50 EJ of global total primary energy supply, and it serves mostly for households cooking and heating that have a significant impact on environment and health. Modern bioenergy used in building sector around 5 EJ in 2012, while around 8 EJ were used in industrial sector. A total of 370 TWh of bioenergy electricity were produced in the year 2012 that corresponds

Correspondence: Ilhan Ozturk

e-mail: ilhanozturk@cag.edu.tr

Table 1 Biofuels generation

Biofuels generation	Commercial scale	Feed stocks
Fist generation	Sugar and starch-based ethanol Oil-crop based biodiesel and vegetable oil Anaerobic digestion	Sugar cane Sugar beet Starch bearings grains Animal fats Cooking oils
Second and third generation	Cellulosic ethanol Biomass-to-liquids diesel Bio-synthetic gas	Algae-based biofuels Diesel-type biofuels using biological or chemical catalysts

Source: IEA (2013).

to 1.5% of global electricity generation (IEA, 2013). Table 1 shows the first, second, and third generation of biofuels.

There are two major classifications of biofuels, that is, one is the primary biofuels and another one is secondary biofuels. Primary biofuels contained fuel wood, wood chips, and pellets that primarily used in household items including heating, cooking, or electricity production. Secondary biofuels include liquid biofuels (i.e., ethanol and biodiesel) that can be used in vehicles and industrial processes. Bioenergy accounts for more than 80% used in household items, while 18–20% used in industry. There is only a 2% used in transport sector which shows a very limited role of bioenergy in this sector (Green Facts, 2010).

The overall facts indicate the importance of biofuels production that is the way towards the future energy fuel to the developed and developing countries. There are number of studies established the viability of the biofuels production in their nationwide and across the region. Therefore, there is a pressing need to evaluate biofuels production coupled with the sustainable indicators in the panel of diverse countries portfolios including the set of developed and developing countries that gives the sound policy vista for developing biofuels energy as a green source of energy. The effectiveness of biomass and bioenergy found in the study of Hall & Scrase (1998) where they probe that in the future, whether biomass fuel will be environmental friendly? This notion is used by many developed countries and found that biomass fuel is environmental friendly. However, the systems need to be caution in terms of environment and social benefits across the countries. Schaldach et al. (2011) examining the potential impact of biofuel development on land-use change in India and found that rising population and biofuel production will affect on the land-use change over the coming decades in India. Ravindranath et al. (2011) further step

toward the sustainability of India in relation with the biofuel development and its impact on three key environmental factors comprising land use, environment, and food production. The results suggest that technological and policy breakthroughs are prerequisite for promoting sustainable and long-term biofuel production in India. According to Groom et al. (2008, p. 602) 'Conservation biologists can significantly broaden and deepen efforts to develop sustainable fuels by playing active roles in pursuing research on biodiversity-friendly biofuel production practices and by helping define biodiversity-friendly biofuel certification standards'.

Escobar et al. (2009) examined the dynamic linkages between environment, food security, technology, and biofuel production and conclude that there is a substantial need to regulate some international certification mechanism for checking the considerable impact of biofuels on environment, land use, and food security across the globe. Yang et al. (2009) conclude that China's current biofuel development path may affect on nation's food supply, terms of trade and nation's environment. Hoefnagels et al. (2010) traces the footprints of greenhouse gas during different biofuel production systems and found that greenhouse gas performance of biofuels varies depends upon the production method applications and the selected system boundaries. Common practice in life cycle assessment (LCA) of bioenergy has been to assume that any carbon dioxide (CO_2) emission related to biomass combustion equals the amount absorbed in biomass, thus assuming no climate change impacts (Cherubini et al., 2011; Van Zelm et al., 2014). Demirbas (2007) conclude that the main advantage of biofuels production is to minimize the risk of greenhouse gas emissions and unpredictable climate change across the globe. According to Hill et al. (2006, p. 11206), 'Negative environmental consequences of fossil fuels and concerns about petroleum supplies have spurred the search for renewable transportation biofuels. To be a viable alternative, a biofuel should provide a net energy gain, have environmental benefits, be economically competitive, and be producible in large quantities without reducing food supplies'. Birur et al. (2008) examined the impact of biofuels production on global agricultural markets using the Computable general equilibrium model and found that terms of trade in oil sector considerable improves in all of the oil exporting regions on the cost of aggregate agricultural sectors. Nigam & Singh (2011) provide extensive literature review on liquid biofuels and conclude that the process of conversion and chemical transformation of liquid biofuels is too much expensive; therefore, for economic viability of liquid biofuels on large scale is difficult to attain this bioenergy.

There is a considerable need to search the cheap way of conservation and chemical transformation of liquid biofuels for using worldwide. Cremonez *et al.* (2015) recognize the three major outcomes that inclined from the production of biofuels, that is, economic impact, environmental preservation, and societal benefits in Brazil. The results derive that there is substantial need to improved the production routes of biofuel energy with lowering cost of production, as the raw material used in biofuels production mostly linked with the nitrous oxide emissions that may affect on agricultural productivity loss, threatened wildlife, and intensive water usage that impact on the productivity loss, environmental degradation, and social problems in the society. Spartz *et al.* (2015) conducted the natural experiment and gathered the public perceptions regarding bioenergy and land-use change in the frames of agriculture and forestry. The results indicate the higher uncertainty in both of the frames; however, perception regarding the future energy prices in the forestry frame is highly uncertain. Panichelli & Gnansounou (2015) considered different modeling choices for examining the potential impact of biofuels production on land-use change and greenhouse gas emissions. The results show the favorable outcomes that addresses from biofuels production to mitigate greenhouse gas emissions and land-use change.

The overall studies indicate the significance of biofuels in the sustainable development of the globe. Therefore, this study explore the dynamic linkages between total biofuels production, carbon dioxide emissions, total primary energy consumption, renewable electricity generation, and total population in the panel of 17 developed and developing countries for the period of 2000–2012. The study used dynamic panel heterogeneous econometric modeling coupled with the robust least square regression for robust estimations. The results of the study provide broader insights toward sustainability for policymaking and biofuels effectiveness across the countries.

Materials and Methods

The data of the biofuels and sustainable indicators taken from U.S. Energy Information Administration (EIA, 2014) over the period of 2000–2012 for 17 developed and developing countries namely Canada, United States, Argentina, Brazil, Cuba, Paraguay, Peru, Austria, Czech Republic, Denmark, France, Germany, Italy, Spain, Sweden, China, and India. The variables comprises total biofuels production in thousand barrels per day, total carbon dioxide emissions from the consumption of energy in million metric tons, total population in million, energy intensity – total primary energy consumption per Dollar of GDP [Btu per year 2005 U.S. Dollars (purchasing power parities)], and total renewable electricity net generation in billion kilowatt hours. These variables selected for examining the long-run relationship between biofuels production and its

regressors, that is, sustainable indicators in the panel of selected 17 developed and developing countries. Table 2 shows the list of variables and their expected signs.

Table 2 shows that sustainable indicators have a positive relationship with the biofuels production, as increasing carbon dioxide emissions elevate the importance of biofuels production in the region. Conventional use of combustion of fossil fuels emerged the deleterious impact on the environmental degradation. Thus, there is a need to emphasize on the production of biofuels that have considerable better impact on environment as compare to the combustion of fossil fuels energy. Energy intensity subsequently increases the use of biofuels production, as total primary energy consumption insufficient to fulfill the need of energy supply and demand gap for the region. Renewable energy consumption search for the environmental friendly energy production that minimizes the risks of environmental-related health issues and provide safe and clean air to the rapid population in a region. These sustainable indicators indicate the undeniable significance to promote biofuels production for scoring sustainable agenda reforms across the countries.

The study used following nonlinear regression equation to explore the impact of sustainable indicators on biofuels production:

$$\text{Log(BIOF)}_{i,t} = \alpha_0 + \alpha_1 \text{Log(CO}_2)_{i,t} + \alpha_2 \text{Log(EINT)}_{i,t} \\ + \alpha_4 \text{Log(RENEGEN)}_{i,t} + \alpha_4 \text{Log(POP)}_{i,t} + \varepsilon_{i,t} \quad (1)$$

where, BIOF represents biofuels production, CO_2 represents carbon dioxide emissions, EINT represents energy intensity, RENEGEN represents renewable energy, POP represents total population, 'i' represents cross section identifiers, that is, country 1 to country 17, 't' represents time period from 2000 to 2012, 'Log' represents natural logarithm, and ε represents error term.

Equation (1) shows the empirical relationship between biofuels production and sustainable indicators for assessing the parameter estimates between the variables. This equation tested by number of panel econometric techniques for robust results. The study first evaluated the panel unit root tests by Im, Pesaran and Shin (i.e., Im *et al.*, 2003) for confirming the stationary series of the candidate variables. The study further used Levin *et al.* (2002) panel unit root test for only those variables that does not confirmed their stationary properties by Im *et al.* (2003). In the Levin, Lin and Chu (LLC) approach, it is assumed that, all the α_i have a common value, so that the null hypothesis is formulated as follows:

Table 2 List of variables

Variables	Symbol	Expected Sign
Biofuels production	BIOF	
Regressors		
Carbon dioxide emissions	CO_2	Positive
Energy Intensity	EINT	Positive
Renewable Energy	RENEGEN	Positive
Population	POP	Positive

$$H_0 : \alpha = 0 \text{ vs. } H_1 : \alpha < 0$$

Thus, an estimator of α is absorbing the heteroscadasticity across the time series that make up the panel. While, in Im, Pesaran, and Shin approach (IPS) is also based on the ADF regressions; however, the H_0 and H_1 are considerable different from the LLC approach, that is, the rejection of the null hypothesis implies that all the series are stationary:

$$H_0 : \alpha_1 = \alpha_2 = \ldots = \alpha_N = 0 \text{ vs. } H_1$$
$$\text{: Some but not necessarily all } \alpha_i < 0$$

The study further employed Pedroni (1999) panel co-integration for substantiate the long-run relationship between the variables. Pedroni's panel co-integration test allows the four panel statistics 'within' dimension including panel v-statistic, panel rho-statistic, panel ADF-statistic, and panel PP-statistics, while there are four weighted panel regression statistics also available in 'within dimension'. In addition, there are three group statistics available for substantiating the long-run relationship between the variables including group rho-statistic, group ADF-statistic and group PP-statistic. In majority of the cases, panel v-statistics substantially show positive and greater magnitude, while remaining panel and group statistics shows negative sign. However, for long-run cointegration relationship between the variables, the respective probability values may also have a considerable impact on confirmation of co-integration relationship between the variables.

In majority of the panel econometric techniques, there appears a problem of endogeneity in the model that leads the wrong results. Thus, the study employed dynamic panel approach – Generalized Method of Moments (GMM) that handle the problem of endogeneity problem in the given data set. Equation (2) shows the GMM model specification.

$$\text{Log(BIOF)}_{i,t} = \alpha_0 + \theta_{i,t} + \alpha_1 \text{Log(CO}_2)_{i,t} + \alpha_2 \text{Log(EINT)}_{i,t}$$
$$+ \alpha_4 \text{Log(RENEGEN)}_{i,t} + \alpha_4 \text{Log(POP)}_{i,t} + \varepsilon_{i,t} \quad (2)$$

where θ represents list of instrumental variables including the first lagged value of the explanatory variables.

After correcting the problem of endogeneity from the model, the study detected and adjusted outliers from the given variables data set and presented robust results. For this purpose, the study used robust least square regression apparatus that is less sensitive to outliers. There are three different methods available for robust regressions including M-estimation proposed by Huber (1973) that addresses the dependent variable

outliers; S-estimation as proposed by Rousseeuw & Yohai (1984) that focuses outliers in regressors' variables; and finally, MM-estimation proposed by Yohai (1987) that is the combination of M-estimation and S-estimations.

Results

This section comprises the following sequential order of estimations, that is, descriptive statistics, correlation matrix, panel unit root test, panel co-integration, dynamic panel modeling, and robust least square regression. Table 3 shows the descriptive statistics of the variables.

The descriptive statistics show that total biofuels production has a minimum value of 0.020 thousand barrels per day to maximum value of 971.619 thousand barrels per day, having a mean value of 57.064 thousand barrels per day. Carbon dioxide emissions have an average value of 948.340 million metric tons with standard deviation of 1808.336 million metric tons. Energy intensity has a minimum value of 3285.539 US dollar to maximum value of 32 192.63 US dollar, having a positive skewed distribution and considerable peak of the distribution. Total population is 5.337 million to maximum 1336.990 million, with an average of 193.838 million having standard deviation of 379.074 million, respectively. Finally, the region comprises minimum value of 0.488 total renewable electricity net generations in Billion Kilowatt hours to maximum value of 1003.515 billion kilowatt hours, having a mean value of 133.583 billion kilo watt hours with standard deviation of 170.008 billion kilowatt hours.

Table 4 shows the correlation matrix between the variables. The results show that carbon dioxide emissions have a positive correlation with the biofuels production, as the correlation coefficient value of $r = 0.445$, $P < 0.000$ depicts the significant association between the variables. The results indicate that along with the increase in carbon emissions, the region emphasized on the production of biofuels rather than the fossil fuel energy that increase the carbon dioxide emissions in the region. In addition, combustion of biofuels during

Table 3 Descriptive statistics

	BIOF	CO$_2$	EINT	POP	RENEGEN
Mean	57.06480	948.3409	8082.234	193.8383	133.5839
Maximum	971.6192	8547.746	32192.63	1336.990	1003.515
Minimum	0.020000	3.453520	3285.539	5.337420	0.488100
SD	153.4918	1808.336	5102.211	379.0744	170.0080
Skewness	3.910798	2.472694	2.637874	2.258935	1.900773
Kurtosis	19.37075	7.911910	11.20377	6.440978	6.990724
Observations	221	221	221	221	221
Cross-section	17	17	17	17	17

Table 4 Correlation matrix

	BIOF	CO_2	EINT	POP	RENEGEN
BIOF	1				
CO_2	0.445***	1			
EINT	−0.083	0.045	1		
POP	0.068	0.654***	0.027	1	
RENEGEN	0.537***	0.771***	0.057	0.528***	1

Note: *** indicates significance of the variables at 1 percent level.

production should be carefully handled. Similarly, renewable energy increases the production of biofuels, as the coefficient value indicates the high correlation between them, that is, $r = 0.537$, $P < 0.000$. Renewable energy consumption in the form of electricity supply needs environmental friendly energy production that useful for providing energy supply according to the requirement of energy demand in the region. The result further confirmed the conventional hypothesis, that is, massive population increases carbon dioxide emissions in the region, as the value of correlation coefficient indicates the positive and high correlation between them, that is, $r = 0.654$, $P < 0.000$. In addition, the results endorsed that renewable energy generation in the form of electricity increases carbon dioxide emissions, that is, the coefficient value indicates $r = 0.771$, $P < 0.000$. This relationship confirmed the need of biofuels production that has a minimal impact on environmental degradation. Finally, massive population required more energy, as correlation coefficient value indicates the positive and significant association between renewable energy and population in the region. This all exercise suggests the following points, that is, (i) the caution should be applied when burning the biofuels during the production process; (ii) to fulfill the energy supply-demand gap, there is required biofuels production; and (iii) rapid population hinders in the objective of sustainable

development, therefore, healthy and safe environment prerequisite for sustainable development.

Table 5 shows the results of panel unit root test to check the stationary series of the individual variables.

The results show that the data series of biofuels production and carbon dioxide emissions are more volatile in nature, as both of the variables are nonstationary at level; however, it becomes differenced stationary. The remaining variables, that is, energy intensity and renewable energy generation both are stationary at level when adjusted the time trend in the series. However, without adjusted trend in the data series, both the variables are differenced stationary. Finally, the data set of total population is not captured any stationary series till second differenced in IPS (Im, Pesaran and Shin) approach. Thus, the study used LLC panel unit root approach for confirming the stationary process. The LLC panel unit root test confirmed that population data is stationary at level, although, it does not become significant at first difference. The results confirmed the mixture of order of integration among the candidate variables, therefore, during estimation; it would be cautions to use robust least square regression that are less sensitive to outliers.

Table 6 shows the Pedroni (1999) panel residual co-integration test results for evaluating null hypothesis of no co-integration against the alternative hypothesis of co-integration relationship between the variables.

The results of Table 6 show that both the weighted panel PP-statistic and weighted panel ADF-statistic significant at 1% and 5% level, respectively. Similarly, group PP-statistic and group ADF-statistic also tend to show significant at 1% and 10% level, respectively. The results reject the null hypothesis of no co-integration and accepted the alternative hypothesis of co-integration relationship between the variables. The results established the long-run association between biofuels production and sustainable indicators in the panel of selected developed and developing countries. The study step toward the estimation of variable parameters in a

Table 5 Results of panel unit root test

Variables	Level		First difference	
	Constant	Constant + Trend	Constant	Constant + Trend
BIOF	3.194	−1.344	−6.353***	−4.040***
CO_2	4.706	−0.629	−6.679***	−4.960***
EINT	1.455	−1.738**	−10.071***	−8.098***
RENEGEN	4.696	−2.104**	−9.002***	−6.858***
POP	2.858	4.269	2.913	5.768
Levin, Lin and Chu (LLC) panel unit root test				
LLC test for POP	−2.749***	−5.518***	−1.368	−0.812

Note: *** and ** shows significance at 1% and 5% level respectively.

Table 6 Pedroni residual co-integration results

Cross-sections included: 17

Null Hypothesis: No co-integration
Trend assumption: No deterministic trend
User-specified lag length: 1
Newey–West automatic bandwidth selection and Bartlett kernel
Alternative hypothesis: common AR coefs. (within-dimension)

	Statistic	P	Weighted Statistic	P
Panel v-Statistic	0.571603	0.2838	−1.567849	0.9415
Panel rho-Statistic	2.952751	0.9984	3.072395	0.9989
Panel PP-Statistic	1.485271	0.9313	−3.103340	0.0010
Panel ADF-Statistic	2.463213	0.9931	−2.060473	0.0197

Alternative hypothesis: individual AR coefs. (between-dimension)

	Statistic	P
Group rho-Statistic	4.929505	1.0000
Group PP-Statistic	−6.059985	0.0000
Group ADF-Statistic	−1.530066	0.0630

multivariate setting. Therefore, the study used dynamic heterogeneous panel technique, that is, generalized method of moments (GMM) to adjust the problem of endogeneity in the given model and the results presented in Table 7.

The results of GMM show that carbon dioxide emissions have a positive and significant association with the biofuels production, as the coefficient value indicate that there is less elastic relationship between the variables. The results imply that environmental quality indicator is affected by the combustion of fossil fuels energy due to the high abundance and cheap source of energy in most of the developed and developing countries. Therefore, for caution of carbon free energy, there is

substantially required biofuels production that adjusted the problem of high emissions in the region. The result further depict that along with the increase in the renewable energy generation, there is largely required biofuels production, as if there is one percent increase in renewable energy generation, biofuels production increases by 0.415%. However, the magnitude is far less than the magnitude of carbon dioxide emissions that increases the production of biofuels in the region. The results obviously would facilitate the policy planners to formulate the policies related with the carbon free economy and sustainable development in the region. The remaining variables, that is, energy intensity and total population does not show any significant association during

Table 7 Panel dynamic modeling – generalized method of moments (GMM)

Dependent Variable: LOG(BIOF)

Method: Generalized Method of Moments
Estimation weighting matrix: HAC (Bartlett kernel, Newey-West fixed bandwidth = 5.000)
Standard errors & covariance computed using estimation weighting matrix
Instrument specification: LOG(CO_2(-1)) LOG(POP(-1)) LOG(RENEGEN(-1)) Log(EINT(-1))
Constant added to instrument list

Variable	Coefficient	SE	t-Statistic	P
C	−0.816184	3.740940	−0.218176	0.8275
LOG(CO_2)	0.828618	0.235484	3.518795	0.0005
LOG(POP)	−0.313812	0.244223	−1.284943	0.2002
LOG(RENEGEN)	0.415999	0.155783	2.670370	0.0082
LOG(EINT)	−0.265458	0.414985	−0.639681	0.5231
R-squared	0.533560	Mean dependent var		1.731545
Adjusted R-squared	0.524882	SD dependent var		2.261811
SE of regression	1.559039	Sum-squared resid		522.5796
Durbin–Watson stat	0.351840	J-statistic		1.94E-42
Instrument rank	5			

Table 8 Outliers detection in panel modeling

Model	Number of predictors	Model Fit statistics Stationary R-squared	Ljung-Box Q(18) Statistics	DF	Sig.	Number of outliers
BIOF-Model_1	4	.939	25.739	18	.106	22

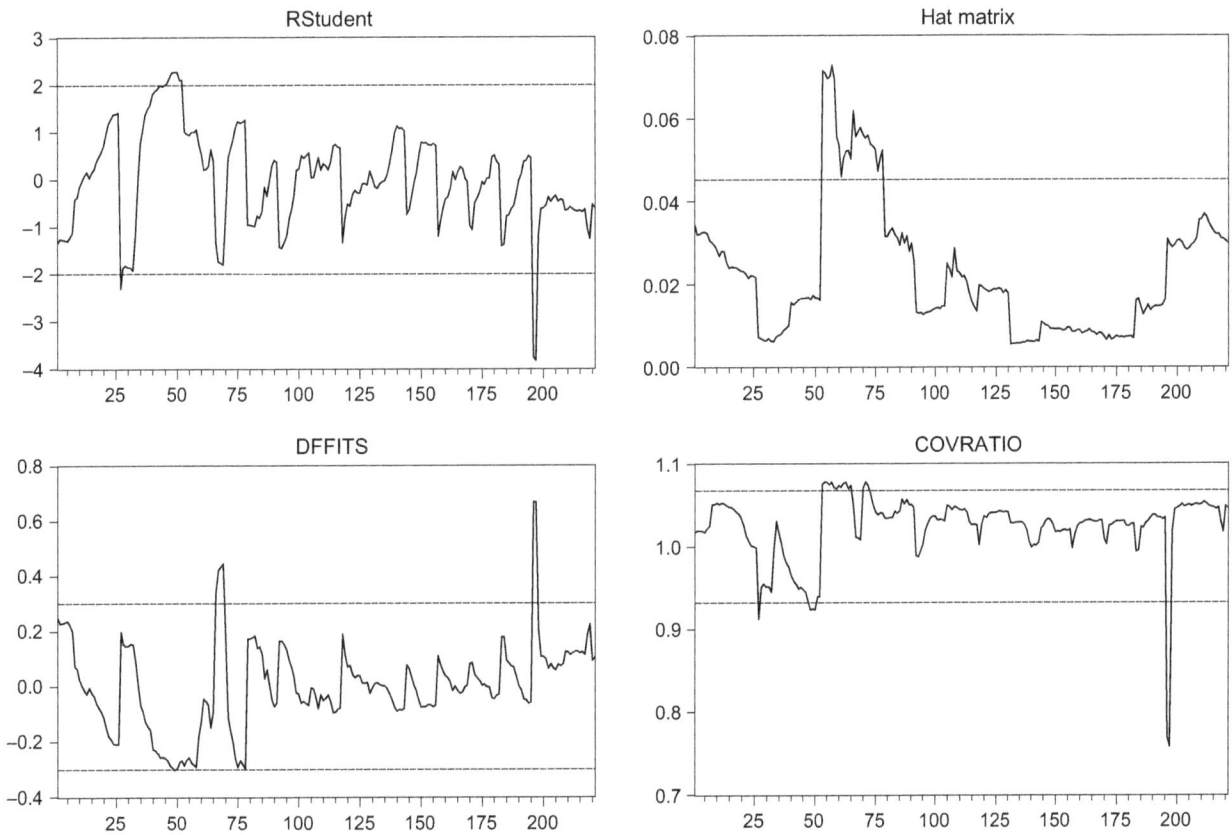

Fig. 1 Influence statistics.

the time period; however, we may not ignore the possible impact of both the variables on biofuels production. The other statistics including adjusted R-squared show that about 52.48% sustainable indicators explained the importance of biofuels production, while the value of J-statistic shows that the problem of endogeneity has been adjusted and the results are free from errors. After adjustment of endogeneity, the study further checked the possible outliers in the panel of countries, as during panel unit root estimation, some of the variables are highly volatile and there is considerable need to adjust the outliers and obtained the robust results. For this purpose, the study used several outliers' detection tests that presented in Table 8, Figs 1 and 2, respectively.

The result confirmed that around 22 outliers presented in the panel of 17 developed and developing

countries which should have to be adjusted before estimation. Figure 1 further shows the influence statistics for closer overview of outliers in the panel of countries.

Figure 1 shows that most of the observations of the given variable series surpass the average values; therefore, there is clear indication that the parameter estimates may be affected by the possible outliers in the given model. All of the four influence statistics confirmed that majority of the variables observations exceeds the average line. Thus, there is required variable by variable to see the observation trends. Figure 2 shows the leverage plots of each variable to observe the outliers in the data series. Figure 2 depict that variable series are fluctuating over the period of time, as one of the reason is obvious the difference of developed and

BIOF vs. variables (Partialled on regressors)

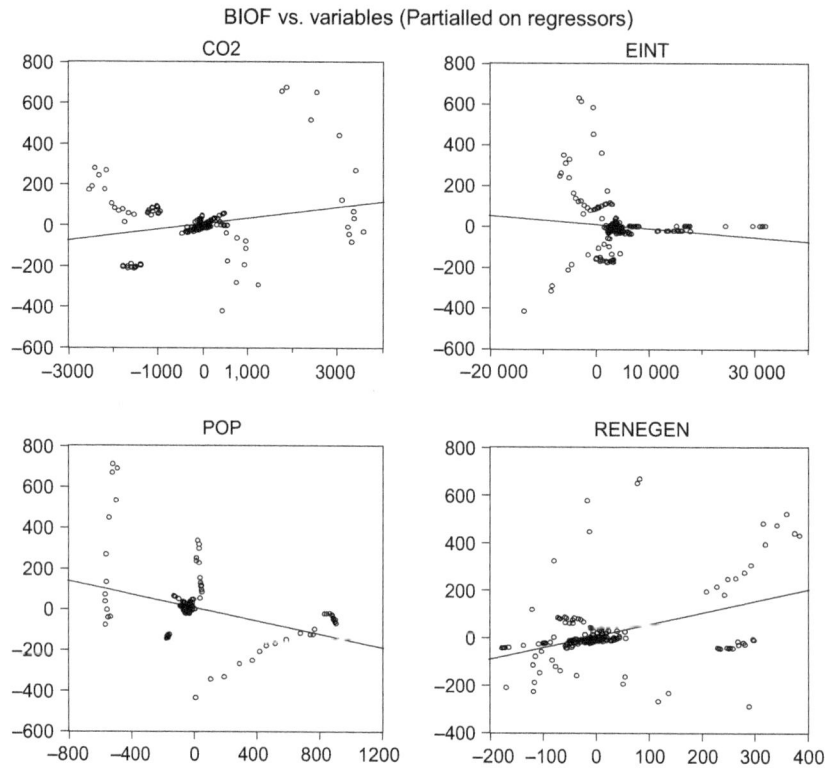

Fig. 2 Stability diagnostics - leverage plots. BIOF vs. Variables (Partialled on Regressors).

developing countries. Developed countries more pronounce toward the biofuels production and environmental indicators while developing countries struggling to achieve the optimum level of energy and carbon emissions.

After careful analysis of above exercise, there is considerable need to adjust the outliers from dependent variable and from its regressors. Therefore, the study used robust least square regression including M-estimations that adjust the outliers of dependent variables, while MM-estimations address both the dependent and independent variables' outliers. Table 9 shows the panel robust least square regression estimated that less sensitive to outliers.

The results of robust least square regression indicate that carbon dioxide emissions and renewable energy generation have a significant and positive impact on increasing biofuels production in the region, while total population significantly decreases the biofuels production. One of the possible implications of this result is that biofuels production sensitizes the total population by moving towards traditional sector to modern era. Therefore, the pace of mechanized world hinders by the rapid population pressure on the cost of supply-demand gap of energy in the region. The robust regression indicated the robust weighted R-squared 59.9% that are greater than the 6.6% higher than the actual

R-squared in GMM estimations. This result adjusted the outliers of dependent variable shows that explanatory variables have only 46.5% impact on biofuels production in the region.

The study further uses MM-estimations of robust least square regression to adjust the model outliers and results are presented in Table 10.

The results show that after adjusting both the dependent and independent variables' outliers, all of the sustainable indications exhibit the significant association with the biofuels production with the greater magnitude and improved statistical significance powers. The results reveal that if there is one percent increase in carbon dioxide emissions, biofuels production increases by 0.934%. The magnitude value is improved in MM-estimations as compared to GMM and M-estimations. Similarly, renewable energy generation along with the energy intensity both exerted the positive relationship with the biofuels production, as the coefficient values are 0.493% and 0.529%, respectively. Finally, total population have a significant and negative relationship with the biofuels production that need to device sound health policy to reduce the pressure of massive populations in order to utilize energy infrastructure across the countries. In both the M-estimation and MM-estimation, R-squared considerably decreasing while robust weighted R-squared increases in a greater extent that

Table 9 Panel robust least square results (M-estimations)

Dependent Variable: LOG(BIOF)

Method: Robust Least Squares
Included observations: 221
Method: M-estimation
M settings: weight = Bisquare, tuning = 4.685, scale = MAD (median centered)
Huber Type I Standard Errors & Covariance

Variable	Coefficient	SE	z-Statistic	P
C	−2.274181	2.086535	−1.089931	0.2757
LOG(CO$_2$)	0.880646	0.112631	7.818870	0.0000
LOG(EINT)	−0.099009	0.225922	−0.438245	0.6612
LOG(POP)	−0.348856	0.125392	−2.782131	0.0054
LOG(RENEGEN)	0.375143	0.081909	4.580000	0.0000
	Robust Statistics			
R-squared	0.474733	Adjusted R-squared		0.465006
Rw-squared	0.599739	Adjust Rw-squared		0.599739
Akaike info criterion	207.9675	Schwarz criterion		227.2109
Deviance	436.9568	Scale		1.477289
Rn-squared statistic	252.0903	Prob(Rn-squared stat.)		0.000000
	Non-robust Statistics			
Mean dependent var	1.729630	SD dependent var		2.256845
SE of regression	1.565053	Sum-squared resid		529.0682

Table 10 Panel robust least square results (MM-estimations)

Dependent variable: LOG(BIOF)

Method: Robust Least Squares
Included observations: 221
S settings: tuning = 0.275, breakdown = 0.90011, trials = 200, subsmpl = 5,
refine = 2, compare = 5
M settings: weight = Bisquare, tuning = 3.78
Random number generator: rng = kn, seed = 1540814231
Huber Type II Standard Errors & Covariance

Variable	Coefficient	SE	z-Statistic	P
C	−7.180870	1.508614	−4.759913	0.0000
LOG(CO$_2$)	0.934792	0.092803	10.07284	0.0000
LOG(EINT)	0.529426	0.160719	3.294115	0.0010
LOG(POP)	−0.712284	0.092219	−7.723788	0.0000
LOG(RENEGEN)	0.493846	0.062015	7.963311	0.0000
	Robust Statistics			
R-squared	0.309064	Adjusted R-squared		0.296269
Rw-squared	0.829145	Adjust Rw-squared		0.829145
Akaike info criterion	442.4348	Schwarz criterion		458.8377
Deviance	184.3613	Scale		0.653386
Rn-squared statistic	328.2296	Prob(Rn-squared stat.)		0.000000
	Non-robust Statistics			
Mean dependent var	1.729630	SD dependent var		2.256845
SE of regression	1.657219	Sum squared resid		593.2168

shows the explanatory powers of the sustainable indicators in the region.

The overall results indicate the importance of biofuels production for a strong policy vista for sustainable bioenergy development across the globe. There is one thing be in a mind that if the economies pertaining to the bioenergy development in a region, there should be keen focused on the land-use changes and greenhouse gas balances for climatic protections. Piccirillo (2012) conclude that biofuel energy is one of the key factors of green development alternative to the standard fossil fuels, as it is produced from renewable energy sources. Besides that the impact of biofuel energy on environmental degradation is observed smaller as compared to the conventional fossil fuel combustion, because the plants used for biofuel production absorb carbon dioxide emissions, that leading to zero net CO_2 emissions.

Discussion

The study focused on the biofuels production that is used as one of the key factor for green sustainable agenda for the panel of developed and developing countries. The study examined the impact of four promising sustainable indicators that impact on the biofuels production in the region. The time series data of 17 developed and developing countries taken from U.S Energy Information Administration (EIA) database for the period of 2000–2012. The study employed number of panel econometric modeling techniques on merit, that is, panel unit root test evaluate stationary series of the variables; panel co-integration test indicates the long-run connection between the variables, panel GMM technique incorporated the endogeneity problem from the model; and robust least square regression technique addresses the outliers in the given model.

The results of panel unit root test indicate the mixture of order of integration, as some of the variables including biofuels, and carbon dioxide emissions are highly volatile data in nature; therefore, it decomposed with differenced stationary. The panel co-integration results rejected the null hypothesis of no co-integration against the alternative hypothesis. Therefore, the study established that there is co-integration relationship between the variables. The results of GMM indicate that both the carbon dioxide emissions and renewable electricity generation have a positive and significant association with the biofuels production in the region. The results of robust least square regression confirmed that all of the sustainable indicators have a dynamic linkage with the biofuels production. Thus, the policy related to the green biofuels production process should be carried out for long-term development in the region.

On the basis of the following results, the study proposed short-term, medium term and long-term policy implications. In the short-term, the policymakers and government officials may have to reduce the burden of trade deficit using the diverse energy mix in their portfolios. However, biofuel energy has a distinct edge on the other energy mix as it may be used as a substitute for high priced petroleum. In addition, biofuels energy using in transport sector may help to mitigate the greenhouse gas emissions to addresses the threat of climate change in the region. In the medium-term plan,

policymakers should have to be devoted a major expenditures on the R & D activities related with the sustainable development of biofuels including the availability of the biofuels, cost effectiveness, ecosystem protection, etc. Finally, in the long-term plan, the policymakers have to promote biofuels as the important component that reduces life cycle CO_2 emissions by replacing fossil fuels. However, it is point of noted that during cultivation and production of biofuels, second-generation biofuels produces less carbon dioxide emissions than the replaced fossil fuel. Biofuels development should be considered as integrated spatial planning with resource efficiency and renewable energy strategies. Therefore, biofuels production is the desirable situation to optimize economic growth that lead to sustainable development across the globe. Future research should focus on different income level country group, longer data series and with more variables to investigate this issue.

References

AENews (2014) Biofuels from engineered tobacco plants? Alternative Energy News. Available at: http://www.alternative-energy-news.info/biofuels-engineered-tobacco-plants/ (accessed 1 March 2015).

Birur D, Hertel T, Tyner W (2008) Impact of biofuel production on world agricultural markets: a computable general equilibrium analysis. Center for Global Trade Analysis. Purdue.

Chauhan SK, Shukla A (2011) Environmental impacts of production of biodiesel and its use in transportation sector. doi: 10.5772/20923

Cherubini F, Peters GP, Berntsen T, Stromman AH, Hertwich E (2011) CO_2 emissions from biomass combustion for bioenergy: atmospheric decay and contribution to global warming. *GGB Bioenergy*, 3, 413–426.

Christopherson S. (2008) What are biofuels? Cornell University Ithaca. Available at: http://www.greenchoices.cornell.edu/energy/biofuels/ (accessed 1 March 2015).

Cremonez PA, Feroldi M, de Oliveira CDJ, Teleken JG, Alves HJ, Sampaio SC (2015) Environmental, economic and social impact of aviation biofuel production in Brazil. *New biotechnology*, 32, 263–271.

Demirbas A (2007) Progress and recent trends in biofuels. *Progress in Energy and Combustion Science*, 33, 1–18.

EIA (2014) U.S. Energy Information Administration, U.S Department of Energy, Washington, D.C. Available at: http://www.eia.gov/cfapps/ipdbproject/IEDIndex3.cfm?tid=79&pid=79&aid=1 (accessed 20 January 2015).

Escobar JC, Lora ES, Venturini OJ, Yáñez EE, Castillo EF, Almazan O (2009) Biofuels: environment, technology and food security. *Renewable and Sustainable Energy Reviews*, 13, 1275–1287.

Green Facts (2010) Liquid Biofuels for Transport Prospects, risks and opportunities. Available at: http://www.greenfacts.org/en/biofuels/l-2/1-definition.htm (accessed 1 March 2015).

Groom MJ, Gray EM, Townsend PA (2008) Biofuels and biodiversity: principles for creating better policies for biofuel production. *Conservation Biology*, 22, 602–609.

Hall DO, Scrase JI (1998) Will biomass be the environmentally friendly fuel of the future? *Biomass and Bioenergy*, 15, 357–367.

Hill J, Nelson E, Tilman D, Polasky S, Tiffany D (2006) Environmental, economic, and energetic costs and benefits of biodiesel and ethanol biofuels. *Proceedings of the National Academy of Sciences*, 103, 11206–11210.

Hoefnagels R, Smeets E, Faaij A (2010) Greenhouse gas footprints of different biofuel production systems. *Renewable and Sustainable Energy Reviews*, 14, 1661–1694.

Huber PJ (1973) Robust regression: asymptotics, conjectures and Monte Carlo. *The Annals of Statistics*, 1, 799–821.

IEA (2013) *Medium-Term Renewable Energy Market Report 2013*. OECD/IEA, Paris.

Im KS, Pesaran MH, Shin Y (2003) Testing for Unit Roots in heterogeneous Panels. *Journal of Economics*, 115, 53–74.

Levin A, Lin CF, Chu CSJ (2002) Unit root tests in panel data: asymptotic and finite sample properties. *Journal of Econometrics*, 108, 1–24.

Nigam PS, Singh A (2011) Production of liquid biofuels from renewable resources. *Progress in Energy and Combustion Science*, **37**, 52–68.

Panichelli L, Gnansounou E (2015) Impact of agricultural-based biofuel production on greenhouse gas emissions from land-use change: key modelling choices. *Renewable and Sustainable Energy Reviews*, **42**, 344–360.

Pedroni P (1999) Critical values for cointegration tests in heterogeneous panels with multiple regressors. *Oxford Bulletin of Economics and Statistics, Special Issue*, **61**, 653–678.

Piccirillo C (2012) Biofuels Production and Greenhouse Gases Emissions. Available at: http://www.decodedscience.com/biofuels-production-greenhouse-gases-emissions/15432 (accessed 2 March 2015).

Ravindranath NH, Lakshmi CS, Manuvie R, Balachandra P (2011) Biofuel production and implications for land use, food production and environment in India. *Energy Policy*, **39**, 5737–5745.

Rousseeuw PJ, Yohai VJ (1984) Robust regression by means of S-estimators. In: *Robust and Nonlinear Time Series Analysis, Lecture Notes in Statistics 26*, (eds Franke J, Hˉardle W, Martin RD), pp. 256–272. Springer Verlag, New York.

Schaldach R, Priess JA, Alcamo J (2011) Simulating the impact of biofuel development on country-wide land-use change in India. *Biomass and Bioenergy*, **35**, 2401–2410.

Spartz JT, Rickenbach M, Shaw BR (2015) Public perceptions of bioenergy and land use change: comparing narrative frames of agriculture and forestry. *Biomass and Bioenergy*, **75**, 1–10.

UNCTAD (2008) Biofuel production technologies: status, prospects and implications for trade and development. United Nations Conference on Trade and Development, New York and Geneva, i – viii, pp. 1-41.

Van Zelm R, Muchada PAN, van der Velde M, Kindermann G, Obersteiner M, Huijbregts MAJ (2014) Impacts of biogenic CO_2 emissions on human health and terrestrial ecosystems: the case of increased wood extraction for bioenergy production on a global scale. *GGB Bioenergy*, **7**, 608–617.

Yang H, Zhou Y, Liu J (2009) Land and water requirements of biofuel and implications for food supply and the environment in China. *Energy Policy*, **37**, 1876–1885.

Yohai VJ (1987) High breakdown-point and high efficiency robust estimates for regression. *The Annals of Statistics*, **15**, 642–656.

13

The influence of feedstock supply risk on location of stover-based bio-gasoline plants

JUAN SESMERO and XIN SUN

Agricultural Economics, Purdue University, 403 West State Street, West Lafayette, IN 47907, USA

Abstract

This study models and quantifies spatially referenced probability distributions of corn residue cost and assesses their influence on comparative advantages of different areas of the Corn Belt to attract biofuel plants. Results suggest that irrigated areas of the Corn Belt, despite their relatively low planting density, may result more attractive than some of their rainfed counterparts in the eastern Corn Belt due to low risk in feedstock cost resulting from stability of yields. Therefore, agricultural districts in the Great Plains of the US may not need to pay high subsidies to compete with those in the eastern Corn Belt to attract biofuel firms. Policy restrictions on irrigation due to concerns over groundwater depletion may, however, diminish the relative comparative advantage of the irrigated Corn Belt for biofuel production.

Keywords: cellulosic biofuels, corn density, feedstock cost, irrigation, stochastic dominance, yield volatility

Introduction

Biofuels policies can be greatly benefited by an understanding of their environmental and economic repercussions. Quantification of environmental and economic implications of stover-based cellulosic biofuels in the US requires knowledge of the spatial pattern of plants' location across the Corn Belt. An important factor influencing plant location is the availability of stover which is determined by corn planting density and yields. These measures vary across space and also randomly over time introducing spatially heterogeneous risk in feedstock cost, and yet, existing economic analyses of stover-based biofuel production have not considered such spatial pattern of risk. This study models and quantifies risk in feedstock cost for different areas of the Corn Belt and their resulting comparative advantage to attract biofuel plants.

Policy background

The transportation sector is one of the main sources of oil consumption in the US (United States Energy Information Administration, 2011). The nonrenewable nature of oil, in combination with rapid economic growth in developing countries, results in predictions of continued price increase over the long run (U.S. Energy Information Administration, Annual Energy Outlook, 2011, http://www.eia.gov/forecasts/aeo/) reaching $150 a

barrel by 2040. Oil consumption also results in emissions of greenhouse gases (GHG) which have the potential to adversely affect the earth's climatic system. The recent imbalance between supply and demand of oil and its deleterious effects on climate have sparked interest in renewable, clean energy resources. Cellulosic biofuels may result in substantial reductions in GHGs emissions relative to regular gasoline (Wang *et al.*, 2011). Development of these sources of fuel can also provide a hedge against future variations in oil prices.

To encourage the production of cellulosic biofuels, the Energy Independence and Security Act of 2007 (EISA) established specific annual mandates for this fuel source which is expected to reach 16 billion gallons of ethanol-equivalent biofuels by 2022. About a third of total advanced biofuels supported by EISA is expected to come from corn stover (National Research Council (US), 2011; Downing *et al.*, 2011). Previous studies (e.g. Petter & Tyner, 2014) have identified the price of biofuel and the cost of feedstock as the most important factors influencing the economic viability of stover-based biofuel plants. Consistently with these findings, firms will try to locate in areas where output prices are high and feedstock cost is low. The Corn Belt is the area in which stover can be procured at the lowest cost but that cost may vary widely within the Corn Belt. On the other hand, the price of gasoline, which is the relevant reference for bio-gasoline, does not vary as much since the entire Corn Belt is contained within the same gasoline district, that is, the PADD 2 district exhibited in Fig. 1. Therefore, we concentrate our attention on the link between spatially heterogeneous stover production conditions

Correspondence: Juan Sesmero
e-mail: jsesmero@purdue.edu

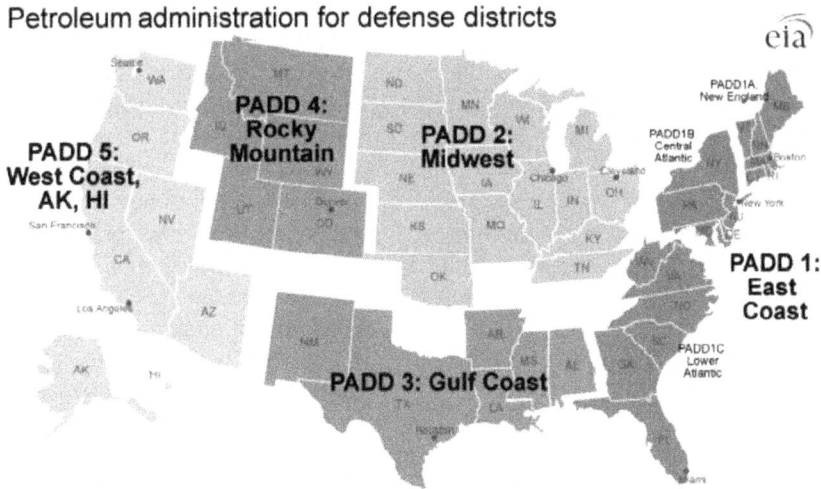

Fig. 1 Geographic aggregation of petroleum markets.

and its influence on feedstock cost and plant profitability.

Environmental implications associated with the development of a cellulosic biofuel industry depend upon the spatial distribution of processing plants. For example, if plants locate in the irrigated part of the Corn Belt, higher pressure on groundwater is to be expected (Sesmero, 2014). If plants locate in the eastern Corn Belt, more water pollution may be observed (e.g. Gramig et al., 2013). Location of processing plants will also determine the spatial pattern of economic rents resulting from policy support. Finally, the comparative advantage of different areas of the Corn Belt regarding the production of cellulosic biofuel will directly inform state-level policies aimed at attracting investment in this industry. Therefore, understanding and quantifying comparative advantages of different regions to attract biofuel plants and the potential spatial pattern of cellulosic biofuel production should be informative of both federal and state-level biofuel policies.

Spatial patterns of feedstock supply

A number of uncertainties surround the assessment of feedstock cost which translates into a wide range of estimates in the scholarly literature. The cost of delivering stover to a processing plant is likely to be influenced by stover production density which changes across space and over time due to yield volatility. Substantial differences in corn planting density (and consequently stover production density) and yield volatility exist across different regions of the Corn Belt as illustrated in Fig. 2. Data in Fig. 2 sample counties from east to west of the Corn Belt along the 41st parallel. States included are Indiana, Illinois, Iowa, and Nebraska. This is an area

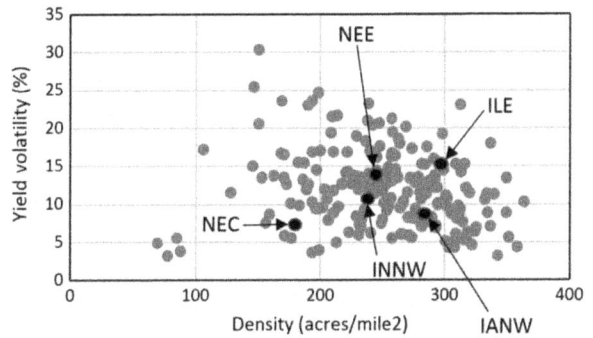

Fig. 2 Density and volatility of stover production across the Corn Belt.

that cuts across the heart of the Corn Belt and that already supports a large number of corn ethanol plants which suggests it is also a promising area for stover-based biofuel production.

Despite vast differences in supply conditions across counties in the Corn Belt, previous economic analyses of stover-based biofuels (Petrolia, 2008a; Petrolia, 2008b; Gallagher et al., 2003; Fiegel et al., 2012; Anex et al., 2010; Brown et al., 2013a,b; Brown & Brown, 2013b; Petter & Tyner, 2014; Jones et al., 2009; Archer and Johnson, 2012) have not explicitly modeled and quantified the link between density and variability of stover production, and the resulting feedstock cost faced by plants in different regions.

This study estimates the probability distribution of stover harvest density in different areas of the Corn Belt and models the cost of feedstock as a function of that distribution. Quantification of the link between stover density and cost under risk permits the assessment of the comparative advantage of different areas of the

Corn Belt to support a cellulosic biofuel plant. We identify interesting risk–return tradeoffs faced by investors trying to decide on location of a plant. We find that firms may prefer to locate in the irrigated Corn Belt, even though it displays a lower corn planting density, because lower volatility in stover yields (attained through irrigation) greatly reduces downside risk. The competitiveness of irrigated areas relative to nonirrigated ones is enhanced by a stronger (negative) link between yields and farmers' willingness to accept for stover.

Literature review

Converting biomass to ethanol may be problematic from a profitability point of view because (1) corn ethanol is less costly and (2) market penetration will be limited by the 'blend wall', which is a technical constraint to the inclusion of ethanol in liquid fuels (Tyner, 2010). The blend wall diminishes the market potential of the biofuel. An interesting option is conversion of biomass into bio-gasoline (also called 'drop-in' biofuel) which has the same chemical properties as regular gasoline and, as such, is not constrained by the blend wall (Tyner, 2011). For these reasons, previous analyses of cellulosic biofuels have focused on drop-ins instead of ethanol. We maintain this premise and analyze biofuel plants producing drop-in cellulosic bio-gasoline.

Numerous studies, in both business and academic realms, have routinely found that a cellulosic bio-gasoline plant built today could have a positive *mean* return on the investment (Cottam & Bridgwater, 1993; Anex et al., 2010; Brown et al., 2013a,b; Brown & Brown, 2013b; Petter & Tyner, 2014; Jones et al., 2009). However, they have also found that there is significant uncertainty around that mean. For instance, Petter & Tyner (2014) found that the probability of economic loss is almost 50%.

Anex et al. (2010) applied techno-economic analysis to measure the returns of a biofuel plant after construction. They analyzed three alternative conversion pathways: pyrolysis, gasification, and biochemical. Breakeven prices of gasoline (the price of gasoline that would set the net present value of this project equal to zero) under each pathway were calculated and compared. Results showed that fast pyrolysis (see US Department of Energy, 2012 and Bridgwater, 2012 for a technological description of the process) is the pathway with the lowest breakeven price ($2/gallon), followed by gasification ($4.5/gallon) and biochemical conversion ($5/gallon). Techno-economic Analysis (TEA) has also been applied to explore the cost of upgrading bio-oil to transportation fuel. According to Wright et al. (2010), purchasing hydrogen to upgrade bio-oil results, for a biofuel plant,

in lower gasoline breakeven price than producing the biofuel ($2.11/gallon vs. $3.09/gallon).

The technical performance of plants is uncertain as technical processes involved in the conversion of biomass to bio-gasoline are relatively new at commercial scale (Meier & Faix, 1999; Meier et al., 2013). Bio-gasoline price is highly uncertain. As a perfect substitute of regular gasoline, the price of bio-gasoline will be affected by the price of oil. As imperfect substitutes, the prices of corn and bio-gasoline may be correlated as well. Prices of oil and corn are volatile, and this volatility will surely transmit to the price of bio-gasoline. Finally, there is also substantial uncertainty regarding the cost at which feedstock can be procured by processing plants.

Petter & Tyner (2014) built on analyses conducted by Brown et al. (2013a,b) and focused on the identification of sources of uncertainty surrounding investment in bio-gasoline plants. The analysis by Petter & Tyner (2014) reveals that technical uncertainty (which includes feedstock cost uncertainty) and output price fluctuation certainly have the potential to influence investment on bio-gasoline plants and that policies addressing such uncertainty may enhance plants' returns and encourage investment.

As suggested by our review of these studies, the volatility in feedstock cost has been incorporated into the economic analysis of bio-gasoline plants in a rather *ad-hoc* way. Significant differences in feedstock cost are likely to emerge across space and over time, and these have received little attention in the literature which limits quantification of regional comparative advantages for cellulosic bio-gasoline production (spatial, deterministic differences in cost across locations are analyzed in Brown et al., 2013a,b). Feedstock costs are determined by on-farm cost (including storage) and transportation cost. Both cost components are influenced by stover harvest density around the plant which is determined by corn planting density and stover yields. The former varies widely across space (variation over time is more limited), and the latter varies both across space and over time.

High corn density (due to high planting density and high yields) reduces the cost of transporting feedstock to the plant (Perlack & Turhollow, 2002; Perrin et al., 2012), and low yield volatility reduces the uncertainty of feedstock supply and cost. Therefore, our hypothesis is that plants will try to locate, all else constant, in areas where corn density (production per square mile) is high and yield volatility is low. However, some areas of the Corn Belt display lower density but also lower volatility than others (e.g. central Nebraska has a lower corn density than north-west Indiana but also a significantly lower yield volatility due to irrigation) creating a potential tradeoff between average return and risk associated with investment in cellulosic biofuels.

Other factors such as infrastructure, community support, and local policies are also very important for biofuel plant location (Parcell & Westhoff, 2006; Richard, 2010; Tyner, 2011). The present analysis abstracts from those factors to concentrate solely on the role of feedstock supply on the economics of bio-gasoline plants. Formalizing such links will help determine regional differences in feedstock cost that can then be combined with other factors to better predict location decisions.

This study formalizes such trade-off and quantifies the relative importance of density and volatility in feedstock cost under alternative scenarios for transportation cost and stover price pass-through. This quantification allows identification of areas of the Corn Belt that dominate others for plant location from a first- and second-order stochastic dominance point of view. This will, in turn, deepen our understanding of spatial comparative advantages which will potentially shape location patterns across the Corn Belt as the industry develops.

Materials and Methods

The plant's cost function and NPV

In line with previous literature (Gallagher et al., 2005; Perrin et al., 2012), we assume that the plant uses a uniform delivered pricing strategy (i.e. the plant pays a uniform price at the farm gate and then takes care of the shipping cost) and that acreage allocated to corn, yield, and harvest practices are homogeneously distributed around the plant. Therefore, additional biomass can only be procured from the extensive margin, that is, from an expanding circle around the delivery point. In contrast with previous literature, however, we allow for yield to be uncertain and model risk though probability distributions. We also allow for yield volatility to affect stover price. Therefore, the uncertainty in feedstock cost is explicitly modeled as a result of uncertainty on yield that can be estimated based on actual data.

Under these assumptions, the quantity harvested within radius R is the amount of stover produced per square mile and the area of a circle of radius R expressed in square miles πR^2. Therefore, $Q = d\pi R^2$, where Q is the amount of biomass procured within the circular fuelshed of radius R, d represents harvest density in tons per square mile, and π is the number pi. Harvest density d is defined as $d = p_d y$, where p_d denotes corn acres harvesting stover around the facility (acres per square mile), and y is the amount of stover harvested per acre. From the expression for Q, it is clear that the radius of the fuelshed of a plant of size Q is $R = \sqrt{\frac{Q}{\pi d}}$.

As shown by Gallagher et al. (2003), the amount of biomass that can be procured from a circle at a distance r from the plant is given by the product of the circumference of the circle ($2\pi r$), the density of residue d, and the width of the ring Δr: $\Delta Q = 2\pi r d\,(\Delta r)$. Therefore, the marginal cost of expanding

this circle by the increment Δr under uniform delivered pricing is $\frac{\partial C}{\partial r} = 2\pi r d(p_s + tr)(\Delta r)$, where $(p_s + tr)$ is the cost of purchasing stover (p_s) plus the cost of transporting it from a point on that circle to the plant (tr). Then, total cost is obtained by taking integral on both sides with respect to r which results in:

$$C = \pi d r^2 \left(p_s + \frac{2}{3} tr \right) \quad (1)$$

Since, for a plant of size Q, $r = \sqrt{\frac{Q}{\pi d}}$, total feedstock cost can be re-expressed as:

$$FC = Q p_s + \frac{2}{3} \frac{t}{\sqrt{\pi d}} Q^{3/2} \quad (2)$$

In addition to feedstock cost, the total cost of producing bio-gasoline includes processing cost and capital cost. Processing cost refers to the cost associated with conversion of one ton of biomass into 1 L of bio-gasoline. Capital cost is the initial investment of the bio-gasoline plant, which are used to purchase land, facilities, and other fixed assets. Detailed data on these components were reported in Tables 2 and 3. The resulting total cost of the plant can be depicted by:

$$TC = p_s Q + \left(\frac{2}{3} \frac{t}{\sqrt{\pi d}} \right) Q^{3/2} + k + bQ \quad (3)$$

where k denotes the capital cost of biomass process, and b represents the cost of processing biomass on a per ton basis.

The above expressions for bio-gasoline production cost clarify the path of risk. Yields are random, which makes density random, which in turn results in random feedstock cost through the second term of total production cost. As depicted in the cost equation, the effect of density on the cost of producing bio-gasoline will increase with a higher transportation coefficient as this component will have a higher weight in total production cost.

In this research, total revenue is influenced by two factors: gasoline price ($ per L) and yearly output of bio-gasoline which is, under the commonly held assumption of fixed proportions technology for biofuel processing (also known as Leontief technology), proportional to the amount of biomass processed:

$$TR = p_g \theta Q \quad (4)$$

where p_g is the bio-gasoline selling price, θ is bio-gasoline yield per unit of biomass processed (liters per metric ton), and Q is the amount of biomass processed as determined by the plant's capacity.

Based on the equations above, net present value (NPV) is calculated as the difference between the present value of future cash inflows, described in Eqn (4), and outflows, described in Eqn (3).

Parameterization

Previous literature has compared alternative pathways for the conversion of biomass into biofuels (Wright et al., 2010; Brown & Brown, 2013a; Petter & Tyner, 2014) on the basis of their technical and economic viability. Technologies discussed in the literature include gasification, hydrolysis, and fast pyrolysis. These analyses reveal that fast pyrolysis is the most cost-competitive process to produce bio-gasoline, so we concentrate our attention in this technological pathway.

Unless otherwise noted, we use technical and cost assumptions as previously used by Brown *et al.* (2013a,b) and Petter & Tyner (2014) for a biomass processing plant using fast pyrolysis. We assume the plant produces 545 000 L of bio-gasoline per day and operates 329 days per year. Regarding feedstock requirements, while Brown *et al.* (2013a,b) assumed that 322 L of bio-gasoline can be obtained per metric dry ton of stover processed, Kior, which is currently the only commercial scale cellulosic bio-gasoline plant, reported a yield of 273 L per metric dry ton (Lane, 2013). Kior's reported value is lower than that in Brown *et al.* (2013a,b) because the latter analyzes a fast pyrolysis pathway, while Kior employs a catalytic fast pyrolysis pathway. To keep consistency between the cost structure and the productivity of our technology, we use Brown *et al.*'s figure of 322 L of bio-gasoline per ton of feedstock. Technical assumptions are summarized in Table 1.

Capital cost was calculated based on financing assumptions reported in Table 2. In this study, capital cost is calculated as the present value of investment cost. The construction period is 3 years. The plant pays back the investment cost with interest in full after 21 years (last year of construction plus 20 years of operation) at a 7.5% interest rate.

Nonfeedstock operating cost is calculated by combining yearly processing cost in Brown *et al.* (2013a,b) and federal taxes. We assume a 20% effective tax rate on net income. Nonfeedstock operating cost is composed of hydrogen cost, and miscellaneous. These costs are reported in Table 3 and were obtained from Brown *et al.* (2013a,b). Total yearly processing cost per liter (for a plant producing 180 million liters per year) is equal to $0.17 per liter. Taxes vary over time, so they are not reported in Table 3 as a constant on a per liter basis.

Inflows described in Eqn (4) are calculated assuming a bio-gasoline price of $0.7 per liter which is the price around which wholesale gasoline price has fluctuated since 2008 (Fig. 3). Outflows are those described in Eqn (3). Future cash flows are discounted at a 10% annual rate.

A critical component of total cost is feedstock cost which is captured by the first two terms of Eqn (3). Given the amount of biomass purchased by the plant, these terms are governed by plant-gate price of stover (first term), and harvest density (second term). Corn plating density and yield are critical for these cost components, and they vary widely across regions of the Corn Belt. We calculate the probability distribution of NPV in five different agricultural districts as defined by USDA: Nebraska Central (NEC), Nebraska East (NEE), Iowa North West (IANW), Illinois East (ILE), and Indiana North West

Table 1 Technical assumptions of biofuel plant

Cost Category	Value
Plant production	180 000 000 L year^{-1}
Plant life	20 years
Feedstock use	659 000 metric tons per year
Bio-gasoline yield	322 L per metric ton of feedstock

Table 2 Assumptions for financing

Parameter	Value	Source
Investment cost	$429 000 000	Brown *et al.* (2013a,b)
Payback	21 years	Author's assumptions
Construction time	3 years	Wright *et al.* (2010)
% of investment in year one	8%	Wright *et al.* (2010)
% of investment in year two	60%	Wright *et al.* (2010)
% of investment in year three	32%	Wright *et al.* (2010)
Interest rate	7.5%	Wright *et al.* (2010)
Liters of bio-gasoline produced per year	212 000 000 L	Author's calculation

Table 3 Processing costs per liter

Cost Category	Value	Source
Hydrogen	$0.13	Brown *et al.* (2013a,b)
Miscellaneous	$0.04	Brown *et al.* (2013a,b)
Total	$0.17	

(INNW). These are important corn-producing agricultural districts which makes them of special interest to our analysis. In addition, they display a wide range of density/volatility combinations allowing for meaningful comparison to assess risk and return tradeoffs.

Combinations of corn planting density and yield volatility for all five agricultural districts are depicted in Fig. 2 so that they can be compared with the broader population of counties across the 41st parallel. As it is apparent from Fig. 2, the heavily irrigated district in the analysis (NEC) displays a substantially lower volatility but also a lower density than all other districts. In the opposite extreme, ILE displays a high volatility but also high density of stover production. On the other hand, IANW seems like a promising location in that it combines a high density with a surprisingly low volatility for a predominantly rainfed district.

According to the National Agricultural Statistical Service (NASS Quick Stats) average corn planting densities in the past 10 years are 190 acres per square mile in NEC, 241 acres per square mile in NEE, 284 acres per square mile in IANW, 299 acres per square mile in ILE, and 237 acres per square mile INNW. Mean stover harvest density in each district (assuming a 1 : 1 relationship with grain yields and 50% removal rate) are 354 metric tons per square mile in NEC, 421 metric tons per square mile in NEE, 544 metric tons per square mile in IANW, 538 metric tons per square mile in ILE, and 410 metric tons per square mile in INNW.

Yields (and, hence, harvest density) vary substantially from year to year especially in nonirrigated agricultural districts (i.e. NEE, IANW, ILE, and INNW). According to data from NASS Quick Stats, 85% of corn land is under irrigation in NEC, 55%

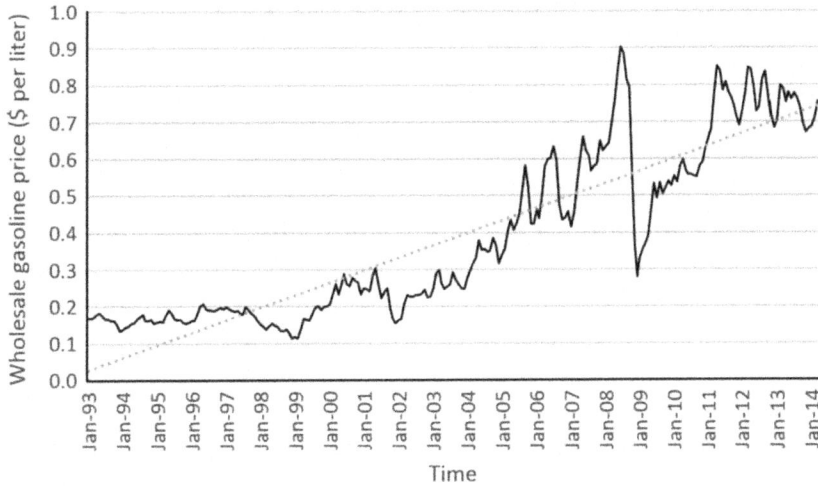

Fig. 3 Average real wholesale gasoline price in the Midwest (PADD, area 2) (Annual Energy Outlook, 2014).

in NEE, and less than 1% of land in the other districts. We have collected county-level data in each agricultural district in the last 10 years and conducted a parametric fit of yield distributions. Descriptive statistics of yield data in each agricultural district are reported in Table S1 of Supporting Information 1, and the estimated probability density functions are plotted in Supporting Information 2. The first row of the Table in Supporting Information 1 denotes the number of observations based on which parametric fitting of distributions was conducted.

Parametric specifications for each district were chosen based on Akaike and Bayesian information criteria. Both the Akaike and the Bayesian information criteria rank different statistical specifications based on the goodness of fit of these specifications (as measured by the value of the log-likelihood function) relative to their complexity (measured by the degrees of freedom of each model). Yields in nonirrigated areas were best approximated by Extreme Value Minimum density functions which belong to the family of Gumbel distributions. In particular, yields in NEE are described by Extreme Value Minimum (172,20), yields in IANW by Extreme Value Minimum (172,20), yields in ILE by Extreme Value Minimum (172,20), and yields in INNW by Extreme Value Minimum (172,20). Interestingly, the best parametric density function for the heavily irrigated district in our sample (NEC) is not described by an asymmetric density function but by a normal distribution, that is, Normal (170,13). The symmetry of yields in this area may be explained by the reduction in downside risk provided by irrigation which eliminates the negative skewness characterizing nonirrigated areas rendering a symmetric probability distribution.

Farm-gate price of stover is determined by harvest cost. A number of studies have obtained estimates of such cost (e.g. Gallagher *et al.*, 2003; Fiegel *et al.*, 2012; Brechbill *et al.*, 2011; Perrin *et al.*, 2012; Brown *et al.*, 2013a,b) and recognized the influence that yield may have on this price. Yet, they have not explicitly modeled variability in yields and the resulting vari-

ability in farm-gate price of stover. Petter & Tyner (2014) do consider uncertainty in feedstock price, but such uncertainty is not linked to variability in yields.

When yields decrease due to weather or other adverse production shocks, on-farm cost of stover increases due to the fact that the same operations (with a fixed cost per acre) result in lower stover yields which increases the cost per ton of biomass harvested. In addition, lower stover production may increase the risk of soil erosion after harvest that may result in an increase in farmers' willingness to accept for biomass. Moreover, high stover yields may increase the benefits of removing stover as excessive residues increase the prevalence of pests and disease and delay planting as soils take longer to warm and dry up in the spring. Therefore, high yields may reduce the farmer's willingness to accept for stover. All of these forces operate to produce, unambiguously, a negative link between stover yield and price. We capture the potential link between stover yield and price by a simple linear relationship:

$$p_s = \alpha - \beta y \qquad (5)$$

where p_s is the farm-gate price of stover, and y is the amount of stover harvested per acre.

In specification (5), the parameter β captures the effect of changes in stover yield on stover price. Cost components measured on a per acre basis can constitute around 50% to 60% of harvest cost depending on technical assumptions and nutrient replacement assumptions. This suggests that a 10% reduction in yield can increase on-farm harvest cost per ton by 7% due to the mere existence of cost components that are fixed per acre. Therefore, a value of 0.7 constitutes an approximate lower bound to the elasticity of stover price with respect to yields. Once soil erosion risk and effects on soil temperature and disease are factored in, the elasticity will be higher. Therefore, we calibrate β so that the elasticity of stover price with respect to yield is equal to 1 at average yield and average plant-gate price in our sample of regions across the Corn Belt. The elasticity is calculated as $\frac{\partial p_s}{\partial y} \frac{y}{p_s} = -\beta \frac{y}{p_s} = 1$, which implies that $\beta = -\frac{p_s}{y}$.

Due to differences in assumptions regarding harvesting operations, removal rate, yields, nutrient replacement, and storage as well as differences in prices and custom rates estimates of on-farm feedstock cost vary widely between \$40 and \$101 (Brechbill *et al.*, 2011; Perrin *et al.*, 2012; Brown *et al.*, 2013a,b), so we assume an average on-farm cost of \$70 (i.e. $p_s = 70$). Average yield across regions considered in this study is 170 bushels per acre which combined with the assumed average on-farm cost yields $\beta = -\frac{70}{170} = -0.41$. This means that a reduction in stover yield of one ton per acre will be associated with a decrease in farm-gate stover price of 41 cents per ton. From Eqn 5, α is then calculated to be 140.

The cost of harvesting stover is spatially heterogeneous. One of the reasons why harvest cost in the irrigated Corn Belt may be different from that in the rainfed Corn Belt is water replacement. Stover removal may increase soil water evaporation and require an increase in irrigation to offset this water loss. The cost of additional irrigation increases the cost of harvesting stover. On the other hand, farmers attain higher yields under irrigation. As a substantial portion of the cost of harvesting stover is fixed on a per acre basis, higher yields in the irrigated Corn Belt translate into a lower mean cost per ton. An additional secondary benefit of the Central Plains irrigated area is low humidity, which means that biomass storage will yield a drier product in this region.

We abstract away from such spatial differences on the mean stover price and focus on the volatility of such price in different regions. It is not the objective of this paper to precisely calculate mean stover price (i.e. mean on-farm cost). Such figure is spatially idiosyncratic in nature, and we do not presume to obtain a representative estimate for each region. Rather the objective of this study is to discuss the influence of feedstock supply volatility on the relative attractiveness of alternative locations across the Corn Belt, *under similar expected (mean) on-farm harvest cost.*

It should be noted that randomness in stover yields introduces risk in plants' cost of production through two channels. First, stover harvest density has been previously defined as the product of corn planting density and stover yield ($d = p_d y$). Consequently, randomness in yield introduces randomness in harvest density and transportation cost [second term of Eqn (3)]. Second, randomness in yield introduces risk in the farm-gate price of stover through Eqn (5). Finally, randomness in plant's cost (3) introduces risk in NPV defined as the pres-

ent value of future streams of revenue (4) and cost (3). The concept of stochastic dominance is employed to compare risky NPVs.

First- and second-order stochastic dominance

To compare risky projects (i.e. projects with random NPVs) and determine the desirability of a location in the Corn Belt relative to another, we utilize two concepts: first-order stochastic dominance and second-order stochastic dominance. Both concepts are illustrated in Fig. 4. Figure 4(Panel a) illustrates first-order stochastic dominance. A location is said to dominate another in a first-order stochastic sense if the cumulative distribution function (CDF) of its NPV is lower than that of the other location in the entire support of the NPVs. In the case illustrated in Fig. 4(a), location Y is preferred to location X for investment in a bio-gasoline plant. Intuitively, the fact that the CDF of NPV in location Y is below that of location X means that the probability of obtaining an NPV below a given level is always lower in location Y. It needs to be kept in mind, however, that it is still possible for the investment in location X to result in higher returns, but it is unlikely. It is this likelihood that the first-order stochastic dominance criterion is based off.

Second-order stochastic dominance is illustrated in Fig. 4(b). The fact that very low NPVs in location Y are less likely (the CDF of NPV in location Y is below that of X up to a certain value of NPV) is offset by the fact that high NPVs are also less likely (the CDF of NPV in location Y is below that of X up to a certain value of NPV). In other words, returns in location Y are less spread out (i.e. are less risky) than returns in location X. Location Y will be preferred to location X in a second-order stochastic sense if area A is larger than the area B. Intuitively, this means that location Y will be preferred to X if the reduction in the probability of extremely bad outcomes associated with location Y is large enough to offset the reduction in the probability of extremely good outcomes also associated with this location.

Results and Discussion

We calculate in this chapter NPV of investment in a bio-gasoline plant located in the five agricultural districts

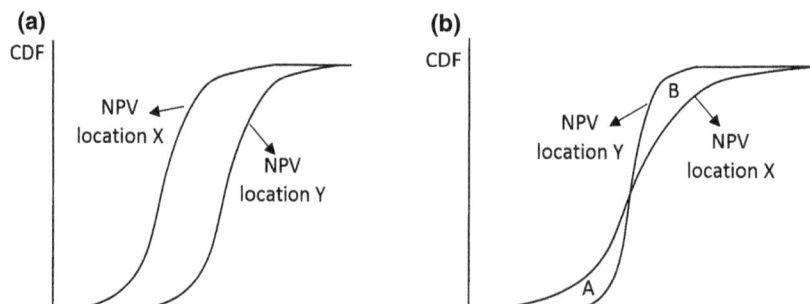

Fig. 4 (a) First-order stochastic dominance. (b) Second-order stochastic dominance.

previously discussed. We run Monte Carlo simulations (2000 iterations were run) based on fitted yield distributions. These simulations result in a probability distribution of NPVs in each location. Results of simulations are plotted in Fig. 5 in the form of CDFs of NPV in each of the five areas considered.

IANW and ILE display the highest mean NPVs (the first two moments of the distributions are reported on the table to the right of the figure), followed by NEC which has a slightly higher mean than INNW and NEE. However, IANW and ILE (especially the latter) display much higher standard deviation than NEC due to the risk-reducing effect of irrigation in the latter. Yet by visual inspection, we can conclude that IANW dominates NEC in a first-order stochastic sense. This means that the increase in mean in IANW is strong enough to offset the increase in downside risk also associated with this location. Therefore, IANW will be preferred to NEC by all investors regardless of their risk preference.

The fact that mean NPV in NEC is higher than that in INNW and NEE is interesting as NEC is the district with the lowest stover harvest density. This fact reveals that the reduction in downside risk associated with irrigation substantially increases mean NPV to the point where NPV performs better, on average, in NEC than it does in INNW and NEE. Results also indicate that ILE, INNW, and NEE cannot be compared to NEC from a first-order stochastic dominance point of view as their CDFs cross. The fact that NEC displays higher mean and lower standard deviation than NEE and INNW implies that NEC dominates the other two locations in a second-order stochastic sense. This means that NEC will be preferred to the other two locations by both risk-neutral and risk-averse investors, though it may not be the preferred location of risk-prone investors.

It can also be concluded by simple visual inspection (areas between curves to the left and right of

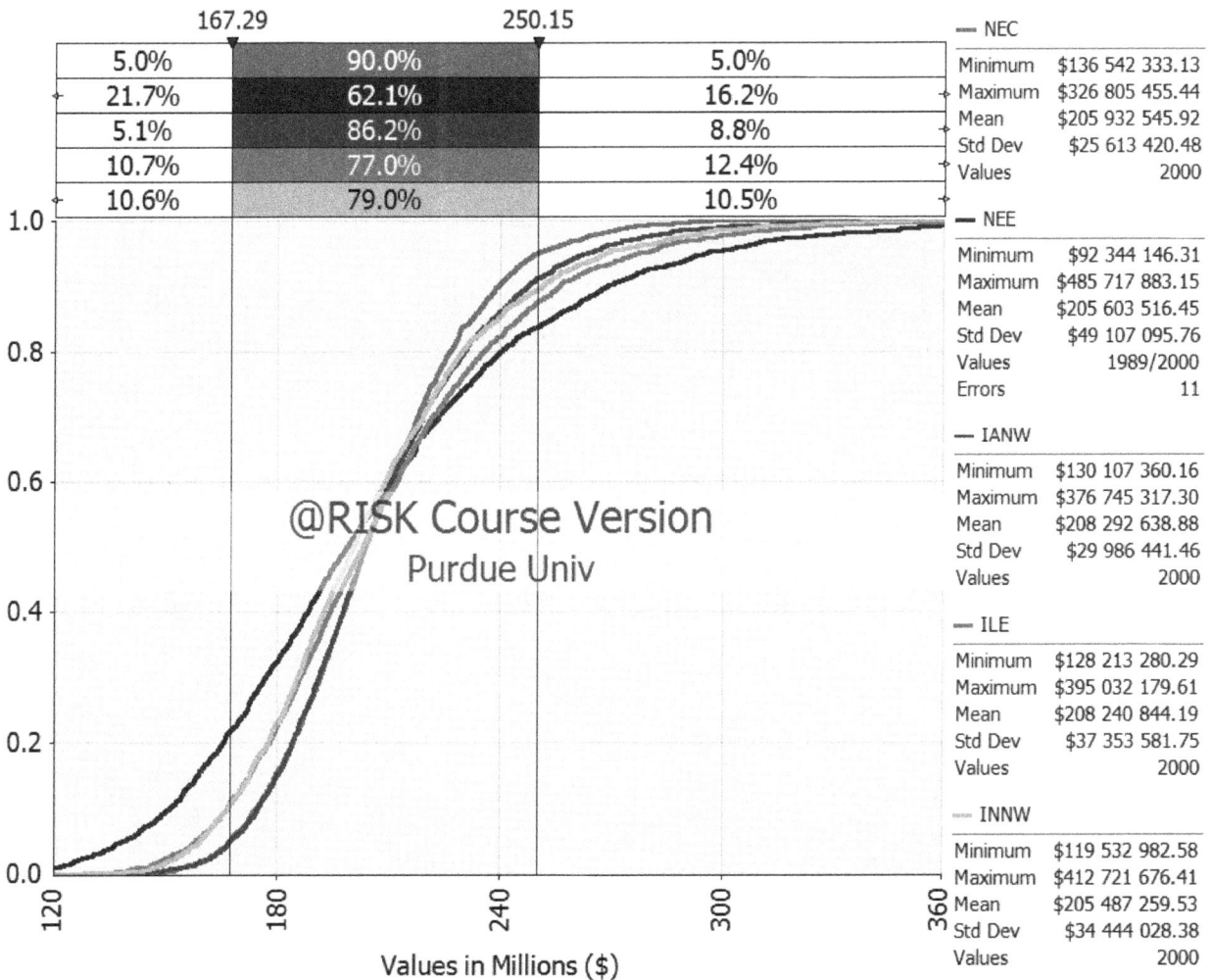

Fig. 5 Cumulative density functions of net present value for five agricultural districts.

the intersection) that IANW dominates INNW and ILE in a second-order stochastic sense. Visual inspection is not enough to compare NEC to ILE as a large difference across areas between curves to the left and right of the intersection point is not apparent. Information on the first two moments of both distributions is also insufficient to determine dominance as ILE displays higher mean but also higher variance than NEC.

Calculation of the areas between ILE's and NEC's CDFs to the left and right of their intersection point requires integration of these CDFs for these subsets of the NPV domain. To integrate these functions, we first fit a parametric approximation to the CDFs and then integrate those specifications on the relevant parts of the domain of NPVs. According to AIC, the best fit for NEC's CDF is a Gaussian function. The CDF for ILE, on the other hand, was best approximated by an Extreme Value distribution function. Areas below approximations to the CDFs to the left and right of the intersection points are calculated as follows:

Nebraska central

Left of intersection point

$$\int_{120\,000\,000}^{207\,500\,000} \left(0.5 + 0.5\,\mathrm{erf}\left(\frac{x - 205\,926\,158}{(2*25\,425\,259)^{0.5}}\right)\right) dx = 1\,094\,720$$

Right of intersection point

$$\int_{207\,500\,000}^{350\,000\,000} \left(0.5 + 0.5\,\mathrm{erf}\left(\frac{x - 205\,926\,158}{(2*25\,425\,259)^{0.5}}\right)\right) dx = 13\,312\,400$$

Illinois east

Left of intersection point

$$\int_{120\,000\,000}^{207\,500\,000} \left(e^{-e^{\frac{190\,981\,926 - x}{30\,746\,973}}}\right) dx = 1\,441\,970$$

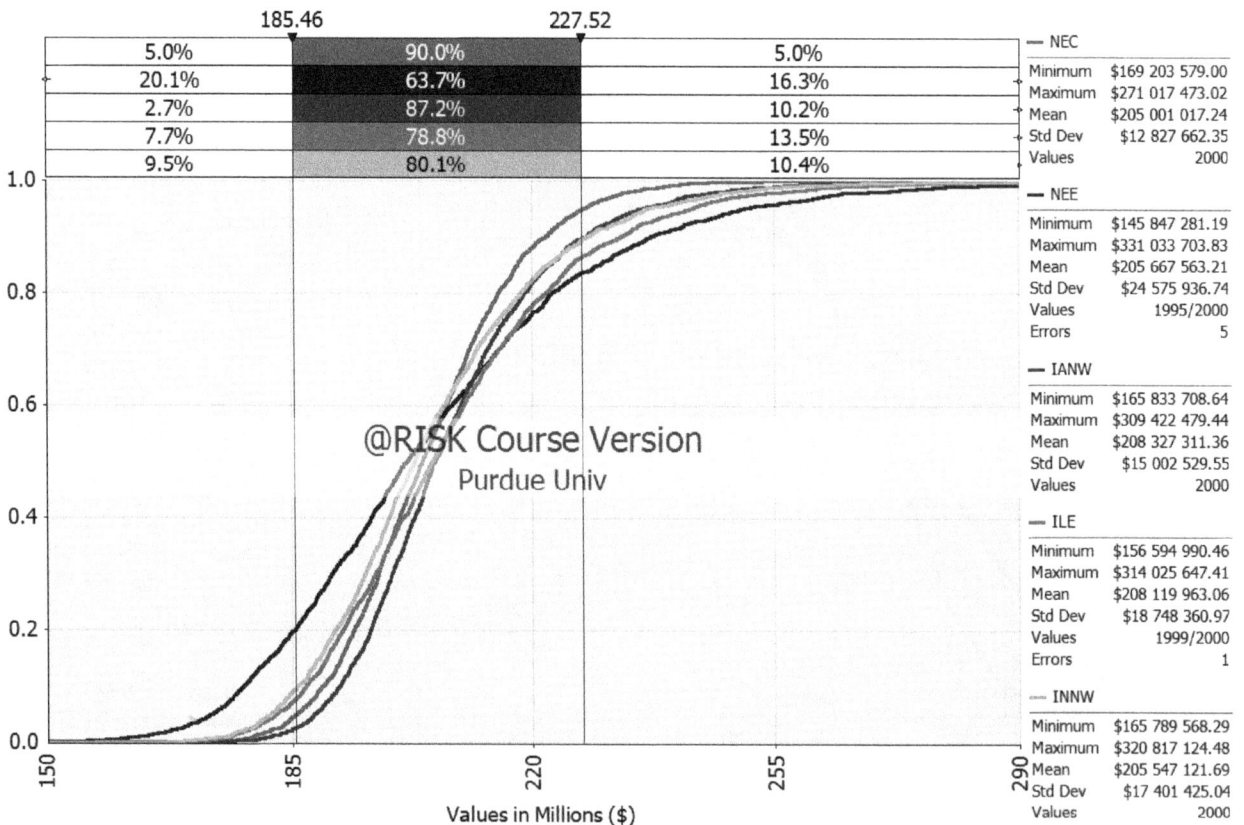

	185.46		227.52	
5.0%		90.0%		5.0%
20.1%		63.7%		16.3%
2.7%		87.2%		10.2%
7.7%		78.8%		13.5%
9.5%		80.1%		10.4%

— NEC

Minimum	$169 203 579.00
Maximum	$271 017 473.02
Mean	$205 001 017.24
Std Dev	$12 827 662.35
Values	2000

— NEE

Minimum	$145 847 281.19
Maximum	$331 033 703.83
Mean	$205 667 563.21
Std Dev	$24 575 936.74
Values	1995/2000
Errors	5

— IANW

Minimum	$165 833 708.64
Maximum	$309 422 479.44
Mean	$208 327 311.36
Std Dev	$15 002 529.55
Values	2000

— ILE

Minimum	$156 594 990.46
Maximum	$314 025 647.41
Mean	$208 119 963.06
Std Dev	$18 748 360.97
Values	1999/2000
Errors	1

— INNW

Minimum	$165 789 568.29
Maximum	$320 817 124.48
Mean	$205 547 121.69
Std Dev	$17 401 425.04
Values	2000

Values in Millions ($)

Fig. 6 Cumulative density functions of net present value in five districts under 0.5 elasticity of price with respect to yield.

Right of intersection point

$$\int_{207\,500\,000}^{350\,000\,000} \left(e^{-e^{\frac{190.981\,926-x}{30\,746\,973}}} \right) dx = 127\,025\,00$$

Results above reveal that NEC does not dominate ILE as the area between CDFs to the left of the intersection point (14 419 700–10 947 200) is smaller than the area between CDFs to the right of the intersection point (133 124 000–127 025 000). This means that risk-neutral investors may prefer to locate in ILE instead of NEC. Finally, risk-averse investors may prefer NEC to ILE due to the downside risk-reducing effect of irrigation.

The sensitivity of our results can be evaluated with respect to a number of assumed parameter values. We will conduct sensitivity analyses with respect to three of the most important and contentious parameters: (1) the elasticity of farm-gate price of stover with respect to stover yield, (2) the price of gasoline, and (3) transportation cost. The robustness of stochastic dominance results with respect to differences in mean stover prices across irrigated and nonirrigated areas is also evaluated and presented in Supporting information 3.

We now proceed to conduct sensitivity analysis of our results to the elasticity of farm-gate stover price with respect to stover yield. Figs 6 and 7 report the CDFs of the NPV of a cellulosic bio-gasoline plant located in the five agricultural districts considered in this study under low elasticity (a 50% reduction relative to the baseline of elasticity unitary elasticity) and high elasticity (a 50% increase relative to the baseline of unitary elasticity), respectively. Comparison of these figures with curves in Fig. 5 reveals that lower responsiveness of stover price to yields favors areas with high density and high volatility (Fig. 6).

Under a 0.5 elasticity of price with respect to yields, the first-order stochastic dominance of IANW over NEC (irrigated district) is enhanced as is the position of ILE relative to NEC. Furthermore, NEC's mean NPV is no longer higher than NEE. In fact, as indicated by simple visual inspection of areas, NEC no longer dominates INNW in a second-order stochastic sense (i.e. the area between curves for NPVs lower than the intersection point is smaller than the area between curves above this value) which is in contrast to the scenario with a unitary elasticity. Therefore, a certain degree of risk aversion would be required on the part of the investor for NEC to be preferred to

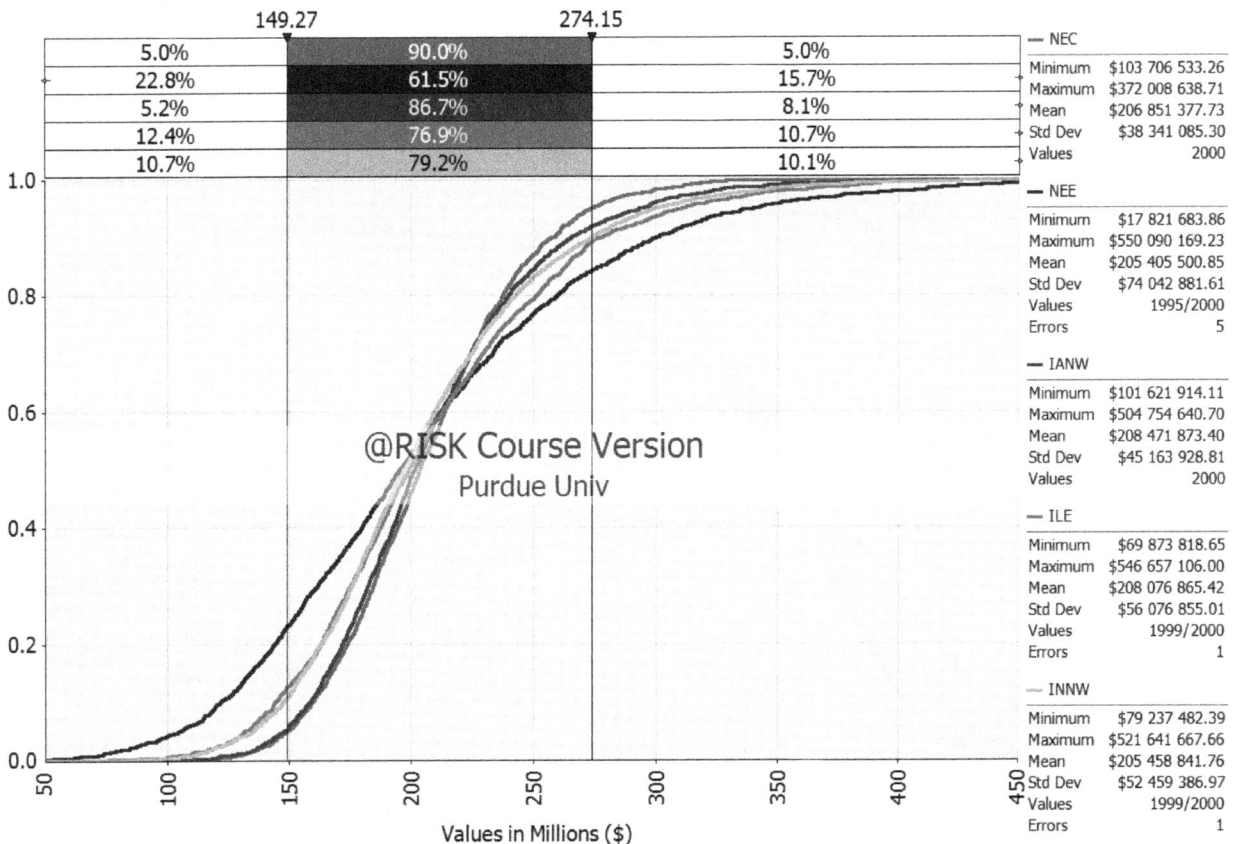

Fig. 7 Cumulative density functions of net present value in five districts under 1.5 elasticity of price with respect to yield.

INNW under this elasticity scenario. Figure 7 reveals that an increase in the elasticity of farm-gate stover price with respect to yields favors the irrigated district over all others.

Figures 8 and 9 reveal that NPV is highly sensitive to gasoline price in all districts. In fact, a 5% reduction (increase) in gasoline price reduces (increases) mean NPV by almost 30% in all districts. As indicated in Fig. 8, a 5% reduction in gasoline price increases the probability of lower economic returns (e.g. lower than $200 million) by orders of magnitude in all agricultural districts. In contrast, a 5% increase in output price greatly enhances NPV (Fig. 9) in all districts. The relative position of CDFs and hence an investor's preference over them is unaffected by a change in gasoline price.

An increase in transportation cost may raise the volatility of nonirrigated areas relative to irrigated ones favoring NEC over other districts. On the other hand, increased transportation cost also favors high density districts as average transportation distances are smaller. Comparison of results presented in Fig. 10 with those in Fig. 5 reveals the net effect of these opposing forces. A doubling of transportation cost increases standard deviation of districts only slightly, while it affects their mean NPV substantially.

The effect on mean NPV is not, as expected, symmetric across districts. While an increase in transportation cost reduces mean NPV by 11% in NEC, it only reduces it by 8% in IANW, the highest density district in the sample. The combination of the effect of increased transportation cost on mean and variance of districts' CDFs, diminishes the attractiveness of NEC relative to all other districts. In fact under high transportation cost, no district is dominated by NEC on a second-order stochastic sense. Therefore, higher transportation costs would make risk-neutral investors more likely to locate in the rainfed districts.

Conclusion and policy implications

Despite its potential importance for firm profitability and location decisions, previous economic analyses of stover-based biofuels have not explicitly modeled and quantified the link between density and variability of stover production, and feedstock cost faced by plants in different regions. Areas of the eastern and central Corn Belt are typically considered better candidates than the western Corn Belt due to higher corn planting density. However, Monte Carlo simulations based on estimated probability density functions of yield in the eastern and western regions of the Corn Belt reveal

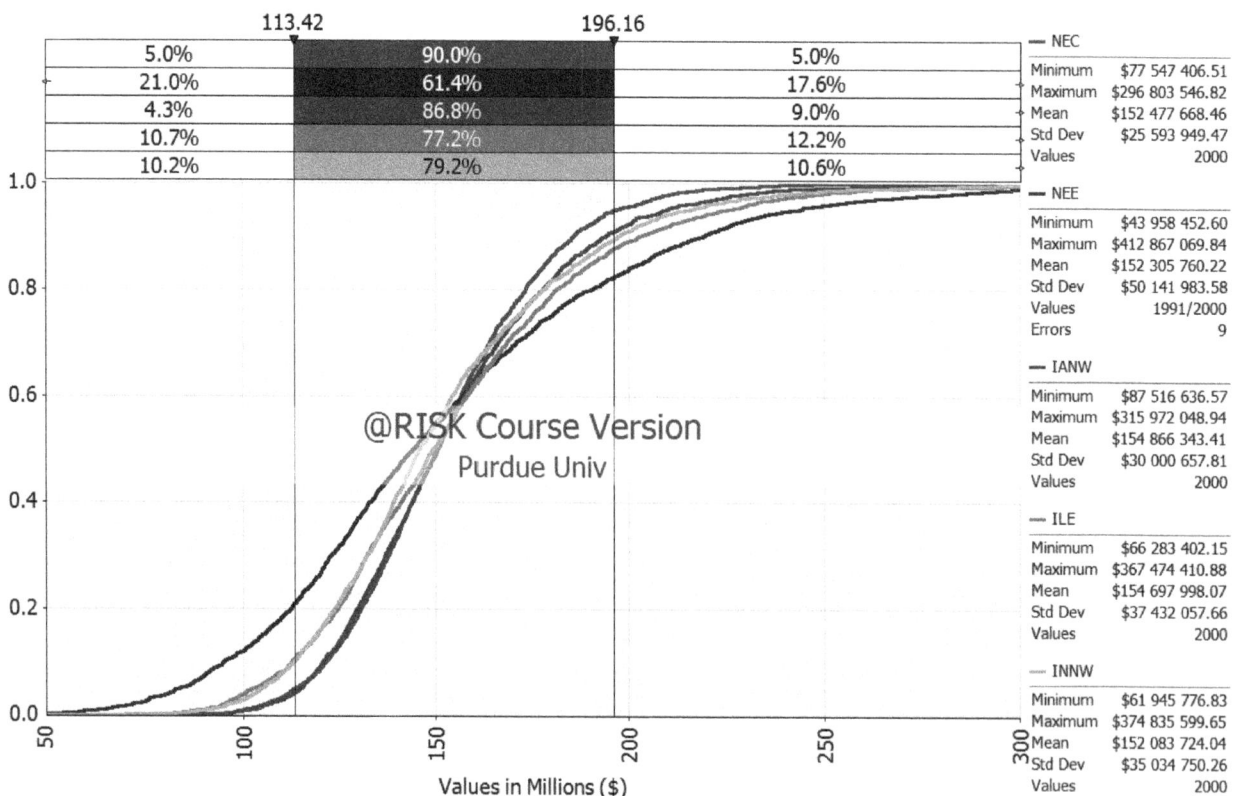

Fig. 8 Cumulative density functions of net present value in five districts under low ($0.665 L^{-1}) gasoline price.

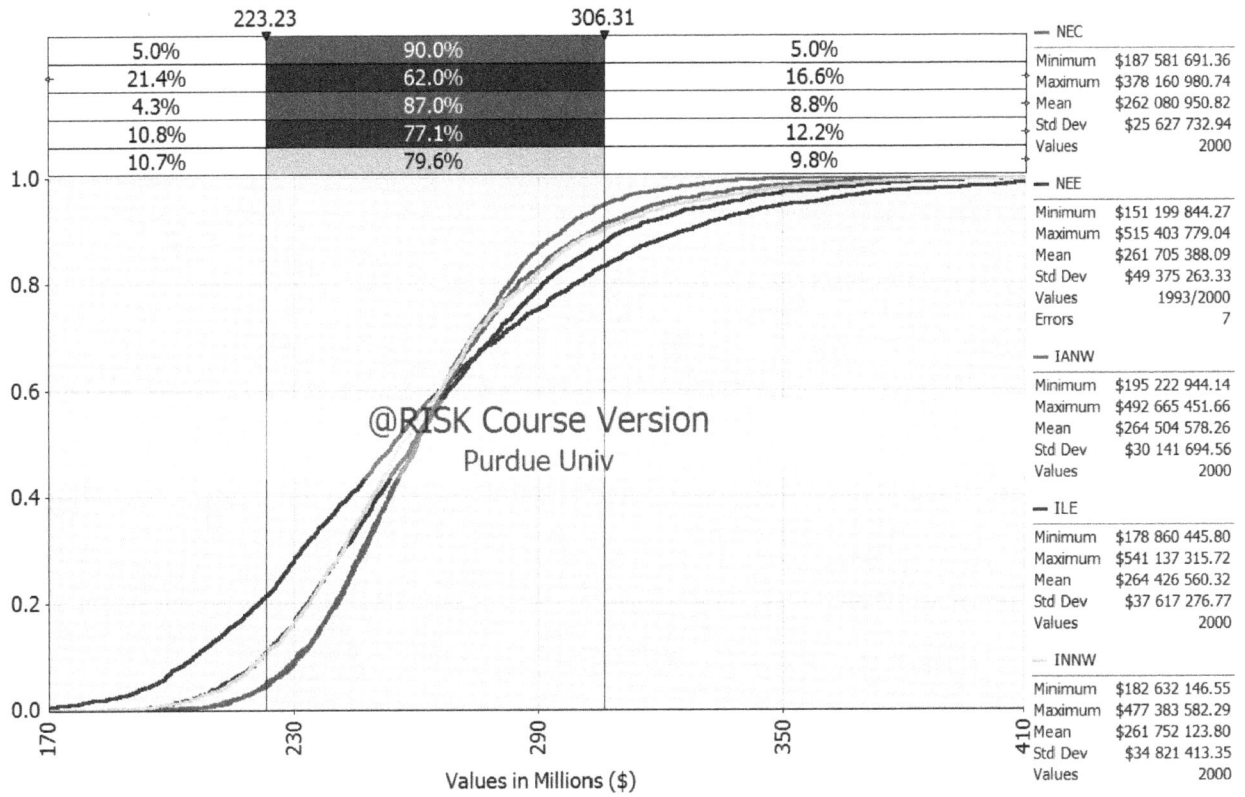

Fig. 9 Cumulative density functions of net present value in five districts under high ($0.735 L^{-1}) gasoline price.

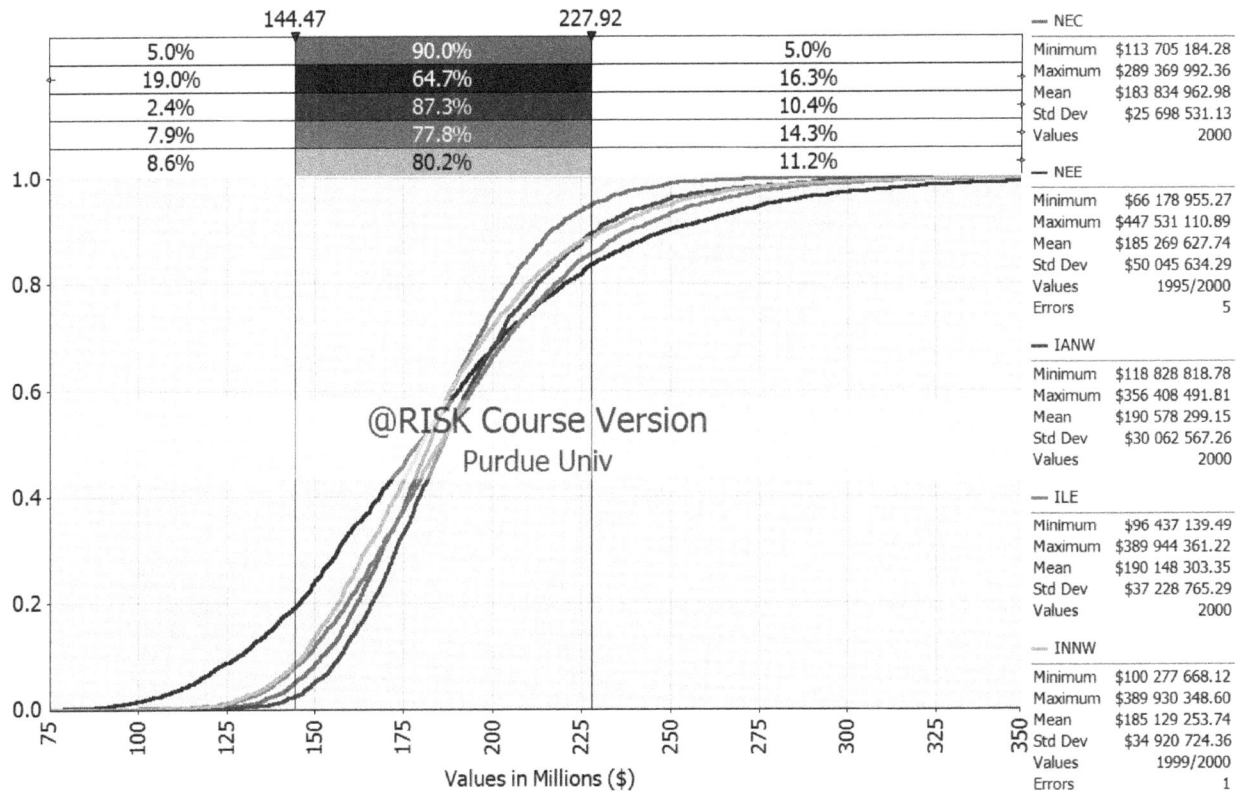

Fig. 10 Cumulative Density Functions of Net Present Value in five districts under high transportation cost.

that irrigation in the west greatly reduces negative skewness of the yield distribution. This, when stover price is affected by local supply conditions, results in significant reduction in the spread of feedstock cost and investment returns.

Our results reveal an interesting risk–return tradeoff associated with irrigated and nonirrigated regions of the Corn Belt. While INNW displays a higher corn planting density than NEC, higher mean and lower volatility in yields due to irrigation in NEC result in dominance of NEC over INNW in a second-order stochastic sense. NEC also dominates NEE due to this risk-reduction effect of irrigation. While NEC does not dominate ILE from a second-order stochastic point of view, it does display a significantly lower downside risk. Therefore, under conditions assumed in this study, risk-averse investors may prefer NEC over ILE. Finally, IANW dominates NEC in a first-order stochastic sense and all other locations in a second-order stochastic sense.

These results have policy implications both at the federal and state levels. First, our results suggest that biofuel policies at the federal level may trigger entry of plants in the irrigated Corn Belt. This may increase pressure on groundwater resources. This translates into two main insights regarding state-level policies. First, states in the irrigated Corn Belt (Nebraska, Kansas, and to a lesser degree Texas) may not need to put in place strong incentives to entice potential investors as the comparative advantage provided by the risk-reducing effect of irrigation makes them attractive for stover-based bio-gasoline producers. This prediction is consistent with anecdotal evidence as Abengoa Bioenergy is building one of the first cellulosic biofuel plants in Hugoton, KS. Second, as these states tighten constraints on irrigation, if such constraints result in deficit irrigation as many have envisioned (e.g. Klocke *et al.*, 2011), the comparative advantage of these regions will be greatly diminished as the risk-reducing effect of irrigation will be weakened.

This study is not without limitations. First, insights from this study may not apply to irrigated areas where agro-climatic conditions differ from those in the Central Plains. Second, the effect of stover harvest density on price is a critical driver of our results. Yet there is no information based on which this relationship can be estimated. There is, hence, a great deal of uncertainty on the link between stover supply and price. Different assumptions on this link (captured in our model by β) can then be made, and they will yield different results. Finally, yield volatility and price effect are the only two factors that we assume are uncertain in this paper.

However, uncertainty may come from multiple sources and these sources may in fact interact with stover yields in driving NPV. This seems like a promising research avenue. Additionally, on-farm harvest cost may be affected by idiosyncratic regional agronomic conditions other than yield and those spatial differences are not explicitly quantified here (although sensitivity analysis was presented in Supporting Information 3).

Acknowledgements

This study received support from Purdue University through its 'Agricultural Research Program'.

References

Anex RP, Aden A, Kazi FK *et al.* (2010) Techno-economic comparison of biomass-to-transportation fuels via pyrolysis, gasification, and biochemical pathways. *Fuel*, **89**, S29–S35.

Annual Energy Outlook (2014) *U.S. Energy Information Administration Office of Integrated and International Energy Analysis*. U.S. Department of Energy, Washington, DC. Available at www.eia.gov/forecasts/aeo (accessed 7 June 2015).

Archer DW, Johnson JMF (2012) Evaluating local crop residue biomass supply: economic and environmental impacts. *Bioenergy Research*, **5**, 699–712.

Brechbill SC, Tyner WE, Ileleji KE (2011) The economics of biomass collection and transportation and its supply to Indiana cellulosic and electric utility facilities. *Bioenergy Research*, **4**, 141–152.

Bridgwater AV (2012) Review of fast pyrolysis of biomass and product upgrading. *Biomass and Bioenergy*, **38**, 68–94.

Brown TR, Brown RC (2013a) A review of cellulosic biofuel commercial-scale projects in the United States. *Biofuels, Bioproducts and Biorefining*, **7**, 235–245.

Brown TR, Brown RC (2013b) Techno-economics of advanced biofuels pathways. *RSC Advances*, **3**, 5758–5764.

Brown TR, Thilakaratne R, Brown RC, Hu G (2013a) Techno-economic analysis of biomass to transportation fuels and electricity via fast pyrolysis and hydroprocessing. *Fuel*, **106**, 463–469.

Brown TR, Thilakaratne R, Brown RC, Hu G (2013b) Regional differences in the economic feasibility of advanced biorefineries: fast pyrolysis and hydroprocessing. *Energy Policy*, **57**, 234–243.

Cottam ML, Bridgwater AV (1993) Techno-economics of pyrolysis oil production and upgrading. In: *Advances in Thermochemical Biomass Conversion* (eds Bridgwater AV), pp. 1343–1357. Springer, Netherlands.

Downing ME, Eaton LM, Graham RL *et al.* (2011) *US Billion-Ton Update: Biomass Supply for a Bioenergy and Bioproducts Industry*. No. ORNL/TM-2011/224. Oak Ridge National Laboratory (ORNL), Oak Ridge, TN.

Energy Independence and Security Act (2007) http://www.afdc.energy.gov/laws/eisa.html (accessed 6 June 2015).

Fiegel J (2012) *Development of a Viable Corn Stover Market: Impacts on Corn and Soybean Markets*. MS thesis, Purdue University.

Gallagher P, Dikeman M, Fritz J, Wailes E, Gauthier W, Shapouri H (2003) Supply and social cost estimates for biomass from crop residues in the United States. *Environmental and Resource Economics*, **24**, 335–358.

Gallagher P, Wisner R, Brubacker H (2005) Price relationships in processors' input market areas: testing theories for corn prices near ethanol plants. *Canadian Journal of Agricultural Economics*, **53**, 117–139.

Gramig BM, Reeling CJ, Cibin R, Chaubey I (2013) Environmental and economic trade-offs in a watershed when using corn stover for bioenergy. *Environmental Science & Technology*, **47**, 1784–1791.

Jones SB, Valkenburg C, Walton CW *et al.* (2009) *Production of Gasoline and Diesel from Biomass via Fast Pyrolysis, Hydrotreating and Hydrocracking: A Design Case*. Pacific Northwest National Laboratory, Richland, WA.

Klocke NL, Currie RS, Tomsicek DJ, Koehn J (2011) Corn yield response to deficit irrigation. *Trans ASABE*, **54**, 931–940.

Lane J (2013) KiOR mulls "Columbus II" facility to accelerate path to profits, as 2013 production forecast is cut. *Biofuels Digest*. Available at: http://www.biofuelsdigest.

com/bdigest/2013/08/12/kior-mulls-columbus-ii-facility-to-accelerate-path-to-pro fits-as-2013-production-forecast-is-cut/ (accessed 7 June 2015).

Meier D, Faix O (1999) State of the art of applied fast pyrolysis of lignocellulosic materials—a review. *Bioresource Technology*, **68** , 71–77.

Meier D, van de Beld B, Bridgwater AV, Elliott DC, Oasmaa A, Preto F (2013) State-of-the-art of fast pyrolysis in IEA bioenergy member countries. *Renewable and Sustainable Energy Reviews*, **20**, 619–641.

National Research Council (US) (2011). *Committee on Economic and Environmental Impacts of Increasing Biofuels Production*. Renewable Fuel Standard: Potential Economic and Environmental Effects of US Biofuel Policy. National Academies Press, Washington, DC.

Parcell JL, Westhoff P (2006) Economic effects of biofuel production on states and rural communities. *Journal of Agricultural and Applied Economics*, **38** , 377.

Perlack RD, Turhollow AF (2002) Assessment of options for the collection, handling, and transport of corn stover. ORNL/TM-2002/44, Report to the US Department of Energy, Office of Energy Efficiency and Renewable Energy, Biomass Program, http://bioenergy. ornl. gov/pdfs/ornltm-200244. pdf (accessed 9 April 2010).

Perrin R, Sesmero J, Wamisho K, Bacha D (2012) Biomass supply schedules for Great Plains delivery points. *Biomass and Bioenergy*, **37**, 213–220.

Petrolia DR (2008a) The economics of harvesting and transporting corn stover for conversion to fuel ethanol: a case study for Minnesota. *Biomass and Bioenergy*, **32**, 603–612.

Petrolia DR (2008b) An analysis of the relationship between demand for corn stover as an ethanol feedstock and soil erosion. *Review of Agricultural Economics*, **30**, 677–691.

Petter R, Tyner WE (2014) Technoeconomic and policy analysis for corn stover biofuels. *ISRN Economics*, **2014**, 1–13.

Richard TL (2010) Challenges in scaling up biofuels infrastructure. *Science (Washington)*, **329**, 793–796.

Sesmero JP (2014) Cellulosic biofuels from crop residue and groundwater extraction in the US Plains: the case of Nebraska. *Journal of Environmental Management*, **144**, 218–225.

Tyner WE (2010) The integration of energy and agricultural markets. *Agricultural Economics*, **41** , 193–201.

Tyner W (2011) Description of 2011 biofuels policy alternatives. *Global Policy Research Institute (GPRI) Policy Briefs*, **1** , 1.

U.S. Energy Information Administration, Annual Energy Outlook (2011) http://www.eia.gov/aeo (accessed 6 June 2015)

Wang MQ, Han J, Haq Z, Tyner WE, Wu M, Elgowainy A (2011) Energy and greenhouse gas emission effects of corn and cellulosic ethanol with technology improvements and land use changes. *Biomass and Bioenergy*, **35**, 1885–1896.

Wright MM, Daugaard DE, Satrio JA, Brown RC (2010) Techno-economic analysis of biomass fast pyrolysis to transportation fuels. *Fuel*, **89**, S2–S10.

14

What can and can't we say about indirect land-use change in Brazil using an integrated economic – land-use change model?

JUDITH A. VERSTEGEN[1], FLOOR VAN DER HILST[1], GEERT WOLTJER[2],
DEREK KARSSENBERG[3], STEVEN M. DE JONG[3] and ANDRÉ P. C. FAAIJ[4]

[1]Faculty of Geosciences, Copernicus Institute for Sustainable Development, Utrecht University, Heidelberglaan 2, 3584 CS, Utrecht, The Netherlands, [2]LEI, Wageningen University & Research Centre, Alexanderveld 5, 2502 LS Den Haag, The Netherlands, [3]Department of Physical Geography, Faculty of Geosciences, Utrecht University, Heidelberglaan 2, 3584 CS, Utrecht, The Netherlands, [4]Energy and Sustainability Research Institute Groningen, University of Groningen, Blauwborgje 6, PO Box 9700 AE, Groningen, The Netherlands

Abstract

It is commonly recognized that large uncertainties exist in modelled biofuel-induced indirect land-use change, but until now, spatially explicit quantification of such uncertainties by means of error propagation modelling has never been performed. In this study, we demonstrate a general methodology to stochastically calculate direct and indirect land-use change (dLUC and iLUC) caused by an increasing demand for biofuels, with an integrated economic – land-use change model. We use the global Computable General Equilibrium model MAGNET, connected to the spatially explicit land-use change model PLUC. We quantify important uncertainties in the modelling chain. Next, dLUC and iLUC projections for Brazil up to 2030 at different spatial scales and the uncertainty herein are assessed. Our results show that cell-based (5×5 km^2) probabilities of dLUC range from 0 to 0.77, and of iLUC from 0 to 0.43, indicating that it is difficult to project exactly where dLUC and iLUC will occur, with more difficulties for iLUC than for dLUC. At country level, dLUC area can be projected with high certainty, having a coefficient of variation (cv) of only 0.02, while iLUC area is still uncertain, having a cv of 0.72. The latter means that, considering the 95% confidence interval, the iLUC area in Brazil might be 2.4 times as high or as low as the projected mean. Because this confidence interval is so wide that it is likely to straddle any legislation threshold, our opinion is that threshold evaluation for iLUC indicators should not be implemented in legislation. For future studies, we emphasize the need for provision of quantitative uncertainty estimates together with the calculated LUC indicators, to allow users to evaluate the reliability of these indicators and the effects of their uncertainty on the impacts of land-use change, such as greenhouse gas emissions.

Keywords: biofuel, Brazil, error propagation, indirect land-use change, land-use change, modelling, Monte Carlo spatio-temporal, sugar cane, uncertainty

Introduction

Governments throughout the world have set mandatory biofuel targets for the transport sector, aiming at mitigating climate change, improving energy security, and stimulating rural development (Sorda et al., 2010). Currently, one of the central problems in the biofuel arena is the premise of biofuel-induced land-use change (IPCC, 2011; Creutzig et al., 2012; Finkbeiner, 2014; Warner et al., 2014). These land-use changes can have negative impacts such as carbon stock loss, rising food prices, loss of biodiversity, and water scarcity, reducing the eligibility of the feedstock as a sustainable source for biofuels. An increased demand for biofuel feedstocks can lead to direct land-use change (dLUC): land use is changed from some previous use to the biofuel feedstock. This, in turn, can lead to indirect land-use change (iLUC): a change of land use outside the biofuel feedstock cultivation area, induced by a change in use or production quantity of that biofuel feedstock. This can happen either when the agricultural land-use type converted to the biofuel feedstock is displaced to elsewhere, in order to continue to meet the demand for its agricultural products, or when the direct conversion triggers a change in the price of agricultural products, causing land to be taken into (or out of) production elsewhere (Wicke et al., 2012). The question to be tackled is to what extent the global increase in demand for biofuels (Broch et al., 2013) leads

Correspondence: Judith Verstegen
e-mail: J.A.Verstegen@uu.nl

to dLUC and iLUC and how the negative effects can be minimized.

Direct land-use changes are unambiguously visible in both historical data and spatial land-use change model results. DLUC takes place wherever a bioenergy crop field appears and consequently displaces the previous land use. On the contrary, iLUC cannot be directly observed (Finkbeiner, 2014), because if, for example, pasture displaces forest in the presence of an expansion of bioenergy cropland over pasture, this does not necessarily mean that the pasture displacement is caused by the expansion of bioenergy cropland. The pasture might have caused deforestation for a reason unrelated to bioenergy. In other words, the indirect effects of a particular demand increase cannot be identified from historical data because the effects are intertwined with a wide range of processes from which the effects are also present in these data (O'Hare et al., 2011; Overmars et al., 2011). Separate identification is only possible by comparing all land-use changes with and without the demand increase for bioenergy, which can be performed using a simulation model (Creutzig et al., 2014).

The processes governing dLUC and iLUC range from global to local scale. For example, the impact of the biofuel targets on demands for feedstocks in different parts of the world is a global market issue. On the other hand, at which location the land-use changes and which previous land use is replaced is primarily steered by local factors, such as accessibility and biophysical conditions (Meyfroidt et al., 2013). Likewise, the impacts of the land-use change are highly location-dependent (e.g. van der Hilst et al., 2014). Therefore, a sound approach to model iLUC is by using a global economic model coupled to a spatially explicit land-use change (LUC) model to take both the global- and local-scale level into account, as, for example, demonstrated by Lapola et al. (2010).

It is commonly recognized that there is a large uncertainty in modelled iLUC (Mathews & Tan, 2009; Wicke et al., 2012; Malins, 2013; Creutzig et al., 2014; Finkbeiner, 2014). The uncertainties arise from model structure uncertainty (Refsgaard et al., 2006; Verstegen et al., 2015), from data (inputs, calibration data set and initial system state) (Dendoncker et al., 2008), and from model coupling (Ray et al., 2012). For iLUC in particular, uncertainty in reported values also stems from the fact that the assumptions, the employed models, and the validity of these models are often not clearly communicated (Mathews & Tan, 2009). Information quantifying uncertainty in iLUC is critical to evaluate whether or not iLUC indicators are reliable enough to be included in legislation, to identify which parts of the modelling chain have the highest priority for improvement, that is cause most uncertainty, and to

assess how this uncertainty propagates to the impacts of iLUC, such as greenhouse gas (GHG) emissions (e.g. Plevin et al., 2015). Uncertainty information can be obtained by (1) being explicit about the applied models, the processes included in these models, and the parameter settings used, as well as the uncertainty in the various model components and the performance of these models (Mathews & Tan, 2009; Broch et al., 2013), and (2) assessment of the magnitude of the output uncertainty by, for example, doing Monte Carlo analyses of iLUC (Wicke et al., 2012, 2015; Nelson et al., 2014; Warner et al., 2014; Plevin et al., 2015). Uncertainty should be assessed at different spatial scales because different types of impacts play a role at different scales and it is known that uncertainty is highly scale-dependent (e.g. Pontius Jr. and Spencer; Verstegen et al., 2012). Yet, such information is currently scarcely reported for iLUC; a status we aim to improve with this study.

We have set up a model study with the global Computable General Equilibrium (CGE) model MAGNET (e.g. Kavallari et al., 2014; Woltjer & Kuiper, 2014), integrated with the spatially explicit land-use change model PLUC (e.g. Verstegen et al., 2012). With this integrated model, we project land-use change caused by an increasing demand for biofuels up to 2030 for Brazil, one of the main bioethanol producers in the world. As Brazil holds the world's major potential for agricultural expansion (Alexandratos & Bruinsma, 2012), production and export of bioethanol are likely to increase in the future (IEA, 2013; OECD/Food and Agriculture Organization, 2014, Walter et al., 2014). Yet, the country also maintains the largest area of natural remnants, with high carbon stocks and high levels of biodiversity, stressing the need to assess potential negative impacts. For this case study, we seek to answer the following research questions: (1) What are the dLUC and iLUC projections for Brazil up to 2030 at different spatial scales and what is the uncertainty herein? (2) What are the sources of uncertainty for each step in the model chain and how do these uncertainties influence dLUC and iLUC projections? (3) What is the contribution of the economic and land-use change model to the uncertainty in dLUC and iLUC at the different spatial scales?

The next section introduces the Brazilian case study, presents the LUC model, the CGE model, and the way they are coupled, describes the calibration method, defines the projection scenario for the increased demand for biofuels, and explains how iLUC is derived from the results. Section three illustrates the results for the three research questions. The final section discusses these results in the light of the research questions and gives suggestions for further research.

Materials and methods

Overview

The projection of dLUC and iLUC in Brazil caused by an increasing demand for biofuels and the uncertainty herein is performed using MAGNET (Woltjer & Kuiper, 2014), a global Computable General Equilibrium (CGE) model, connected to the land-use change model PLUC (e.g. Verstegen *et al.*, 2012), tailored to Brazil (Fig. 1). For 2006, an initial land-use map is created by combining tabular area data per land-use type and land-use maps with satellite data. This map is used as the initial system state for PLUC. Next, PLUC is calibrated from 2007 until 2012 based on trends per land-use type from agricultural statistics databases. To project the dLUC and iLUC effects of the biofuel mandates, we define both a 'biofuel scenario' that includes these mandates and a 'reference scenario' that does not include them. For both scenarios, MAGNET determines the supply and demand of all commodities in all world regions up to 2030 and, related to that, the areas they occupy. This 2013–2030 time series of land area demands per land-use type for Brazil is then input for the spatially explicit land-use change projection up to 2030 by PLUC. The PLUC outputs are a time series of land-use maps. By comparison of the maps of the two scenarios, dLUC and iLUC are assessed.

In the model chain, uncertainties in the inputs, calibration data set, initial system state, and model structure are quantified, part of which propagates through the model coupling (Fig. 1). To quantify uncertainty in MAGNET, it is run with two different parameter sets, resulting in an upper and a lower demand limit. PLUC, including the generation of the initial land-use map, the calibration, and the demand coming from MAGNET, is used stochastically by running it in Monte Carlo mode (Fig. 1).

Case study

Brazil has been producing bioethanol from sugar cane since the beginning of the 20th century and has been exporting the etha-

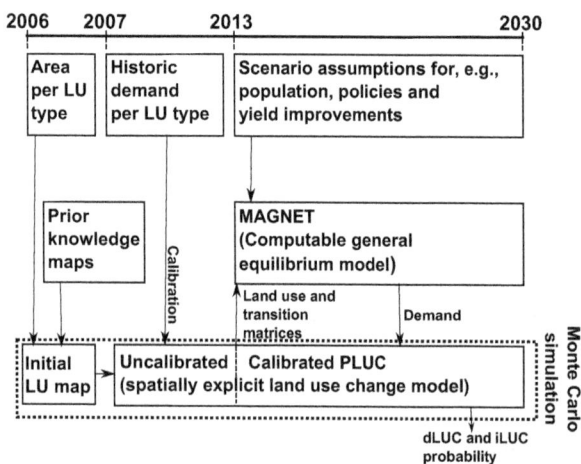

Fig. 1 Overview of the modelling chain and model run-time frame to simulate the probability of dLUC and iLUC in Brazil up to 2030.

nol since 1989 (Andrade de Sá *et al.*, 2013). Sugar cane currently occupies the third largest area of all crops in Brazil, topped only by soya and maize (although a large quantity of the maize is cultivated as second crop) (IBGE, 2013b). The main sugar cane production areas are the Central South region and the north-east region. Recent expansion has mainly taken place in the Central South region: in the past decade, the total area dedicated to sugar cane cultivation has more than doubled in that region (Rudorff *et al.*, 2010). It expected that future expansion will also predominantly occur in the Central South region (Nassar *et al.*, 2008; Lapola *et al.*, 2010). According to Adami *et al.* (2012), over 99% of all sugar cane expansion in the last decade has taken place over existing agricultural land, signifying that the direct effect of increasing ethanol demand on deforestation is negligible. However, deforestation can still take place through iLUC, which is also shown by others (Lapola *et al.*, 2010; e.g., de Souza Ferreira Filho & Horridge, 2014).

Initial land-use map and land-use change model

We distinguish 11 different land-use types n, where $n = 1, 2, \ldots, 11$: urban, water, natural forest, rangeland, crops (excluding sugar cane), grass and shrubs, sugar cane, planted forest, planted pasture, bare soil, and abandoned agricultural land. Planted pasture and natural pasture (rangeland) are modelled separately because the extensively managed, naturally vegetated rangelands have a stocking rate of about 70% lower than the intensively managed planted pastures (IBGE, 2006, Aguiar & d'Athayde, 2014). Cropland includes both annual and permanent crops. Sugar cane is modelled as a separate land-use type to be able to evaluate where sugar cane expands in reaction to the increased ethanol demand and which other land uses it replaces.

PLUC (PCRaster Land Use Change model) (van der Hilst *et al.*, 2012, 2014; Verstegen *et al.*, 2012; Diogo *et al.*, 2014) is founded on the separation between the quantity of change per land-use type and the spatial allocation of this change, like many other land-use change models (Pontius Jr. and Neeti, 2010). The quantity of land demanded per land-use type n is called 'demand' $d_{n,t}$, in which t is the time step in years, with $t = 1, 2, \ldots, T$. The total area per land-use type in the demand time series (tabular area data from agricultural statistics) for the initial year of the simulation should match the total area per land-use type in the initial land-use map, that is the initial system state of the model. If the time series and initial map are coming from different sources, which is likely, a perfect match is obviously never going to be the case. You & Wood (2005) provide a deterministic method to create a land-use map that matches the time series, by spatially disaggregating land-use areas per administrative region from the time series into raster cells within that region, using prior knowledge maps. We apply this procedure, using municipalities as administrative regions (5566 in total for Brazil), to create an initial land-use map for Brazil with a cell size of 5×5 km^2 for the year 2006. This year is chosen because it was the year in the recent past (to have a calibration period) with the best data availability for both the tabular and prior knowledge map data. Compared to You & Wood (2005), we do a few things differently, most importantly adding a method to make a stochastic map instead of a deterministic map, in order to include uncertainty arising from errors in the initial land-use map into the model chain, as explained in Methods S1.

Of the eleven land-use types considered in PLUC, five are assumed to respond to changes in the economy by expanding or contracting: rangeland, planted forest, crops, sugar cane, and planted pasture. These *active* land-use types are demand-driven (Table 1). The other six land-use types do not have demands. They are either *passive*, meaning that they can contract or expand due to the dynamics of the active land-use types, or *static*, meaning that they cannot change and are thus fixed on the map. Passive land-use types are natural forest, grass and shrubs, bare soil, and abandoned agricultural land. Abandoned land originates when an active land-use type contracts; it is not present in the initial land-use map. Static land-use types are urban and water.

The demands for the five dynamic land-use types over time in Brazil have been subdivided into six regions (Fig. 2), corresponding to the macroregions defined by the Brazilian Institute of Geography and Statistics (IBGE). We added one region by splitting the north-eastern macroregion into two regions, as suggested by Nassar *et al.* (2010), because the north-east coast differs significantly from the north-east Cerrado (savannah) in terms of agricultural production.

In PLUC, the spatial allocation is regulated by spatial attributes that serve as proxies for important drivers of location, that is processes that determine where a land-use type expands or contracts. These are called suitability factors k, with $k = 1, 2, \ldots, K_n$ (each active land-use type n can have a different number of suitability factors). For each n defined as active, a weighted sum of these suitability factors forms the total suitability map. In one model time step, representing 1 year, the demands of the active land-use types are allocated sequentially for each macroregion, as follows. For the first active land-use type n, the total suitability map is sorted, and cells are allocated to n,

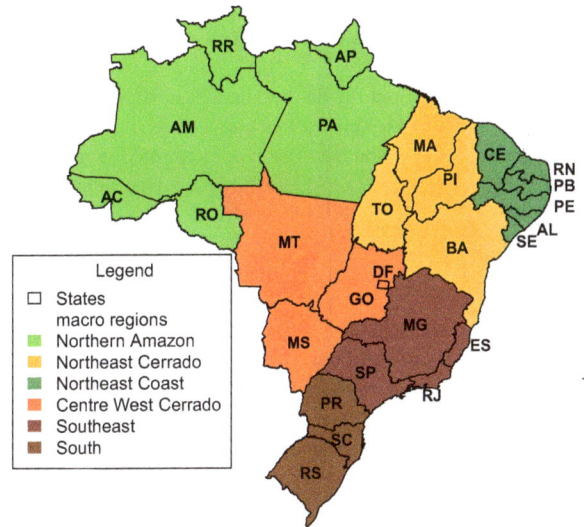

Fig. 2 The six macroregions in Brazil (six different colours) used as demand input units in PLUC and the 27 states (black lines), or in fact 26 states and one federal district, used as calibration units. The state name abbreviations are as follows: AC, Acre; AL, Alagoas; AM, Amazonas; AP, Amapá; BA, Bahia; CE, Ceará; DF, Distrito Federal; ES, Espírito Santo; GO, Goiás; MA, Maranhão; MG, Minas Gerais; MS, Mato Grosso do Sul; MT, Mato Grosso; PA, Pará; PB, Paraíba; PI, Piauí; PR, Paraná; RJ, Rio de Janeiro; RN, Rio Grande do Norte; RO, Rondônia; RR, Roirama; RS, Rio Grande do Sul; SC, Santa Catarina; SE, Sergipe; SP, São Paulo; TO, Tocatins.

starting with the cell with the highest suitability value that is not yet of type n, until $d_{n,t}$ is fulfilled. Next, the same is performed for the second land-use type in the sequence, with the exception that cells occupied by the first land-use type cannot

Table 1 Suitability factors, k, of the active land-use types, n, for the Brazilian case study

n	Land-use type	k	Process represented	Suitability factor
4	Rangeland	1	Economies of scale	n in the neighbourhood
		2	Transportation costs	Distance to roads
		3	Potential profits per hectare	Potential yield of n
5	Planted forest	1	Economies of scale	n in the neighbourhood
		2	Transportation costs	Distance to roads
		3	Potential profits per hectare	Potential yield of n
6	Crops	1	Economies of scale	n in the neighbourhood
		2	Transportation costs	Travel time to hubs for n
		3	Potential profits per hectare	Potential yield of n
		4	Costs to make the land cultivatable	Conversion elasticity
		5	Double-cropping potential	Growing season length
8	Sugar cane	1	Economies of scale	n in the neighbourhood
		2	Transportation costs	Travel time to hubs for n
		3	Potential profits per hectare	Potential yield of n
		4	Costs to make the land cultivatable	Conversion elasticity
9	Planted pasture	1	Economies of scale	n in the neighbourhood
		2	Transportation costs	Distance to hubs for n
		3	Potential profits per hectare	Potential yield of n

be changed. This procedure continues until the demands of all active land-use types in all macroregions have been allocated (see also Methods S2).

The suitability factors for the Brazilian case study are given in Table 1. To represent economies of scale ($k = 1$), the number of neighbours of the same land-use type is counted in a square window of 5 by 5 cells (25×25 km^2). For transportation costs ($k = 2$), the travel time to hubs is used as a proxy. This is the time it takes to transport the products originating from the land-use type to the nearest production facility. For planted forest, we have no data about the location of hubs (e.g. saw mills), and for rangelands, we believe that the livestock hubs are of lower importance, because livestock from rangeland is often 'finished' elsewhere before being slaughtered. Therefore, for these two land uses, we apply distance to roads as the proxy for transportation costs. Potential profits per hectare ($k = 3$) are represented by potential yield maps, using IIASA's GAEZ data (Tóth et al., 2012). As, to our knowledge, no potential yield map exists for woody biomass, we use IIASA's map of the length of the growing season as a proxy for the potential yield of planted forest. The costs to make the land cultivable ($k = 4$) are estimated using a conversion elasticity, that is a fraction indicating the ease with which a certain land-use type can be transformed into the land-use type that implements the suitability factor, especially relevant for crops. Double-cropping potential ($k = 5$) is an important suitability factor in Brazil, indicated by the rapid increase in double-cropped area or even triple-cropped area over the last decade (Galford et al., 2008, Conab, 2014). We do not have a map of double-cropping potential, so we use the growing season length as a proxy, which is supported by an analysis of the relation between these two by Arvor et al. (2014).

The no-go map, that is areas where expansion is not allowed, is an overlay of military areas, areas of indigenous people, and federal and state conservation units (Gurgel et al., 2009). Conservation policies or initiatives which have historically not been well enforced, such as the Forest Act (Sparovek et al., 2012), the soy moratorium (Rudorff et al., 2011), and the sugar cane zoning (Padua Junior et al., 2012), are not taken into account in this simulation. We are preparing another study, in which we include more scenarios with, among other things, stricter nature conservation rules (F. van der Hilst, J.A. Verstegen, G. Woltjer, E.M.W. Smeets, A.P.C. Faaij, unpublished results).

We use a Monte Carlo simulation with 5000 realizations. The weights of the suitability factors and the order of allocation are modelled stochastically. Their prior probability distributions are uninformed (see Methods S2).

Calibration

The aim of the calibration phase, 2007 to 2012, is to narrow the probability distributions of all stochastic elements: the order of the land-use types and all weights of the suitability factors (Table 1). The model calibration is performed using a Bayesian data assimilation technique, the sequential importance resampling (SIR) particle filter (van Leeuwen, 2009). In short, the SIR particle filter compares the land-use system simulated by PLUC and observations of the land-use system from the real world, taking into account the uncertainty in these observations. Next, it updates the Monte Carlo ensemble in such a way that well-performing realizations are progressed and poorly performing realizations are discarded. An extensive explanation of this model structure identification and calibration method for a case study in the São Paulo state is provided by Verstegen et al. (2014).

For calibration, a time series of land-use/cover data is required as observational data. For Brazil, we use time series of areal data per land-use type per state (Fig. 2). These time series are derived using information from agricultural statistics databases (IBGE, 2013a,b, ABRAF, 2013, see F. van der Hilst, J.A. Verstegen, G. Woltjer, E.M.W. Smeets, A.P.C. Faaij, unpublished results). These observational data are not error free. Between the yearly (IBGE, 2013b (for crops), IBGE, 2013a (for livestock)), and the 10-yearly (IBGE, 2006) census data sources, areas and area increases differ from zero up to more than 100%. As one cannot calculate a standard deviation based on two values, we make an educated guess of the average error based on these data sources. Under the assumption that the observational errors are uncorrelated over space and time, we assign an observation error to the observed increase in area with a standard deviation of 20% of the observed increase in that time step.

After calibration, a land-use matrix, summarizing the total areas per land-use type in 2012, is computed per macroregion, representing the initial system state for MAGNET (Fig. 1). In addition, a land transition matrix is calculated per macroregion, to be used for the calibration of MAGNET. These six land transition matrices show the average area of conversion from every land-use type to every other land-use type derived from PLUC over the whole calibration period.

As a measure of model performance, we calculate root-mean-squared error (RMSE), the root of the summed squared differences between the median of the modelled area and observed area over all states. We determine the reduction in RMSE (%) for the results of the calibrated model, that is with uncertainty reduced by the SIR particle filter, compared to the non-calibrated model. To evaluate the effect of calibration, we apply a split-sample approach: PLUC is calibrated using data from 2007 to 2009, and the model reduction in RMSE is evaluated from 2010 to 2012. This split-sample approach is used only to evaluate the effect of calibration. The model parameters we use for the projection, integrated with the MAGNET model, are calibrated based on all available observational data (2007–2012).

Economic (CGE) model

The growing demand for food, feed, fibre, and bioenergy requires an increased agricultural output. This can be reached by raising inputs such as fertilizers, machinery and labour (bound by technological limitations), that is expansion at the intensive margin, or by converting new land to agriculture, that is expansion at the extensive margin (Hertel, 2011), which can result in iLUC. At what ratio both alternatives are applied in face of a growing demand depends on, for example, land availability, prices, and policies that vary worldwide. To evaluate how demand grows over time and to assess to what extent this demand is fulfilled by expansion at the intensive and extensive margins, we use a global Computable General Equilibrium (CGE) model (Rose, 1995). Key parameters in CGE models are the elasticities, simulating behavioural responses, for example the response of the demand for a commodity to a change in price or the response of consumption to a change in GDP per capita.

The CGE model used is MAGNET (Modular Applied GeNeral Equilibrium Toolbox) (see for an extensive explanation Woltjer & Kuiper, 2014). This is the modularized and improved version of LEITAP (e.g. Banse et al., 2011; Hoefnagels et al., 2013). MAGNET uses the GTAP database version 8 (Narayanan et al., 2012), in an extended and adaptable form. For this case study, we use the database with 42 sectors (including various ethanol sectors that take into account co- and by-products such as molasses and electricity, and a difference between planted pasture and rangeland), 45 commodities and 15 regions, of which Brazil is one. Brazil has been subdivided into six regions, matching the input macroregions for PLUC (Fig. 2). These six macroregions are a subdivision in MAGNET in terms of agricultural production and land area only; for international trade, Brazil is considered as one region. Total land availability per macroregion is calculated from the no-go map.

To model land cover change, a regional land transition approach has been developed that is inspired on the work of de Souza Ferreira Filho & Horridge (2014) and further developed by Woltjer (2013). Herein, the area of land that is changed from one particular land-use type n to another one m depends on the land transition elasticity $e_{n,m}$. Using expert knowledge and trial and error, we test for all combinations of n and m for what values of $e_{n,m}$ MAGNET can best reproduce the 2012 system state given by the land-use matrix from PLUC, and the transitions given by the land transition matrix.

To assess the uncertainty related to the key parameters in the economic model, two runs are performed, one with considerably higher (200%) and one with considerably lower (25%) land transition elasticities $e_{n,m}$ than the values found by the procedure above. This results in two demand time series per land-use type, one for the upper land transition elasticities, $d_{u,n,t}$, and one for the lower land transition elasticities, $d_{l,n,t}$, where all potential lines between these time series are assumed to have equal likelihood:

$$d_{n,t} = d_{l,n,t} + Z_d \cdot (d_{u,n,t} - d_{l,n,t}), \text{ with } Z_d \sim U(0,1), \quad (1)$$

for each active n in each t.

Equation 1 shows that the demand input of PLUC $d_{n,t}$ in the projection phase has an error model based on a uniform distribution between $d_{u,n,t}$ and $d_{l,n,t}$.

Projection

In the projection from 2013 to 2030, the socio-economic developments are based on the Shared Socioeconomic Pathways (SSPs) (O'Neill et al., 2014). The SSPs quantify global drivers of the energy–economy–land-use system such as demographics and economic development. In these pathways, projections are

included on population and GDP growth. We use SSP2, the Middle of the Road pathway with some additional assumptions on, for example, the agricultural intensification over time (see F. van der Hilst, J.A. Verstegen, G. Woltjer, E.M.W. Smeets, A.P.C. Faaij, unpublished results).

Using SSP2 and these assumptions, MAGNET is run up to 2030, providing total land areas occupied by all land-use types for all world regions and the six macroregions in Brazil for the years 2013, 2015, 2020, 2025, and 2030. Yearly demand time series for the six macroregions to serve as an input for PLUC are obtained by a linear interpolation between these years and an aggregation of the areas of all individual crops, except sugar cane, into the single class cropland.

To evaluate the future dLUC and iLUC effects caused by current and planned ethanol mandates worldwide, we define both a 'biofuel scenario' including these mandates and a 'reference scenario' excluding them. This does not mean that there is no increase in the demand for sugar cane in the reference scenario, only that there is no (additional) increase originating from the increased ethanol demand. All other inputs and parameters of both models are kept the same as in the biofuel scenario.

Direct land-use change (dLUC) and indirect land-use change (iLUC)

Normally, direct land-use change can be assessed using one scenario, as the difference between current and projected land use. In our case, however, we want to assess dLUC from sugar cane caused by the biofuel mandates, that is, only sugar cane expansion for ethanol. Therefore, we want to exclude sugar cane expansion that is a result of an increased demand for sugar over time. Hence, both dLUC and iLUC originating from the mandates are assessed through the difference between the reference and the biofuel scenario (Table 2) in 2030. A grid cell that is sugar cane in the biofuel scenario, and something else in the reference scenario, is considered dLUC, that is sugar cane expansion

Table 2 Classification of differences in land use between the reference and the biofuel scenario that are considered undesirable effects of increasing ethanol demand (dLUC and iLUC, dark grey), and the opposite effects (neg_dLUC and neg_iLUC, light grey). The class 'other agriculture' includes rangeland, planted forest, crops, and planted pasture. The class 'nature' includes natural forest, grass and shrubs, bare soil, and abandoned agricultural land, thereby assuming that land will eventually become nature when left abandoned. Zero stands for no difference, that is neither (neg_)dLUC nor (neg_)iLUC

Reference scenario	Biofuel scenario		
	Sugar cane	Other agriculture	Nature
Sugar cane	0	neg_dLUC	neg_dLUC
Other agriculture	dLUC	0	neg_iLUC
Nature	dLUC	iLUC	0

resulting from the biofuel mandates. A grid cell that is nature in the reference scenario and agricultural land but not sugar cane is considered iLUC. The opposite effects exist as well. A grid cell that is sugar cane in the reference scenario and something else in the biofuel scenario is negative dLUC (neg_dLUC), and a grid cell that is agriculture in the reference scenario and nature or abandoned land in the biofuel scenario is negative iLUC (neg_iLUC).

Especially for iLUC, this opposite effect might appear in the real world. If, for example, an area of 10 000 ha of wheat fields is present, and 80% of this area is taken over by sugar cane for ethanol, then the remaining 20% of wheat land might be abandoned because the advantages of economies of scale have disappeared. The 8000 ha of displaced wheat land and the 2000 ha of wheat land now grown elsewhere make 10 000 ha of iLUC. In our methodology, we count the abandoned land as −2000 ha of iLUC [and therefore, we call it neg_iLUC (Table 2)], coming to a total of 8000 ha iLUC, which was indeed the area of land shifted by sugar cane.

To compare outcomes at different spatial scales, we focus our analysis on local, regional, and national level, calculated from output of PLUC. At the regional level, we use 250×250 km^2 blocks. We do not use administrative levels, like states, because these differ in size and are thus problematic to compare. The coefficient of variation (cv) (standard deviation of dLUC or iLUC area over all Monte Carlo realizations divided by the mean of dLUC or iLUC area over all Monte Carlo realizations) is used as the measure of uncertainty. As this measure of uncertainty is standardized by the mean, the cv is comparable between dLUC and iLUC and between regions with different magnitudes of dLUC or iLUC. As the local level, we use probabilities of dLUC and iLUC in single cells (5×5 km^2).

Contribution of the two models to total output uncertainty

We compare the contribution of the two models to the total output uncertainty, by running the projection until 2030 three times, all three with 5000 realizations. One Monte Carlo run is with both models stochastic (the default run used in all analysis described above). One run is with only PLUC stochastic (including the uncertainty in the initial land-use map and calibration time series). In this run the demand $d_{n,t}$ is fixed at the mean between the upper and lower time series, by setting Z_d (equation 1) to 0.5 for all Monte Carlo realizations to exclude uncertainty from MAGNET. The uncertainty in the output of this run is thus caused by uncertainty in PLUC only. The final run is with only MAGNET stochastic. In PLUC, the weights, the order of allocation, and the land-use map for 2012 are fixed by taking the medians hereof from the calibrated model, to exclude uncertainty from PLUC. This run results in information about output uncertainty caused by MAGNET. For the three runs, we compare the mean and the coefficient of variation in dLUC and iLUC area at the different spatial scales.

Results

Sources of uncertainty for each step in the model chain and their influence on dLUC and iLUC projections

Initial land-use map. For the initial year, 2006, a land-use map was created for each Monte Carlo realization to serve as the initial system state (Fig. 3). The total area per land-use type per macroregion is the same for all realizations and also the locations of individual patches within the macroregion are the same, but the shape of these patches differs slightly, see, for example, the patch of sugar cane at the bottom of the map view in Fig. 3. The patches of land-use types that were assumed to be known precisely, being urban, water, and bare soil (see Methods S1) always have the same shape, see, for example, the shape of the city Natal, in the north-east of the map. We can conclude that the uncertainty in the initial land-use map is very local, important only at cell level. The effect for projected iLUC will mainly be that when sugar cane expands in a certain grid cell, uncertainty in the initial land-use map makes that in some realizations, it expands over agricultural land, which may result in iLUC through displacement (depending on the demand trend for the displaced agricultural land-use type), and in other realizations over nature, not resulting in iLUC, because there is no displacement effect.

Land-use change model, calibration. The input demand time series that were constructed from agricultural statistics for calibration are shown in Fig. 4 (2007–2012, indicated by an arrow). After calibration using this demand and the observations, of the 120 possible sequences (see Methods S2) for the order of allocation of the land-use types, 72 obtain a posterior probability of zero, that is they are not present anymore in the ensemble. So, 48 unique sequences remain, with posterior probabilities ranging between 0.002 and 0.19. The land-use sequence with the highest posterior probability is planted pasture–planted forest–sugar cane–rangeland–crops. An analysis of all other sequences and their posterior probabilities reveals that there is a dichotomy in this most common sequence. Planted pasture, planted forest, and sugar cane usually (in about 80% of the realizations) come in the first part of the sequence, and rangeland and crops in the last part, but the order among them fluctuates.

Sugar cane can only displace land-use types coming after it in the sequence. So, in 80% of the realizations, it predominantly replaces crops and rangeland. An important consequence of this calibration result with regard to iLUC is that the iLUC within a macroregion will originate mainly from the displacement of crops and

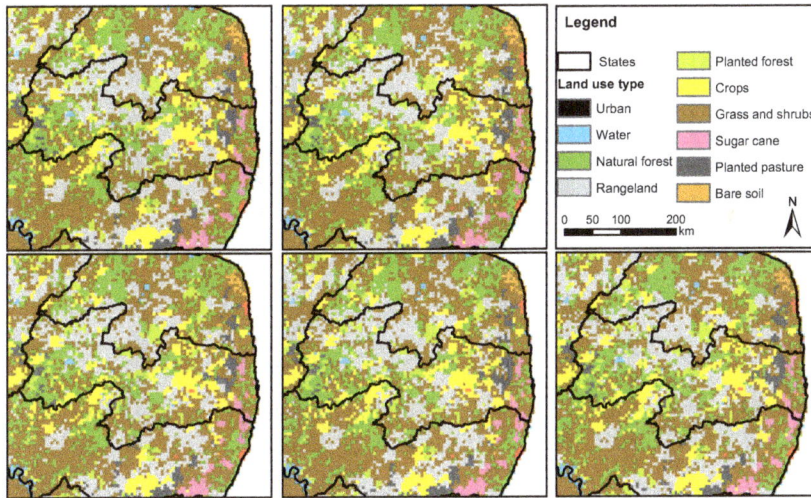

Fig. 3 Five randomly selected realizations out of the total of 5000 realizations of the initial land-use map (year 2006) zoomed in to the state Paraíba, in the north-east coast region of Brazil (see Fig. 2).

rangeland, which is in line with the findings of Lapola *et al.* (2010).

The weights of the suitability factors have been calibrated as well (Table 3). In general, suitability factor $k = 1$, representing n in the neighbourhood, obtains a high weight. This means that a land-use type is likely to expand in regions in which it is already cultivated. This factor has the highest median posterior weight for rangeland, sugar cane, and planted pasture. Accordingly, dLUC will take place close to existing sugar cane patches. For cropland, the double-cropping potential ($k = 5$) is the most important suitability factor for expansion. Galford *et al.* (2008) have found in a case study in Matto Grosso (an important expansion region, see Fig. 2) by means of remote sensing that newly established cropland is usually single cropped, but is converted to double cropping after 2–3 years. The high weight for the double-cropping potential factor indicates that this potential already plays a role at the establishment of the cropland, while the actual implementation of double cropping takes place a few years later. As a consequence, the location of iLUC in the case of displaced cropland is likely to be a location with a high double-cropping potential. In conclusion, the calibrated land-use change model mainly influences the location of dLUC and iLUC within the macroregion,

that is distribution between and also within states in a macroregion.

The results above were based on calibration over 2007–2012. To show the effect of calibration, we have applied a split-sample approach, with calibration only from 2007 to 2009, to allow a comparison with observational data in the validation period from 2010 to 2012. We compare the modelled against observed area of cropland (Fig. 5), because this land-use type gives the most information on model performance as it both expands and contracts in the calibration period. For most states without a clear break in their trend, for example Roirama (RR), Ceará (CE), and Mato Grosso (MT), the modelled median remains good in the validation period. However, the areas of states that do show a trend break, for example Maranhão (MA) and Rio de Janeiro (RJ), are poorly simulated, although at least for Maranhão, the observed cropland area falls within the 95% confidence interval of the modelled area.

To summarize the effect of calibration for all land-use types, we compare the root-mean-squared error (RMSE) in area summed over all states of the calibrated and noncalibrated model (Table 4). For crops, sugar cane, and planted pasture, a considerable RMSE reduction is achieved. The highest reduction is achieved for crops, with a maximum of 53% in 2010. For sugar cane, the

Fig. 4 Demand for the five dynamic land-use types in the six macroregions and Brazil as a whole for the initial year, for the calibration period [using data from IBGE (2013a,b) and ABRAF (2013)], and for two of the five output years in the projection period (output from the MAGNET model). The ranges of the y-axes differ between macroregions to improve the visibility of trends. In the projection period, the hatched bar is the reference scenario and the nonhatched bar is the biofuel scenario. The thick box on top of the bars indicates the uncertainty in the output, that is the difference between $d_{u,n,t}$ (elasticities set to 200%) and $d_{l,n,t}$ (elasticities set to 25%). In the case of a filled box, $d_{l,n,t}$ is higher than $d_{u,n,t}$, and in case of an unfilled box, $d_{l,n,t}$ is lower.

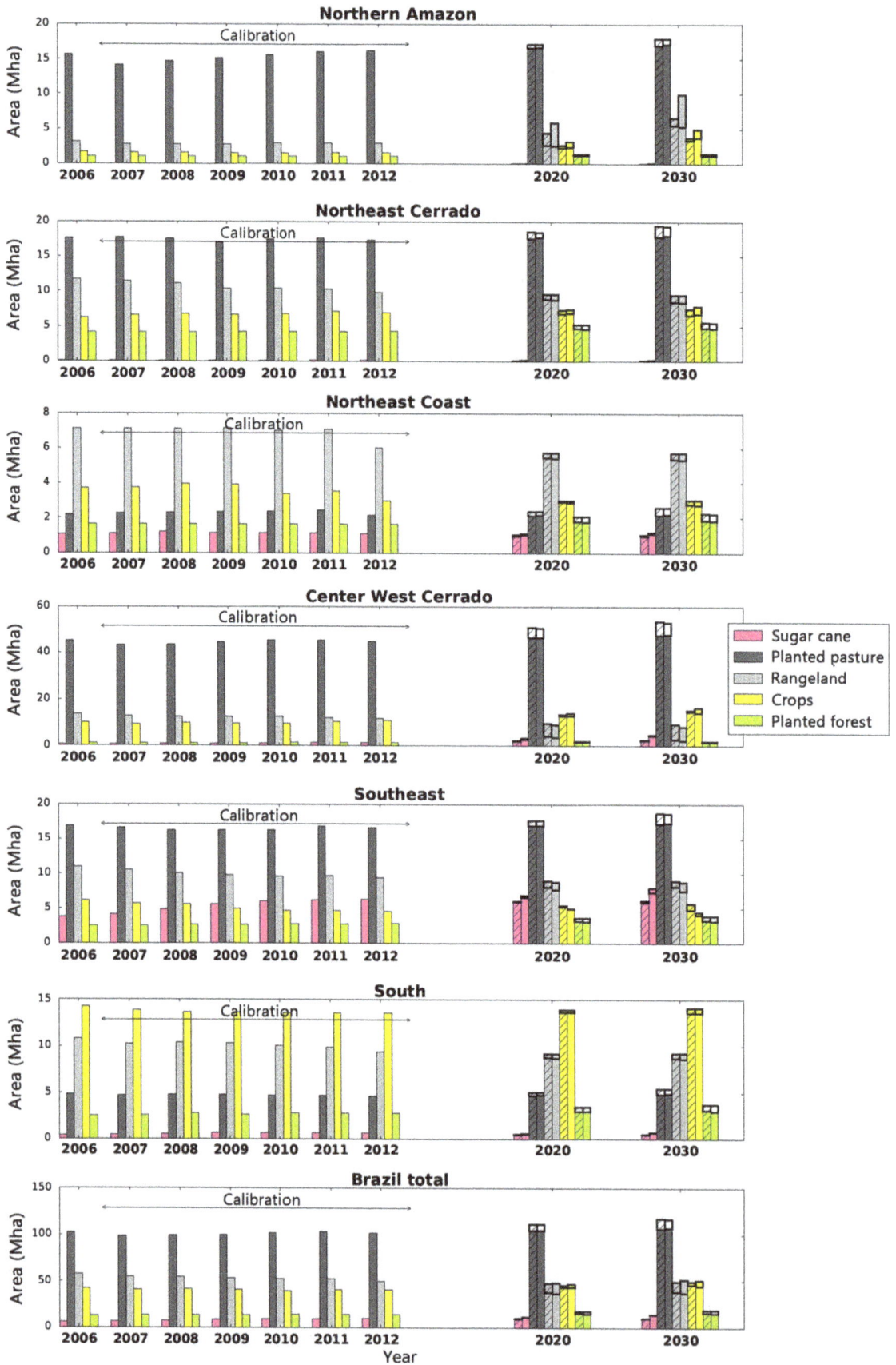

Table 3 The mean, first quartile, and third quartile of the weights of the suitability factors k of all active land-use types n resulting from the calibration

n	Name	k	Suitability factors	First quartile	Median	Third quartile
4	Rangeland	1	n in the neighbourhood	0.35	0.46	0.60
		2	Distance to roads	0.24	0.35	0.45
		3	Potential yield of n	0.03	0.19	0.28
5	Planted forest	1	n in the neighbourhood	0.19	0.29	0.36
		2	Distance to roads	0.27	0.34	0.37
		3	Potential yield of n	0.32	0.37	0.51
6	Crops	1	n in the neighbourhood	0.11	0.22	0.36
		2	Travel time to hubs for n	0.05	0.14	0.22
		3	Potential yield of n	0.05	0.11	0.20
		4	Conversion elasticity	0.12	0.23	0.33
		5	Growing season length	0.21	0.30	0.36
8	Sugar cane	1	n in the neighbourhood	0.23	0.29	0.36
		2	Travel time to hubs for n	0.21	0.28	0.33
		3	Potential yield of n	0.15	0.22	0.26
		4	Conversion elasticity	0.17	0.21	0.24
9	Planted pasture	1	n in the neighbourhood	0.40	0.53	0.66
		2	Distance to hubs for n	0.33	0.45	0.56
		3	Potential yield of n	0.00	0.02	0.03

average reduction is 24%. A significant reduction for sugar cane is important, as it is the land-use type of main interest. Being able to correctly project the location of sugar cane expansion connotes correct modelling of dLUC, which is the first step in also correctly projecting iLUC as the two are chained. For rangeland and planted forest, the calibration does not bring the modelled median area per state closer to the observed area. The modelled median even becomes worse, although not significantly, only a few percentage. The reason why PLUC cannot find weights for the suitability factors that result in a correct projection is probably the poor data availability for these two land-use types. For example, for the initial land-use map, no good prior knowledge maps were available (see Methods S1), and for the suitability factors, we have no information about the locations of the hubs for these land-use types.

Economic model, projection. Demands are projected by MAGNET per land-use type for 2013, 2015, 2020, 2025, and 2030. To illustrate the trend, the demands for 2020 and 2030 for the reference and the biofuel scenario and the uncertainty herein are shown in Fig. 4. An interesting result is that the uncertainty within a scenario is often higher than the difference between the scenarios. This indicates that it can be problematic to draw conclusions about the effect of, for example, a policy by means of comparing scenarios from the CGE model. If the land transition elasticities are uncorrelated between the two scenarios, the large uncertainty makes that the policy effects might be negative as well

as positive. Yet, we believe that although the elasticities are uncertain, they are correlated between the two scenarios, as these scenarios represent the same system, as long as the difference between scenarios is not too large. Others doubt this; a discussion that is known in economic modelling as the Lucas critique. Lucas (1976) argues in his work that the parameters in economic models are not policy-invariant and that they would therefore change when a policy is implemented. This discussion is interesting, but goes beyond the scope of this study. Nevertheless, we should be aware, that if Lucas is correct, the uncertainties in dLUC and iLUC shown in the next sections might be significantly higher.

In the reference scenario, sugar cane mainly expands in the Centre West Cerrado and the South-east (together called the Central South). The extra demand for sugar cane for ethanol from the mandates (biofuel scenario) also mainly ends up in these two regions. In the biofuel scenario, the total area of sugar cane in the Centre West Cerrado almost triples by 2030 compared to 2012.

The difference between the reference scenario and the biofuel scenario for the other land-use types within Brazil is the largest in the South-east (Fig. 4). In this macroregion, the areas of crops and rangeland are significantly smaller in the biofuel scenario than in the reference scenario. As the productivity of all land-use types are roughly the same in these two scenarios, this decrease in area means that MAGNET assumes that these areas of crops and rangeland are displaced by

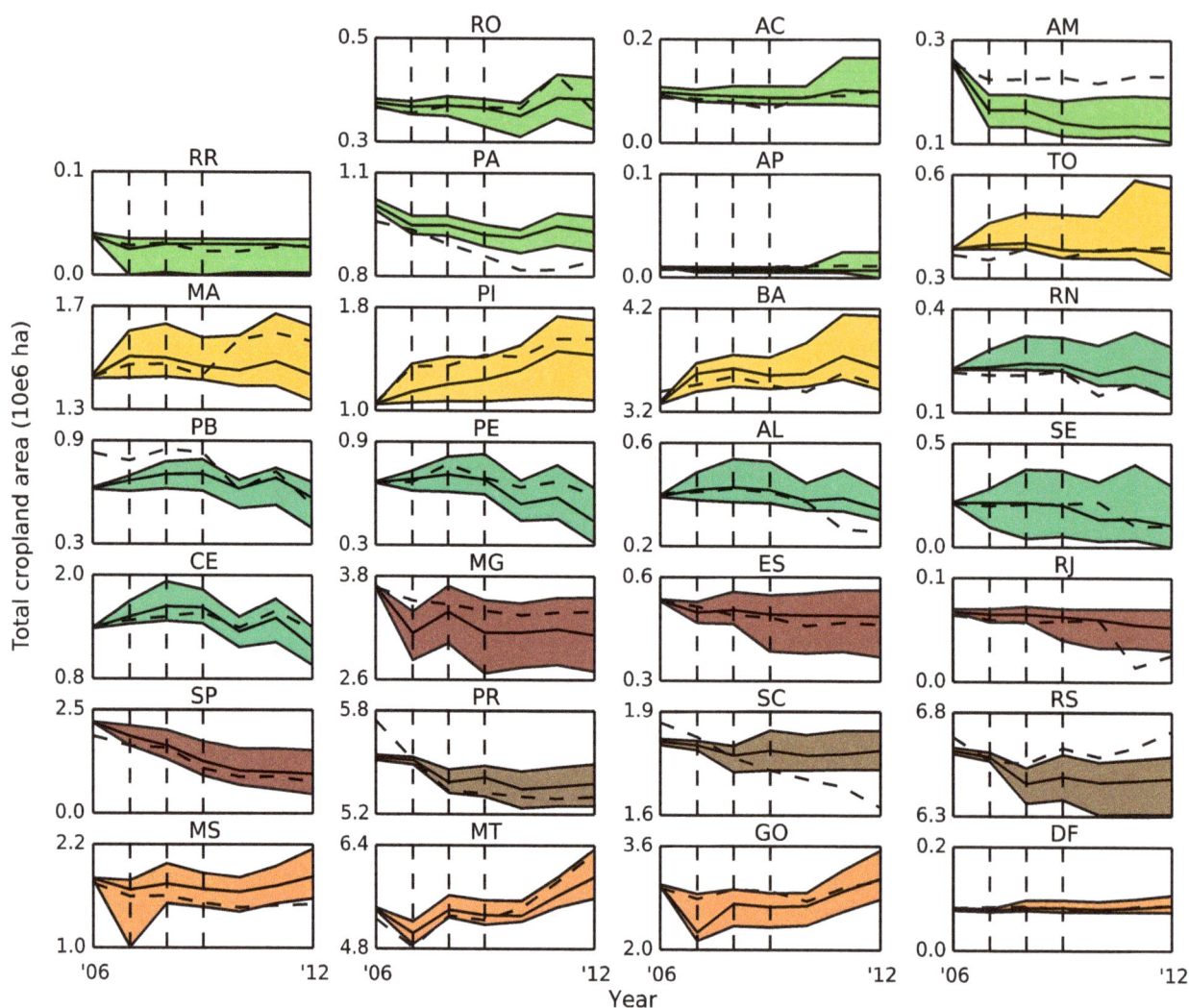

Fig. 5 Modelled and observed areas per state, for the land-use type cropland as an example. Vertical dashed lines: calibration years, dashed lines: observed area, solid lines: modelled median area, coloured planes: 95% confidence interval of the modelled area, where the colour corresponds to the colour of the macroregion in Fig. 2, to be able to quickly see which states belong to the same macroregion. The state name abbreviations are as follows: AC, Acre; AL, Alagoas; AM, Amazonas; AP, Amapá; BA, Bahia; CE, Ceará; DF, Distrito Federal; ES, Espírito Santo; GO, Goiás; MA, Maranhão; MG, Minas Gerais; MS, Mato Grosso do Sul; MT, Mato Grosso; PA, Pará; PB, Paraíba; PI, Piauí; PR, Paraná; RJ, Rio de Janeiro; RN, Rio Grande do Norte; RO, Rondônia; RR, Roirama; RS, Rio Grande do Sul; SC, Santa Catarina; SE, Sergipe; SP, São Paulo; TO, Tocatins.

sugar cane. The displaced land uses are shifted to the north-east Cerrado and the Northern Amazon: here, crops and rangeland occupy a larger area in the biofuel scenario than in the reference scenario (Fig. 4, difference between the hatched and plain bars).

Conceptual differences between the two models. Despite the 'shared' conversion matrix between MAGNET and PLUC and despite the fact that MAGNET provides the demand as an input for PLUC, the conversion dynamics between the two models differ because of conceptual differences between the models. A result of this is that the area of iLUC for the whole of Brazil calculated from

MAGNET differs from the area calculated by PLUC (further discussed later on), although ideally these two would be the same. This problem does not occur for dLUC, only for iLUC, and in the following we explain why.

The origin of the problem is that in PLUC sugar cane expands in the projection period, besides over cropland and rangeland, also often over planted pasture; a displacement also observed in other studies (e.g. Rudorff *et al.*, 2010; Adami *et al.*, 2012). In MAGNET, however, the area of conversion from planted pasture to sugar cane is negligible. This displacement of planted pasture in PLUC, not present in MAGNET, has two effects. One

Table 4 Reduction in root-mean-squared error (%) of the median area of land-use type n for the calibrated model compared to the reference case (Monte Carlo run without particle filter), summed over all states, given per year for the validation period (2010–2012)

n	Land-use type	Year		
		2010	2011	2012
4	Rangeland	−1	−2	−4
5	Planted forest	−6	0	2
6	Crops	53	50	42
8	Sugar cane	26	21	24
9	Planted pasture	33	35	25

is that in PLUC, the area of planted pasture is decreased in a region, such that, in that same year, planted pasture should expand (in addition to the expansion caused by a potential increase in demand already given by MAGNET) elsewhere in that region in order to make up for the lost acreage. This causes iLUC in the region, not anticipated by MAGNET. Another effect is that the areas of rangeland and/or crops in PLUC are larger than dictated by the demand in MAGNET for that year, so that these land uses will contract, resulting in abandoned land. This causes negative iLUC, which by definition never occurs in MAGNET. It can be debated which of the two models, if any, is correct. But the most important implication for our study is that uncertainty in iLUC projections does not only stem from uncertainties of parameters and model structure within one model component, but also from the dissimilarity in model concepts between the two models within the integrated model chain.

DLUC and iLUC projections for Brazil up to 2030 at different spatial scales and the uncertainty herein

The direct land-use change as a result of an increased ethanol production from 2013 to 2030 mainly takes place in the Central South region. The highest $(5 \times 5 \text{ km}^2)$ cell-based probabilities, up to 0.77, exist in the south of Mato Grosso do Sul and the west of São Paulo (Fig. 6, frame 1). The highest probabilities of indirect land-use change, with a maximum of 0.43, occur in the Amazonian states Rondônia, Amapá, and Roirama (Fig. 6, frame 2). Probabilities in these three small states are high because they are the only places in the northern Amazon where any agricultural land-use type can expand, as the rest of the northern Amazon has very few roads, almost no existing agriculture and thus few hubs, and many protected areas. Implementation of new roads in the Amazon could change the spatial distribution drastically, but this is not included into the

model due to limited spatial planning data availability. In the other macroregions, there are more options for expansion, and there is more variation in the suitability maps (best locations for expansion) between the different land-use types and between the individual Monte Carlo realizations, that is more uncertainty. In these other macroregions, iLUC locations with high probabilities are the frontier of the sugar cane expansion area (Goiás, Mato Grosso do Sul, and Mato Grosso) as well as the 'arc of deforestation', the transition area from cultivated land to mainly natural vegetation (Mato Grosso, Pará, and Rondônia).

As expected, there are only very few cells experiencing negative dLUC, and with negligibly low probabilities, with a maximum of 0.07 (Fig. 6, frame 3). Conversely, negative iLUC (land abandonment in the biofuel scenario and not in the reference scenario, Fig. 6, frame 4) does appear, with probabilities up to 0.48, mainly in Espírito Santo, Minas Gerais, and the Pantanal, which is the wetland area in the west of Mato Grosso do Sul and the south of Mato Grosso. These are areas where the suitability for most agriculture is low, resulting in land abandonment when the demand in the biofuel scenario is lower than in the reference scenario (see also the discussion in the previous section). With lower probabilities, up to 0.1, this effect also occurs in the rest of the Central South.

In the following, we add up dLUC and negative dLUC, and iLUC and negative iLUC, obtaining net dLUC and net iLUC. Scaling up to $250 \times 250 \text{ km}^2$ blocks (Fig. 7), we can calculate the coefficient of variation (cv), indicating relative uncertainty in dLUC and iLUC. Clearly, the uncertainty in iLUC is generally larger than in dLUC. The median cv over all selected blocks for dLUC is 0.91, while for iLUC, it is 1.61. This is caused by the fact that dLUC is affected by the dynamics of sugar cane only, while iLUC is an effect of the interplay of all land-use types, thereby being subjected to the uncertainties in all weights of all suitability factors (Table 3) and the order of allocation. The maximum cv value of dLUC is 4. The maxima occur at the expansion frontier of sugar cane, through Mato Grosso, Goiás, and Minas Gerais. The maximum cv value of iLUC is 6. This means that, when considering the 95% confidence interval, mean iLUC values might be as much as 13 times as high or as low. In a nutshell, the uncertainty in these blocks is so high that we can say practically nothing about expected iLUC there, except when mean iLUC is (very close to) zero (13 times zero is still zero). Coefficients of variation in the arc of deforestation generally range from 1 to 3, which is a bit better, but still very uncertain.

Comparing Figs 6 and 7, it becomes apparent that blocks of maxima in cv of iLUC area correspond to

(a)   Direct land use change (dLUC)

(b)   Indirect land use change (iLUC)

(c)   Negative direct land use change (neg_dLUC)

(d)   Negative indirect land use change (neg_iLUC)

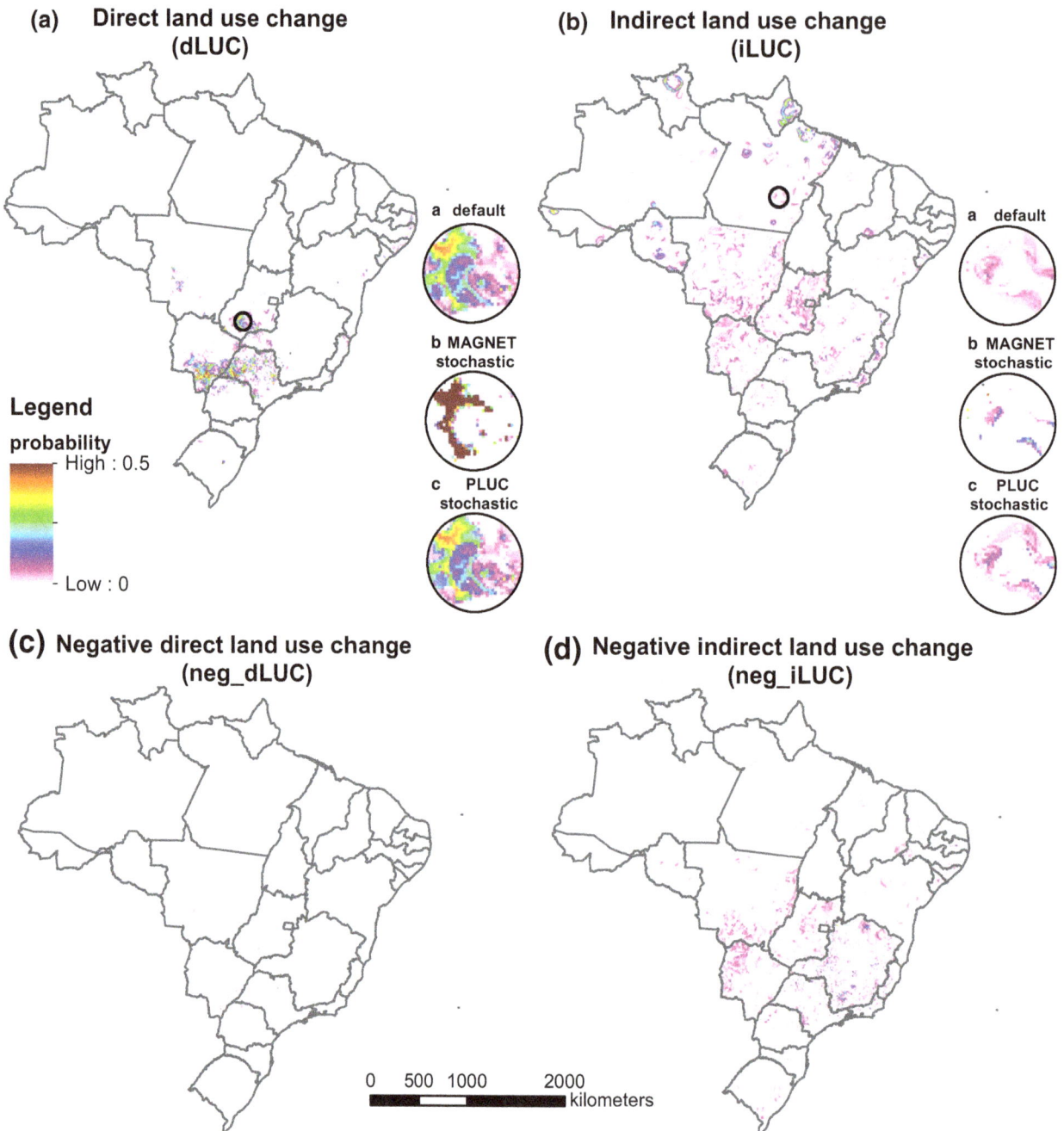

Fig. 6 Probability of (1) dLUC, (2) iLUC, (3) negative dLUC, and (4) negative iLUC per grid cell. Probabilities are shown at a common scale from 0 to 0.5. Only for dLUC, a few cells with higher probabilities exist, up to 0.77, but stretching the scale up to 0.77 reduces discernibility between different cells with low probability in all four maps. For dLUC and iLUC map, detail frames at a location in the expansion area are provided, showing probabilities for the three runs: (a) the default with both models stochastic, (b) with only PLUC stochastic, and (c) with only MAGNET stochastic.

regions where both iLUC and negative iLUC might appear. When some Monte Carlo realizations have negative iLUC values and others positive iLUC values, the standard deviation is large, and correspondingly the cv. In these blocks, the net iLUC effect might be positive as well as negative, so that the impacts on, for example, biodiversity, might be negative as well as positive, respectively.

When looking at the cv for dLUC and iLUC area for Brazil as a whole (Table 5, both models stochastic), the values are many times smaller. The total amount for the whole of Brazil can be determined about nine times as

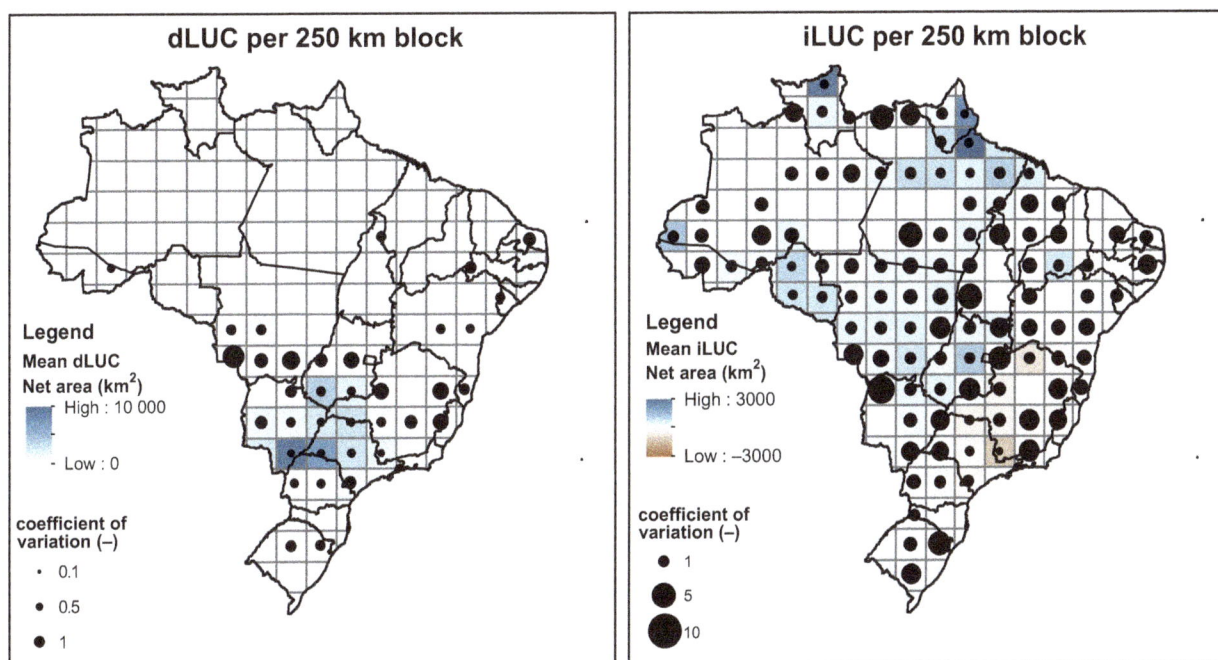

Fig. 7 Mean net area (km^2) (colour of the block) and the coefficient of variation (cv) (–) (size of the red circle) of dLUC (left) and iLUC (right), per 250 × 250 km^2 block. For the display of the cv, blocks smaller than 31 250 km^2 (half of a 250 × 250 km^2 block, occurring at the map edges) are filtered out, as the cv is heavily influenced by the support size of the block. Also blocks with mean dLUC or iLUC smaller than 25 km^2 (one cell) are filtered out, because when the mean goes to zero, the cv becomes infinite.

Table 5 Total area (Mha), standard deviations (SD) (Mha), and coefficients of variation (cv) (–) of dLUC and iLUC for Brazil for three different runs: (1) both the land-use change model and the economic model stochastic, (2) land-use change model stochastic and the economic model deterministic, (3) the land-use change model deterministic and the economic model stochastic

Run	Mean dLUC	SD dLUC	cv dLUC	Mean iLUC	SD iLUC	cv iLUC
Both models stochastic	4.20	0.10	0.02	3.13	2.25	0.72
PLUC stochastic	4.21	0.00	0.00	3.15	0.61	0.06
MAGNET stochastic	4.20	0.10	0.02	2.62	2.19	0.84

precise for dLUC and about two times as precise for iLUC compared to the median of the 250 × 250 km^2 blocks, because small-scale errors balance each other out when aggregating.

Contribution of the economic and land-use change model to the uncertainty in dLUC and iLUC at different spatial scales

The cv of the dLUC area at national level is for 100% caused by MAGNET (Table 5), which is logical, as MAGNET determines the total demand for sugar cane, and PLUC only allocates it within the macroregions. For iLUC area, this is not the case. The cv value of iLUC area for the run with only MAGNET stochastic is about sixteen times higher than cv value of iLUC area for the run with only PLUC stochastic, so about 93% of uncertainty in iLUC stems from MAGNET. Yet, the exact contribution of both models cannot be determined,

because errors from the two models partly compensate each other (the cv values of the two runs do not add up to the cv value of 0.72 found in the default run). The reason that uncertainty in iLUC at national level is not fully determined by MAGNET is that in PLUC iLUC can occur within a macroregion that is additional to the iLUC between macroregions from MAGNET.

At the grid cell level, many cells have some probability of experiencing dLUC or iLUC when both models are stochastic (Fig. 6, detail frame 1a). With only PLUC stochastic and MAGNET deterministic, there is in general not much difference with the results of the default run (Fig. 6, panel 1c), although somewhat fewer cells have a probability above zero on dLUC and iLUC. This indicates that only a small part of the uncertainty at cell level is caused by MAGNET. With only MAGNET stochastic and PLUC deterministic, compared to the default run less than half of the cells have a probability above zero on dLUC and iLUC, and the ones that have, have a relatively high probability, indicating much lower uncertainty (Fig. 6, panel 1b). The uncertainty now mainly exists at the edges of the expansion patches, caused by the variation in demand from MAGNET. For iLUC (Fig. 6, panels 2a–c), the same reasoning applies. In conclusion, uncertainty at grid cell level is mainly caused by uncertainty in PLUC, for both dLUC and iLUC.

Discussion

In this study, we have demonstrated a general methodology to calculate direct and indirect land-use change (dLUC and iLUC) stochastically with an integrated economic – land-use change model, taking into account the important uncertainties in all components of the modelling chain. The proficiencies of this methodology were shown for a case study of land-use change in Brazil up to 2030, steered by the current and planned ethanol mandates worldwide. Here, we shortly discuss the answers to our three research questions and give recommendations for further studies.

What are the dLUC and iLUC projections for Brazil up to 2030 at different spatial scales and what is the uncertainty herein?

Cell-based $(5 \times 5 \text{ km}^2)$ probabilities of dLUC range from 0 to 0.77, and of iLUC from 0 to 0.43. Thus, given our scenario assumptions, there is no single cell in Brazil for which it can be said with certainty that dLUC or iLUC will take place up to 2030. So, it is difficult to project exactly where dLUC and iLUC will occur, but it is certain that it will occur (there are no Monte Carlo realization without dLUC or iLUC effects). Yet, overall locations of iLUC are in line with the locations projected by Lapola *et al.* (2010). For dLUC, our study shows some locations with high probabilities in Mato Grosso do Sul and Goiás, where Lapola *et al.* (2010) do not project dLUC. As there are the 'new' expansion areas, this inconsistency is likely caused by the fact that their projection is for 2020, while we project up to 2030. Also, in our projections, there are many cells for which it can be concluded with certainty that no dLUC or iLUC will take place in 2030, which is surely relevant information. In $250 \times 250 \text{ km}^2$ blocks, the coefficient of variation (cv) ranges from 0 to 4 for dLUC and from 0 to 6 for iLUC. Large cv values for dLUC occur at the frontier of sugar cane expansion. High cv values for iLUC occur where both iLUC and the opposite effect (agriculture in the reference scenario is abandoned land in the biofuel scenario), introduced in this study, might take place.

The uncertainty in iLUC area and location is generally higher than in dLUC, because iLUC is caused by the interplay of various land-use types that each have their uncertain model parameters, while dLUC is mainly affected by the parameters for sugar cane. Uncertainty in dLUC and iLUC is lower at higher aggregation levels. For iLUC, the decrease in uncertainty by aggregation is smaller. At country level, the cv for iLUC is 36 times higher than for dLUC in our case study. At this level, dLUC can be projected with high certainty, having a cv of only 0.02, while iLUC is still uncertain, having a cv of 0.72. Thus, to answer the question posed in the title, what we can and cannot say about iLUC: we can merely say things about iLUC with high uncertainties. Estimated iLUC areas, even at country level, might as well be 2.4 times as high or as low, given the 95% confidence interval.

What are the sources of uncertainty for each step in the model chain and how do these uncertainties influence dLUC and iLUC projections?

Uncertain components in the land-use change model are (1) the initial land-use map, causing uncertainty at cell level, (2) the order of allocation of the land uses, causing uncertainty in especially iLUC, and (3) the selection and weights of the suitability factors for allocation of the land-use types, causing mainly uncertainty at intermediate aggregation levels, like states. The reduction in root-mean-squared error in the modelled median land-use areas per state by model calibration, compared to a noncalibrated model is on average 20%. Poor-performing land-use types are rangeland and planted forest, probably due to poor data availability for the drivers of location of land-use change.

For the economic model, we have assessed only the effect of one of the most critical parameters: the land transition elasticities that simulate the likelihood of particular land transitions. The uncertainty caused by varying these elasticities mainly plays a role at national level. At the cell level, it only causes uncertainty at the edge of the patch of expansion.

The final aspect generating uncertainty in the output is the difference in the conceptual model between the economic and the land-use change model concerning the reclaiming of abandoned land. This conceptual difference affects the total amount of iLUC and the opposite iLUC effect.

What is the contribution of the economic and land-use change model to the uncertainty in dLUC and iLUC at the different spatial scales?

At the cell level, uncertainty is primarily determined by the land-use change model. Going to higher aggregation levels, the influence of the uncertainty in the Computable General Equilibrium (CGE) model on output uncertainty increases. At national level, the cv of dLUC is caused by the CGE model for 100%. The contribution of the economic model to the cv of iLUC at this level is about 93%, although this cannot be determined precisely, because errors from the two models partly compensate each other.

Implications and recommendations

From the above, we can conclude that projected iLUC areas and locations are highly uncertain. Based on the case study, our opinion is that threshold evaluation for iLUC indicators should not be implemented in legislation. Thresholds (cf. Malins, 2013) have no use when the model, used to check whether an indicator for a specific case is above or below this threshold, gives an output confidence interval that straddles the threshold. This is likely to happen considering the high uncertainties found in our study. As most iLUC (or LUC) indicators in legislation are provided in terms of greenhouse gas (GHG) emissions generated, the impacts of the uncertainty in dLUC and iLUC projections on GHG emissions should be assessed to underpin our conclusion. Error propagation assessment for other impacts, such as biodiversity and water availability, is also desirable. Our opposition to thresholds for iLUC factors in legislation does not mean we favour negligence of biofuel-induced land-use change. We propose, in line with, for example, Finkbeiner (2014) and Mathews & Tan (2009), a change of focus from quantifying iLUC to taking proactive measures to mitigate iLUC, even though the effectiveness of these measures might be difficult to quantify.

Our quantification of the sources of uncertainty allows identification of the parts of the modelling chain having the highest priority for improvement. If one wants better estimates of dLUC and iLUC at cell level for a given case study, for example to be able to better quantify local GHG emissions caused by the biofuel targets, one should focus on improving the land-use change model. Spatially explicit input data could be improved, especially for land-use types that are problematic to derive from remote sensing: rangeland (problematic to distinguish from natural savannah) and planted forest (problematic to distinguish from natural forest). And data that are now only included at an aggregate level, such as land management and yield level, could be included spatially to account for spatial variation. Also, better information on data accuracy would be helpful. Due to the lack of accuracy information we had to make strong assumptions on the errors in the maps used to create the initial land-use map and the observational data used for calibration.

If one wants better estimates of dLUC and iLUC at country level, one should focus on improving the economic model. Our current estimates of uncertainty in the CGE model might be underestimated, because we have evaluated the uncertainty from the and transition elasticities only, while land-use changes might be sensitive to other parameters as well (Kavallari *et al.*, 2014). Yet, making other parameters stochastic could also reduce uncertainty, when they cancel out each other's errors. It would be good if making parameters stochastic and running Monte Carlo simulations would become common practice in economic modelling. Another option for better country level estimates might be the usage of a whole different type of model or tool, obviously also stochastic, although our current study gives cannot ascertain whether and to what extent that could reduce uncertainty.

One thing that could improve iLUC estimates at all spatial scales is a better match between the economic and land-use change model. The best solution would be to link the economic and the land-use model with a hard link that includes a feedback, as also suggested by Wicke *et al.* (2015). However, there is an inherent risk that this feedback loop is infinite, meaning that the land-use dynamics cannot be resolved, and there are many technical obstacles that complicate hard linking.

Yet, even if improved models, or improved model connections are used, in all cases, we strongly advise to provide quantitative uncertainty estimates together with the calculated dLUC, iLUC, or LUC indicators so that users of these indicators can evaluate the reliability of the indicators.

Acknowledgements

This work was carried out within the BE-Basic R&D Program, which was granted a FES subsidy from the Dutch Ministry of Economic Affairs, Agriculture, and Innovation (EL&I). We thank two anonymous reviewers for their valuable comments.

References

ABRAF (2013) *Yearbook Statistical ABRAF 2013, Base Year 2012*. ISSN: 1980–8550, Brazilian Association of Forest Plantation Producers (ABRAF), Brasilia.

Adami M, Rudorff BFT, Freitas R, Aguiar DA, Sugawara LM, Mello MP (2012) Remote sensing time series to evaluate direct land use change of recent expanded sugarcane crop in Brazil. *Sustainability*, **4**, 574–585.

Aguiar GAM, d'Athayde HP (2014) *Information about Cattle Production and Agriculture in Brazil*. SCOT Consultoria, Bebedouro, São Paulo, Brazil.

Alexandratos N, Bruinsma J (2012) World Agriculture Towards 2030/2050: The 2012 Revision. ESA Working paper No. 12-03. Food and Agriculture Organization of the United Nations, Agricultural Development Economics Division, Rome.

Andrade de Sá S, Palmer C, di Falco S (2013) Dynamics of indirect land-use change: empirical evidence from Brazil. *Journal of Environmental Economics and Management*, **65**, 377–393.

Arvor D, Dubreuil V, Ronchail J, Simões M, Funatsu BM (2014) Spatial patterns of rainfall regimes related to levels of double cropping agriculture systems in Mato Grosso (Brazil). *International Journal of Climatology*, **34**, 2622–2633.

Banse M, van Meijl H, Tabeau A, Woltjer G, Hellmann F, Verburg PH (2011) Impact of EU biofuel policies on world agricultural production and land use. *Biomass and Bioenergy*, **35**, 2385–2390.

Broch A, Hoekman SK, Unnasch S (2013) A review of variability in indirect land use change assessment and modeling in biofuel policy. *Environmental Science and Policy*, **29**, 147–157.

Conab (2014) Séries Históricas Relativas às Safras 1976/77 a 2013/14 de Área Plantada, Produtividade e Produção.

Creutzig F, Popp A, Plevin R, Luderer G, Minx J, Edenhofer O (2012) Reconciling top-down and bottom-up modelling on future bioenergy deployment. *Nature Climate Change*, **2**, 320–327.

Creutzig F, Ravindranath NH, Berndes G et al. (2014) Bioenergy and climate change mitigation: an assessment. *Global Change Biology Bioenergy* (Early view).

de Souza Ferreira Filho JB, Horridge M (2014) Ethanol expansion and indirect land use change in Brazil. *Land Use Policy*, **36**, 595–604.

Dendoncker N, Schmit C, Rounsevell M (2008) Exploring spatial data uncertainties in land-use change scenarios. *International Journal of Geographical Information Science*, **22**, 1013–1030.

Diogo V, van der Hilst F, van Eijck J et al. (2014) Combining empirical and theory-based land use modelling approaches to assess future availability of land and economic potential for sustainable biofuel production: Argentina as a case study. *Renewable & Sustainable Energy Reviews*, **34**, 208–224.

Finkbeiner M (2014) Indirect land use change – Help beyond the hype? *Biomass and Bioenergy*, **62**, 218–221.

Galford GL, Mustard JF, Melillo JM, Gendrin A, Cerri CC, Cerri CEP (2008) Wavelet analysis of MODIS time series to detect expansion and intensification of row-crop agriculture in Brazil. *Remote Sensing of Environment*, **112**, 576–587.

Gurgel HC, Hargrave J, França F et al. (2009) Unidades de conservação e o falso dilema entre conservação e desenvolvimento. *Boletim regional, urbano e ambiental*, **3**, 109–119.

Hertel TW (2011) The global supply and demand for agricultural land in 2050: a perfect storm in the making? *American Journal of Agricultural Economics*, **93**, 259–275.

Hoefnagels R, Banse M, Dornburg V, Faaij A (2013) Macro-economic impact of large-scale deployment of biomass resources for energy and materials on a national level-A combined approach for the Netherlands. *Energy Policy*, **59**, 727–744.

IBGE (2013a) Pesquisa Pecuária Municipal 2006–2012. Available at: http://www.ibge.gov.br/home/estatistica/economia/ppm/2013/ (accessed 7 May 2015).

IBGE (2013b) Produção Agrícola Municipal 2006–2012. Available at: http://www.ibge.gov.br/home/estatistica/economia/pam/2013/ (accessed 7 May 2015).

IBGE (2006) Censo Agropecuário 2006. Available at: http://www.ibge.gov.br/home/estatistica/economia/agropecuaria/censoagro/2006/ (accessed 7 May 2015).

IEA (2013) *World Energy Outlook 2013*. International Energy Agency, Paris, France.

IPCC (2011) *IPCC Special Report on Renewable Energy Sources and Climate Change Mitigation*. Cambridge University Press, Cambridge, UK and New York, NY, USA.

Kavallari A, Smeets E, Tabeau A (2014) Land use changes from EU biofuel use: a sensitivity analysis. *Operational Research*, **14**, 261–281.

Lapola DM, Schaldach R, Alcamo J, Bondeau A, Koch J, Koelking C, Priess JA (2010) Indirect land-use changes can overcome carbon savings from biofuels in Brazil. *Proceedings of the National Academy of Sciences of the United States of America*, **107**, 3388–3393.

Lucas RE Jr (1976) Econometric policy evaluation: a critique. American Elsevier (Ed.), Carnegie-Rochester Confer. Series on Public Policy New York 1, 19–46.

Malins C (2013) A model-based quantitative assessment of the carbon benefits of introducing iLUC factors in the European Renewable Energy Directive. *Global Change Biology Bioenergy*, **5**, 639–651.

Mathews JA, Tan H (2009) Biofuels and indirect land use change effects: the debate continues. *Biofuels, Bioproducts and Biorefining*, **3**, 305–317.

Meyfroidt P, Lambin EF, Erb K, Hertel TW (2013) Globalization of land use: distant drivers of land change and geographic displacement of; land use. *Current Opinion in Environmental Sustainability*, **5**, 438–444.

Narayanan G, Badri AA, McDougall R (2012) Global Trade, Assistance, and Production: The GTAP 8 Data Base.

Nassar AM, Antoniazzi LB, Moreira MR, Chiodi L, Harfuch L (2010) An Allocation Methodology to Assess GHG Emissions Associated with Land Use Change. Institute for International Trade Negotiations. Available at: http://www.iconebrasil.com.br/publication/study/details/649 (accessed 7 May 2015).

Nassar AM, Rudorff BFT, Antoniazzi LB, Aguiar DA, Bacchi MRP, Adami M (2008) Chapter 3: Prospects of the sugarcane expansion in Brazil: impacts on direct and

indirect land use changes. In: *Sugarcane Ethanol: Contributions to Climate Change Mitigation and the Environment* (eds Zuurbier P, van de Vooren J), pp. 63–112. Wageningen Academic Publishers, Wageningen.

Nelson GC, van der Mensbrugghe D, Ahammad H et al. (2014) Agriculture and climate change in global scenarios: why don't the models agree? *Agricultural Economics (United Kingdom)*, **45**, 85–101.

OECD/Food and Agriculture Organization of the United Nations (2014) *OECD-FAO Agricultural Outlook 2014*. OECD Publishing, Paris, France.

O'Hare M, Delucchi M, Edwards R et al. (2011) Comment on "Indirect land use change for biofuels: testing predictions and improving analytical methodologies" by Kim and Dale: Statistical reliability and the definition of the indirect land use change (iLUC) issue. *Biomass and Bioenergy*, **35**, 4485–4487.

O'Neill BC, Kriegler E, Riahi K et al. (2014) A new scenario framework for climate change research: the concept of shared socioeconomic pathways. *Climatic Change*, **122**, 387–400.

Overmars KP, Stehfest E, Ros JPM, Prins AG (2011) Indirect land use change emissions related to EU biofuel consumption: an analysis based on historical data. *Environmental Science and Policy*, **14**, 248–257.

Padua Junior AL, Costa Pasini AC, Comatsu CE et al. (2012) Agro-enviromental Zoning - Green Ethanol - Environmental System for São Paulo - Government of São Paulo.

Plevin RJ, Beckman J, Golub AA, Witcover J, O'Hare M (2015) Carbon accounting and economic model uncertainty of emissions from biofuels-induced land use change. *Environmental Science and Technology*, **49**, 2656–2664.

Pontius RG Jr, Neeti N (2010) Uncertainty in the difference between maps of future land change scenarios. *Sustainability Science*, **5**, 39–50.

Pontius RG Jr, Spencer J (2005) Uncertainty in extrapolations of predictive land-change models. *Environment and Planning B: Planning and Design*, **32**, 211–230.

Ray DK, Pijanowski BC, Kendall AD, Hyndman DW (2012) Coupling land use and groundwater models to map land use legacies: assessment of model uncertainties relevant to land use planning. *Applied Geography*, **34**, 356–370.

Refsgaard JC, van der Sluijs JP, Brown J, van der Keur P (2006) A framework for dealing with uncertainty due to model structure error. *Advances in Water Resources*, **29**, 1586–1597.

Rose A (1995) Input-output economics and computable general equilibrium models. *Structural Change and Economic Dynamics*, **6**, 295–304.

Rudorff BFT, Adami M, Aguiar DA et al. (2011) The soy moratorium in the Amazon biome monitored by remote sensing images. *Remote Sensing*, **3**, 185–202.

Rudorff BFT, Aguiar DA, Silva WF, Sugawara LM, Adami M, Moreira MA (2010) Studies on the rapid expansion of sugarcane for ethanol production in São Paulo State (Brazil) using landsat data. *Remote Sensing*, **2**, 1057–1076.

Sorda G, Banse M, Kemfert C (2010) An overview of biofuel policies across the world. *Energy Policy*, **38**, 6977–6988.

Sparovek G, Berndes G, Barretto A, Klug ILF (2012) The revision of the Brazilian Forest Act: increased deforestation or a historic step towards balancing agricultural development and nature conservation? *Environmental Science & Policy*, **16**, 65–72.

Tóth G, Kozlowski B, Prieler S, Wiberg D (2012) *Global Agro-ecological Zones (GAEZ v3.0)*. IIASA, Laxenburg, Austria and FAO, Rome, Italy.

van der Hilst F, Verstegen JA, Karssenberg D, Faaij APC (2012) Spatio-temporal land use modelling to assess land availability for energy crops - illustrated for Mozambique. *Global Change Biology Bioenergy*, **4**, 859–874.

van der Hilst F, Verstegen JA, Zheliezna T, Drozdova O, Faaij AP (2014) Integrated spatiotemporal modelling of bioenergy production potentials, agricultural land use, and related GHG balances; demonstrated for Ukraine. *Biofuels, Bioproducts and Biorefining*, **8**, 391–411.

van Leeuwen PJ (2009) Particle filtering in geophysical systems. *Monthly Weather Review*, **137**, 4089–4114.

Verstegen JA, Karssenberg D, van der Hilst F, Faaij APC (2015) Detecting systemic change in a land use system by Bayesian data assimilation. *Environmental Modelling & Software*, Early view.

Verstegen JA, Karssenberg D, van der Hilst F, Faaij APC (2014) Identifying a land use change cellular automaton by Bayesian data assimilation. *Environmental Modelling & Software*, **53**, 121–136.

Verstegen JA, Karssenberg D, van der Hilst F, Faaij APC (2012) Spatio-temporal uncertainty in spatial decision support systems: a case study of changing land availability for bioenergy crops in Mozambique. *Computers, Environment and Urban Systems*, **36**, 30–42.

Walter A, Galdos MV, Scarpare FV et al. (2014) Brazilian sugarcane ethanol: developments so far and challenges for the future. *Wiley Interdisciplinary Reviews: Energy and Environment*, **3**, 70–92.

Warner E, Zhang Y, Inman D, Heath G (2014) Challenges in the estimation of green-house gas emissions from biofuel-induced global land-use change. *Biofuels, Bio-products and Biorefining*, **8**, 114–125.

Wicke B, van der Hilst F, Daioglou V *et al.* (2015) Model collaboration for the improved assessment of biomass supply, demand, and impacts. *Global Change Biology Bioenergy*, **7**, 422–437.

Wicke B, Verweij P, Van Meijl H, Van Vuuren DP, Faaij APC (2012) Indirect land use change: review of existing models and strategies for mitigation. *Biofuels*, **3**, 87–100.

Woltjer GB (2013) Forestry in MAGNET: a new approach for land use and forestry modelling. WOt-werkdocument 320, Statutory Research Tasks Unit for Nature & the Environment (WOT Natuur & Milieu), Wageningen, The Netherlands.

Woltjer GB, Kuiper MH (2014) The MAGNET Model: Module description. LEI Report 14-057, LEI Wageningen UR (University & Research centre), Wageningen, The Netherlands. Available at: http://edepot.wur.nl/310764 (accessed 19 January 2015).

You L, Wood S (2005) Assessing the spatial distribution of crop areas using a cross-entropy method. *International Journal of Applied Earth Observation and Geoinforma-tion*, **7**, 310–323.

15

Elevated atmospheric [CO_2] stimulates sugar accumulation and cellulose degradation rates of rice straw

CHUNWU ZHU[1], XI XU[1†], DAN WANG[2], JIANGUO ZHU[1], GANG LIU[1] and SAMAN SENEWEERA[3]

[1]State Key Laboratory of Soil and Sustainable Agriculture, Institute of Soil Science, Chinese Academy of Sciences, NO.71 East Beijing Road, Nanjing 210008, China, [2]International Center for Ecology, Meteorology and Environment, School of Applied Meteorology, Nanjing University of Information Science and Technology, Nanjing 210044, China, [3]Centre for Systems Biology, University of Southern Queensland, Toowoomba, Qld 4350, Australia

Abstract

Rice straw can serve as potential material for bioenergy production. However, the quantitative effects of increasing atmospheric carbon dioxide concentration [CO_2] on rice straw quality and the resulting consequences for bioenergy utilization are largely unknown. In this study, two rice varieties, WYJ and LY, that have been shown previously to have a weak and strong stimulatory response to rising [CO_2], respectively, were grown with and without additional CO_2 at China free-air carbon dioxide enrichment (FACE) platform. Qualitative and quantitative measurements in response to [CO_2] included straw biomass (including leaf, sheath, and stem), the concentration of nonstructural and structural carbohydrates, the syringyl-to-guaiacyl (S/G) ratio of lignin, glucose and xylose release from structural carbohydrate, total sugar release by enzymatic saccharification, and sugar yield and the ratio of cellulose and hemicellulose degradation. Elevated [CO_2] significantly increased straw biomass and nonstructural carbohydrate contents while enhancing the degraded ratio of structural carbohydrates as indicated by the decreased lignin content and increased S/G ratio. Overall, total sugar yield (g m^{-2}) in rice straw significantly increased by 27.1 and 57% for WYJ and LY at elevated [CO_2], respectively. These findings, while preliminary, suggest that rice straw quality and potential biofuel utilization may improve as a function of rising [CO_2].

Keywords: biofuel, elevated [CO_2], rice, saccharification, straw, sugar release

Introduction

Climate change and energy security have driven renewable energy production to the top of global agendas (Karp & Shield, 2008). Potentially, plant-based sources of bioenergy (e.g., ethanol) could lower CO_2 emissions and help mitigate climate change impacts (Cuevas et al., 2010; Erdei et al., 2010).

Rice is the dominant source of calories for a large portion of the human population (Shimono & Bunce, 2009). It is cultivated globally in about 160 million hectares; in addition to a grain production about 740 million tons, it also produces about ~730 million tons of straw as a by-product (Wang et al., 2011; FAO, 2014). As demand for rice is expected to increase in many countries, the availability of rice straw will also increase (Yoswathana et al., 2010; Lim et al., 2012).

At present, excess rice straw is often subject to open-field burning following harvest (Oanh et al., 2011). Open-field burning wastes energy while resulting in environmental and public health concerns. Alternatively, rice straw represents a potential resource that could be used for biofuel and energy production (Domínguez-Escribá & Porcar, 2010; Lim et al., 2012). For example, it has been estimated that rice straw has the potential to produce 205 billion liters of bioethanol per year, equivalent to 5% of current fossil fuel energy production (Yoswathana et al., 2010).

Rising [CO_2] in addition to its role as a greenhouse gas is the sole source of carbon for photosynthesis, and its increase has been shown to effect rice growth and grain yield (Zhu et al., 2014). However, the impact of rising [CO_2] on quantitative and/or qualitative changes and the subsequent consequences for utilization of rice straw as a source of biofuel have, heretofore, not been investigated.

Higher levels of [CO_2] could alter rice straw quantity by stimulating growth, with the degree of stimulation being cultivar specific (Ziska et al., 1996). Rising CO_2

Correspondence: Chunwu Zhu

e-mail: cwzhu@issas.ac.cn

†The equal contribution to the first author.

could also alter rice straw quality. For example, elevated [CO$_2$] can increase the ratio of carbon to nitrogen (C : N), alter carbon partitioning between nonstructural and structural carbohydrates (Liu *et al.*, 2009; Zhu *et al.*, 2012), and alter the amount of sugar release from structural carbohydrates (Studer *et al.*, 2011).

A fundamental understanding of how rising [CO$_2$] affects those biological parameters that, in turn, alter rice straw production and quality will be of obvious interest in determining the future utility of rice straw as a potential biofuel. To identify and quantify these parameters, we grew two rice cultivars that differ in [CO$_2$] sensitivity at ambient and elevated [CO$_2$] under free-air carbon dioxide enrichment (FACE) conditions. The objectives of this study were to determine (1) whether, and to what extent, rising [CO$_2$] affects rice carbohydrate accumulation and sugar release efficiency for biofuel production and (2) whether any observed [CO$_2$] effect on biomass and biofuel potential differed between rice cultivars.

Materials and methods

Site description

The study was conducted at the FACE facility in Zongcun village (32°35′5″N, 119°42′0″E), Jiangdu city, Jiangsu Province. This facility is situated in the Yangtze River Delta region, where rice is typically grown in a rice–wheat rotation. The region is typical of a north subtropical monsoon climate. Soil is classified as Shajiang Aquic Cambiosol with a sandy loam texture. Soil properties at a depth of 0–15 cm are as follows: bulk density 1.16 g cm^{-3}, soil organic carbon 18.4 g kg^{-1}, total nitrogen 1.45 g kg^{-1}, available phosphorous 10.1 mg kg^{-1}, available potassium 70.5 mg kg^{-1}, and pH 6.8.

FACE system

Details about the FACE facility have been described previously (Okada *et al.*, 2001; Zhu *et al.*, 2008). In brief, three rectangular paddy fields were used due to their uniformity in growth and yield. Within each field, a FACE plot was paired with an ambient control, and plot centers were 90 m apart to avoid movement of additional [CO$_2$] to the ambient plots. Each FACE plot was encircled with an octagonal ring (14 m in diameter) with emission tubes that injected pure CO$_2$ at 30 cm above the plant canopy. Emission tubes were raised as the canopy grew to maintain the [CO$_2$] set point at the top of the plant canopy. Ambient control plots did not receive any supplemental CO$_2$. The CO$_2$ set point in FACE plots was 200 μmol mol^{-1} above that of ambient control plots. Carbon dioxide release was controlled by a computer program with an algorithm based on wind speed and direction to keep the target CO$_2$ concentration within the FACE plot. During the 2012 and 2013 seasons, average daytime [CO$_2$] at canopy height during the experiment was 378 and 374 for the ambient rings, and 571 and

584 μmol moL^{-1} for elevated FACE rings, respectively. The average temperature during the growing stage was ranging from 24.4 °C to 24.8 °C, respectively.

Rice cultivation and sample pretreatment

Two rice (*Oryza sativa* L.) varieties, Wuyunjing21 (WYJ, Japanese inbred) and Liangyou084 (LY, Indica hybrid), were selected. Seeds of each line were sown on May 20, and seedlings were transplanted on June 21 in 2012 and 2013. The spacing of the hills was 16.7 cm × 25 cm (equivalent to 24 hills m^{-2}). The heading dates of LY and WYJ, respectively, were Aug 21 and Aug 25 in 2012, and Aug 20 and Aug 24 in 2013. Both lines were harvested on October 10 and October 17 in 2012 and 2013, respectively. Yield was measured from a 2-m^2 patch (excluding plants in the borders) for each subplot (Yang *et al.*, 2009; Zhu *et al.*, 2014).

Phosphorus and potassium (9 g m^{-2}) were applied as basal fertilizers before transplanting. Total nitrogen fertilizer was 22.5 g m^{-2}, with 40%, 30%, and 30% of the total amount applied before transplanting, tillering, and heading, respectively. Paddy fields were submerged with water from 13 June to 10 July, drained several times from 11 July to 4 August, and then flooded with intermittent irrigation from 5 August to 10 days before harvest. Herbicide and pesticide were applied as follows for the 2012 and 2013 seasons: prevention of rice stem borer, rice blast and stripe disease using chlorpyrifos, tricyclazole and imidacloprid; prevention of rice sheath blight, rice blast, Cnaphalocrocis medinalis and Chilo suppressalis using Fiponil, chlorpyrifos and sheath blight bane on; prevention of Cnaphalocrocis medinalis, panicle neck disease, rice plant hopper and leaf blight using Armure and Fiponil, validamycin and buprofezin; and prevention of ear disease and rice planthopper using tricyclazole, fenobucarb and chlorpyrifos. As climates between years were similar, and management practices remained the same in 2012 and 2013, samples collected from 2012 to 2013 were combined for analysis.

Nonstructural and structural carbohydrates

Sucrose, free glucose, and fructose were determined using a carbohydrate kit (Sigma-Aldrich, USA) and starch was measured using the starch (HK) assay kit. β-1, 3-1, 4-glucan was measured with glucan (mixed linkage) assay kit (Megazyme international, Ireland). Cellulose, hemicellulose, and lignin were measured as previously described (He *et al.*, 2008). After enzymatic hydrolysis, the released glucose and xylose were measured using the glucose assay kit and monosaccharides kit (Sigma-Aldrich).

S/G ratio

Syringyl-to-guaiacyl (S/G) ratio of lignin was analyzed as described before (Studer *et al.*, 2011). Briefly, ~4 mg of ground straw material was pyrolyzed for 2 min at 500 °C (CDS Pyroprobe 5200, Australia). Pyrolysis vapors were entrained in helium flowing at 2 L min^{-1} to a mass spectrometer (Agilent 5975C, USA). Spectra were read over a mass-to-charge

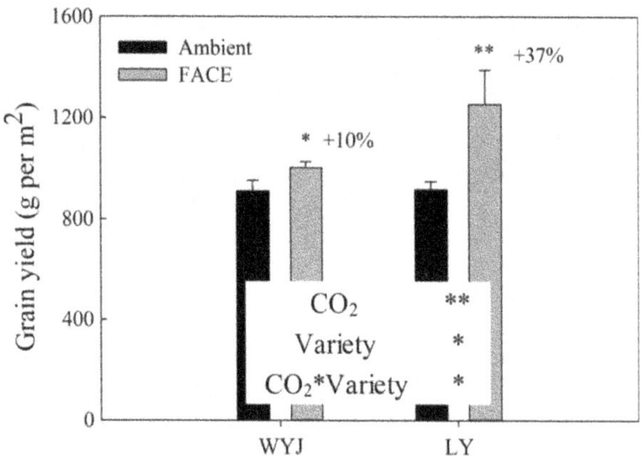

Fig. 1 The average grain yield of 2012 and 2013 seasons for rice cultivars WYJ and LY under ambient and FACE conditions. The mean was the average of 3 replications ($n = 3$) $\pm$ SD. $**P < 0.01$, $*P \leq 0.05$, $^{\dagger}P \leq 0.1$, $^{ns}P > 0.1$.

ratio (m/z) range from 30 to 450 using 22.5-eV electron impact ionization. S/G ratio of lignin was determined by summing up the intensity of the peaks at 154, 167, 168, 182, 194, 208, and 210 and dividing the sum of intensity of guaiacyl peaks at 124, 137, 138, 150, 164, and 178.

Pretreatment and enzymatic hydrolysis

For quantification of nonstructural carbohydrates, samples of rice straw (300 mg) were placed in plastic tubes and 3 mL distilled water added. The tubes were heated to 100 °C for 10 min while agitating every 2 min with a vortex mixer. Sodium acetate buffer (3 mL, pH 4.8) with amyloglucosidase (1 mg, 60 units mg^{-1}, Sigma, USA) and β-glucosidase (0.5 mg, 30 units mg^{-1}, Solarbio, China) was then added and the tube incubated on a shaker at 50 °C for 50 rpm. After 4 h of digestion, the tubes were centrifuged at 12 000 g for 10 min and the liberated glucose was measured. The total recovery of glucose after enzymatic hydrolysis was estimated as the amount of liberated glucose plus the amounts of free glucose, fructose, and sucrose.

After the remaining supernatant was removed, 4 mL pure water was added twice to wash the remaining soluble sugars. Then, sodium acetate buffer (5 mL, pH 4.8) with cellulose (60 mg, 0.93 U mg^{-1}, Sigma) and hemicellulose (40 mg, 2.50 U mg^{-1}, Sigma) was added. Tubes were agitated for 5 min with a vortex mixer, then incubated in a shaker at 50 °C for 48 h at 50 rpm, and then centrifuged at 12 000 g for 10 min (Park et al., 2010). Supernatant was used to determine glucose and xylose concentration in the sample. Each sample was measured twice.

During the saccharifying progress, cellulose ($C_6H_{10}O_5$, molecular weight: 162) was hydrolyzed to glucose ($C_6H_{12}O_6$, molecular weight: 180). The hemicellulose ($C_5H_8O_4$, molecular weight: 132) was hydrolyzed into the xylose ($C_5H_{10}O_5$, molecular weight: 150). The degradation ratio of structural carbohydrates was calculated according to equations (1) and (2) as described (Poornejad et al., 2013).

Table 1 Component (each component per dry weight) of rice straw at mature stage of WYJ and LY under ambient (AMB) and FACE conditions. Mean was the average of replications ($n = 3$)

| CO$_2$ | Variety | Nonstructural carbohydrates | | | | | Sum‡ (%) | Cellulose (%) | Hemicellulose (%) | Lignin (%) |
		Starch (%)	Sucrose (%)	Free glucose (%)	Free fructose (%)	β-1, 3-1, 4-glucan (%)				
AMB	WYJ	4.03 ± 0.40	4.39 ± 0.16	0.29 ± 0.04	0.30 ± 0.03	0.26 ± 0.01	9.27 ± 0.36	27.97 ± 0.95	23.71 ± 1.61	11.41 ± 0.52
FACE	WYJ	4.89 ± 0.24	5.43 ± 0.49	0.37 ± 0.04	0.38 ± 0.05	0.33 ± 0.04	11.40 ± 0.63	23.25 ± 1.46	23.19 ± 1.45	10.53 ± 0.47
AMB	LY	1.73 ± 0.40	6.39 ± 0.29	0.36 ± 0.03	0.36 ± 0.03	0.24 ± 0.04	8.84 ± 0.33	27.02 ± 1.25	20.59 ± 0.67	12.91 ± 0.61
FACE	LY	2.25 ± 0.40	7.51 ± 0.30	0.43 ± 0.03	0.44 ± 0.03	0.31 ± 0.02	10.92 ± 0.35	24.65 ± 1.25	20.62 ± 1.74	10.63 ± 0.44
CO$_2$		**	**	**	**	**	**	**	ns	**
Variety		**	**	*	*	ns	ns	ns	**	*
CO$_2$*Variety		ns	ns	ns	ns	ns	ns	ns	ns	†

$**P < 0.01$; $*P \leq 0.05$; $^{\dagger}P \leq 0.1$, $^{ns}P > 0.1$.
‡The amount of soft carbohydrates was calculated as the sum of starch, sucrose, free glucose, free fructose, and β-1,3-1,4-glucan.

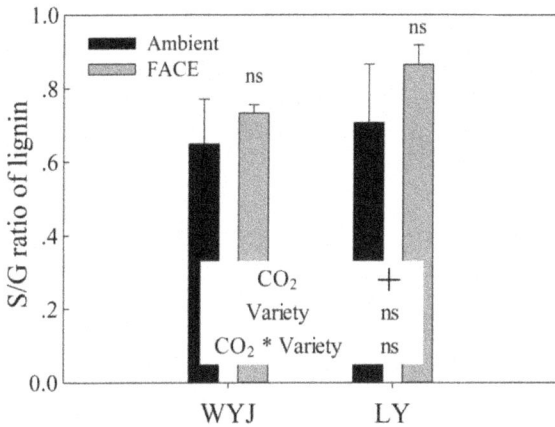

Fig. 2 Syringyl-to-guaiacyl (S/G) ratio of lignin within rice straw for rice cultivars WYJ and LY at maturity under ambient and FACE conditions for the 2012 and 2013 seasons. The mean was the average of 3 replications $(n = 3) \pm$ SD. $^{**}P < 0.01$, $^{*}P \le 0.05$, $^{\dagger}P \le 0.1$, $^{ns}P > 0.1$.

Ratio of cellulose degradation (%)
$$= \text{Glucose produced} / (\text{Cellulose in sample} \times 1.111) \times 100\%$$
(1)

Ratio of hemicellulose degradation (%)
$$= \text{Xylose produced} / (\text{Hemicellulose in sample} \times 1.136) \times 100\%$$
(2)

Total sugar release was determined as follows:

Total sugar release (g per g straw)
$$= \text{Sugars release from nonstructural cabohydrate}$$ (3)
$$+ \text{cellulose} + \text{hemicellulose(g per g straw)}$$

Sugar yield of straw was determined as follows:

Total Sugar yield (g per m2) = Rice straw biomass (g per m2)
$$+ \text{Total sugar release (g per g straw)}$$
(4)

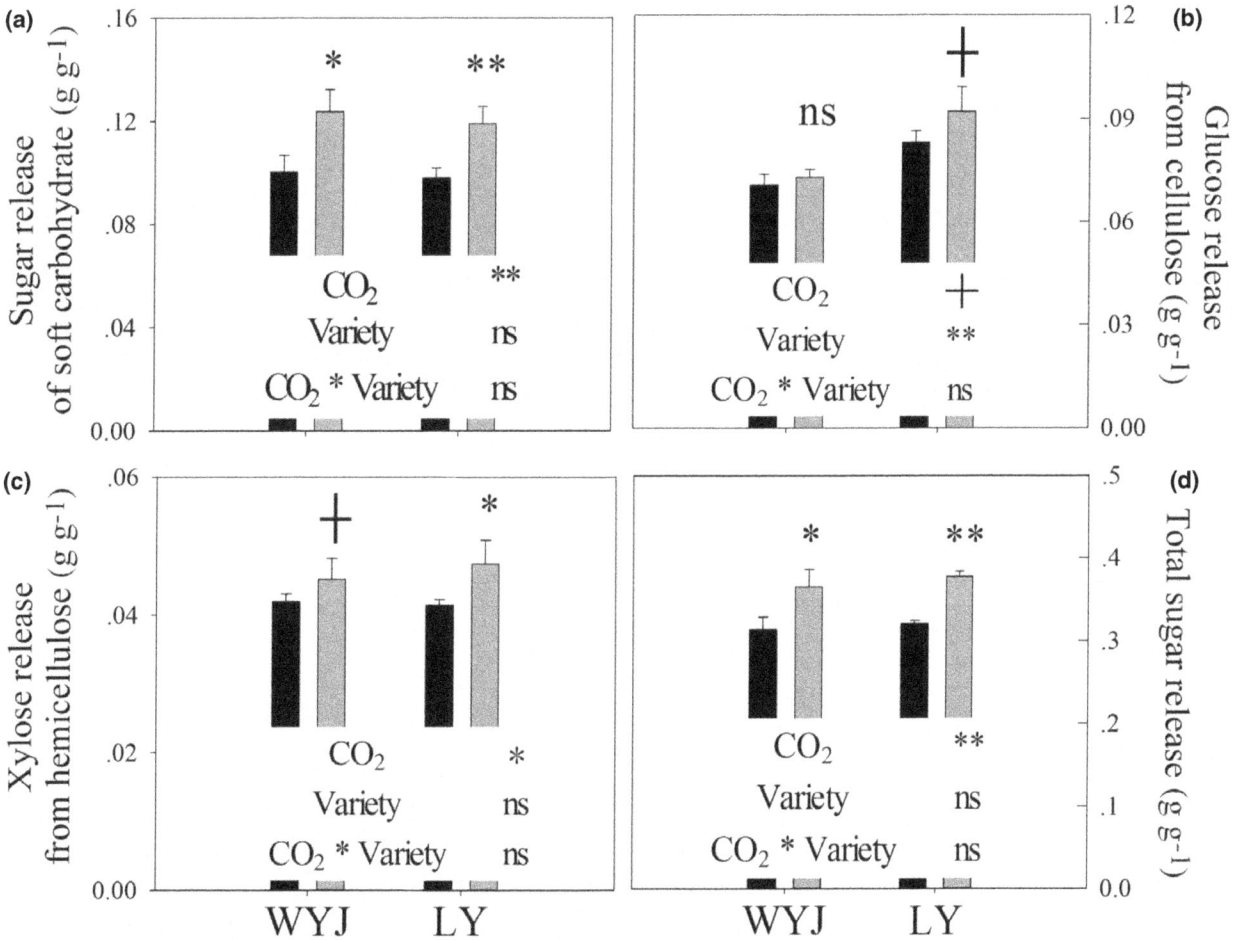

Fig. 3 Sugar release from nonstructural carbon (NSC) (a), glucose release from cellulose (b), xylose release from hemicellulose (c), and total sugar release from nonstructural and structural carbon (d) within rice straw for cultivars WYJ and LY 10 8 at maturity for ambient and FACE conditions for both 2012 and 2013 seasons. The mean was the average of 3 replications $(n = 3) \pm$ SD. $^{**}P < 0.01$, $^{*}P \le 0.05$, $^{\dagger}P \le 0.1$, $^{ns}P > 0.1$.

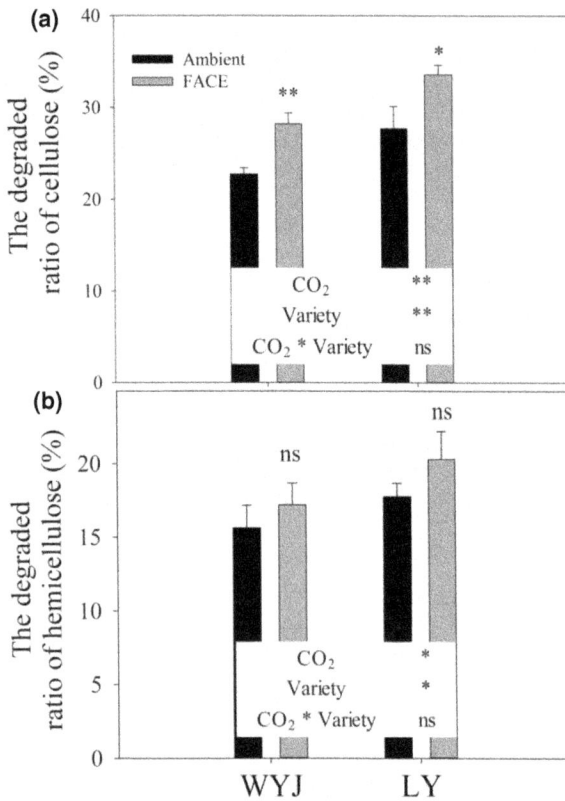

Fig. 4 Changes in the ratio of cellulose (a) and hemicellulose (b) degradation for rice straw biomass for rice cultivars WYJ and LY under ambient and FACE conditions for both 2012 and 2013 seasons. The mean was the average of replications ($n = 3$) ± SD. $**P < 0.01$, $*P \leq 0.05$, $^{\dagger}P \leq 0.1$, $^{ns}P > 0.1$.

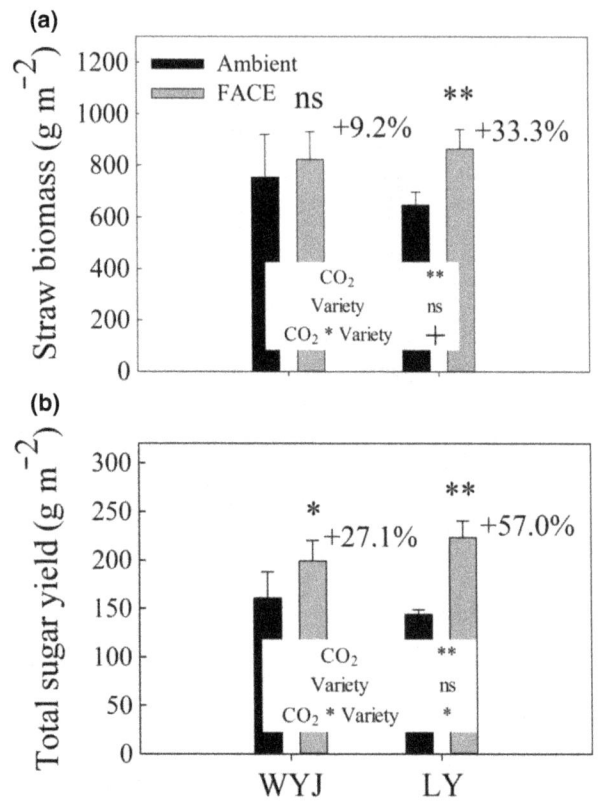

Fig. 5 Average rice straw biomass (a) and total sugar yield [straw biomass (g m^{-2}) × total sugar release (g g^{-1})] by saccharification (b) for rice cultivars WYJ and LY under ambient and FACE conditions for both 2012 and 2013 seasons. $n = 3$ ± SD. $**P < 0.01$, $*P \leq 0.05$, $^{\dagger}P \leq 0.1$, $^{ns}P > 0.1$.

Statistical analyses

The experiment design was a split-plot factor arranged within a randomized complete block design with 3 replications (three rectangular paddy fields) to test for the effect of [CO$_2$] on the rice carbohydrate accumulation and sugar release efficiency for biofuel productions. Subplots were blocked by variety levels to test the difference between the two varieties. For statistical analysis, [CO$_2$] was treated as the fixed-effect whole-plot factor, variety as the split-plot factor, and block as the random effect factor. The statistics were derived using a mixed linear model procedure (SPSS statistical software 19.0, SPSS Inc., USA) to test the effect of [CO$_2$], variety, and their interactions. The linear relationship between sugar releases, the ratio of cellulose and hemicellulose degradation to lignin content, and S/G ratio of lignin was determined using the linear regression model.

Results

Yield

Elevated [CO$_2$] significantly enhanced the grain yield for both varieties (Fig. 1). Consistent with previous studies, the stimulation of yield was larger for LY (37%) than WYJ (10%). A significant [CO$_2$] × cultivar interaction was observed.

Nonstructural and structural carbohydrates and S/G ratio

As shown in Table 1, a large amount of nonstructural carbohydrates are represented by straw, with starch and sucrose as the major components in both varieties. At ambient [CO$_2$], WYJ had similar starch and sucrose contents, whereas LY had more sucrose relative to starch (Table 1). At the elevated [CO$_2$] treatment, starch, sucrose, free glucose, fructose, and β-1, 3-1, 4-glucan contents were significantly increased, and the total nonstructural carbohydrate content was increased from 9.27% to 11.40% for WYJ and from 8.4% to 10.92% for LY in response (Table 1). Conversely, cellulose and lignin were significantly reduced at elevated [CO$_2$] for both varieties, while hemicellulose content did not change. There was a marginally significant interactive effect of CO$_2$ and variety on lignin content (Table 1). S/G ratio of lignin was reduced at elevated [CO$_2$] (Fig. 2).

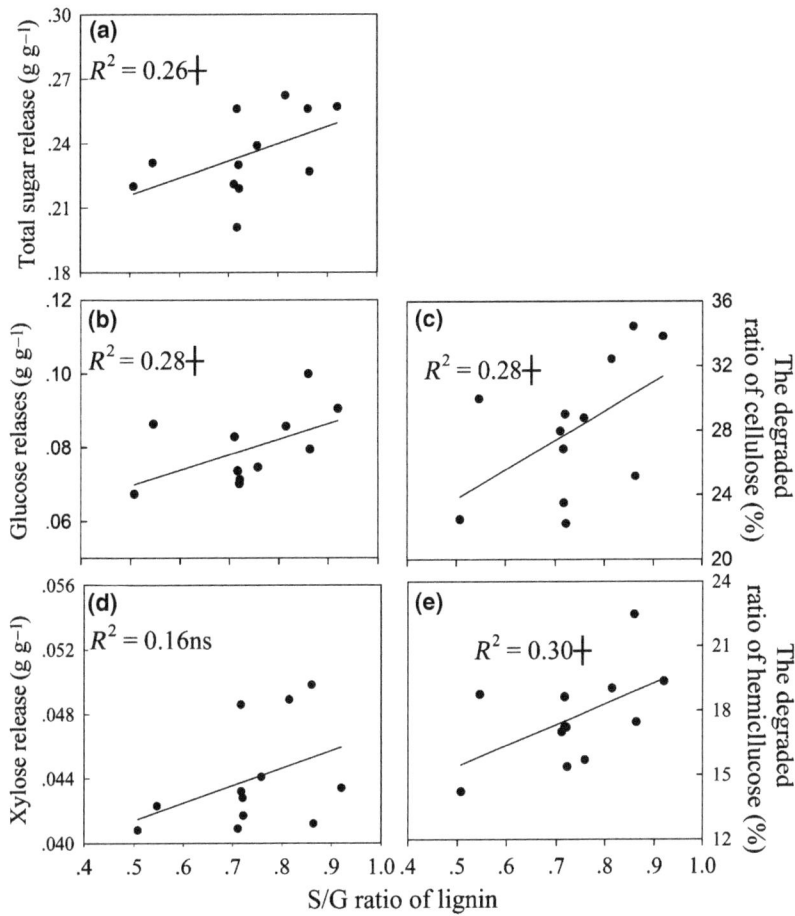

Fig. 6 Total sugar release (a), glucose release from cellulose (b), degraded cellulose ratio (c), xylose release from hemicelluloses (d), and degraded hemicellulose ratio (e), and their relationship with S/G ratio of lignin within rice straw for rice cultivars WYJ and LY under ambient and FACE conditions for both 2012 and 2013 seasons. $n = 12$ [2 CO_2 × 2 varieties × 3 replicates], $*P \leq 0.05$, $^{\dagger}P \leq 0.1$, $^{ns}P > 0.1$.

Sugar release from nonstructural and structural carbohydrates

Elevated [CO_2] significantly increased the total sugar release from nonstructural and structural carbohydrates for both varieties (Fig. 3). Elevated [CO_2] tended to increase the glucose release from cellulose, but there was only a marginal increase in sugar release for LY. Elevated [CO_2] also increased the amount of xylose release from hemicellulose for both varieties. Cellulose degradation was significantly increased for both varieties; however, the degradation of hemicellulose was only slightly enhanced under elevated [CO_2] (Fig. 4).

Straw biomass and total sugar yield from straw

Elevated [CO_2] increased straw biomass by 9.2% and 33.3% for WYJ and LY, respectively (Fig. 5). Similar cultivar-specific increases in total sugar yield were also observed (27.1% and 57.0% for WYJ and LY, respectively).

The relationship between S/G and lignin content to sugar release

Total sugar release, glucose release from cellulose, and the ratio of cellulose and hemicellulose degradation were marginally ($P < 0.10$) positively correlated with S/G ratio (Fig. 6a,b,c,e). Total sugar release was marginally ($P < 0.10$) negatively correlated with increasing lignin; however, a significant correlation was observed for xylose release as lignin content increased (Fig. 7a,e).

Discussion

The current study indicates that elevated [CO_2], by stimulating vegetative biomass, could enhance the

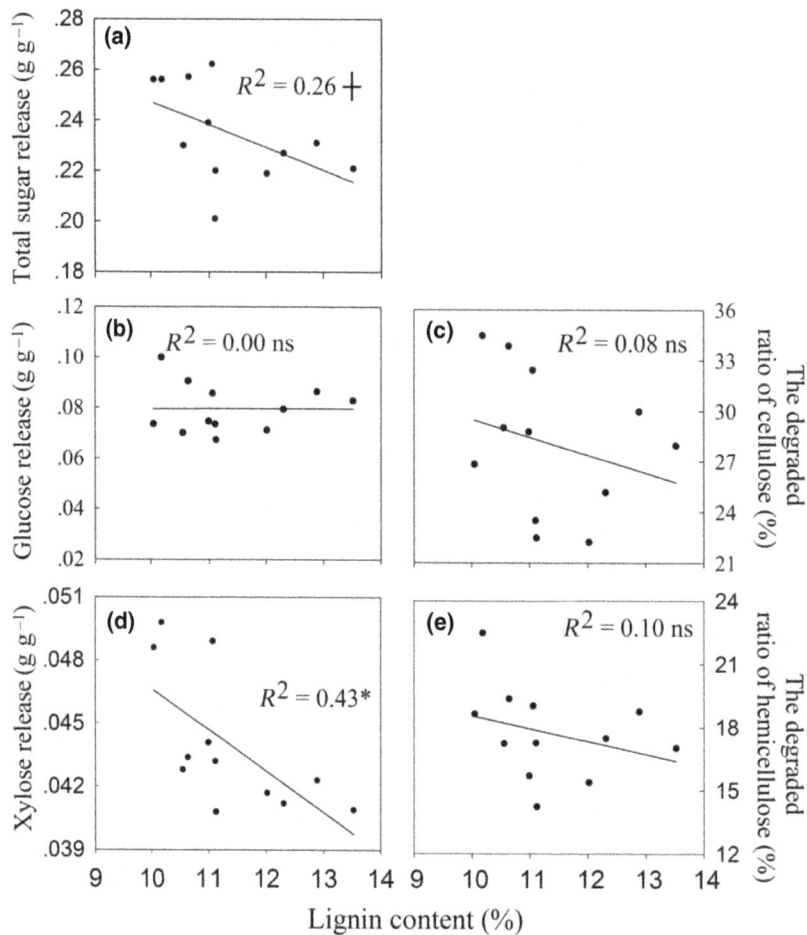

Fig. 7 Total sugar release (a), glucose release from cellulose (b), degraded cellulose ratio (c), xylose release from hemicellulose (d), and degraded hemicellulose ratio (e), and their relationship with lignin content within rice straw of WYJ and LY at mature stage under ambient and FACE conditions for both 2012 and 2013 seasons. $n = 12$ [2 CO_2 × 2 varieties × 3 replicates], *$P \leq 0.05$, †$P \leq 0.1$, ns$P > 0.1$.

potential of rice straw as a bioethanol source in the future. In addition, the degree of stimulation by elevated [CO_2] appeared to be cultivar specific. For example, in the current study, the hybrid LY showed a higher shoot biomass response (+33.3%) than the conventional WYJ variety (+9.2%). Such variation is consistent with other studies that have shown that [CO_2] could enhance shoot biomass from 5% to 39% among rice lines (Yang *et al.*, 2006; Liu *et al.*, 2008; Shimono *et al.*, 2009). This variation, in turn, could be considered in selecting rice lines that could show a stronger vegetative response to elevated [CO_2] (Shimono *et al.*, 2009).

Elevated [CO_2] can not only stimulate the amount of biomass, but also the quality of the biomass produced. For example, Henning *et al.* (1996) and Booker *et al.* (2005) reported that elevated [CO_2] increased the lignin concentration in sorghum stems; Billings *et al.* (2003) showed that elevated [CO_2] had no effect on cellulose

and lignin concentration for soybean and four shrub species; Hall *et al.* (2005) found that rising [CO_2] did not change the content of cellulose, hemicellulose, and lignin across plant species within a scrub oak community. Alternatively, Newman *et al.* (2003) found that rising [CO_2] decreased lignin concentrations of tall fescue. It has been previously documented that elevated [CO_2] can increase nonstructural carbohydrate accumulation in stems and leaf blades of C_3 crops, including rice (Seneweera *et al.*, 2002; Ainsworth *et al.*, 2004). It has also been demonstrated that elevated [CO_2] significantly increased the content of starch through increasing the size and number of starch in C_3 tissues (Teng *et al.*, 2006).

For this study, elevated [CO_2] also affected straw quality. Nonstructural carbohydrate was significantly increased, but structural carbon content was significantly reduced, with a subsequent decline in cellulose and lignin. Under elevated [CO$_2$], the cellulose content was reduced, but the degradation of cellulose was increased as was the glucose release from cellulose, especially for LY. Interestingly, the hemicellulose content was unaffected by [CO$_2$], but the ratio of degradation of hemicelluloses was enhanced. The amount of xylose release from hemicellulose was also significantly enhanced with elevated [CO$_2$]. Previous studies have demonstrated the possible mechanism of cell wall traits, anatomy, and biochemistry associated with resistance to carbohydrate degradation. A common assumption was that high lignin content adversely affected enzymatic hydrolysis (Chang & Holtzapple, 2000; Dien et al., 2006). Furthermore, S-rich lignin is more reactive to be degraded (Stewart et al., 2009; Sannigrahi et al., 2010). In this study, decreased lignin content and increased S/G ratio under elevated [CO$_2$] could have also altered the rate of straw degradation. However, the differential response of hemicellulose content relative to cellulose and lignin content within rice straw in response to elevated [CO$_2$] requires further study.

Overall, in contrast to biomass and straw production, similar enhancement of total sugar release from straw was observed for both lines (e.g., 16.4% for WYJ and 17.8% for LY) in response to [CO$_2$]. Selection for total sugar release may be possible in response to elevated [CO$_2$], but a larger range of cultivars will need to be evaluated.

To the best of our knowledge, this is the first study that has quantified [CO$_2$]-induced changes in rice straw in the context of its utility as a biofuel source. Although there is merited interest in the impact of rising [CO$_2$] on grain yield, less is known with respect to how [CO$_2$] can alter bioethanol production. Yet, given the importance of rice cultivation globally, rice straw could, potentially, represent a large, potential energy source. In this context, we would argue that it is important to evaluate how rising [CO$_2$] could influence growth of other major crops and the subsequent potential of those crops for biofuel. We would suggest that other crop FACE studies (soybean and maize-FACE of USA; barley, wheat, and maize-FACE of Germany, wheat-FACE of Australia, rice and wheat-FACE of China, and rice-FACE of Japan) could be used to accumulate a database on how elevated [CO$_2$] alters biofuel production of major crops globally.

Acknowledgements

This work was supported by the National Basic Research Program (973 Program, 2014CB954500), the National Natural Science Foundation of China (Grant No. 31370457, 41301209, 31261140364, 31201126, 41101232), and Natural Science Foundation of Jiangsu Province in China (Grant No. BK20131051, BK20140063). The FACE system instruments were supplied by the National Institute of Agro-Environmental Sciences and the Agricultural Research Center of Tohoku Region (Japan). We also thank Dr. Lewis H. Ziska of USDA-ARS for his review and suggestions to the manuscript.

Reference

Ainsworth EA, Rogers A, Nelson R et al. (2004) Testing the "source-sink" hypothesis of down-regulation of photosynthesis in elevated CO$_2$ in the field with single gene substitutions in Glycine max. Agricultural and Forest Meteorology, 122, 85–94.

Billings SA, Zitzer SF, Weatherly H et al. (2003) Effects of elevated carbon dioxide on green leaf tissue and leaf litter quality in an intact Mojave Desert ecosystem. Global Change Biology, 9, 729–735.

Booker FL, Prior SA, Torbert HA et al. (2005) Decomposition of soybean grown under elevated concentrations of CO$_2$ and O$_3$. Global Change Biology, 11, 685–698.

Chang VS, Holtzapple MT (2000) Fundamental factors affecting biomass enzymatic reactivity. Applied Biochemistry and Biotechnology, 84–86, 5–37.

Cuevas M, Sanchez S, Bravo V et al. (2010) Determination of optimal pre-treatment conditions for ethanol production from olive-pruning debris by simultaneous saccharification and fermentation. Fuel, 89, 2891–2896.

Dien BS, Jung HJG, Vogel KP et al. (2006) Chemical composition and response to dilute-acid pretreatment and enzymatic saccharification of alfalfa, reed canary-grass, and switchgrass. Biomass and Bioenergy, 30, 880–891.

Domínguez-Escribá L, Porcar M (2010) Rice straw management: the big waste. Biofuels, Bioproducts and Biorefining, 4, 154–159.

Erdei B, Barta Z, Sipos B et al. (2010) Ethanol production from mixtures of wheat straw and wheat meal. Biotechnology for Biofuels, 3, 16.

FAO (2014) FAOSTAT: Production-Crops, 2012 data.

Hall MC, Stiling P, Moon DC et al. (2005) Effects of elevated CO$_2$ on foliar quality and herbivore damage in a scrub oak ecosystem. Journal of Chemical Ecology, 31, 267–286.

He YF, Pang YZ, Liu YP et al. (2008) Physicochemical characterization of rice straw pretreated with sodium hydroxide in the solid state for enhancing biogas production. Energy & Fuels, 22, 2775–2781.

Henning FP, Wood CW, Rogers HH et al. (1996) Composition and decomposition of soybean and sorghum tissues grown under elevated atmospheric carbon dioxide. Journal of Environmental Quality, 25, 822–827.

Karp A, Shield I (2008) Bioenergy from plants and the sustainable yield challenge. New Phytologist, 179, 15–32.

Lim JS, Manan ZA, Alwi SRW et al. (2012) A review on utilisation of biomass from rice industry as a source of renewable energy. Renewable and Sustainable Energy Reviews, 16, 3084–3094.

Liu HJ, Yang LX, Wang YL et al. (2008) Yield formation of CO$_2$-enriched hybrid rice cultivar Shanyou 63 under fully open-air field conditions. Field Crops Research, 108, 93–100.

Liu J, Han Y, Cai ZC (2009) Decomposition and products of wheat and rice straw from a FACE experiment under flooded conditions. Pedosphere, 19, 389–397.

Newman JA, Abner ML, Dado RG et al. (2003) Effects of elevated CO$_2$, nitrogen and fungal endophyte-infection on tall fescue: growth, photosynthesis, chemical composition and digestibility. Global Change Biology, 9, 425–437.

Oanh NTK, Ly BT, Tipayarom D et al. (2011) Characterization of particulate matter emission from open burning of rice straw. Atmospheric Environment, 45, 493–502.

Okada M, Lieffering M, Nakamura H et al. (2001) Free-air CO$_2$ enrichment (FACE) using pure CO$_2$ injection: system description. New Phytologist, 150, 251–260.

Park JY, Arakane M, Shiroma R et al. (2010) Culm in rice straw as a new source for sugar recovery via enzymatic saccharification. Bioscience Biotechnology and Biochemistry, 74, 50–55.

Poornejad N, Karimi K, Behzad T et al. (2013) Improvement of saccharification and ethanol production from rice straw by NMMO and BMIM OAc pretreatments. Industrial Crops and Products, 41, 408–413.

Sannigrahi P, Ragauskas AJ, Tuskan GA (2010) Poplar as a feedstock for biofuels: a review of compositional characteristics. Biofuels Bioproducts & Biorefining, 4, 209–226.

Seneweera SP, Conroy JP, Ishimaru K et al. (2002) Changes in source-sink relations during development influence photosynthetic acclimation of rice to free air CO$_2$ enrichment (FACE). Functional Plant Biology, 29, 945–953.

Shimono H, Bunce JA (2009) Acclimation of nitrogen uptake capacity of rice to elevated atmospheric CO$_2$ concentration. Annals of Botany, 103, 87–94.

Shimono H, Okada M, Yamakawa Y et al. (2009) Genotypic variation in rice yield enhancement by elevated CO_2 relates to growth before heading, and not to maturity group. *Journal of Experimental Botany*, **60**, 523–532.

Stewart JJ, Akiyama T, Chapple C et al. (2009) The effects on lignin structure of over-expression of ferulate 5-hydroxylase in hybrid poplar. *Plant Physiology*, **150**, 621–635.

Studer MH, DeMartini JD, Davis MF et al. (2011) Lignin content in natural Populus variants affects sugar release. *Proceedings of the National Academy of Sciences of the United States of America*, **108**, 6300–6305.

Teng N, Wang J, Chen T et al. (2006) Elevated CO_2 induces physiological, biochemical and structural changes in leaves of Arabidopsis thaliana. *New Phytologist*, **172**, 92–103.

Wang F, Hu GH, Xiao JB et al. (2011) Improvement in the productivity of xylooligosaccharides from rice straw. *Archives of Biological Science Belgrade*, **63**, 161–166.

Yang LX, Huang JY, Yang HJ et al. (2006) Seasonal changes in the effects of free-air CO_2 enrichment (FACE) on dry matter production and distribution of rice (Oryza sativa L.). *Field Crops Research*, **98**, 12–19.

Yang LX, Liu HJ, Wang YX et al. (2009) Yield formation of CO_2-enriched inter-subspecific hybrid rice cultivar Liangyoupeijiu under fully open-air field condition in a warm sub-tropical climate. *Agriculture Ecosystems and Environment*, **129**, 193–200.

Yoswathana N, Phuriphipat P, Treyawutthiwat P et al. (2010) Bioethanol production for rice straw. *Energy & Fuels*, **1**, 26–31.

Zhu CW, Zeng Q, Ziska L et al. (2008) Effect of nitrogen supply on carbon dioxide–induced changes in Competition between Rice and Barnyardgrass (Echinochloa crus-galli). *Weed Science*, **56**, 66–71.

Zhu CW, Ziska L, Zhu JG et al. (2012) The temporal and species dynamics of photosynthetic acclimation in flag leaves of rice (Oryza sativa) and wheat (Triticum aestivum) under elevated carbon dioxide. *Physiologia Plantarum*, **145**, 395–405.

Zhu CW, Zhu JG, Cao J et al. (2014) Biochemical and molecular characteristics of leaf photosynthesis and relative seed yield of two contrasting rice cultivars in response to elevated [CO_2]. *Journal of Experimental Botany*, **65**, 6049–6056.

Ziska LH, Manalo PA, Ordonez RA (1996) Intraspecific variation in the response of rice (Oryza sativa L.) to increased CO_2 and temperature: growth and yield response of 17 cultivars. *Journal of Experimental Botany*, **47**, 1353–1359.

16

Enhancing tree belt productivity through capture of short-slope runoff water

RICHARD G. BENNETT[1], DANIEL MENDHAM[2], GARY OGDEN[1] and JOHN BARTLE[3]

[1]Ecosystem Sciences, CSIRO, Underwood Avenue, Floreat, WA 6014, Australia, [2]Ecosystem Sciences, CSIRO, College Road, Sandy Bay, Tas. 7025, Australia, [3]Department of Parks and Wildlife, Kensington, WA, Australia

Abstract

A selection of multi-stemmed, drought-tolerant mallee eucalypts, planted in belt form and integrated with crops in dryland agricultural areas of Australia, may be able to produce biomass as a commercially attractive feedstock for biofuel production. This study aimed to determine if small (40–50 cm high) bunds along mallee belts could trap otherwise underutilized surface water runoff within paddocks, thereby increasing water available to the mallee trees and their growth rates. An experiment was established in 5 year-old *Eucalyptus polybractea* (RT Baker) mallee belts near the town of Narrogin in the central wheatbelt area of Western Australia. Bunds led to significant (12%) increases in biomass accumulation after about 2 years and 35% increases at around 3 years. Bunds also led to significant increases in predawn leaf water potential and significant decreases in soil water deficit within 12 months, which persisted for the remainder of the 39 month trial. We suggest that the increase in biomass accumulation was largely due to increased water availability, but that increased nutrient supply from run-on and trapping of organic residues may have also had some effect on bunded plots, despite our attempts to mitigate this effect by experimentally adding nutrients to all treatments. Results show that installing bunds along mallee belts would be a cost-effective investment at sites where within-paddock runoff is likely (i.e. gently sloping and with a loamy sand or heavier soil texture). Installation costs should be offset by improved biomass production within a few years and ongoing improvements in growth over the long term.

Keywords: 2nd generation biofuels, lignocellulosic, sustainable water use, water efficiency

Introduction

The international focus for development of bioenergy production is shifting to second-generation (lignocellulosic) biomass feedstocks to alleviate concerns around the 'food or fuel' conflict (Tilman *et al.*, 2009). In Australia, this focus has spurred development of an agroforestry system that is able to supply woody biomass from fast-growing, coppicing *Eucalyptus* species adapted to the low and medium rainfall agricultural regions of southern Australia.

Australia is richly endowed with many species of drought- and fire-tolerant, multi-stemmed, coppicing eucalypts commonly called mallees (Brooker, 2002). Several of these species are being domesticated for use as narrow belts (mallee belts) integrated with dryland agriculture and harvested on a 3–5 year cycle (Bartle & Abadi, 2010). Although there is currently no market for the product, economic modelling of this system has shown that farmers could profitably produce mallee biomass at prices that are likely to be commercially

attractive to processors (Abadi *et al.*, 2012). Mallees readily coppice after harvest, so initial establishment costs produce a dividend in biomass harvest over several decades. The integration of mallee belts with traditional crop or pasture agricultural systems conveys biophysical advantages to both components. Most importantly, mallees can better capture and use natural resources like water and nutrients effectively lost to traditional agriculture components, so the mallees can grow more productively than if they were growing in a conventional forestry plantation form. To ensure a commercially viable return from mallee belts, the trees need to be managed to both maximize productivity but also to ensure their productivity is sustained over many harvest cycles. A detailed understanding of nutrient and water dynamics in mallee-crop systems is required to properly inform this management.

The annual water use of mallees in nonenergy-limited environments can far exceed that supplied by annual rainfall where additional water is available (Knight *et al.*, 2002; Wildy *et al.*, 2004; Carter & White, 2009) and the productivity of mallee belts is heavily reliant on water availability (Brooksbank *et al.*, 2011). After early clearing of the deep-rooted native vegetation for annual

Correspondence: Richard G. Bennett
e-mail: richard.bennett@csiro.au

cropping systems, the deep soils of WA have typically accumulated a substantial amount of water (Peck & Williamson, 1987), often equating to 2–3 times the annual rainfall (Sudmeyer & Goodreid, 2007). Robinson *et al.* (2006) found that roots of mallees can access this stored soil water to at least 10 m depth within 7 years of planting and Knight *et al.* (2002) found roots of 4 year-old mallees were using water to 5 m depth and had extracted 399 mm (rainfall equivalent) of stored subsoil water. Despite this rapid exploitation of stored soil water, little consideration has been given to the implications for declining productivity of mallee belts in the longer term, as they deplete this stored soil water.

Apart from incident rainfall, water sources available to mallee belts potentially include surface runoff, shallow subsurface flows, lateral spread of roots and access to deep groundwater (Cooper *et al.*, 2005). Management intervention which allows mallee belts greater access to surface runoff and shallow subsurface flows could increase biomass productivity in the short term and help maintain productivity in the long term. Such intervention could include application of water interception or detention devices. This could also be complementary to agriculture and have a positive impact on the natural environment in south-western Australia, where excess surface and subsurface flow can potentially lead to negative environmental outcomes, including erosion, waterlogging and salinity associated with rising groundwater tables. Mallee belts are also well placed to use summer rainfall events, which can be intense but irregular, and are not reliable enough for agricultural cropping. Cooper *et al.* (2005) advocated the installation of water retaining bunds or banks adjacent to mallee belts to trap, retain and encourage infiltration of overland flow where the mallees have need for water. The use of bunds to retain surface water flows for the benefit of crops is a long-standing agricultural practice (Yair, 1983), but examples where the effect of bunds on adjacent crops has been tested are rare, particularly in dryland agricultural systems. Makurira *et al.* (2009) present one such study and demonstrate positive effects on soil water and increased agricultural productivity due to man-made bunds. Modelling by McGrath *et al.* (2012) has also demonstrated great potential for capture of additional moisture by installation of bunds on amenable sites.

This study aimed to understand the impacts of bunds on soil water availability under mallees and the associated impacts on the water status and growth of mallees. We tested the hypotheses that bunds adjacent to mallee belts would (i) capture and retain water from overland flows, (ii) increase the soil water under mallee belts and (iii) have positive effects on the water status and biomass accumulation of mallees. We installed bunds

around experimental plots of 5 year-old mallees and measured water collection, soil water status, leaf water potential and biomass over 3 years.

Materials and methods

Trial site, management and experimental conditions

The site chosen for this experiment was located approximately 7.5 km south-east of the town of Narrogin in Western Australia's medium rainfall wheatbelt (33.0°S, 117.2°E). It was on a midslope position below a laterite breakaway on a slope of approximately 4–6% and approximately 2 km long and had a duplex soil profile (sandy gravel A horizon over sandy clay subsoil) that previously carried wandoo and marri (*Eucalyptus wandoo* and *Corymbia calophylla*) woodland. The site had been cleared for over 50 years. The average annual rainfall between 1981 and 2010 for Narrogin was 460 mm (COA, 2013). Soil chemistry was measured in six samples taken from alleys between mallee belts in the experimental area during July 2011 (Table 1).

The mallee belts were planted to *Eucalyptus polybractea* (blue mallee) in 2005 in a four-row belt configuration. Spacing between the four rows was 2, 4, and 2 m (except in the uppermost belt, which were all spaced at 2 m) and between tree spacing along rows was 1.8 m. Four adjacent belts were used in the experiment. The belts were spaced evenly with approximately 60 m between centres and were approximately parallel to the contour of the paddock. The experiment ran between May 2010 and July 2013.

During the course of the experiment, the alleys between tree belts were managed as a typical annual pasture, with the dominant species being annual ryegrass (*Lolium rigidum*) and subterranean clover (*Trifolium subterraneum*) grazed intermittently during the experiment by sheep. The mallee trees were not grazed by livestock. No fertilizer was applied to the pasture during the experiment except where required for the experimental treatments.

The experiment compared plots with or without bunds. There were a total of six replicates of each treatment which were installed over four adjacent belts within the paddock to capture a larger cross section of the inherent site variation. We placed one replicate of each treatment on three of the belts used in the experiment (those lower in the landscape) and the three remaining replicates of each treatment were installed on the fourth (highest) belt. Earthen bunds, 30 m in length were installed during May 2010 along the downhill side of bunded treatments. Returns (90° into the tree belt) were installed at

Table 1 Summary of surface soil chemistry at the experimental site. Standard deviation is shown in parentheses (*n* = 6)

	0–10 cm depth	10–20 cm depth
Organic C (%)	1.97 (0.35)	0.56 (0.24)
Total N (%)	0.16 (0.08)	0.04 (0.01)
Conductivity (dS m^{-1})	0.06 (0.02)	0.03 (0.01)
pH (CaCl$_2$)	5.1 (0.36)	4.8 (0.24)

each end of the bund to prevent water running around the bund ends, effectively creating small dams around each bunded plot (Fig. 1). The gross plots were 30 m long, and an inner measure plot (20 m length) was marked out in each of the plots, leaving a 5 m buffer between measured trees and end of the treated plots. An extra 5 m buffer was also retained between bund returns and the start of the next gross plot.

To provide a firm footing for the earthen bunds, a grader was used to clear the previous year's crop residues along the downhill side of the mallee belts extending from approximately 1 m–5 m from the downhill row of trees. Approximately 10 m^3 of gravelly loam removed from an area adjacent to the experimental site was formed into a bund with the top approximately level at 0.5 m high, 30 m long and located approximately 2 m from the lowermost row of trees using the grader. As the tree belts were not precisely parallel with the elevation contour, in some cases, there was a small grade in elevation within the bunded plots. This grade was never more than 0.2%, or 60 mm along the length of the bund. A spillway was constructed at 0.4 m above the lowermost point of each plot, which was reinforced with rubber matting to reduce erosion. Over time, the height of the bunds decreased as the material settled and animal traffic caused some attrition. At the conclusion of the experiment, the lowest point on most bunds had reduced by approximately 0.1 m to leave the overflow point at approximately 0.3 m. Such a bund, 0.3 m at the lowest point on a 4% grade, should in theory, detain 1.125 kl of water m^{-1} of bund when full, which translates into a rainfall equivalent of 94 mm over a belt of 12 m width (in the case of the lower three belts) or 112 mm on a belt 10 m wide (top belt).

Prior to the start of the experiment, it was observed that surface water movement from previous rainfall events had transported organic matter (animal droppings and crop residues) onto the mallee belts. Therefore, we added nutrients to all treatments to alleviate nutrient limitations that the trees may have been experiencing and thereby reduce the potential confounding effect of additional nutrients captured by bunds. Fertilizer was applied during June 2011 and July 2012 to all treatments. In both years, the fertilizer was applied on two occasions around 2–3 weeks apart at half the total rate to minimize the risk of the granular fertilizer moving off the application area

with surface water flow. In 2011, nutrients were applied at rates of 69 kg N, 41 kg P, 42 kg K, 32 kg Ca, 31 kg S, 1.1 kg Mg, 11.3 kg Fe, 1.9 kg Mn, 0.75 kg Zn, 0.75 kg Cu, 0.15 kg B, and 0.038 kg Mo, all per hectare. In 2012, 69 kg N ha^{-1} was applied, but the other elements were omitted. Nutrients were applied as urea, triple superphosphate, potassium sulphate and 'Micromax' (Scotts International). All fertilizers were broadcast by hand over an area of 600 m^2 for each plot encompassing the 30 m length of plots and a width of 20 m centred over the belt.

Measurement of biomass

The biomass of mallees in plots was measured by relating tree heights and canopy dimensions to above-ground green biomass through an allometric relationship. These relationships were built through destructive harvest of 25 trees of various sizes representative of those in the experimental plots. The harvested trees were growing in the same plantation as the experimental trees but outside the experimental area. Prior to harvesting these trees, the basal area, length of canopy (along the row), width of canopy (perpendicular to the row) and tree height were measured. Trees were then cut at ground level and green biomass was measured using a large balance mounted on a covered trailer as described by Peck et al. (2012). Figure 2 shows the regression between measured biomass and biomass predicted from canopy dimensions using Eqn (1).

$$\ln(\text{above-ground biomass}) = 1.908.\ln(\text{height}) + 0.67.\ln(\text{length}) + 0.527.\ln(\text{width}) - 0.477 \quad (1)$$

where above-ground biomass is measured in green kg; height is distance from ground level to tallest part of tree canopy (m); length is greatest dimension of tree canopy along rows (m) and; width is greatest dimension of tree canopy across rows (m).

Canopy dimensions of all mallees within the measured plots were assessed seven times during the experiment (June 30 2010, April 14 and October 10 2011, January 16, April 17 and October 25 2012 and August 20 2013). The biomass of plots was converted to an area basis, using the convention that the belt

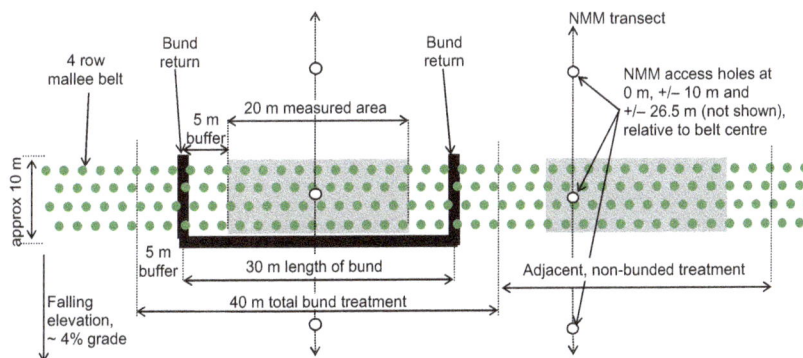

Fig. 1 Schematic showing dimension and location of earthen bunds constructed around bund treatments, the location of buffers in bunded and unbunded plots and the location of neutron moisture metre (NMM) access tubes arranged in a transect through each plot.

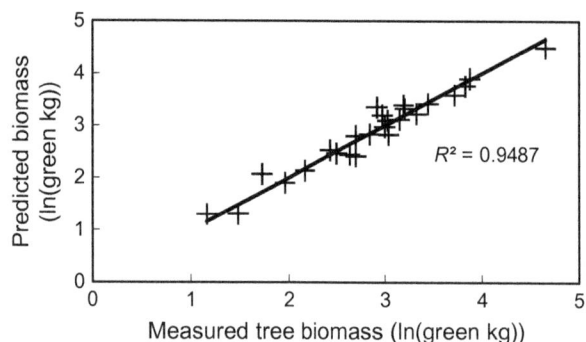

Fig. 2 Relationship between predicted biomass based on canopy dimensions [Eqn (1)] and measured biomass in 25 sampled trees.

area occupies an additional 2 m on either side of each outer row (Peck *et al.*, 2012). The effects of bunds on the estimated absolute biomass at each measure and the growth increment expressed as a percentage of initial plot biomass [100*(biomass at t_t – biomass at t_0)/biomass at t_0] were then tested using repeated two-way ANOVAS for each date (with treatment and replicate as factors) in R version 3.0.0 (R Core Team, 2013). The effect of bunds on the growth increment over time was also analysed using a repeated measures ANOVA analysis in Genstat 16.1 (VSN International, 2011) with per cent increase in biomass as the dependent variable, blocked by replicates, and with treatment and measurement date as the independent variables.

Rainfall

Hourly rainfall at the experimental site was measured using a tipping bucket rainfall gauge with 0.2 mm resolution supplied by ESIS and connected to a SL5 μSmart logger system between November 2010 and May 2013. On some occasions, it was necessary to augment this data with daily rainfall records from the Bureau of Meteorology weather station at Narrogin (COA, 2013), approximately 8 km away. Bureau data were used for the periods April 2010 to October 2010, 18 April 2013 to 16 March 2013 and for June and July 2013. No data were collected on canopy interception or evaporation.

The total rainfall at the site during the experimental period May 2010 and July 2013 was 1436 mm, of which 188 mm fell from May to December 2010, 656 mm fell in 2011, 370 mm fell in 2012 and 222 mm fell from January to July 2013.

Leaf water potential

To assess when mallees were suffering moisture stress and to investigate the effect of bunds on this stress, regular measurements of predawn leaf water potential (LWP) were made using a Scholander-style pressure chamber (Turner, 1988) on five mallees in each plot which were randomly selected at the beginning of the experiment. Predawn LWP was measured on all six replicates on April 4, July 19 and September 29 2011, January 10, April 11, July 4 and October 10 2012 and January 23 and May 17 2013. To increase the number of dates when predawn LWP was measured without adding excessively to

workload, predawn LWP measures were also made on a subset of three replicates on January 27, April 19, May 5, May 26, July 19, August 4, August 25, October 11, October 26 and November 18 2011, January 17, January 31, February 23, April 11, May 2, May 25 and November 8 2012 and February 5 and June 6 2013. For each measurement date, a one-way ANOVA was conducted on plot averages to identify significant ($P < 0.05$) treatment effects in R version 3.0.0 (R Core Team, 2013).

Soil water measures

To assess the effect of bunds on soil water availability, 40 mm PVC neutron moisture metre access tubes were installed, and a CPN 503DR Hydroprobe neutron moisture metre (NMM) was used at regular intervals to assess soil water. The NMM access tubes were established in transects across the plots, with 5 NMM access tubes per transect, arranged such that two were located in the middle of the (approximately) 53 m upper and lower alleys between belts (approximately 26.5 m from the nearest row of trees), two were located in the upper and lower alley at 6 m away from the nearest row of trees (approximately 10 m from the centreline of belts), and the last access tube was located as close as possible to the centreline of the belt. These locations were named alley above, alley below, interface above, interface below and belt respectively. NMM access tubes were installed to 8.25 m or impenetrable layer (whichever was the shallowest), and backfilled with a slurry of kaolin, cement and water. Soil water was measured at 0.2 m intervals between 0.3 m and 1.5 m deep and then at 0.5 m intervals from 2 m to 8 m. Soil water was assessed 27 times between November 2010 and June 2013. The NMM counts were converted to volumetric soil water using a prior calibration for the same instrument on a similar soil type (Mendham *et al.*, 2011), and soil water deficit for each access point was calculated as the difference between the measured soil water and the soil at its wettest point in time observed at that depth within each of the transects over the whole measurement time. A lineal soil water deficit (m^3 of water per m of belt) was calculated assuming that each of the transect positions represented a strip of land, 12 m wide in the alley and belt positions, and 13 m wide in the interface positions (for a total of 62 m). The treatment effect was assessed as the difference in soil water deficit between the bunded and nonbunded plots and subject to one-way ANOVA analyses in R version 3.0.0 (R Core Team, 2013). The cumulative effect of bunds on soil water was calculated by summing the difference between the wettest and driest measurements of each wetting and drying cycle in both treatments, and then calculating the difference between treatments.

Water depth in the bunded areas

To measure the detention of water behind the bunds, capacitance water level probes (Dataflow systems, Christchurch, New Zealand) were installed in bunded treatments inside slotted, 40 mm diameter PVC tubes. The water level was logged hourly from August 4 2011 to October 31 2012.

A simple analysis of how much water was collected by bunds compared to the rainfall records was undertaken for two reasons. First, the analysis provided insight into the variation in water detention in bunds of different replicates, as their small plot design, rather than along an entire belt, meant they were affected by small differences in topography. Second, we sought to clearly demonstrate that bunds were able to capture significant amounts of water which could then infiltrate into the dry soil under the mallee belts. The depth of water recorded behind each bund was transformed into a volume per metre of belt length, assuming a constant grade of 4% inside the bunded area. To reduce processing effort and increase the reliability of the relationship, records where bunds collected less than 0.05 m^3 water m^{-1} belt were removed (approximately equivalent to 65 mm depth of water inside the bund). The results generally showed sets of contiguous data for each filling event where volumes increased as rainfall occurred, but also showed the volumes decreasing as water infiltrated once rainfall ceased. To isolate the effect of rainfall on bund filling, the first peak volume collected from each filling event was identified and then all following readings in that contiguous data set were removed. This left, for each filling event, the volumes leading up to and including the first peak volume, which was plotted against the total rainfall occurring in the last 3 hours. Subsequent filling events were only included if the water volume dropped below 0.05 m^3 m^{-1} belt, which avoided confounding effects of relatively small rainfall events showing an artificially large effect on volumes by adding to water already collected in the bunds. The relationship between rainfall events and water collected by bunds was summarized using an exponential line of best fit.

Results

Mallees in bunded treatments had a significantly ($P < 0.05$) higher cumulative increase in biomass (expressed as a percentage increase over the starting biomass in each plot) after bunds had been installed for around 24 months (Fig. 3) which continued to be significant for the remainder of the experiment. Over those periods where significant differences in cumulative biomass increase were found, the bunds increased productivity by between 12.7% and 35.4%. By the end of the experiment (around 40 months after bunds were installed), mallees in bunded treatments had increased their initial biomass by 266%, compared to 197% for mallees in unbunded treatments. The repeated measures ANOVA analysis revealed that treatment ($P = 0.022$), measurement date ($P < 0.001$) and the interaction between treatment and date ($P < 0.001$) were all significant effects, indicating that bunds did significantly increase biomass and that this effect increased over time, as demonstrated by the increasingly divergent lines in Fig. 3a. While significant differences were found in the percentage increase in biomass compared to the original biomass [$100*$(biomass at t_t − biomass at t_0)/biomass at t_0], there was no significant difference between absolute biomass of bunded and

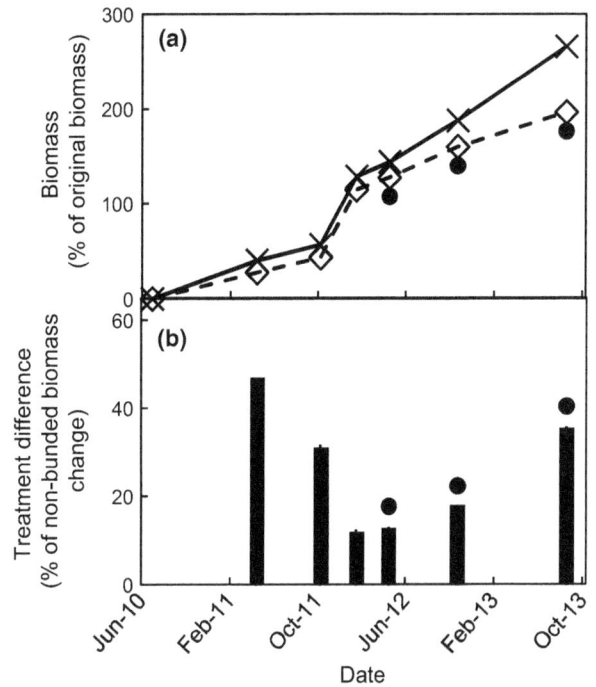

Fig. 3 (a) Cumulative biomass increase in trees in bunded (solid line) and unbunded (dashed line) treatments expressed as a per cent of biomass on 30-6-2010 (approximately 1 month after bunds were installed). (b) Difference in cumulative biomass increase expressed as % of the biomass increase in nonbunded treatments. Black circles near measures indicate dates where treatment were significant ($P < 0.05$).

unbunded mallees at any time during the experiment (Table 2).

Mallees in bunded treatments had significantly less negative predawn LWP ($P < 0.05$) at 7 of the 28 LWP measurement times. The mallees showed a typical seasonal pattern of water stress (Fig. 4), with higher predawn LWP during the winter and spring seasons

Table 2 Total biomass (green tonnes ha^{-1}) in bunded and unbunded treatments, the percentage difference between treatments (treatment effect) and the significance of the treatment effect calculated using ANOVA on seven dates during the experiment

Measurement date	Biomass		Treatment effect	Significance (P-value)
	No bunds	Bunds		
30/06/2010	30.4	31.0	2.1%	0.92
14/04/2011	39.5	43.6	10.4%	0.70
10/10/2011	44.0	48.6	10.0%	0.68
16/01/2012	65.2	70.3	7.7%	0.72
17/04/2012	69.1	74.7	8.1%	0.69
25/10/2012	79.5	88.9	11.8%	0.61
20/08/2013	90.1	111.5	23.8%	0.28

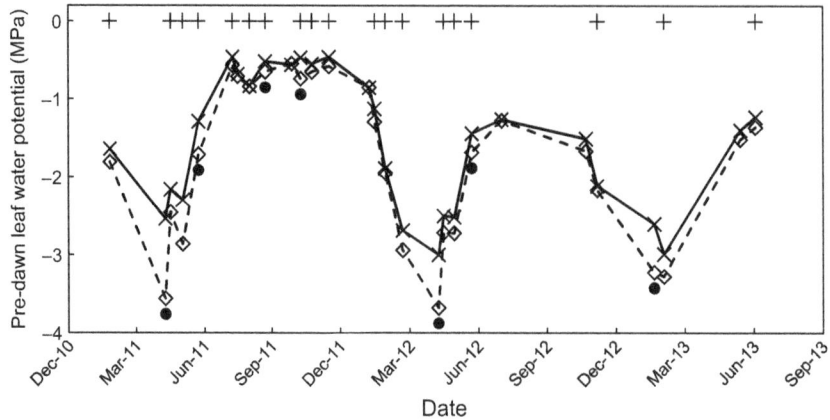

Fig. 4 Predawn leaf water potential of trees in bunded (solid line) and unbunded (dashed line) treatments over the course of the experiment. Results include dates where only three of the six replicates were sampled (indicated by + symbol at $y = 0$). Solid black circles beneath measurements indicate dates at which the bunded treatments had significantly ($P < 0.05$) less negative water potentials than unbunded treatments.

(approximately June to November) and seasonally low predawn LWP during the drier summer and autumn seasons (particularly from February to May). Three of the significant differences in LWP occurred in the summer and autumn months when the trees were displaying the lowest predawn LWP in the annual cycle (April 12 2011, April 11 2012 and January 23 2013). Annual differences in the cycle were also observed, with the predawn LWP levels not recovering to the same extent during the winter of 2012, compared to 2011.

The variability between replicates in the amount of water captured by bunds (Fig. 5) reflects local-scale topography upslope from the plots, i.e. slightly concave topography would concentrate runoff more effectively than slightly convex topography. For example, replicate

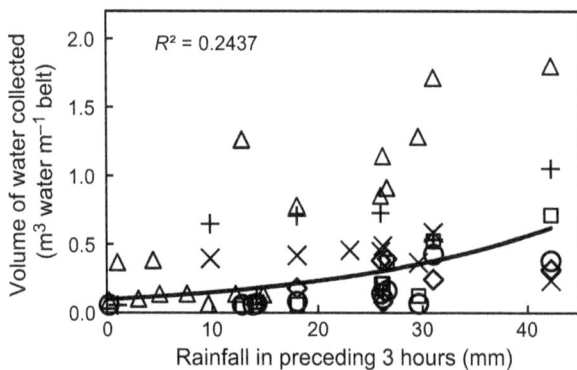

Fig. 5 Volume of water collected by bunds plotted as a response to the rainfall in the preceding 3 h. Replicates are indicated separately: replicate 1, plus symbols; replicate 2, open diamonds; replicate 3, open squares; replicate 4, cross symbols; replicate 5, open circles; replicate 6, open triangles. Trendline based on all data is fitted using an exponential relationship $[y = 0.0954*e^{(0.0443*x)}]$.

6 showed the largest response to rainfall events, while replicate 5 showed the least response and visual assessment indicated that the upslope topography was a factor contributing to the range of variability in water capture. Several of the bunds overflowed during the experiment. Replicate 6 recorded a water volume greater than 1.5 m³ m⁻¹ belt after two rainfall events. Using the relationship of water collected in bunds in response to rainfall (Fig. 5), 10, 20, 30 and 40 mm of rainfall over 3 h would be sufficient to generate approximately 0.15, 0.23, 0.36 and 0.56 m³ water m⁻¹ belt pooling in the bunds respectively. Our rainfall records showed that rainfall in the previous 3 h reached between 10 and 20 mm 10 times, 20 and 30 mm twice, 30 and 40 mm twice and over 40 mm once during the experiment. All of the five rainfall events over 20 mm in 3 h occurred in November or December.

The bund installed around replicate 6 overflowed several times during the experiment. On three occasions, the overflow was vigorous enough to wash away a portion of the earthen bund and repairs were required to maintain its integrity. On December 12 and December 13 2012, a rainfall event of more than 120 mm over approximately 48 h occurred and caused four of the bunds to overflow, all of which required repair. Although replicate 1 also received large inflows, the paddock-scale topography apparently ensured that any overflow was gentle enough to not severely damage the bund. This bund was, therefore, a useful case study to help understanding inflow and infiltration rates. A rainfall event of approximately 25 mm over the course of 1 h was sufficient to fill the bund from empty to 200 mm water depth (equating to approximately 0.5 m³ water m⁻¹ belt). A further 75 mm of rainfall fell over 17 h, which maintained this level or higher of water in the bund. Once rainfall ceased around 3:00 hours on December 14 2012, the water level in the bund did not return to normal (no water) for approximately 72 h.

Comparisons of soil water deficit calculated from NMM measurements in transects through the bunded and unbunded plots (Fig. 6) demonstrate that the bunds had a significant effect ($P < 0.05$) on reducing soil water deficit on several occasions during the experiment. These significant effects were particularly large (often around 5 m^3 water m^{-1} belt) at the 'belt' and 'interface below' transect positions, whereas significant effects seen in the 'interface above' and in the 'alley below' positions were relatively small impacts (generally less than 2 m^3 water m^{-1} belt).

The cumulative water detention by the bunds (Fig. 7, assessed as the difference between the successive peaks

and troughs in soil water) showed that much of the water detained by the bunds was retained within the belt (cumulative value of around 30 m^3 water m^{-1} of belt), but there was also substantial quantities of water at both interface positions. The bunds also significantly influenced water accumulation in the 'alley below' position, but the effect on the 'alley above' position was not significant.

Discussion

The primary conclusions to be drawn from this experiment are that bunds on the lower side of mallee belts could collect and retain surface water, have measurable impacts on soil water availability, significantly improve the water status of mallees and, most importantly, significantly increase biomass productivity.

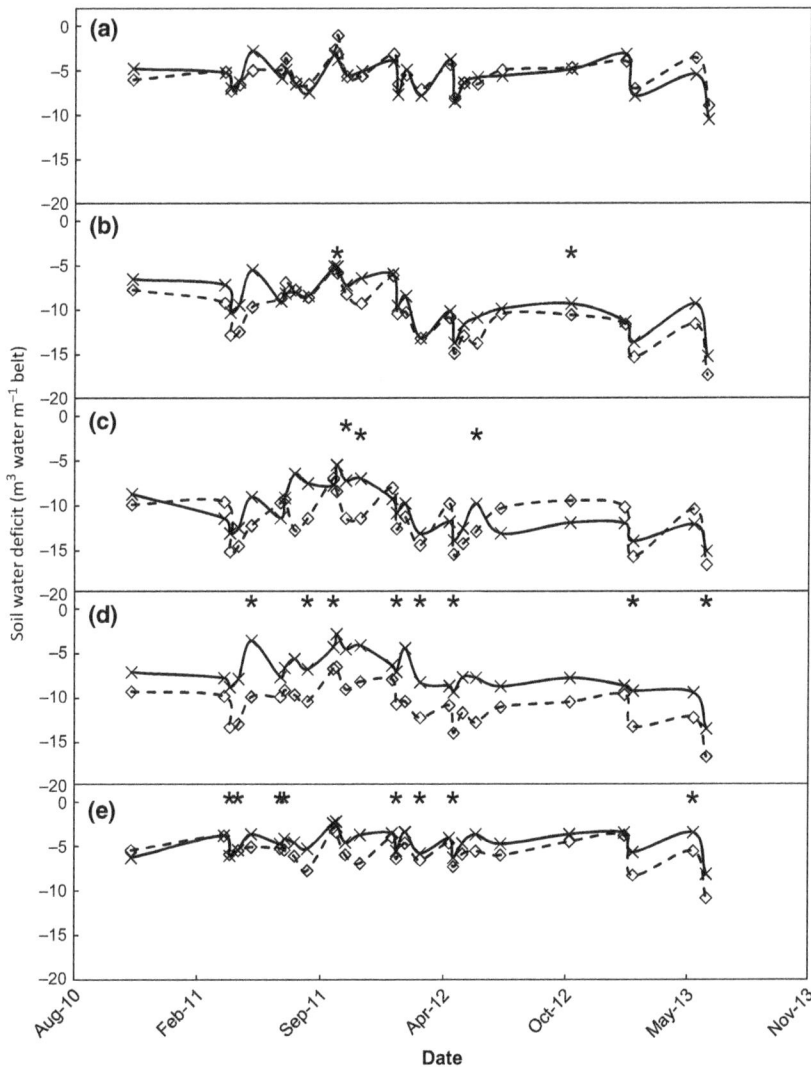

Fig. 6 Soil water deficit (m^3 water lineal m^{-1} of belt) in the whole profile over time in the alley above the belt (a), interface above the belt (b), within the belt (c), interface below (d) and alley below (e). Asterisks show significant differences (*$P < 0.05$).

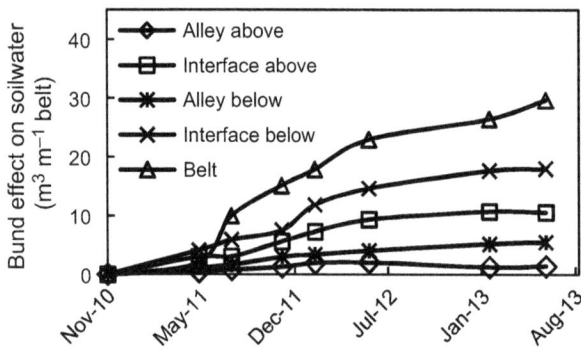

Fig. 7 Cumulative effect of bunds on soil water detention along transect through tree belt.

Proportional to their initial biomass, mallees growing in bunded treatments had significant improvements of between 12% and 35% greater biomass accumulation compared to unbunded treatments after bunds had been installed for approximately 24 months. Treatment effects earlier in the trial were of a similar magnitude, but the absolute biomass differences were smaller and so the significance of these results was masked by inherent variation in initial biomass and the variable effectiveness of the narrow experimental bunds in capturing water in different parts of the landscape, as illustrated in Fig. 5. The importance of our results is highlighted when the young age and uncoppiced nature of the mallees used in this experiment are considered, as these the mallees were unlikely to have fully depleted the soil water accrued from the agricultural practice preceding their planting (Peck & Williamson, 1987; Knight et al., 2002; Robinson et al., 2006). Had the mallees been older or in later coppice rotations, the effect of additional water from bunds may have been even more substantial.

Mallees growing in bunded treatments produced approximately 20 green tonnes ha^{-1} more biomass than those without bunds over the three and a quarter years of the experiment. The construction cost of the small, dam-style bunds used in this experiment is not relevant to the construction cost of paddock-scale bunds which will vary considerably with labour cost, equipment availability and topography. Nevertheless, a brief economic assessment of the benefits of bunds will be useful for land managers who are best placed to account for these variables to determine a construction cost. We can assume that the price paid for delivered biomass will be approximately $45 green tonne^{-1} which includes harvest and delivery costs of $26 green tonne^{-1} (taken from Bartle & Abadi, 2010) leaving a value for delivered biomass of $19 green tonne^{-1}. The additional biomass seen in our experiment is therefore worth approximately $380 hectare^{-1}. Although this figure alone should compare favourably with the initial cost of bund

construction, the real value of constructing bunds will be in their long-term impacts. Even conservative improvements in biomass accumulation of 10–15% will offer profound improvements to the economic viability of mallee belt biomass production systems when compounded over the long term, particularly as the once-off cost of installing bunds will be amortized over multiple harvest cycles. The increased rate of biomass accumulation may also reduce the time taken for a coppice cycle to reach sufficient maturity for harvesting, improving cash flow of mallee production and possibly reducing competition impact on the adjacent crop or pasture (Peck et al., 2012).

The improved biomass accumulation seen in bunded treatments was related to increased supply of water to the mallee belts and reduced water stress in the bunded trees (Figs. 3 and 4). A lack of water availability is known to be a limitation to biomass accumulation in oil mallee agroforestry systems (Cooper et al., 2005; Carter et al., 2006; Brooksbank et al., 2011) and the bunds in this experiment caused additional water to pool and infiltrate on the tree belts, significantly reduced soil water deficits, and significantly increased predawn LWP in mallees growing with bunds, particularly at times of the year when leaf water potentials were at their most negative.

The increased LWP in bunded mallees is evidence that the bunded trees suffered reduced water stress and could sustain growth for longer during the dry seasons. Turgor loss point (TLP) is a metric which is both an (inversely correlated) indicator of the drought tolerance of plants (Bartlett et al., 2012) and, beyond which, cell expansion and growth of plants is impossible (Cosgrove, 1986). To our knowledge, the TLP of E. polybractea, the species used in this experiment, has not been measured. However, the TLP of E. kochii ssp. borealis (-3.5 MPa) (Carter & White, 2009) is likely to be lower than that of E. polybractea, as E. kochii ssp. borealis is considered more drought tolerant. If we use −3.5 MPa as a conservative estimate of TLP in E. polybractea, our results show that predawn LWP of trees in unbunded treatments were below TLP twice during the experiment and their growth would have ceased due to low water availability at these times. Crucially, the trees in bunded treatments had significantly higher predawn LWP (above −3 MPa) on both these occasions and so should have been able to maintain critical growth processes like gas exchange and carbon uptake for longer. Apart from those dates where predawn LWP of unbunded trees were measured at less than −3.5 MPa, it is reasonable to expect that the LWP of unbunded mallees would have fallen below −3.5 MPa at other times during summer and autumn, especially during the day when the trees were exposed to high vapour pressure deficit, and that trees with bunds would have also experienced less moisture stress at these times.

The quantity of additional water provided to the mallees due to bunds is best assessed using the soil water results. Using a belt width of 12 m and even distribution of soil water across the tree belt, the increase in soil water content due to bunds of 30 m^3 m^{-1} belt under the tree belt equates to 2500 mm additional rainfall being provided to the belt area over the 39 month course of the experiment. This estimation suggests the bunds are more than doubling the moisture available to the trees from the 1436 mm of rainfall that fell during the course of the trial, comparable to estimates from a theoretical analysis of runoff potential by McGrath et al. (2012). Naturally, the quantity of moisture available to mallees is not limited to the rainfall incident on the belt area alone, as mallees are known to access water and nutrients through roots many metres beyond the planted area (Robinson et al., 2006; Sudmeyer et al., 2012). However, our results also show that bunds had profound effects on soil water content at the interface above and interface below transect positions where the soil water deficit was reduced by 10 and 18 m^3 m^{-1} belt respectively. These results are evidence that additional water supplied by bunds may remove much of the concern regarding the sustainability of water supply to the mallees in long-term coppice harvest regimes.

Fertilizer was applied to both treatments in this experiment in an attempt to ensure the mallees were not nutrient limited and thereby isolate the effect that additional water collected by bunds had on the productivity of mallees. Nevertheless, it is possible that some of the increased growth that we observed in bunded mallees was a result of additional nutrients deposited onto the belts by surface runoff. If employed on the landscape scale, additional nutrients transported into mallee belts by surface water movement and trapped there by bunds may prove to be a major advantage of installing bunds. We observed that there was an obvious increase in organic matter deposition (mostly crop stubble and animal manure) inside the bunds following large rainfall events, compared with unbunded mallee belts. In addition to the extra nutrients provided by imported organic matter, increased moisture in the topsoil may also improve the ability of trees to access soil nutrients otherwise locked up in dry soils. The effect of bunds on the soil nutrient levels was not the focus of this study and, therefore, was not measured, however, nutrient effects from bunds may improve the long-term nutrient balance of the mallee belts and could be highly desirable as repeated coppice rotations remove nutrients in the above-ground biomass (Grove et al., 2007; Peck et al., 2012) and should be more thoroughly investigated in future experiments.

In this experiment, bunds were in the form of downslope contour walls with short returns at the ends, effectively creating small dams as a simulation of larger scale water interception structures. However, some of our experimental bunds were occasionally filled by large rainfall events, leading to overflowing, erosion of the bund walls and loss of water that may have otherwise contributed to biomass production. The effects of our small bunds on tree biomass were also quite variable as the quantity of runoff water they received depended on their position in relation to upslope localscale topography. In practice, bunds will need to extend along the entire mallee belt and be aligned to contours, but with a small elevation gradient to allow water to spread evenly along the mallee belt and to optimize infiltration. Bartle et al. (2012) have further discussed the major design objectives of water interception structures and we expect that larger scale bunds with practical design would minimize the overtopping problems seen in this experiment and result in more consistent biomass benefits along the belts.

Our experiment placed bunds on the lower side of mallee belts as this would allow water to infiltrate directly into the most water-depleted area under the trees and the drier, more littered and more uneven soil under trees should maximize infiltration rates (Ellis et al., 2006, 2007; McGrath et al., 2012). However, it may be necessary to place bunds on the upper side of mallee belts to provide better access for machinery necessary to clean out material deposited in the bund and to repair any damage in the long term. Our results show that bunds significantly increased soil water in both the 'interface below' and 'alley below' transect positions (in addition to directly under the trees), suggesting that water trapped by bunds placed above the belts should ultimately move downhill into the mallee trees' root zone. Nevertheless, placing bunds below belts remains preferable in terms of hydrological principals if longterm maintenance requirements can be met.

Despite the overflowing problems due to the experimental bund design discussed above, our results show that bunds of 0.4–0.3 m high were sufficient to produce both additional biomass productivity and reduce the soil water deficit. Modelling by McGrath et al. (2012) indicated that increasing the height of bunds beyond 0.4 m would not significantly improve infiltration of surface water in their simulation of annual rainfall, but that taller bunds would probably be required to fully capture runoff from extreme rainfall events. In our experiment, the largest rainfall events occurred during summer, when additional water would provide the most productivity gain as soil water deficit was highest and leaf water potential was most negative. Therefore, we believe that bund height should aim to maximize capture of these intense summer rainfall events, but some compromise will be necessary as building higher bunds would also cost more. Further modelling could be used to better inform bund height and design.

McGrath *et al.* (2012) also investigated which sites were likely to benefit most from bunds. They found that sandy soil types would not benefit from bunds unless water repellence was an issue, but that sites with loamy or clayey topsoils would. They also found that sites with wider spacing of tree belts benefited more than those with narrow spacing and that, when bunds were installed, the degree of slope did not have a large bearing on infiltration once beyond a grade of around 1%. In this context, our site had relatively permeable topsoils of gravelly loam, moderate belt spacing of around 60 m and a suitable grade of approximately 4%. Better results might be expected on sites with heavier soil types or with larger belt spacing.

Despite significant interannual variation in rainfall during the study, the total rainfall during the experiment (1436 mm) is close to what might be expected for a 39 month period based on the mean annual rainfall (460 mm) from 1981 to 2010 for the nearest Bureau of Meteorology site at Narrogin (COA, 2013), an indication that the results in this trial should be representative of those expected under average conditions. However, the effectiveness of bunds depends on runoff water which occurs under heavy rainfall over the course of minutes to hours. A thorough analysis of rainfall event frequency is beyond the scope of this study, but a superficial analysis suggests that the rainfall conditions during our study were slightly unusual in the context of the historical rainfall of the region. We recorded five instances where rainfall exceeded 20 mm in a 3 hour period during the (approximately) 3 year long experiment, all of which occurred during November or December. Over the 100 years since 1913, Bureau of Meteorology records for Narrogin show daily rainfall exceeded 20 mm only 122 times during the wider summer–autumn period (November to April) (COA, 2013). The Bureau of Meteorology records are limited to daily data, so many of these rainfall totals greater than 20 mm would not have been confined to a 3 h period. In addition, the largest daily rainfall event for November or December over the last 100 years occurred during the course of this experiment (84.4 mm on December 13, 2011) (COA, 2013). Nevertheless, projections indicate that the proportion of annual rainfall occurring in summer and autumn is likely to increase in the wheatbelt of WA (COA, 2013), so our results may be a better indication of the effects of bunds in future climates, compared to historical climate.

In summary, our experiment has demonstrated that installation of bunds on the downhill side of mallee tree belts significantly increased the rate of biomass accumulation by around 35% after 3 years. Given that the bunded treatments generally had higher predawn leaf water potentials, particularly during summer and autumn when leaf water potential measurements were lowest, and significantly more soil water, it is likely that the increased biomass was largely due to increased water supply. Installing bunds along mallee belts would be a cost-effective investment at many sites as initial costs should be readily offset by biomass benefits that continue into the long term through more efficient use of within-paddock resources.

Acknowledgements

The authors are very grateful for research funding provided by the Australian Government through the Second Generation Biofuels Research and Development Grant Program and the Cooperative Research Centre for Future Farm Industries. The experimental site was kindly hosted by the Skerritt family and we appreciate their tolerance of our disruption to their farming systems. We also thank our colleagues Paul Turnbull, Mark Tibbett, Adam Peck and Kim Brooksbank for useful advice at project conception and throughout. Georg Wiehl, Damien Priest and Tammi Short all provided time and assistance with site setup, maintenance and measurements, thank you. Finally, we appreciate the time and effort that reviewers provided and for their useful and insightful comments to improve the manuscript.

References

Abadi A, Bartle J, Giles R, Thomas Q (2012) Supply and delivery of mallees. In: *Bioenergy in Australia: Status and Opportunities* (eds Stucley C, Schuck S, Sims R, Bland J, Marino B, Borowitzka M, Abadi A, Bartle J, Giles R, Thomas Q) pp. 140–172. Bioenergy Australia, St Leonards, New South Wales.

Bartle JR, Abadi A (2010) Toward sustainable production of second generation bioenergy feedstocks. *Energy & Fuels*, **24**, 2–9.

Bartle J, Abadi A, Giles R, Mazanec R (2012) Design of commercial mallee biomass production systems. In: *Management of Mallee Belts for Profitable and Sustained Production*. A report compiled by CSIRO, WA DEC and UWA, supported in part by the Australian Government through the Second Generation Biofuels Research and Development Grant Program, and the CRC for Future Farm Industries (eds Mendham D, Bartle J, Peck A, Bennett R, Ogden G, McGrath G, Abadi A, Vogwill R, Huxtable D, Turnbull P), pp. 111–135. CRC for Future Farm Industries, Perth, Australia.

Bartlett MK, Scoffoni C, Sack L (2012) The determinants of leaf turgor loss point and prediction of drought tolerance of species and biomes: a global meta-analysis. *Ecology Letters*, **15**, 393–405.

Brooker MIH (2002) Botany of the eucalypts. In: *Eucalyptus: The Genus Eucalyptus* (ed. Coppen JJW), pp. 3–36. Taylor & Francis, London.

Brooksbank K, Veneklaas EJ, White DA, Carter JL (2011) Water availability determines hydrological impact of tree belts in dryland cropping systems. *Agricultural Water Management*, **100**, 76–83.

Carter JL, White DA (2009) Plasticity in the Huber value contributes to homeostasis in leaf water relations of a mallee Eucalypt with variation to groundwater depth. *Tree Physiology*, **29**, 1407–1418.

Carter JL, Veneklaas EJ, Colmer TD, Eastham J, Hatton TJ (2006) Contrasting water relations of three coastal tree species with different exposure to salinity. *Physiologia Plantarum*, **127**, 360–373.

COA, (2013) Commonwealth of Australia, Bureau of Meteorology.

Cooper D, Olsen G, Bartle J (2005) Capture of agricultural surplus water determines the productivity and scale of new low-rainfall woody crop industries. *Australian Journal of Experimental Agriculture*, **45**, 1369–1388.

Cosgrove DJ (1986) Biophysical control of plant cell growth. *Annual Review of Plant Physiology*, **37**, 377–405.

Ellis TW, Leguedois S, Hairsine PB, Tongway DJ (2006) Capture of overland flow by a tree belt on a pastured hillslope in south-eastern Australia. *Australian Journal of Soil Research*, **44**, 117–125.

Ellis T, Potter N, Hairsine P *et al.* (2007) *Using Banded Tree-Agriculture Systems to Meet Surface Water Targets - a Design Framework*. Rural Industries Research and Development Corporation Barton, Australian Capital Territory.

Grove TS, Mendham DS, Rance SJ, Bartle J, Shea S (2007) *Nutrient Management of Intensively Harvested Oil Mallee Tree Crops*. Rural Industries Research and Development Council, Kingston ACT.

Knight A, Blott K, Portelli M, Hignett C (2002) Use of tree and shrub belts to control leakage in three dryland cropping environments. *Australian Journal of Agricultural Research*, **53**, 571–586.

Makurira H, Savenije HHG, Uhlenbrook S, Rockstroem J, Senzanje A (2009) Investigating the water balance of on-farm techniques for improved crop productivity in rainfed systems: a case study of makanya catchment, Tanzania. *Physics and Chemistry of the Earth*, Parts A/B/C, **34**, 93–98.

McGrath GS, Vogwill R, Hipsey M, Bartle J, (2012) An analysis of the potential of passive and active capture of overland flow by tree-belt systems. In: *Management of Mallee Belts for Profitable and Sustained Production*. A report compiled by CSIRO, WA DEC and UWA, supported in part by the Australian Government through the Second Generation Biofuels Research and Development Grant Program, and the CRC for Future Farm Industries (eds Mendham D, Bartle J, Peck A, Bennett R, Ogden G, McGrath G, Abadi A, Vogwill R, Huxtable D, Turnbull P), pp. 35–66. CRC for Future Farm Industries, Perth, Australia.

Mendham DS, White DA, Battaglia M, McGrath JF, Short TM, Ogden GN, Kinal J (2011) Soil water depletion and replenishment during first- and early second-rotation *Eucalyptus globulus* plantations with deep soil profiles. *Agricultural and Forest Meteorology*, **151**, 1568–1579.

Peck AJ, Williamson DR (1987) Effects of forest clearing on groundwater. *Journal of Hydrology*, **94**, 47–65.

Peck A, Sudmeyer R, Huxtable D, Bartle J, Mendham D, (2012) Productivity of mallee agroforestry systems under various harvest and competition management regimes, RIRDC Project No PRJ-000729. Available at: https://rirdc.infoservices.com.au/downloads/11-162 (accessed 18 February 2014).

R Core Team (2013) R: a language and environment for statistical computing, R Foundation for Statistical Computing, Vienna, Austria. Available at: http://www.R-project.org (accessed 30 August 2013).

Robinson N, Harper RJ, Smettem KRJ (2006) Soil water depletion by *Eucalyptus* spp. integrated into dryland agricultural systems. *Plant and Soil*, **286**, 141–151.

Sudmeyer RA, Goodreid A (2007) Short–rotation woody crops: a prospective method for phytoremediation of agricultural land at risk of salinisation in southern Australia? *Ecological Engineering*, **29**, 350–361.

Sudmeyer RA, Daniels T, Jones H, Huxtable D (2012) The extent and cost of mallee–crop competition in unharvested carbon sequestration and harvested mallee biomass agroforestry systems. *Crop and Pasture Science*, **63**, 555–569.

Tilman D, Socolow R, Foley JA *et al.* (2009) Beneficial biofuels – the food, energy, and environment trilemma. *Science*, **325**, 270–271.

Turner N (1988) Measurement of plant water status by the pressure chamber technique. *Irrigation Science*, **9**, 289–308.

VSN International (2011) GenStat for windows 16th edition. VSN International, Hemel Hempstead, UK. Available at: GenStat.co.uk (accessed 30 June 2011).

Wildy DT, Pate JS, Bartle JR (2004) Budgets of water use by *Eucalyptus kochii* tree belts in the semi-arid wheatbelt of Western Australia. *Plant and Soil*, **262**, 129–149.

Yair A (1983) Hillslope hydrology water harvesting and areal distribution of some ancient agricultural systems in the northern Negev Desert. *Journal of Arid Environments*, **6**, 283–301.

17

A global meta-analysis of forest bioenergy greenhouse gas emission accounting studies

THOMAS BUCHHOLZ[1,2], MATTHEW D. HURTEAU[3], JOHN GUNN[4] and DAVID SAAH[1,5]

[1]Spatial Informatics Group, LLC, 3248 Northampton Ct., Pleasanton, CA 94588, USA, [2]Gund Institute for Ecological Economics, University of Vermont, 617 Main Street Burlington, Vermont 05405, USA, [3]Department of Ecosystem Science and Management, The Pennsylvania State University, Bigler Road, State College, PA 16803, USA, [4]Spatial Informatics Group, Natural Assets Laboratory, 11 Pond Shore Drive, Cumberland, ME 04021, USA, [5]Department of Environmental Science, University of San Francisco, 2130 Fulton Street, San Francisco, CA 94117, USA

Abstract

The potential greenhouse gas benefits of displacing fossil energy with biofuels are driving policy development in the absence of complete information. The potential carbon neutrality of forest biomass is a source of considerable scientific debate because of the complexity of dynamic forest ecosystems, varied feedstock types, and multiple energy production pathways. The lack of scientific consensus leaves decision makers struggling with contradicting technical advice. Analyzing previously published studies, our goal was to identify and prioritize those attributes of bioenergy greenhouse gas (GHG) emissions analysis that are most influential on length of carbon payback period. We investigated outcomes of 59 previously published forest biomass greenhouse gas emissions research studies published between 1991 and 2014. We identified attributes for each study and classified study cases by attributes. Using classification and regression tree analysis, we identified those attributes that are strong predictors of carbon payback period (e.g. the time required by the forest to recover through sequestration the carbon dioxide from biomass combusted for energy). The inclusion of wildfire dynamics proved to be the most influential in determining carbon payback period length compared to other factors such as feedstock type, baseline choice, and the incorporation of leakage calculations. Additionally, we demonstrate that evaluation criteria consistency is required to facilitate equitable comparison between projects. For carbon payback period calculations to provide operational insights to decision makers, future research should focus on creating common accounting principles for the most influential factors including temporal scale, natural disturbances, system boundaries, GHG emission metrics, and baselines.

Keywords: biomass, carbon accounting, carbon payback period, classification and regression tree analysis, climate change, life cycle assessment, meta-analysis, wildfire

Introduction

The greenhouse gas (GHG) benefits of displacing fossil energy with biofuels are driving policy development in the absence of complete information. Getting the accounting correct is particularly important given the recent heavy emphasis on use of biomass energy to meet national and regional emissions reduction goals. For example, by 2020, between 8% and 11% of the UK's primary energy supply should be from biomass (United Kingdom, 2012; see Beurskens & Hekkenberg, 2011 for renewable energy projections of other EU states). The initial assumption regarding biomass energy was that of 'carbon neutrality', whereby a biologically based energy feedstock does not contribute to a net increase in atmospheric CO_2 relative to a defined fossil-fuel energy

Correspondence: Thomas Buchholz
e-mail: tbuchhol@uvm.edu

baseline (Searchinger et al., 2009). The carbon neutrality of forest biomass is a source of considerable debate because of the complexity of dynamic forest ecosystems, varied feedstock types, and multiple energy production pathways. The evaluation of forest biomass carbon neutrality requires a defined set of criteria that capture initial forest conditions, *in situ* carbon dynamics (e.g. fluxes), energy conversion efficiency, and a well-defined fossil energy source for comparison, among others (Walker et al., 2013; Mika & Keeton, 2014). Much of the research to date has focused on the appropriate choice of baseline (Gunn et al., 2012; Lamers & Junginger, 2013; Walker et al., 2013) or leakage (Gan & McCarl, 2007).

Defining a baseline for carbon stocks in a forest ecosystem has been the focus of considerable research and policy debate, because it is the carbon benchmark against which the effect of biomass energy development is evaluated and therefore influences the carbon neutrality of a project (Zanchi et al., 2012). A 'reference point' baseline uses the carbon stock on a given land area at a given point in time as the benchmark (EPA, 2014, Fig. 1). A 'dynamic' (or 'anticipated future') baseline requires defining a business-as-usual (BAU) condition that is projected without any new use of biogenic feedstocks for energy. Carbon stock changes under bioenergy scenarios are then compared to a fossil-fuel energy scenario to quantify the overall emissions effect from fuel switching (Fig. 2). In this case, the choice of baseline directly influences the determination of carbon neutrality.

Leakage, defined as activity shifting in the presence of a biomass project (Henders & Ostwald, 2012), has the potential to drive forest harvest outside the project area to continue meeting *a priori* economic demand for biomass (e.g. wood products). Although the leakage concept has been well defined, it is challenging to quantify because of the varying size and global nature of markets for different forest products (Gan & McCarl, 2007; Chen, 2009; Fankhauser & Hepburn, 2010).

There are many other attributes that can influence the length of the carbon payback period or point at which the biomass energy produced becomes carbon neutral from an atmospheric perspective (Lamers & Junginger, 2013; Vanhala *et al.*, 2013). These attributes are often project specific and can include biomass feedstock source or type, forest type, fossil-fuel source replaced, and life cycle analysis boundaries, among others (Lamers & Junginger, 2013; Walker *et al.*, 2013). Given the range of attributes influencing biomass projects, the carbon benefits of any given project can be influenced by site-specific aspects and decisions made by researchers in establishing the parameters for comparison.

Previous efforts to synthesize the literature on this topic have generally focused on part of the system (Mann, 2011; Muench & Guenther, 2013), had a small sample size ((Holtsmark, 2013; Sedjo, 2013), or the methods chosen relied on a descriptive analytical framework restricting the authors' deductions to very general conclusions (Helin *et al.*, 2013; Lamers & Junginger, 2013;

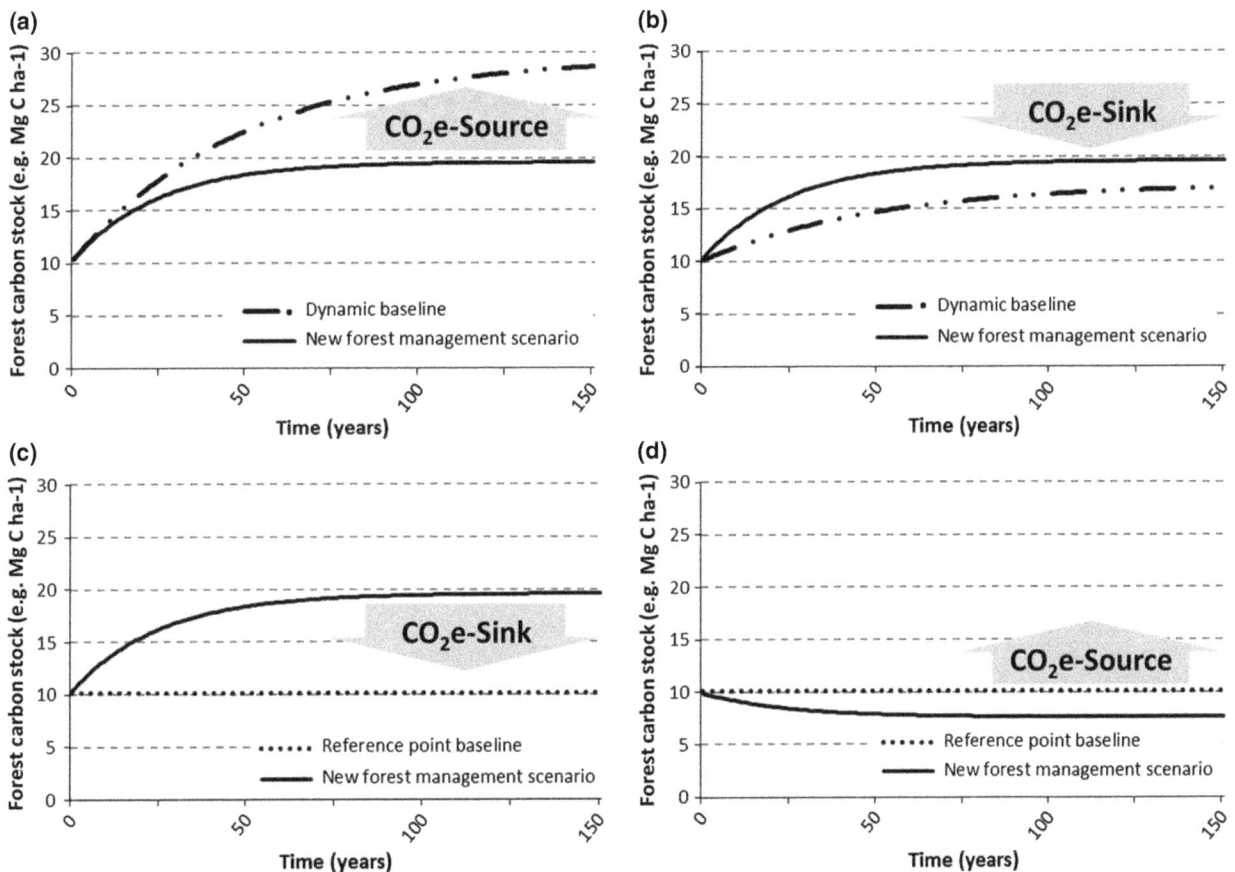

Fig. 1 With a dynamic or anticipated future baseline, future emissions are compared to a modeled baseline that assumes a given trend in forest carbon pools in the absence of the bioenergy activity (a, b). A reference point baseline is defined by the forest carbon stock in a given area at a given point in time. With a reference point baseline, future emissions are compared to this static point in time (c, d). The carbon balance of a particular bioenergy can change as a function of baseline type.

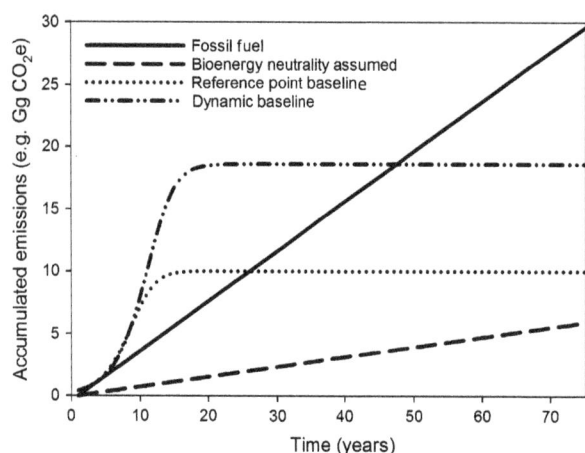

Fig. 2 Baseline choices influence carbon payback when comparing bioenergy alternatives with fossil-fuel emissions. In this hypothetical case, the reference point baseline assumes a scenario where forest carbon stocks briefly decrease followed by a recovery compared to a reference point in time. The dynamic baseline assumes a project scenario where forest stocks decrease compared to business as usual and require a longer time span to recover.

Miner *et al.*, 2014), which all limit the ability to identify system-wide influential factors. Our objective was to identify the attributes of bioenergy GHG emissions analysis that exert the strongest influence over the length of the carbon payback period using an exhaustive review of the literature on this topic paired with a quantitative analytical approach.

Materials and methods

We conducted a literature review using Scopus, searching for the keywords *carbon accounting, forest biomass, greenhouse gas emissions,* and *bioenergy* in studies published between 1991 and 2014. We identified 59 peer-reviewed studies that investigated the carbon neutrality of forest-based bioenergy systems on a temporal scale, as well as seven influential studies in the gray literature (see supporting information). When a particular study included multiple scenarios such as a range of forest eco-systems, benchmark fossil energy sources (e.g. coal, mix, natural gas, oil), or energy conversion efficiency (e.g. electricity, liquid transportation fuel, combined heat and power, heat), we divided the study into separate cases. If the overall results of a single, multi-case study were not directly attributable to specific cases, we associated each case with the overall result. The 59 studies utilized in this analysis included a total of 149 cases.

We identified twenty attributes to classify the publications (Table 1). The baseline assumption referred to authors' choice of assuming carbon neutrality or applying a dynamic or reference point baseline for the forest ecosystem carbon stocks (see Fig. 1). Author clusters described a set of authors that published frequently together or were located at the same institution and using a common set of assumptions or models. Wildfire refers to the inclusion or exclusion of wildfire dynamics in the study's methodology. The stochastic nature of wildfire dynamics (e.g. frequency, size, etc.) can alter source-sink dynamics, adding additional uncertainty to ecosystem model results. The GHG impact of biomass removal from forests to reduce wildfire severity or risk is currently not settled in the scientific community and might rely largely on model assumptions, site conditions, and analytical system boundaries (Camp-

Table 1 Attributes included in the classification and regression tree analysis used to identify the most influential factors for carbon payback period

Attribute	Definition
Author clusters	Authors are from same institution or publish together
Publication year	
Carbon payback period	Study result in upper and lower bounds of carbon payback period in years
Geographic Region	Africa, Australia, Canada, Europe, South America, US, Global
Climatic zone	Tropical dry, temperate, cold; based on Köppen classification
Geographic Scale	County, forest, state, national, regional, global
Spatial unit	Stand, forest, landscape
Temporal scale	Total years considered in analysis
Data source	Hypothetical, regional, field data
Baseline assumption	Reference point, dynamic or neutrality assumed for forest ecosystem carbon stock (see Fig. 1)
Forest type	Natural forest, plantation, or both
Biomass source	Additional harvests or current logging residue only
Wildfire	Inclusion of wildfire dynamics
LCA pools	Number of LCA carbon pools included
LCA boundaries	Comparable system boundaries for fossil-fuel and bioenergy systems or imbalanced (e.g. more detailed bioenergy analysis)
Energy types compared	Electricity, transportation fuel, heating fuel, combined heat and power
Fossil fuel replaced	Coal, energy mix, natural gas, oil product
Wood products	Inclusion of wood product LCA (upstream emissions associated with processing and disposal)
Product substitution	Substitution of wood products for alternative fossil-fuel emission intensive products
Leakage	Accounted for leakage with project implementation

bell *et al.*, 2011; Hurteau *et al.*, 2012). While some authors argue that the carbon stock reduction associated with biomass removal is compensated for by reduced fire severity and risk (Hurteau *et al.*, 2008, 2014a), other studies suggest the opposite (Mitchell *et al.*, 2012). Despite a considerable wildfire risk in large areas of the world's forests, the inclusion of wildfire dynamics when calculating carbon payback period is not commonplace in fire-prone regions (e.g. Jonker *et al.*, 2014). We also screened each case to determine which forest and nonforest carbon pools were considered within each analysis (attribute 'LCA pools'). Seven carbon pools were restricted to the forest ecosystem (Above ground live biomass, Aboveground standing dead biomass, Belowground live biomass, Belowground dead biomass, Forest floor, Merchantable timber, Harvest residue), four carbon pools described the processing of material (Forest treatment operations, Recovery of biomass in the forest, Transport, Mill residue), while two carbon pools described product fate (Wood products in use, Wood products in landfill), and two described indirect effects (Leakage, Product substitution). The studies evaluated were characterized by a very inconsistent inclusion of carbon pools, ranging from the inclusion of 1–16 carbon pools, with an average of nine pools. Leakage was considered in only eight cases, and product substitution was only considered in 21 cases of the 149 total cases.

We analyzed the cases using classification and regression tree analysis. Classification and regression tree (CART) analysis is a nonparametric test where algorithms for constructing decision trees usually work top-down, by choosing a numeric or categorical variable at each step that best splits the set of items (De'Ath & Fabricius, 2000), making it a useful tool for meta-analyses (Dusseldorp *et al.*, 2013). The goal is to create a model that predicts the value of a target variable based on several input variables. Variables used for the first splits are considered the most predictive ones, explaining the highest amount of variance in the dependent variable. Using the JMP PRO 10.0.0 software (SAS, Cary, NC, USA), we validated the model with a randomized binary variable to assess the optimum number of splits based on R^2 values. As not all of the 59 studies analyzed used carbon payback period as a carbon emissions metric, only those 38 studies that incorporated a calculation of a carbon payback period and covering 123 cases were included in the CART analysis.

Results

The CART model validation resulted in a minimum number of eight splits ($R^2 = 0.87$). We validated the stability of the CART model by including and excluding studies represented by disproportionally high cases or long payback periods. We concluded that the outliers did not change the number of splits required nor the attribute ranking based on their predictive power. The hierarchical ranking of these attributes based on their effect on carbon payback period for forest biomass projects indicated that the single largest determinant of carbon payback period length was the inclusion of wildfire dynamics (Fig. 3). Studies that included wildfire

dynamics had a mean carbon payback period of 856 years (SD = 1299), while those that did not had a mean carbon payback period of 51 years (SD = 75). This initial level of classification had a significant influence on the importance of subsequent factors, such that there was no overlap between influential attributes following this highest level classification (Fig. 3). Studies having the shortest carbon payback period (μ = 5 years, SD = 15) did not account for wildfire dynamics or leakage, were from an author group other than authors who were at some point associated with the Joanneum Research Forschungsgesellschaft mbH (Graz, Austria), included a fossil energy source other than natural gas, and did not use a dynamic baseline. Studies having the longest carbon payback period (μ = 2945 years, SD = 1082) were a subgroup of studies that included wildfire dynamics but also considered a wood products LCA, utilized electricity generation as the dominant technology, and were conducted in natural forests (Fig. 3). The total range of payback periods covered a span from 0 to 4500 years, with the largest ranges occurring in studies that included wildfire dynamics or wood products LCA (Fig. 4). The three attributes commonly identified as important for evaluating the carbon neutrality of biomass projects (baseline, leakage, and product substitution) were less influential overall. In studies where wildfire dynamics were considered, leakage and baseline were not influential in the first four levels of classification (Fig. 3). In studies where wildfire dynamics were not considered, leakage was the second level and baseline the fifth level of classification (Fig. 3). In these cases, including leakage increased the carbon payback period such that the interquartile range exceeded that of cases that did not include leakage (Fig. 4b) and the type of baseline had little influence over the carbon payback period.

Discussion

CART analysis

Project baseline and leakage are two attributes consistently used in the quantification of forest carbon projects and in quantifying the atmospheric greenhouse gas effects of a forest bioenergy project (Guest *et al.*, 2013). While these attributes are important for forest carbon offset projects (Hurteau *et al.*, 2012), our results suggest they are less informative for evaluating the carbon benefits of forest biomass projects. Interestingly, the choice of baseline type (dynamic or reference point) was only influential in 33% of cases, and only after studies had been segregated based on four other attributes (Fig. 3).

The inclusion of wildfire dynamics was the attribute with the greatest influence over carbon payback period

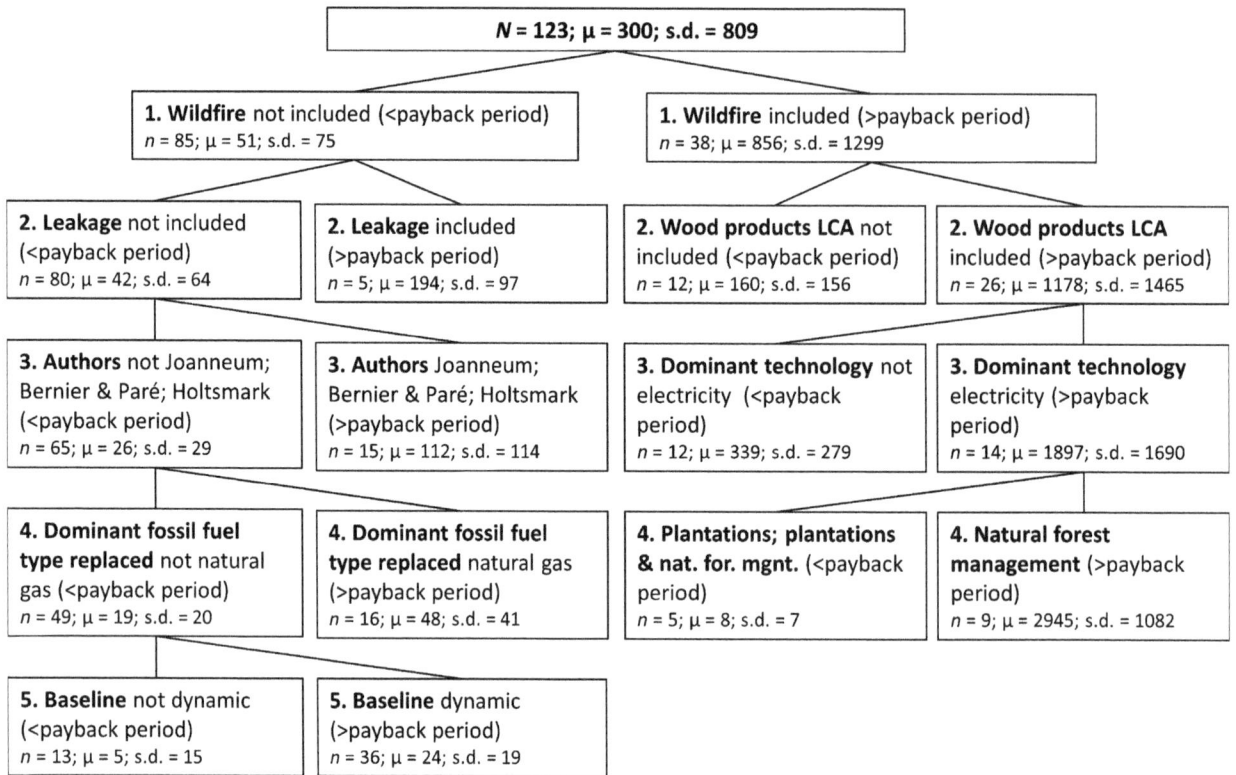

Fig. 3 Classification and regression tree (CART) analysis on the influence of different variables on carbon payback period in years for forest bioenergy. CART ranks independent variables based on predictive power with the variable that explains the highest amount of variance in the dependent variable on top. A total of eight splits resulted in a R^2 of 0.87, additional splits did not produce meaningful increases in R^2.

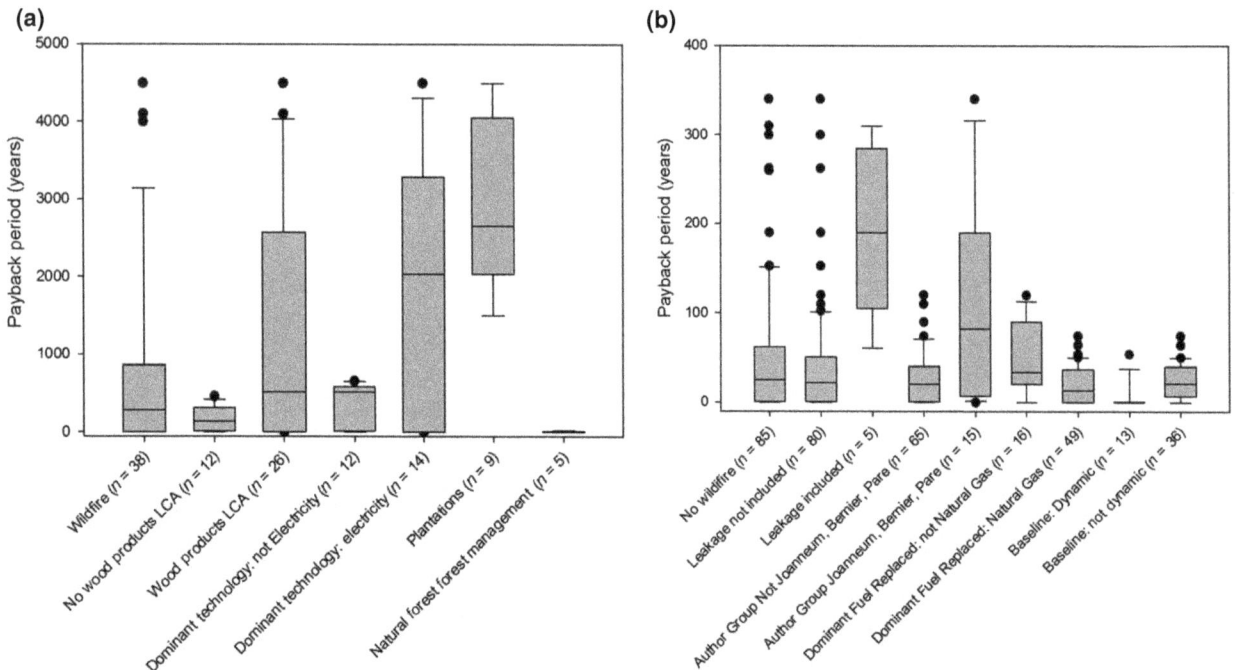

Fig. 4 Carbon payback periods based on variables with high predictive power as indicated by classification and regression tree analysis (Fig. 3). Figure a and b exhibit carbon payback periods for variables including and excluding wildfire dynamics, respectively.

length, suggesting that the role of natural disturbance within a system exerts strong control (Fig. 3). Unlike other natural disturbances (e.g. hurricanes, ice storms), wildfire risk can be managed. In biomass projects where fuel-reduction treatments are considered, the original driver for the treatments needs to be clearly defined because it directly influences the appropriate baseline condition. For example, if biomass is only a by-product of thinning that is already occurring, GHG emissions and derived carbon payback periods for bioenergy scenarios differ compared to scenarios where the presence of a biomass market triggers a decision to implement a fuel-reduction thinning (Walker *et al.*, 2013). The few studies in fire-prone regions where open burning of fuel-reduction treatment residues is common practice conclude that using the material for bioenergy results in immediate carbon benefits (Jones *et al.*, 2010; Springsteen *et al.*, 2011). Where a market facilitates the decision to thin, the carbon payback period is influenced by a suite of factors. The carbon costs associated with treatments (e.g. thinning and prescribed burning) have the potential to reduce mortality and emissions from subsequent wildfire when compared with the untreated forest condition (Hurteau *et al.*, 2008; North & Hurteau, 2011). However, the potential benefits (in terms of short payback periods) to be gained from reduced wildfire emissions following treatment are dependent on the probability of occurrence, size, and severity of wildfire as well as the growth response of trees retained during treatment (Campbell *et al.*, 2011; Hurteau *et al.*, 2014a). Given the influence of projected changes in climate on forest growth (Silva & Anand, 2013) and disturbance frequency and effect (Westerling *et al.*, 2011; Moritz *et al.*, 2012; Hurteau *et al.*, 2014b), disturbance dynamics are likely to become even more influential in evaluating biomass energy projects over meaningful temporal scales. Therefore, while simulating stochastic disturbance adds additional challenges to modeling efforts, in disturbance-prone areas, it is an integral component of both baseline and project scenario conditions.

Other influential attributes in determining carbon payback period can be broadly classified into decision criteria and regional market influences. Decision criteria attributes, including leakage and wood products LCA, require clearly defining the study boundary and present an opportunity for standardization of evaluation criteria. When the effects of wood harvest displacement to meet market demands are absent, the influence of market forces is left unaccounted and the actual effects of a project on the global carbon cycle are neglected. Likewise, accounting for the use and disposal of wood products can strongly influence conclusions about the carbon benefits of forest management (Lippke *et al.*, 2011). Creating a framework in which there is consensus on the

specific boundaries for evaluation or inclusion of a range of boundaries will facilitate comparison across studies.

Attributes related to geographic location and local markets (e.g. dominant technology, fossil-fuel source) exert influence over the carbon payback period and pose a challenge for equitable comparison of forest biomass energy across large spatial scales. The dominant technology and its influence on carbon payback period are functions of conversion efficiency and are highly sensitive to the fossil-fuel source (McKechnie *et al.*, 2010). In our evaluation of dominant technology, electricity production vs. other technologies such as combined heat and power or heat only was the defining factor. This result was not surprising given the slightly low conversion efficiencies associated with producing electricity only from woody biomass combustion over fossil-fuel consuming systems (Schlamadinger & Marland, 1996). When other technologies are employed, such as combined heat and power, the overall conversion efficiency of woody biomass combustion systems increases (Richter *et al.*, 2009) and approaches that of fossil-fuel consuming combined heat and power systems; therefore, the carbon payback period is reduced. While decisions regarding dominant technology are in part influenced by location, the replacement fuel comparison is entirely a function of geographic location. Power sources and the emissions per unit of power generated vary by region (Chen, 2009). If the regional power mix is comprised primarily of natural gas, woody biomass energy will have a considerably longer carbon payback period ($\mu = 82$ years, SD = 83, $n = 21$). However, if the regional energy mix is primarily from coal combustion, the carbon payback period is reduced ($\mu = 36$ years, SD = 48, $n = 21$).

Author group was an influential attribute for classifying carbon payback period. The partitioning based on author groups is most likely attributable to the repeated application of modeling frameworks and software used within a confined circle of researchers. Models are a representation on how authors understand the system to be analyzed. Providing a host of results using various models is a common characteristic of complex systems where scientific consensus has not been reached. An example is the inclusion of 41 different climate models in the fifth assessment report of the Intergovernmental Panel for Climate Change (Flato *et al.*, 2013). This result validates how models are consistent within their applications but also how they can create 'half-predictable' outcomes based on their assumptions. This finding further reinforces the need to establish a common set of criteria for evaluation. In particular, specifying model components such as stochastic disturbances in general and wildfire in particular is a case in point.

Additional insights

Calculating a carbon payback period as a metric to describe the GHG impact of alternative scenarios is becoming standard practice outnumbering other metrics frequently employed such as tons of carbon displaced per energy unit of biomass fuel (e.g. (Hall *et al.*, 1991; Schmidt *et al.*, 2011), carbon emissions for various scenarios over a given timescale (e.g. (Domke *et al.*, 2008), or a carbon neutrality factor that measures GHG emissions in percent of a baseline scenario over a given period of time (e.g. Schlamadinger *et al.*, 1995; US Forest Service, 2009; Zhang *et al.*, 2010; Kilpeläinen *et al.*, 2012; Winford & Gaither, 2012). Carbon payback period was the principal metric in 26 of the 59 studies while nine studies used GHG savings in % over a fossil-fuel scenario over a given time. Other metrics such as CO_2 savings per ha (e.g. Dwivedi *et al.*, 2014) or CO_2 savings per MWh (e.g. Kilpeläinen *et al.*, 2011) were infrequent. A conversation on the advantages and disadvantages of one metric over the others is largely absent.

For the majority of studies, we observed a high trust in models that was exhibited by the willingness of authors to report in 100+ year timespans as well as a frequent absence of uncertainty metrics when reporting results. We also observed no consistent pattern in the use of temporal scales for modeling. The temporal scale of analysis for all studies analyzed ranged from 20 (e.g. (Hudiburg *et al.*, 2011) to 10 000 years (Mitchell *et al.*, 2012) with a median of 240 years. The lowest temporal scale was applied by (Hudiburg *et al.*, 2011) to avoid the risk of 'overstretching data', that is owing to data uncertainty. No neutrality was achieved over these 20 years in this study. All other authors seemed to have enough confidence in their assumptions, datasets and models to investigate carbon fluxes over longer time scales although only a few cases included episodic carbon pulses that occur on large temporal and spatial scales such as wildfire (included in 8% or 14% of all studies), insect outbreaks or storm events. Most studies used hypothetical data (35% or 59% of all studies), only seven studies (12%) used field data. Among those studies that modeled neutrality over time on temporal scales surpassing 100 years, the share of studies using hypothetical data was even higher (67% or 30 of 45 studies). Uncertainties affecting other system elements such as baselines (Buchholz *et al.*, 2014a), product substitution (York, 2012; Bird, 2013), soil carbon (Buchholz *et al.*, 2014b), or market effects (Sedjo, 2013) were frequently underreported or excluded.

Setting assessment boundaries provide a major challenge when comparing bioenergy GHG emission studies and can result in incomplete accounting. For instance, we confirmed the observation of (Muench & Guenther, 2013) that most studies did not account for all upstream fossil-fuel emissions such as building machinery and facilities. Notably, a broader set of metrics to assess GHG implications using bioenergy systems was largely absent. The inclusion of non-CO_2 GHG relevant emissions (other reactive gases, biogenic aerosols, and factors such as methane or atmospheric particles), surface albedo only considered by (Guest *et al.* (2013), evapotranspiration or discounting approaches to account for the release of GHG emissions along a temporal scale (e.g. Cherubini *et al.*, 2011; Pingoud *et al.*, 2012) was not common practice. Nevertheless, our CART analysis suggests that a focus on top-priority system attributes such as wildfire dynamics, leakage, or wood products LCA can substitute for a more complete assessment that includes a maximum set of (ultimately less influential) attributes. This insight is supported by an observation of Holtsmark (2012), finding that complex global warming potential decay functions 'did not change the results fundamentally' compared to a model that used a simple accumulation model of CO_2 in the atmosphere.

While the CART analysis suggests some influence of plantation vs. natural forest management practices on carbon payback periods, this was only true for a small subset of cases. The full sample revealed no apparent differences in carbon payback periods between the two management types (Fig. 5). Reducing rotation lengths to increase profitability can be a major advantage of plantation over natural forest regimes (Cubbage *et al.*, 2010). Our results do not show that a switch from natural forest management regimes to plantation forestry provides

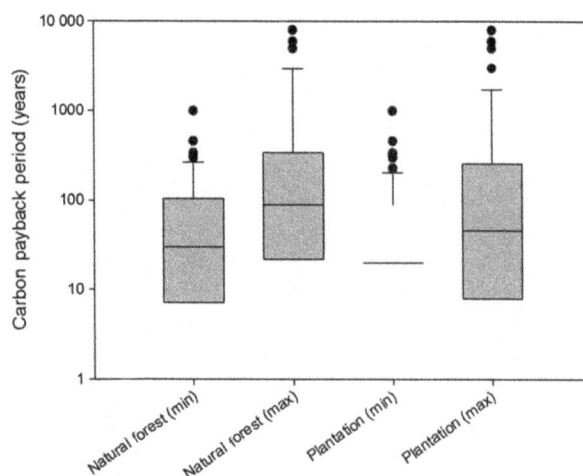

Fig. 5 Range of minimum and maximum carbon payback periods for natural and plantation forests.

a strong argument to reduce carbon payback period. Similarly, (Pyörälä *et al.*, 2012) also concluded for boreal forests that shorter rotations do not always automatically produce more favorable emission balances on behalf of bioenergy. This result challenges the generalization by (Lamers & Junginger, 2013) that shorter rotations result in shorter carbon payback periods.

Recommendations

In summary, for carbon payback period calculations to provide operational insights to decision makers, future research should focus on creating consistent accounting principles including the consideration of stochastic disturbance, temporal scales, quantifying and reporting uncertainties, standardization of carbon pools evaluated, GHG emission metrics considered and baseline definition.

Acknowledgements

Partial support for this research came through Placer County Air Pollution Control District and a Joint Venture Agreement between the Manomet Center for Conservation Sciences and the USDA Forest Service Northern Research Station. We are grateful for conversations with Anna Mika on methodological implications using CART and the thoughtful feedback from three anonymous reviewers.

References

Beurskens LWM, Hekkenberg M (2011) *Renewable Energy Projections as Published in the National Renewable Energy Action Plans of the European Member States Covering all 27 EU Member States*. ECN-E-10-069, European Environment Agency, Openhagen, Denmark, 244p.

Bird DN (2013) *Estimating the Displacement of Energy and Materials by Woody Biomass in Austria*. Joanneum Research, Graz, Austria.

Buchholz T, Prisley S, Marland G, Canham C, Sampson N (2014a) Uncertainty in projecting greenhouse gas emissions from bioenergy. *Nature Climate Change*, **4**, 1045–1047.

Buchholz T, Friedland AJ, Hornig CE, Keeton WS, Zanchi G, Nunery J (2014b) Mineral soil carbon fluxes in forests and implications for carbon balance assessments. *GCB Bioenergy*, **6**, 305–311.

Campbell JL, Harmon ME, Mitchell SR (2011) Can fuel-reduction treatments really increase forest carbon storage in the western US by reducing future fire emissions? *Frontiers in Ecology and the Environment*, **10**, 83–90.

Chen Y (2009) Does a regional greenhouse gas policy make sense? A case study of carbon leakage and emissions spillover. *Energy Economics*, **31**, 667–675.

Cherubini F, Peters GP, Berntsen T, StrøMman AH, Hertwich E (2011) CO2 emissions from biomass combustion for bioenergy: atmospheric decay and contribution to global warming: global warming potential of CO2 from bioenergy. *GCB Bioenergy*, **3**, 413–426.

Cubbage F, Koesbandana S, Mac Donagh P *et al.* (2010) Global timber investments, wood costs, regulation, and risk. *Biomass and Bioenergy*, **34**, 1667–1678.

De'Ath G, Fabricius KE (2000) Classification and regression trees: a powerful yet simple technique for ecological data analysis. *Ecology*, **81**, 3178–3192.

Domke GM, Ek AR, Becker DR *et al.* (2008) *Assessment of Carbon Flows Associated with Forest Management and Biomass Procurement for the Laskin Biomass Facility*. Staff Paper Series No. 198, Department of Forest Resources, College of Food, Agricultural and Natural Resource Sciences, University of Minnesota, St. Paul, MN, 39p.

Dusseldorp E, van Genugten L, van Buuren S, Verheijden MW, van Empelen P (2013) Combinations of techniques that effectively change health behavior: evidence from meta-CART analysis. *Health Psychology*, **33**, 1530–1540.

Dwivedi P, Bailis R, Khanna M (2014) Is use of both pulpwood and logging residues instead of only logging residues for bioenergy development a viable carbon mitigation strategy? *BioEnergy Research*, **7**, 217–231.

EPA (2014) *Framework for Assessing Biogenic CO2 Emissions from Stationary Sources*. United States Environmental Protection Agency, Office of Air and Radiation, Office of Atmospheric Programs. Climate Change Division, Washington, DC, USA.

Fankhauser S, Hepburn C (2010) Designing carbon markets, Part II: carbon markets in space. *Energy Policy*, **38**, 4381–4387.

Flato G, Marotzke J, Abiodun B *et al.* (2013) Evaluation of Climate Models. In: *Climate Change 2013: The Physical Science Basis. Contribution of Working Group I to the Fifth Assessment Report of the Intergovernmental Panel on Climate Change* (eds Stocker TF, Qin D, Plattner G-K, Tignor M, Allen SK, Boschung J, Nauels A, Xia Y, Bex V, Midgley PM), pp. 741–866. Cambridge University Press, Cambridge, UK.

Gan J, McCarl BA (2007) Measuring transnational leakage of forest conservation. *Ecological Economics*, **64**, 423–432.

Guest G, Bright RM, Cherubini F, Strømman AH (2013) Consistent quantification of climate impacts due to biogenic carbon storage across a range of bio-product systems. *Environmental Impact Assessment Review*, **43**, 21–30.

Gunn JS, Ganz DJ, Keeton WS (2012) Biogenic vs. geologic carbon emissions and forest biomass energy production. *GCB Bioenergy*, **4**, 239–242.

Hall DO, Mynick HE, Williams RH (1991) *Carbon Sequestration Versus Fossil Fuel Substitution: Alternative Roles for Biomass in Coping with Greenhouse Warming*. Springer, New York, USA.

Helin T, Sokka L, Soimakallio S, Pingoud K, Pajula T (2013) Approaches for inclusion of forest carbon cycle in life cycle assessment – a review. *GCB Bioenergy*, **5**, 475–486.

Henders S, Ostwald M (2012) Forest carbon leakage quantification methods and their suitability for assessing leakage in REDD. *Forests*, **3**, 33–58.

Holtsmark B (2012) Harvesting in boreal forests and the biofuel carbon debt. *Climatic Change*, **112**, 415–428.

Holtsmark B (2013) The outcome is in the assumptions: analyzing the effects on atmospheric CO $_2$ levels of increased use of bioenergy from forest biomass. *GCB Bioenergy*, **5**, 467–473.

Hudiburg TW, Law BE, Wirth C, Luyssaert S (2011) Regional carbon dioxide implications of forest bioenergy production. *Nature Climate Change*, **1**, 419–423.

Hurteau MD, Koch GW, Hungate BA (2008) Carbon protection and fire risk reduction: toward a full accounting of forest carbon offsets. *Frontiers in Ecology and the Environment*, **6**, 493–498.

Hurteau MD, Hungate BA, Koch GW, North MP, Smith GR (2012) Aligning ecology and markets in the forest carbon cycle. *Frontiers in Ecology and the Environment*, **11**, 37–42.

Hurteau MD, Robards TA, Stevens D, Saah D, North M, Koch GW (2014a) Modeling climate and fuel reduction impacts on mixed-conifer forest carbon stocks in the Sierra Nevada, California. *Forest Ecology and Management*, **315**, 30–42.

Hurteau MD, Westerling AL, Wiedinmyer C, Bryant BP (2014b) Projected effects of climate and development on California wildfire emissions through 2100. *Environmental Science and Technology*, **48**, 2298–2304.

Jones G, Loeffler D, Calkin D, Chung W (2010) Forest treatment residues for thermal energy compared with disposal by onsite burning: emissions and energy return. *Biomass and Bioenergy*, **34**, 737–746.

Jonker JGG, Junginger M, Faaij A (2014) Carbon payback period and carbon offset parity point of wood pellet production in the South-eastern United States. *GCB Bioenergy*, **6**, 371–389.

Kilpeläinen A, Alam A, Strandman H, KellomäKi S (2011) Life cycle assessment tool for estimating net CO2 exchange of forest production: life cycle assessment tool. *GCB Bioenergy*, **3**, 461–471.

Kilpeläinen A, Kellomäki S, Strandman H (2012) Net atmospheric impacts of forest bioenergy production and utilization in Finnish boreal conditions. *GCB Bioenergy*, **4**, 811–817.

Lamers P, Junginger M (2013) The "debt"is in the detail: a synthesis of recent temporal forest carbon analyses on woody biomass for energy. *Biofuels, Bioproducts and Biorefining*, **7**, 373–385.

Lippke B, Oneil E, Harrison R, Skog K, Gustavsson L, Sathre R (2011) Life cycle impacts of forest management and wood utilization on carbon mitigation: knowns and unknowns. *Carbon Management*, **2**, 303–333.

Mann M (2011) *Biomass Power Feedstock and Greenhouse Gas Emissions. Presentation to the Energy Foundation Biopower Meeting, Feb 22–23, 2011 Minneapolis, MN*. National Renewable Energy Laboratory, Washington, DC.

McKechnie J, Colombo S, Chen J, Mabee W, MacLean HL (2010) Forest bioenergy or forest carbon? Assessing trade-offs in greenhouse gas mitigation with wood-based fuels. *Environmental Science and Technology*, **45**, 789–795.

Mika AM, Keeton WS (2014) Net carbon fluxes at stand and landscape scales from wood bioenergy harvests in the US Northeast. *GCB Bioenergy*, in press, see DOI: 10.1111/gcbb.12143.

Miner RA, Abt RC, Bowyer JL *et al.* (2014) Forest carbon accounting considerations in US bioenergy policy. *Journal of Forestry*, **112**, 591–606.

Mitchell SR, Harmon ME, O'Connell KEB (2012) Carbon debt and carbon sequestration parity in forest bioenergy production. *GCB Bioenergy*, **4**, 818–827.

Moritz MA, Parisien M-A, Batllori E, Krawchuk MA, Van Dorn J, Ganz DJ, Hayhoe K (2012) Climate change and disruptions to global fire activity. *Ecosphere*, **3**, art49.

Muench S, Guenther E (2013) A systematic review of bioenergy life cycle assessments. *Applied Energy*, **112**, 257–273.

North MP, Hurteau MD (2011) High-severity wildfire effects on carbon stocks and emissions in fuels treated and untreated forest. *Forest Ecology and Management*, **261**, 1115–1120.

Pingoud K, Ekholm T, Savolainen I (2012) Global warming potential factors and warming payback time as climate indicators of forest biomass use. *Mitigation and Adaptation Strategies for Global Change*, **17**, 369–386.

Pyörälä P, Kellomäki S, Peltola H (2012) Effects of management on biomass production in Norway spruce stands and carbon balance of bioenergy use. *Forest Ecology and Management*, **275**, 87–97.

Richter D deB, Jenkins DH, Karakash JT, Knight J, McCreery LR, Nemestothy KP (2009) Wood energy in America. *Science*, **323**, 1432–1433.

Schlamadinger B, Marland G (1996) The role of forest and bioenergy strategies in the global carbon cycle. *Biomass and Bioenergy*, **10**, 275–300.

Schlamadinger B, Spitzer J, Kohlmaier GH, Lüdeke M (1995) Carbon balance of bioenergy from logging residues. *Biomass and Bioenergy*, **8**, 221–234.

Schmidt J, Leduc S, Dotzauer E, Schmid E (2011) Cost-effective policy instruments for greenhouse gas emission reduction and fossil fuel substitution through bioenergy production in Austria. *Energy Policy*, **39**, 3261–3280.

Searchinger TD, Hamburg SP, Melillo J *et al.* (2009) Fixing a critical climate accounting error. *Science*, **326**, 527–528.

Sedjo R (2013) *Comparative Life Cycle Assessments: Carbon Neutrality and Wood Biomass Energy*. Resources for the Future, Washington, DC, USA, 21p.

Silva LCR, Anand M (2013) Probing for the influence of atmospheric CO2 and climate change on forest ecosystems across biomes. *Global Ecology and Biogeography*, **22**, 83–92.

Springsteen B, Christofk T, Eubanks S, Mason T, Clavin C, Storey B (2011) Emission reductions from woody biomass waste for energy as an alternative to open burning. *Journal of the Air & Waste Management Association*, **61**, 63–68.

United Kingdom D of T (2012) *UK Bioenergy Strategy*. Department of Transport, London, UK, 86p.

US Forest Service (2009) *Biomass to Energy: Forest Management For Wildfire Reduction, Energy Production, and Other Benefits*. Pacific Southwest Research Station, Albany, CA, USA.

Vanhala P, Repo A, Liski J (2013) Forest bioenergy at the cost of carbon sequestration? *Current Opinion in Environmental Sustainability*, **5**, 41–46.

Walker T, Cardellichio P, Gunn JS, Saah DS, Hagan JM (2013) Carbon accounting for woody biomass from Massachusetts (USA) managed forests: a framework for determining the temporal impacts of wood biomass energy on atmospheric greenhouse gas levels. *Journal of Sustainable Forestry*, **32**, 130–158.

Westerling AL, Turner MG, Smithwick EAH, Romme WH, Ryan MG (2011) Continued warming could transform greater Yellowstone fire regimes by mid-21st century. *Proceedings of the National Academy of Sciences*, **108**, 13165–13170.

Winford EM, Gaither JC (2012) Carbon outcomes from fuels treatment and bioenergy production in a Sierra Nevada forest. *Forest Ecology and Management*, **282**, 1–9.

York R (2012) Do alternative energy sources displace fossil fuels? *Nature Climate Change*, **2**, 441–443.

Zanchi G, Pena N, Bird N (2012) Is woody bioenergy carbon neutral? A comparative assessment of emissions from consumption of woody bioenergy and fossil fuel. *GCB Bioenergy*, **4**, 761–772.

Zhang Y, McKechnie J, Cormier D, Lyng R, Mabee W, Ogino A, MacLean HL (2010) Life cycle emissions and cost of producing electricity from coal, natural gas, and wood pellets in Ontario, Canada. *Environmental Science & Technology*, **44**, 538–544.

18

Influence of corn, switchgrass, and prairie cropping systems on soil microbial communities in the upper Midwest of the United States

EDERSON DA C. JESUS[1,2], CHAO LIANG[3,4], JOHN F. QUENSEN[1], ENDANG SUSILAWATI[1,5], RANDALL D. JACKSON[3], TERESA C. BALSER[3] and JAMES M. TIEDJE[1]

[1]Center for Microbial Ecology and DOE Great Lakes Bioenergy Research Center, Michigan State University, 540 Plant and Soil Sciences Building, East Lansing, MI 48824-1325, USA, [2]Embrapa Agrobiologia, BR 465, km 7, Seropédica, Rio de Janeiro 23890-000, Brazil, [3]Department of Agronomy and DOE Great Lakes Bioenergy Research Center, University of Wisconsin-Madison, 1575 Linden Drive, Madison, WI 53706, USA, [4]State Key Laboratory of Forest and Soil Ecology, Institute of Applied Ecology, Chinese Academy of Sciences, Shenyang 110164, China, [5]Department of Chemical Engineering and Applied Chemistry, University of Toronto, Toronto, ON M5S 3E5, Canada

Abstract

Because soil microbes drive many of the processes underpinning ecosystem services provided by soils, understanding how cropping systems affect soil microbial communities is important for productive and sustainable management. We characterized and compared soil microbial communities under restored prairie and three potential cellulosic biomass crops (corn, switchgrass, and mixed prairie grasses) in two spatial experimental designs – side-by-side plots where plant communities were in their second year since establishment (i.e., *intensive* sites) and regionally distributed fields where plant communities had been in place for at least 10 years (i.e., *extensive* sites). We assessed microbial community structure and composition using lipid analysis, pyrosequencing of rRNA genes (targeting fungi, bacteria, archaea, and lower eukaryotes), and targeted metagenomics of *nifH* genes. For the more recently established intensive sites, soil type was more important than plant community in determining microbial community structure, while plant community was the more important driver of soil microbial communities for the older extensive sites where microbial communities under corn were clearly differentiated from those under switchgrass and restored prairie. Bacterial and fungal biomasses, especially biomass of arbuscular mycorrhizal fungi, were higher under perennial grasses and restored prairie, suggesting a more active carbon pool and greater microbial processing potential, which should be beneficial for plant acquisition and ecosystem retention of carbon, water, and nutrients.

Keywords: bacterial communities, biofuel crops, fungal communities, lipid analysis, *nifH*, pyrosequencing

Introduction

To help reduce dependence on fossil fuels, there is great interest in using plant biomass for energy. Significant efforts are underway to understand what biomass crops should be grown, where they should be grown, and how they can be managed in sustainable ways (Kim *et al.*, 2012; Gao *et al.*, 2013; Werling *et al.*, 2014). Annual crops such as corn continue to be the most readily abundant and available crops for biofuel production in the United States (US-DOE, 2011), while canola and soybeans have been used to produce biodiesel (IEA, 2007). However, these crops are monocultures that require

Correspondence: James M. Tiedje
e-mail: tiedjej@msu.edu

high-energy inputs to maintain, lose soil and nutrients, and serve as key food crops for humans and/or livestock. Their use for biofuel may increase food costs and may encourage more land to be converted to agriculture, which has negative ramifications for ecosystem carbon balance, wildlife habitat, and a host of other ecosystem services (Fargione *et al.*, 2008).

The disadvantages listed above have encouraged a focus on perennial grasses such as *Panicum virgatum* (switchgrass) for lignocellulosic ethanol production (IEA, 2007). Switchgrass produces high amounts of biomass and is a native species of North America that should require less intensive agricultural management than annual crops (Wright & Turhollow, 2010). Biologically diverse, low-input systems consisting of mixtures of native grasses are also candidates for cellulosic biomass crops because they can be highly productive, while conserving nutrients (Jach-Smith & Jackson, 2015), can positively affect the diversity of other groups of organisms (Werling *et al.*, 2014), and result in lower greenhouse gas emissions than annuals and perennial monocultures (Tilman *et al.*, 2006; Oates *et al.*, 2015). Theoretically, more diverse plant communities will improve sustainability by requiring fewer inputs than monocultures, although support for this hypothesis in agricultural production settings is scant (but see Tilman *et al.*, 2006; Webster *et al.*, 2010). In more productive soils, it may be more important to have a particular plant community, for example, one that includes the most productive taxa, to minimize inputs, while improving stability and resilience (Adler *et al.*, 2009).

All agricultural activity affects biodiversity, soil fertility, and water resources (Groom *et al.*, 2008), and these factors should be assessed when evaluating the sustainability of biofuel cropping systems. Considering the impacts on microbial soil communities is an important component of this assessment because most soil-based ecosystems services such as organic matter degradation, nitrogen fixation, nitrification, denitrification, soil aggregation, and water retention are driven by microbial activity (Swift *et al.*, 2004). As has been shown for other crops, the cultivation of biofuel crops can be expected to influence soil microbial communities, thus affecting the key ecosystem processes and the services they provide (Groom *et al.*, 2008; Liang *et al.*, 2012). Corn is presently the major crop used for ethanol production in the United States, and soil microbial communities under corn and prairie have often been contrasted, showing differences in microbial community composition and improved carbon storage and soil aggregation under prairie (Bailey *et al.*, 2002; Allison *et al.*, 2005; Bach *et al.*, 2010; Fierer *et al.*, 2013; Murphy & Foster, 2014). The growing interest in using switchgrass and other grasses for cellulosic biomass production has sparked similar investigations contrasting soil microbial communities under these grasses with those under corn (Jesus *et al.*, 2010; Mao *et al.*, 2011, 2013; Liang *et al.*, 2012). The ways and extent to which these crops have been found to influence soil microbial communities varied according to the methods used, spatial sampling schemes, soil and environmental variables, land management and land history, but a general finding has been that cultivation of perennial grasses stimulates communities more similar to those under prairies (Liang *et al.*, 2012). This is a desirable outcome, because systems with soil microbial communities similar to those under prairies should require fewer external inputs and, for this reason, be more sustainable. But for the most part, these studies have been performed in local settings and the need to carry out studies at larger geographic scales, including

sites with different management types, different times since crop establishment, and for a range of soil conditions is necessary to better examine shifts in microbial communities.

For this reason, we used a more holistic approach for our study. Our initial hypothesis was that the cultivation of switchgrass and mixed grasses would lead to microbial communities more similar to those under prairie species, implying a more sustainable system. To test this hypothesis, we sampled over a larger geographic scale, including sites in two states with a range of soil conditions, with different times since crop conversion, and under two different sampling strategies. We compared microbial communities in soils cultivated with three potential biofuel crops (corn, switchgrass, and mixed grasses) and with prairie species, and in two spatial experimental designs. One design consisted of side-by-side plots where plant communities were in their second year since establishment (i.e., *intensive* sites), and the other consisted of regionally distributed fields where plant communities had been in place for at least 10 years (i.e., *extensive* sites). We assessed the microbial communities using three different methods: lipid analysis, pyrosequencing of ribosomal genes (that target fungi, bacteria, archaea, and lower eukaryotes), and targeted metagenomics of a gene important for a key ecological function, *nifH* coding for nitrogen reductase (N_2 fixation). Our main questions were as follows: (i) how do the different biofuel crops affect soil microbial communities, that is, are soil microbial communities under switchgrass and mixed grasses more similar to those under prairie, (ii) how are any effects modified by location and soil type, and (iii) how do alternative soil microbial assay methods compare in revealing community differences?

Materials and methods

Site description and soil sampling

Soil samples were collected from sites in southern Michigan and southern Wisconsin under two different designs that have been used for other studies by the Great Lakes Bioenergy Research Center (GLBRC) (Fig. S1). These two designs are *intensive* and *extensive*.

The intensive plots were located at the Kellogg Biological Station (KBS) in Michigan and at the Arlington Agricultural Research Station (AARS) in Wisconsin. The plots were arrayed in randomized complete blocks designs consisting of five replicated 30 × 40 m plots of each of four plant communities – corn, switchgrass, mixed grasses, and restored prairie – and were harvested annually for biomass. We sampled from three of the five blocks 2 years after their establishment.

The extensive sites were fields located on working farms or reserves in Michigan and Wisconsin and were selected from among those studied by Werling *et al.* (2014) to cover the range of soil types and conditions of the southern regions of both states. Nine fields were sampled in each state – three in corn, three in switchgrass, and three in restored prairie – but only the corn fields were harvested. All sites had been under their respective vegetation for at least 10 years.

Three composite samples were taken at random from each of the sampling units. Each composite sample consisted of five soil cores taken to a depth of 10 cm. All samples were transported on ice to the laboratory and then stored at −20 °C until processing.

Soil analysis

Soil samples were analyzed for elemental composition (Al, B, Ca, Cu, Fe, K, Mg, Mn, Na, P, S, and Zn), total C, total N, pH, and soil texture as previously described (Liang *et al.*, 2012).

Lipid analysis

Microbial community composition was determined using a hybrid procedure of phospholipid fatty acid (PLFA) and fatty acid methyl ester (FAME) analysis as previously described (Liang *et al.*, 2012). The total biomass of bacteria (B), fungi (F), and protozoa was estimated. Bacteria were further subdivided into Gram-positive (Gm$^+$) and Gram-negative (Gm$^-$) categories, and the fungi, into arbuscular mycorrhizal fungi (AMF) and saprophytic fungi (SF) (Liang *et al.*, 2012).

DNA extraction

DNA was extracted from each well-mixed 500 mg soil sample using MoBio's Power Soil DNA Isolation Kit (Mobio Laboratories Inc., Carlsbad, CA, USA) according to the manufacturer's instructions. The DNA was quantified with a Nanodrop ND-1000 spectrophotometer (Nanodrop Technology, Wilmington, DE, USA) and stored at −20 °C until use.

Preparation of 16S/18S (SSU) rRNA gene amplicon libraries for pyrosequencing

The V6-V8 region of smal subunit (SSU) rRNA was amplified from the template DNA using primers 926F (5′- cct atc ccc tgt gtg cct tgg cag tct cag AAA CTY AAA KGA ATT GRC GG- 3′) and 1392R (5′ - cca tct cat ccc tgc gtg tct ccg act cag - <XXXXX> - ACG GGC GGT GTG TRC - 3′). Primer sequences were modified by the addition of 454 A or B adapter sequences (lower case). In addition, the reverse primer included a 5 bp bar code (designated by <XXXXX> above) for multiplexing of samples during sequencing. Twenty microliter PCRs were performed in duplicate and pooled to minimize PCR bias using 0.4 μl Advantage GC 2 Polymerase Mix (Advantage-2 GC PCR Kit, Clonetech, Mountain View, CA, USA), 4 μl 5× GC PCR buffer, 2 μl 5 M GC Melt Solution, 0.4 μl 10 mM dNTP mix (MBI Fermentas, Amherst, MA, USA), 1.0 μl of each 25 nM primer, and 10 ng sample DNA. The thermal cycler protocol was 95 °C for 3 min, 25 cycles of

95 °C for 30 s, 50 °C for 45 s, and 68 °C for 90 s, and a final 10-min extension at 68 °C. PCR amplicons were purified using SPRI Beads and quantified using a Qubit fluorometer (Thermo Fisher Scientific Inc., Waltham, MA, USA). Samples were diluted to 10 ng μl^{-1} and mixed in equal concentrations. Emulsion PCR and sequencing of the PCR amplicons were performed following the Roche 454 GS FLX Titanium technology manufacturer's instructions. Sequencing tags were analyzed using the software tool PYROTAGGER (Kunin & Hugenholtz, 2010) using a 180 bp sequence length threshold as described in Engelbrektson *et al.* (2010).

16S/18S rRNA gene nucleotide sequences were deposited in the European Nucleotide Archive (http://www.ebi.ac.uk/ena) as part of study PRJEB6704 under accession numbers ERR571396 through ERR571438.

Preparation of 28S (LSU) rRNA gene amplicon libraries for pyrosequencing

PCR amplification of template DNA was also performed using the primers LR3 (5′-CCGTGTTTCAAGACGGG-3′) and LR0R (5′-ACCCGCTGAACTTAAGC-3′) (Liu *et al.*, 2012). These primers target a 625 bp fragment of the large subunit (LSU) rRNA gene in fungi. Detailed amplification and purification protocols are given in Penton *et al.* (2013). Adapters and bar codes were ligated to the amplicons prior to sequencing at Utah State University using Lib-L kits.

28S rRNA gene nucleotide sequences were deposited in the European Nucleotide Archive (http://www.ebi.ac.uk/ena) as part of study PRJEB6704 under accession numbers ERR571439 through ERR571456.

Preparation of nifH amplicon libraries for pyrosequencing

The extracted DNA also served as template to prepare *nifH* gene libraries as described in Wang *et al.* (2013). The primers were based on those of Poly *et al.* (2001), which target an approximately 320 bp region of the *nifH* gene. *NifH* gene libraries were sequenced by the Research Technology Support Facility (RTSF) at Michigan State University (East Lansing).

NifH nucleotide sequences were deposited in the European Nucleotide Archive (http://www.ebi.ac.uk/ena) as part of study PRJEB6704 under accession numbers ERR571353 through ERR571395.

Data analysis

Principal component analysis (PCA) was used to display distances between sites based on their soil attributes. For these analyses, the arc sin transformation was first applied to percentages of sand and silt, and then all soil variables were standardized to zero mean and unit variance prior to PCA on the correlation matrix using the R (R Core Team, 2012) package VEGAN's (Oksanen *et al.*, 2012) *rda* function. To aid interpretation, vectors for the soil variables were added to the PCA plots using vegan's *envfit* function.

Total carbon concentrations were compared by analysis of variance (ANOVA) using the *lm* function of the R package stats and a cutoff value of $\alpha = 0.05$.

Lipid data were used to estimate microbial biomass, and the following ratios were calculated: fungi/bacteria (F/B), arbuscular mycorrhizal fungi/saprophytic fungi (AMF/SF), and Gram-positive bacteria/Gram-negative bacteria (Gm^+/Gm^-). Data were displayed per crop, and mean and errors bars were calculated for each of the experiments types. Treatment differences were tested in the same manner as total carbon. When ANOVA was significant, treatment contrasts were made with the *TukeyHSD* function of the R package stats.

All 28S LSU rRNA sequences were first processed through RDP's pyrosequencing initial processing tool (http://pyro.c-me.msu.edu/). Because the amplicons were ligated with the adapters and bar codes, both primer sequences were entered in the forward primer box. Filter parameters were 0 mismatches to the forward primer, 250 bp length filter, maximum number of N's = 0, and minimum quality score of 20. Because some sequences were read from each direction, it was not possible to align them. The sequences were therefore classified directly using the Ribosomal Database Project (RDP) classifier (Wang *et al.*, 2007) with a manually curated LSU gene training set v1 (Liu *et al.*, 2012) also used in Penton *et al.* (2013) which provides additional detail. Sequences were binned by genus if identified with confidence of 0.5 or greater, or otherwise to the lowest rank category for which confidence was at least 0.5, resulting in 639 categories.

All 16S/18S rRNA gene sequences that passed the quality controls of the GL FLX software were uploaded on the PYROTAGGER pipeline (Kunin & Hugenholtz, 2010). Raw sequences were sorted by bar code, trimmed, filtered to remove sequences of low quality (10% threshold), and aligned. The minimum sequence length allowed was 150 bp. Potential chimeras were identified and excluded from downstream analysis. Sequences were clustered at the level of 97% identity, and the best hit in Greengenes (for prokaryotes) and Silva (for eukaryotes) databases was determined for each cluster. The output $OTU_{0.03}$ tables were used for statistical analysis.

All *nifH* sequences also were initially processed using the pyrosequencing pipeline tools on RDP's Web site. Reads passing the initial filters were frame shift corrected and translated into NifH protein sequences using the RDP FrameBot tool (Wang *et al.*, 2013). About 15% of the sequences had frame shift errors detected and corrected by FrameBot, such that more than 99% of the sequences were retained for analysis. The protein sequences were aligned using the HMMER3 aligner, clustered at 95% identity, and the representative sequences for each cluster classified using the FunGene Pipeline (Fish *et al.*, 2013) to find the nearest match among 675 protein sequences in a curated reference set (Wang *et al.*, 2013).

Good's coverage (Good, 1953) was calculated as a percentage for each sample from the 16S/18S rRNA, 28S rRNA, and NifH data as 100 times the quantity one minus the number of singletons divided by the total number of sequences.

PCA was used to display distances between sites based on all four data types (lipid, 16S/18S rRNA, 28S rRNA, and NifH). For these analyses, the Hellinger transformation (Legendre & Gallagher, 2001) was applied to the OTU count data using

vegan's *decostand* function prior to PCA on the variance–covariance matrix using vegan's *rda* function. Multivariate analysis of variance by permutation (PMANOVA) was used to test for significant differences in dispersion among groups and for differences between group centroids (Anderson, 2001, 2006) using vegan's *betadisper* and *adonis* functions. The factors considered in these analyses were state (location), crop, and the interaction between them. Here, 'state' is actually a proxy for several correlated soil attributes, differing between intensive and extensive experiments, as explained in the Results section.

The sequences contributing most significantly to the ordinations were identified using Biodiversity. R's *ordiequilibriumcircle* function (Kindt & Coe, 2005), and indicator group analysis (Dufrene & Legendre, 1997) was used to identify OTUs whose occurrences were linked to specific crop types using labdsv's *indval* function (Roberts, 2012) and the package QVALUE (Dabney *et al.*, 2012) to assign statistical significance. Procrustes analysis (Cox & Cox, 2001) was applied to determine whether there were significant correlations between ordinations based on the four types of data using vegan's *protest* function.

Results

Soil chemical and physical analysis

Michigan and Wisconsin intensive sites differed markedly in their physical and chemical soil attributes (Fig. S2, Tables S1 and S2). Samples from the two states were separated along the first PCA axis, which explained 75.1% of the variance and represent a sand/silt gradient (Fig. S2). Except for pH and Cu, the measured soil variables, which are all linked to nutrient concentrations in the soil, were positively correlated with higher percentages of silt. Of the variables measured, pH had the highest projection on the second PCA axis, which explained only 8.4% of the variance.

A sand/silt gradient also separated the extensive sites by state, but fewer of the soil variables were strongly correlated with this gradient (Fig. S3). Indeed, Mg, Ca, and pH were orthogonal to the gradient, while K, Na, S, and total C were nearly so. However, none of these soil variables separated sites by crop.

There were important differences between states and experiments in total soil carbon. For the intensive sites, total soil carbon was higher in the Wisconsin samples for each of the studied crops (ANOVA, $P < 0.001$, Fig. 1b). For the extensive sites, total carbon tended to be higher in the Wisconsin samples for each crop, but the differences were less pronounced and not significant (ANOVA, $P = 0.15$).

Lipid analysis

In most cases, small sample size and high variance precluded detection of statistically significant among

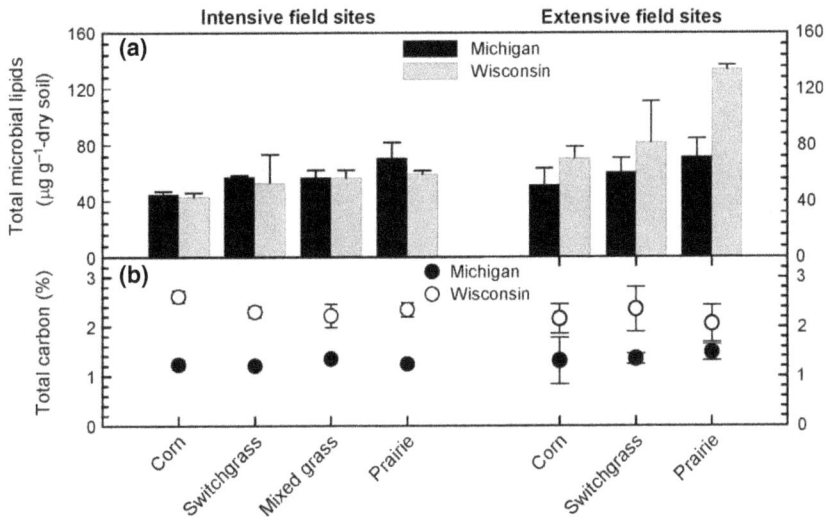

Fig. 1 Total lipid biomass (a) and total carbon (b) under different biofuel cropping systems at intensive and extensive sites in Wisconsin and Michigan. Bars represent ± 1 standard error.

treatment differences in lipids characterizing various microbial groups. For this reason, we are limited to discussing trends in the data, the strengths of which may be judged from the relative error bars in Figs 1–3.

For the intensive sites, total lipid biomass per treatment did not differ between states (Fig. 1a). We

observed a trend for higher total microbial biomass under perennial species than under corn for both states. For Michigan, this modest difference was accounted for by increases of both bacteria and fungi under the perennial grasses, with a slightly higher proportion of AMF in mixed grasses and prairie (Fig. 2a,b). For Wisconsin,

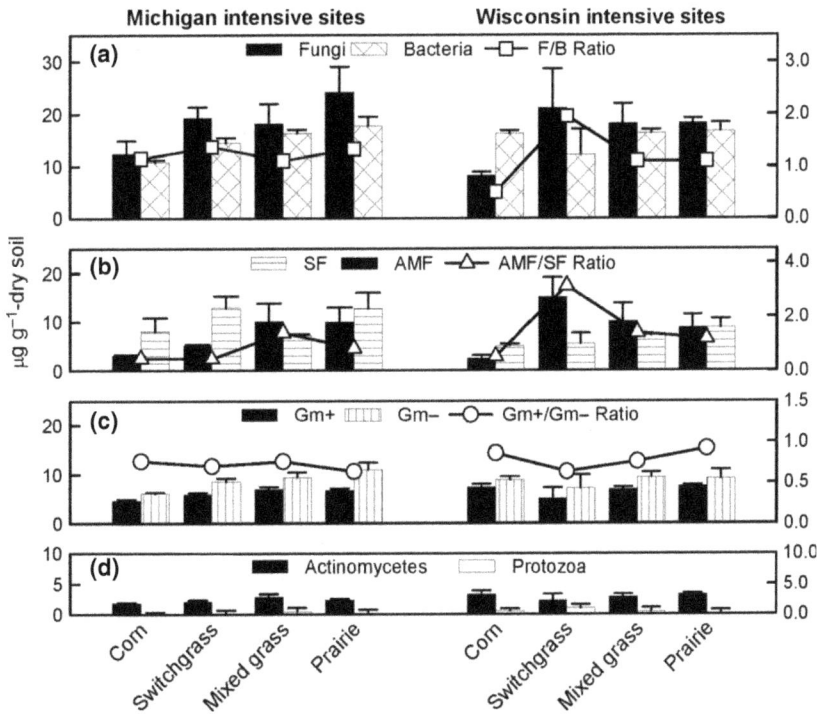

Fig. 2 Abundance (left axis) of fungi and bacteria (a), arbuscular mycorrhizal (AMF) and saprophytic fungi (SF) (b), Gram-positive (Gm$^+$) and Gram-negative bacteria (Gm$^-$) (c), actinomycetes and protozoa (d), and the F/B, AMF/SF and Gm$^+$/Gm$^-$ ratios (right axis) of microbial communities under corn, switchgrass, mixed grasses, and prairie at intensive sites in Michigan and Wisconsin. Bars represent ± 1 standard error.

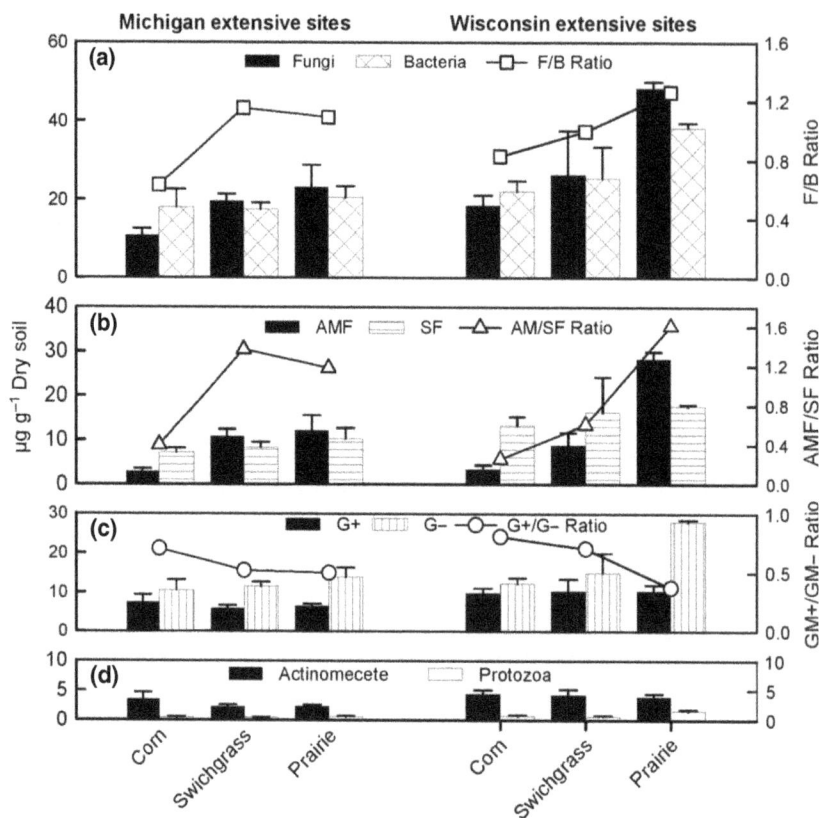

Fig. 3 Abundance (left axis) of fungi and bacteria (a), arbuscular mycorrhizal (AMF) and saprophytic fungi (SF) (b), Gram-positive (Gm$^+$) and Gram-negative bacteria (Gm$^-$) (c), actinomycetes and protozoa (d), and the F/B, AMF/SF and Gm$^+$/Gm$^-$ ratios (right axis) of microbial communities under corn, switchgrass, and prairie at the extensive sites in Michigan and Wisconsin. Bars represent ± 1 standard error.

this increased total biomass under the perennials was accounted for mainly by the fungi, and of the fungi, mainly by the AMF, especially in switchgrass. There were no differences in actinomycetes or protozoa lipid markers among treatments for either state (Fig. 2d).

The extensive sites exhibited much greater differences in total biomass between the two states (Fig. 1a). Microbial biomass per treatment was higher for the extensive sites in Wisconsin than for those in Michigan, but markedly so only for the prairie sites. It also tended to increase among treatments from corn to switchgrass to prairie, especially in Wisconsin. Wisconsin extensive sites exhibited higher microbial biomass per treatment than corresponding intensive sites, especially for the corn and prairie sites.

For the extensive sites in Wisconsin, the higher biomass under perennial grasses was due to an increase in the biomass of both fungi and bacteria, and more so to fungi, but for those sites in Michigan, it was due to fungi only (Fig. 3a). For both states, the Gram-positive/Gram-negative (Gm$^+$/Gm$^-$) ratio was lower (Fig. 3c) and the AMF/SF ratio was greater (Fig. 3b) under the perennial grasses. AMF increased from 3.5 ± 0.9 μg g^{-1}

in Wisconsin and 2.9 ± 0.6 μg g^{-1} in Michigan under corn to 28.3 ± 1.6 μg g^{-1} and 12.1 ± 3.4 μg g^{-1} in the prairie. Actinomycetes and protozoa markers were more abundant in Wisconsin; however, there were no apparent differences in these two groups among treatments for either state (Fig. 3d).

28S rRNA gene pyrosequencing

We obtained 124 654 28S rRNA gene sequences for the 16 extensive sites samples retained in the study, with an average read length of 446 bp. Samples MIE.Co.16 and WIE.Co.2, extensive corn sites in MI and WI, respectively, were excluded for yielding too few sequences. Sequences not identified as fungi by the RDP classifier with the confidence filter set at 0.5 were removed, leaving a total of 119 793 sequences in 632 categories, 94 of which were universal singletons. Sequences per sample ranged from 1979 to 13 280. Good's coverage was high, ranging from 97.2 to 99.6 with a mean of 98.9%.

The three most abundant identifiable phyla were Ascomycota (67% of total sequences), Basidiomycota

(15%), and Chytridiomycota (4%) (Fig. S4). Fungi unclassified at the phylum level made up 13% of the sequences. Basidiomycota were most abundant at the Michigan prairie, and Chytridiomycota were most abundant at the Wisconsin prairie.

By IndVal analysis, the genera *Ascobolus*, *Podospora*, *Coprinellus*, *Ascodesmis*, and *Byssonectria* characterized the corn sites, with *Ascobolus* being the most abundant (Table S3). Of the many genera characteristic of the prairie, unclassified Helotiales, *Clavaria*, and *Tricladium* were the most abundant. *Beauveria* was characteristic of the switchgrass sites, but weakly so because it was not abundant and was also found at prairie sites.

As arbuscular mycorrhizal fungi (AMF) were identified by lipids analysis as an abundant group, we sought to identify fungal sequences belonging to this group in the 28S rRNA gene pyrosequencing data. We found that only 0.13% of the sequences were classified to Glomeromycota. Of these, *Paraglomus* sequences were by far the most abundant in the dataset. Most of the AMF sequences were recovered from prairie soils, which also presented the highest detected richness, with a combination of at least four genera per site. In contrast, just one or two genera could be found in soils cultivated with switchgrass, with a predominance of *Paraglomus*. The same applies to soils cultivated with corn in Wisconsin. No AMF sequences were recovered from soils cultivated with corn in Michigan.

16S/18S rRNA gene pyrosequencing

We obtained 167 848 16S/18S rRNA gene sequences with an average of 7570 ± 894 sequences per sample and a minimum read length of 150 bp. These samples were aligned and clustered into 10 092 clusters (OTUs) at a distance of 3%; 6628 of these clusters were global singletons with Good's estimated sample coverage of 52 to 92% and averaging 85%.

Prokaryote sequences accounted for 86.3% of the recovered sequences and eukaryote sequences accounted for 9.2% (Fig. S5). Unassigned sequences accounted for 4.5%. Prokaryote sequences (86.3%) were mostly bacterial with only 0.02% belonging to Archaea. The most abundant bacterial phyla in the libraries were *Proteobacteria*, *Actinobacteria*, and *Acidobacteria*, at both intensive and extensive sites. Fungi, Metazoa, and Cercozoa were the more abundant eukaryotic phyla at both intensive and extensive sites.

No significant differences among locations or treatments were evident at the phylum level. Although such differences were observed for OTUs, interpretation was problematic due to the large number of clusters, most containing few sequences, and due to the poor identification of representative sequences, with many not being identified past the phylum level.

NifH gene pyrosequencing

We obtained 195 385 NifH sequences for the 41 samples retained in the study, with a mean of 4765 sequences per sample and a standard deviation of 811. The average read length was 320 bp. After frame shift correction and translation into amino acids, they were clustered at a distance of 0.05 yielding 2799 OTUs. Of these, 773 were global singletons, with Good's estimated sample coverage varying among samples from 93.1 to 98.4 with a mean of 96.4%. FrameBot, included in the Fungene Pipeline (Fish *et al.*, 2013), was used to match representative sequences from each cluster to 187 of 782 unique NifH reference sequences. These 187 matches fell into 100 genera.

More than 95% of the recovered NifH sequences were assigned to *Proteobacteria* (Fig. S6). Within this phylum, closest matches to the *Alphaproteobacteria* and *Betaproteobacteria* were generally more abundant and *Gammaproteobacteria* least abundant. Variances were large, but there was a tendency for *Alphaproteobacteria* affiliates to be higher in the extensive sites and *Betaproteobacteria* to be higher in the intensive sites, the latter especially for switchgrass. Matches to the *Deltaproteobacteria* were most abundant in the Michigan extensive switchgrass sites, dominating all three replicates. Unidentified environmental sequences and sequences known in *Actinobacteria*, *Bacteroidetes*, *Chlorobi*, *Cyanobacteria*, *Euryarchaeota*, *Firmicutes*, *Fusobacteria*, *Nitrospirae*, *Spirochaetes*, *Synergistetes*, and *Verrucomicrobia* were also detected.

For the NifH data, sequences contributing the most significantly to ordination of the intensive sites were closest matches to *Azospirillum*, *Bradyrhizobium*, *Rubrivivax*, *Leptothrix*, *Dechloromonas*, and *Geobacter* (Fig. S7). Of these, *Geobacter*-like sequences were present in all samples, but they were especially abundant in Michigan, representing more than 20% of the sequences in soils under prairie, mixed grasses, and switchgrass. *Bradyrhizobium* and *Rubrivivax*-related sequences were more characteristic of Wisconsin: *Rubrivivax*-like sequences accounted for more than 17% of the sequences from soils under corn, prairie, and switchgrass in that state. *Azospirillum*, *Dechloromonas*, and *Leptothrix*-like sequences did not distinguish samples by state, being more related to the pH gradient.

Genera contributing most significantly to the ordination of the extensive sites were *Geobacter*- and *Hyphomicrobium*-related (Fig. S8). *Geobacter*-like sequences represented 60% of the sequences from Michigan switchgrass samples, while *Hyphomicrobium*-like sequences were most abundant in prairie soil from both states.

Microbial community structure

Microbial data (lipid, 16S/18S rRNA, and NifH data) from the intensive sites were analyzed by PCA and PMA-NOVA (Table 1 and Fig. 4) and for differences in dispersion among factors. PMANOVA indicated that the 16S/18S rRNA and NifH data separated the samples by location only, and centroids are drawn for location only in Fig. 4b,c. The only significant difference in dispersion ($\alpha = 0.05$) was for the NifH data by location (Fig. 4c), indicating that the Wisconsin samples were also more variable. For the lipid data, however, the interaction term was significant, indicating a crop effect differing by state. Centroids drawn for all treatment combinations (Fig. 4a) depict no separation of Michigan samples by crop, but do reveal a separation of Wisconsin samples by crop. In particular, corn and to a lesser extent mixed grasses are separated from prairie and switchgrass.

In contrast to what was observed for the intensive sites, all four data types (lipid, 28S rRNA, 16S/18S rRNA, and NifH) separated communities by crop in the extensive sites (Table 2 and Fig. 5), and there were no significant differences in dispersion ($\alpha = 0.05$). The 28S rRNA gene data separated prairie and corn sites from each other, but samples from the switchgrass sites overlapped both (Fig. 5a). The lipid data separated corn from switchgrass and prairie (Fig. 5b). The 16S/18S rRNA gene data separated all three crops (Fig. 5d). For the NifH data, there was some overlap between the corn and Wisconsin switchgrass samples, but otherwise crops were separated (Fig. 5e). Additionally, the lipid and NifH data separated the samples by location (Table 2 and Fig. 5c,f).

Procrustes analysis

We performed Procrustes analyses to determine whether there were significant correlations between ordinations based on the four types of data (lipid, 28S

rRNA, 16S/18S rRNA, and NifH). For the intensive sites, ordination by the lipid data was correlated with those by 16S/18S rRNA and by NifH, but ordinations by 16S/18S rRNA and NifH differed significantly. The difference was primarily due to two Michigan corn samples having a greater distance from their centroid by 16S/18S rRNA gene data than by NifH data (Fig. 4b, c), but this did not influence interpretation of results. For the extensive sites, ordinations were correlated with the exceptions of 16S/18S rRNA vs. NifH data and 16S/18S rRNA vs. 28S rRNA.

Discussion

The three methodological approaches we used to characterize microbial communities provided complimentary insights. Lipid analysis provided general taxonomic information coupled to biomass estimates that gave insight into ecosystem function (Kirk *et al.*, 2004). Moreover, the lipid data proved more sensitive to cropping system treatments showing the importance of management on ecosystem processes. Pyrosequencing the rRNA gene provided in-depth taxonomic information (Roesch *et al.*, 2007), and gene-targeted metagenomics provided information on a subset of the community responsible for a certain function (Iwai *et al.*, 2010, 2011). In our case, we targeted the *nifH* gene, which codes for dinitrogenase reductase, a component of nitrogenase, the enzyme responsible for N_2 fixation. N_2-fixing bacteria were chosen as a model to test the effect of cultivation on an important functional group as opposed to information provided by taxonomic markers. Indeed, there is evidence that perennial grasses with potential for biofuel production, such as *Miscanthus*, may be associated with N_2-fixing bacteria (Tjepkema & Burris, 1976; Davis *et al.*, 2010; Mao *et al.*, 2013; Keymer & Kent, 2014), which points to N_2-fixing microorganisms as an important target group. Additionally, we expected NifH to be less conserved than ribosomal genes, thus giving us a contrast to the highly conserved rRNA gene.

All three approaches revealed similar differences among cropping systems in community structure. This agreement between approaches indicates that similar factors are shaping the structure of bacteria, fungi, and N_2-fixing communities under our studied conditions and that disparate taxa are being affected similarly by cultivation, soil type, and land use. Differences in community structure could be linked to both treatment and environmental factors, but the relative importance of the linkages differed between intensive and extensive sites.

Our initial hypothesis that soil microbial communities under switchgrass and mixed grasses would be more

Table 1 Results of multivariate analysis of variance by permutation (function *adonis* in VEGAN package) of sequence data for the intensive sites

Source of variation	Lipids		rRNA		NifH	
	df†	F statistics	df	F statistics	df	F statistics
Location	1	5.2***	1	2.9***	1	6.3***
Crop	3	2.2*	3	1.2ᴺˢ	3	1.3ᴺˢ
Interaction	3	2.7**	3	1.1ᴺˢ	3	1.0ᴺˢ
Residuals	13		14		15	
Total	20		21		22	

Significance codes: ***0.001; **0.01; *0.05; NS, non significant.
†Degrees of freedom.

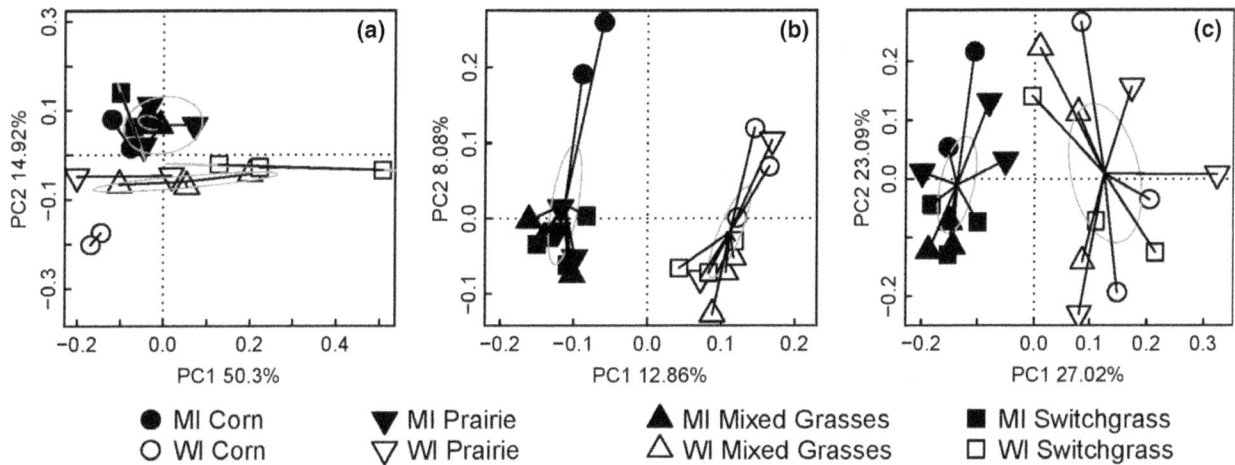

Fig. 4 Principal components analysis of soil microbial communities from Michigan and Wisconsin intensive sites as evaluated by sequencing of SSU rRNA genes (a), lipid analysis, (b) and NifH sequences (c). Ellipses are 95% confidence intervals about centroid means and were drawn to indicate the main factors related to community structure.

Table 2 Results of multivariate analysis of variance by permutation (function *adonis* in VEGAN package) of sequence data for the extensive sites

Source of variation	28S rRNA		Lipids		16S rRNA		NifH	
	df†	F statistics	df	F statistics	df	F statistics	df	F statistics
Location	1	1.5[NS]	1	3.2[*]	1	1.0[NS]	1	2.5[**]
Crop	2	2.2[***]	2	8.0[***]	2	1.8[**]	2	2.6[***]
Interaction	2	1.0[NS]	2	1.7[NS]	2	1.0[NS]	2	1.3[NS]
Residuals	10		11		11		12	
Total	15		16		16		17	

Significance codes: ***0.001; **0.01; *0.05; NS, non significant.
†Degrees of freedom.

like those under prairies was confirmed in the older, extensive sites, but not in the young, intensive sites. In the second case, soil type was a stronger predictor of community structure and composition. Here, soil type is confounded with location (state), meaning that we cannot separate the effects of soil type from the effects of geographical distance. We assume, however, that soil type is the key factor because of the sharp difference in soil texture and fertility between our intensive sites in the two states (Fig. S2). The Michigan soils we studied are sandier and have lower fertility than the Wisconsin soils, which are loess-derived. Our results agree with those of Mao et al. (2013) who also compared microbial soil communities under biofuel crops by pyrosequencing 16S rRNA and *nifH* genes and found that site-to-site variation surpassed variation stemming from plant type.

In contrast, communities at the extensive sites tended to group more strongly by crop, indicating that plant species had a stronger influence on microbial communi-

ties as the plant communities effects on soil microbes accumulated over time. Our results were similar to those of Allison et al. (2005) and Mao et al. (2013) in that communities under corn were separated from those under perennial grasses by lipids and 16S rRNA gene analysis, respectively. Mao et al. (2013), however, were not able to detect differences in N_2-fixing communities between crops, while we did for the extensive sites.

The differential response to crops observed between intensive and extensive experiments is likely related to the length of time the crops had been grown at the sites. Previous experiments by Murphy & Foster (2014) and Buckley & Schmidt (2003) demonstrated that despite changes in plant cover and management, soil microbial communities remained similar even after 6 and 7 years, respectively. In another experiment, Jangid et al. (2011) found an even longer historical effects lasting through 17 years of succession in a previously cultivated field. At the time of our sampling, crops at the intensive sites had been cultivated for only 2 years, which likely was

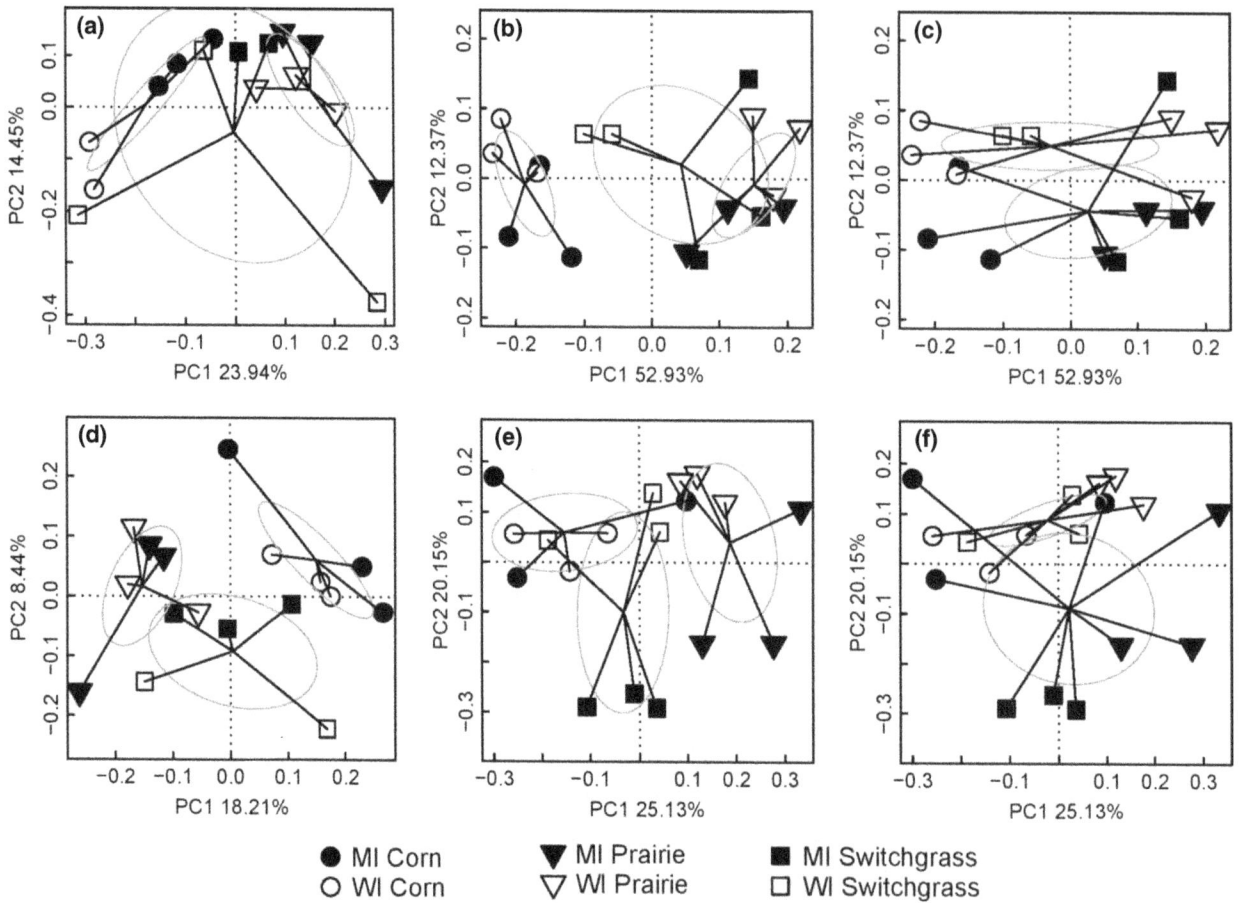

Fig. 5 Principal components analysis of soil microbial communities from Michigan and Wisconsin extensive sites as evaluated by 28S rRNA gene sequences (a), lipid analysis (b and c), SSU rRNA gene sequences (d), and NifH protein sequences(e and f). Ellipses are 95% confidence intervals about centroid means and were drawn to indicate the main factors related to community structure.

not enough time to imprint significant differences on the structure of the communities. In contrast, the extensive sites were much older, >10 years, allowing time for a more pronounced differentiation of community structure according to crop type. Our findings reinforce the previous findings of these authors, but in the context of a larger geographical scale, and including two different cultivation settings and a diverse range of soil types, especially for the extensive sites.

The one exception to this generalization was that the lipid data did reveal a crop effect on microbial community structure for the intensive Wisconsin sites. This may have to do with the differences in relative proportion of AMF biomass between treatments for the two states. Bacterial and SF biomasses were similar among treatments for intensive sites in both states. For the richer Wisconsin soils, however, there was a greater relative difference in AMF biomass between corn and the other crops, especially switchgrass (Fig. 2b), which plots farthest from corn in Fig. 4a. Herzberger *et al.* (2014) also reported total

biomass, and especially AMF biomass, was higher under restored prairie than corn two years after establishment at Wisconsin intensive sites. Community differences due to AMF would not be revealed by the rRNA or NifH data.

As previously observed (Liang *et al.*, 2012), when compared to corn, perennial grasses favored the accumulation of microbial biomass as well as an increase in F/B and AMF/SF ratios in both states, indicating that these grasses favor the accumulation of biomass and fungi, especially AMF. This was especially true among the extensive sites in Wisconsin, which had more time to become established than the intensive sites and had higher C (Fig. 1b) and clay (Table S1) contents than the Michigan sites. Soils with higher C and clay contents are often associated with higher microbial biomass because there is more C available for microbial growth (Bach *et al.*, 2010). On the other hand, the higher abundance of Gram-positive bacteria and actinomycetes under corn indicates a more stressful environment, because these organisms are known for thriving in stressful environments and for producing spores (Yao et al., 2000; Fierer et al., 2003).

The perennial character of switchgrass and prairie assemblages likely contributes to the accumulation of fungal biomass, especially AMF biomass, while hyphae are disrupted by tillage at corn sites. Tillage breaks the hyphae of AMF and increases the decomposition of organic matter, which might decrease the abundance of AMF and favor saprophytic fungi. Mycorrhizal fungi are known to enhance P absorption and utilization, so an increase in AMF implies an improvement in the absorption and utilization efficiency of this nutrient. Additionally, recently reported results showed that mycorrhizae increased the amount of total N in shoots of switchgrass (Schroeder-Moreno et al., 2012), indicating that mycorrhizae can also contribute to the increased utilization of this nutrient. The higher fungal biomass under perennial grasses implies greater potential for C accumulation under these crops (Blanco-Canqui, 2010) because it is assumed that root production is higher (Zan et al., 2001), soil aggregates form more quickly and to a higher degree (Jastrow, 1987; Jarchow & Liebman, 2012) and that fungi produce C compounds more difficult to degrade (Allison et al., 2005). Tilman et al. (2006) also noted that C sequestration was higher under native grassland perennials than under corn, and Bailey et al. (2002) observed larger quantities of C and larger activity ratios in the soil of a restored prairie compared to a neighboring corn farmland.

Fungal species specific to each crop were all saprophytic fungi. Only a few sequences belonging to AMF were identified, but they did reveal some interesting patterns. Among these, there was a greater abundance and richness (number of OTUs) of AMF under prairies for both states, including groups known to have distinct functional traits, such as Paraglomerales and Diversisporales (van der Heijden & Scheublin, 2007). The higher number of OTUs observed under prairie may be a consequence of its higher plant diversity, because a greater diversity of plant hosts creates a greater diversity of niches for AMF. In addition, it is possible that the lower AMF richness under corn stemmed from higher soil disturbance in the cultivation of this crop, because there is evidence that soil disturbance was responsible for reducing the phylotype richness of AMF communities under seminatural grasslands (Schnoor et al., 2011). AMF was linked previously to higher plant productivity (Maherali & Klironomos, 2007), so this greater AMF richness under prairies may positively influence the productivity of the prairie vegetation. It is worth noting that most of the AMF sequences found at our studied sites, and especially those in the prairie soils, belonged to the genus Paraglomus. Previous researchers found Paraglomerales are difficult to detect in roots and soils

and that commonly used primers for AMF fail to amplify Paraglomerales sequences (Lumini et al., 2010; Gosling et al., 2014). Thus, our data indicate that Midwestern prairies may be good places to study the diversity and ecology of this lesser known AMF genus.

There was an inconsistency in the AMF prevalence indicated by the lipid method where AMF were 41% of the fungal biomass and the pyrosequencing methods where only 0.13% of the fungal 28S rRNA sequences were assigned to Glomeromycota. A similarly low fraction of AMF was observed in the 18S rRNA data. Both SSU and LSU primers used were perfect matches to most known strains, so mismatch is not a likely explanation, although other biases have been observed in competitive rRNA gene amplification. One possible explanation is that the biomass (hyphae) measured by lipid may not be filled with protoplasm or with nuclei. Cytoplasmic streaming is known to occur in soil fungi, often resulting in evacuated hyphae because their protoplasm is concentrated at the growing tips (Klein & Paschke, 2004). The reason for this large discrepancy is important to resolve for proper accounting of this important group of soil fungi.

While we could specifically link SSU rRNA gene sequences to crops and soils, and given that these sequences provide better taxonomical resolution, the detection of large numbers of uncultured organisms and taxa with no clearly defined roles limited physiological and functional interpretation. Furthermore, the large number of sequences and OTUs made it difficult to detect relevant indicator organisms based on rRNA pyrosequencing data. This is a problem common to many SSU rRNA sequencing studies, especially for environmental samples. On the other hand, sequence assignment and the identification of indicator organisms were more informative with the NifH sequences, although some horizontal gene transfer may cloud precise taxonomic resolution.

Contrary to what we expected, the faster evolving protein coding gene nifH provided no better resolution than the other methods after conversion of nucleotide sequences to amino acid sequences, which was necessary to correct for sequencing errors as well as to reflect function. The low diversity we found for NifH may be the result of the primers we used not amplifying all nifH variants. The primers used are reported as being selective for Proteobacteria sequences (Diallo et al., 2008). However, it is worth noting that these authors used a previous PCR step, with different primers, which might have introduced extra bias into their PCR. In a more recent study, Gaby & Buckley (2012) reported that no nifH primers were comprehensive for the known nifH genes and that the primers developed by Poly et al. (2001) do exclude certain groups. A trade-off exists between finding primers and conditions that give the best coverage for groups important in the habitat and reliable amplification. We decided the Poly primers were the current best choice because amplification with the broader coverage Zehr & McReynolds (1989) primers was troublesome, probably because of their high degeneracy.

We conclude that location, a proxy primarily for soil type but also including site history, landscape, and climate, was the major factor determining microbial communities in our 2-year-old intensive sites and that these study sites were not under cultivation long enough for the crop to impose a strong signature on the microbial communities. The only exception to this was that the lipid data revealed a crop effect in the richer Wisconsin soil. In contrast, when the same crop had been grown on a site for 10 years or longer, a crop effect was observed, with communities under corn clearly differentiated from those under perennial grasses. Both presence of perennial plants and higher plant diversity likely favored the accumulation of microbial biomass and fungi, especially AMF, under switchgrass, mixed grasses, and prairie, leading to a more stable environment and highlighting that these alternatives to corn for biofuels may improve soil functional stability and sustainability.

Acknowledgements

This work was funded by the DOE Great Lakes Bioenergy Research Center (DOE BER Office of Science DE-FC02-07ER64494). We thank Gregg Sanford and Joe Simmons for managing the intensive cropping systems experiments in Wisconsin and Michigan, respectively. Thanks to Doug Landis and Ben Werling for help with sampling and providing information on the extensive sites in Michigan; Tim Meehan for help with sampling on the extensive sites in Wisconsin; Susanna Tringe, Stephanie Malfatti, Tijana Galvina del Rio at the Joint Genome Institute for SSU rRNA pyrotag sequencing; James Cole from the Ribosomal Database Project for support and comments with sequence data analysis; Harry Read for lipid analysis; and David Duncan for soil physicochemical analysis.

References

Adler PR, Sanderson MA, Weimer PJ, Vogel KP (2009) Plant species composition and biofuel yields of conservation grasslands. *Ecological Applications*, **19**, 2202–2209.

Allison VJ, Miller RM, Jastrow JD, Matamala R, Zak DR (2005) Changes in soil microbial community structure in a tallgrass prairie chronosequence. *Soil Science Society of America Journal*, **69**, 1412–1421.

Anderson MJ (2001) A new method for non-parametric multivariate analysis of variance. *Austral Ecology*, **26**, 32–46.

Anderson MJ (2006) Distance-based tests for homogeneity of multivariate dispersions. *Biometrics*, **62**, 245–253.

Bach EM, Baer SG, Meyer CK, Six J (2010) Soil texture affects soil microbial and structural recovery during grassland restoration. *Soil Biology and Biochemistry*, **42**, 2182–2191.

Bailey VL, Smith JL, Bolton H (2002) Fungal-to-bacterial ratios in soils investigated for enhanced C sequestration. *Soil Biology and Biochemistry*, **34**, 997–1007.

Blanco-Canqui H (2010) Energy crops and their implications on soil and environment. *Agronomy Journal*, **102**, 403–419.

Buckley DH, Schmidt TM (2003) Diversity and dynamics of microbial communities in soils from agro-ecosystems. *Environmental Microbiology*, **5**, 441–452.

Cox TF, Cox MAA (2001) *Monographs on Statistics and Applied Probability. Multidimensional Scaling*. Chapman & Hall/CRC, Boca Raton.

Dabney A, Storey JD, Warnes R (2012) qvalue: Q-value estimation for false discovery rate control. Available at: http://www.bioconductor.org/packages/release/bioc/html/qvalue.html (accessed 15 October 2012).

Davis SC, Parton WJ, Dohleman FG, Smith CM, Del Grosso S, Kent AD, DeLucia EH (2010) Comparative biogeochemical cycles of bioenergy crops reveal nitrogen-fixation and low greenhouse gas emissions in a *Miscanthus x giganteus* agro-ecosystem. *Ecosystems*, **13**, 144–156.

Diallo MD, Reinhold-Hurek B, Hurek T (2008) Evaluation of PCR primers for universal *nifH* gene targeting and for assessment of transcribed *nifH* pools in roots of *Oryza longistaminata* with and without low nitrogen input. *FEMS Microbiology Ecology*, **65**, 220–228.

Dufrene M, Legendre P (1997) Species assemblages and indicator species: the need for a flexible asymmetrical approach. *Ecological Monographs*, **67**, 345–366.

Engelbrektson A, Kunin V, Wrighton KC, Zvenigorodsky N, Chen F, Ochman H, Hugenholtz P (2010) Experimental factors affecting PCR-based estimates of microbial species richness and evenness. *ISME Journal*, **4**, 642–647.

Fargione J, Hill J, Tilman D, Polasky S, Hawthorne P (2008) Land clearing and the biofuel carbon debt. *Science*, **319**, 1235–1238.

Fierer N, Schimel JP, Holden PA (2003) Variations in microbial community composition through two soil depth profiles. *Soil Biology and Biochemistry*, **35**, 167–176.

Fierer N, Ladau J, Clemente JC et al. (2013) Reconstructing the microbial diversity and function of pre-agricultural tall grass prairie soils in the United States. *Science*, **342**, 621–624.

Fish JA, Chai B, Wang Q, Sun Y, Brown CT, Tiedje JM, Cole JR (2013) FunGene: the functional gene pipeline and repository. *Frontiers in Microbiology*, **4**, 291.

Gaby JC, Buckley DH (2012) A comprehensive evaluation of PCR primers to amplify the *nifH* gene of nitrogenase. *PLoS ONE*, **7**, e42149.

Gao J, Thelen KD, Hao XM (2013) Life cycle analysis of corn harvest strategies for bioethanol production. *Agronomy Journal*, **105**, 705–712.

Good IJ (1953) The population frequencies of species and the estimation of population parameters. *Biometrika*, **40**, 237–264.

Gosling P, Proctor M, Jones J, Bending GD (2014) Distribution and diversity of *Paraglomus* spp. in tilled agricultural soils. *Mycorrhiza*, **24**, 1–11.

Groom MJ, Gray EM, Townsend PA (2008) Biofuels and biodiversity: principles for creating better policies for biofuel production. *Conservation Biology*, **22**, 602–609.

van der Heijden MGA, Scheublin TR (2007) Functional traits in mycorrhizal ecology: their use for predicting the impact of arbuscular mycorrhizal fungal communities on plant growth and ecosystem functioning. *New Phytologist*, **174**, 244–250.

Herzberger AJ, Duncan DS, Jackson RD (2014) Bouncing back: plant-associated soil microbes respond rapidly to prairie establishment. *PLoS ONE*, **9**, e115775.

IEA (2007) Biofuel Production - ETE02. Energy Technology Essentials. Available at: http://www.iea.org/techno/essentials2.pdf (accessed 23 March 2012).

Iwai S, Chai BL, Sul WJ, Cole JR, Hashsham SA, Tiedje JM (2010) Gene-targeted-metagenomics reveals extensive diversity of aromatic dioxygenase genes in the environment. *ISME Journal*, **4**, 279–285.

Iwai S, Chai B, Jesus E, Penton CR, Lee TK, Cole JR, Tiedje JM (2011) Gene-targeted metagenomics (GT Metagenomics) to explore the extensive diversity of genes of interest in microbial communities. In: *Handbook of Molecular Microbial Ecology I - Metagenomics and Complementary Approaches* (ed. De Bruijn FJ), pp. 235–243. Wiley-Blackwell, Hoboken, NJ.

Jach-Smith LC, Jackson RD (2015) Switchgrass and prairie response to N fertilizer and harvest time on productive soils of upper Midwest. *Agriculture, Ecosystems and Environment*, **204**, 62–71.

Jangid K, Williams MA, Franzluebbers AJ, Schmidt TM, Coleman DC, Whitman WB (2011) Land-use history has a stronger impact on soil microbial community composition than aboveground vegetation and soil properties. *Soil Biology and Biochemistry*, **43**, 2184–2193.

Jarchow ME, Liebman M (2012) Tradeoffs in biomass and nutrient allocation in prairies and corn managed for bioenergy production. *Crop Science*, **52**, 1330–1342.

Jastrow JD (1987) Changes in soil aggregation associated with tallgrass prairie restoration. *American Journal of Botany*, **74**, 1656–1664.

Jesus ED, Susilawati E, Smith SL et al. (2010) Bacterial communities in the rhizosphere of biofuel crops grown on marginal lands as evaluated by 16S rRNA gene pyrosequences. *Bioenergy Research*, **3**, 20–27.

Keymer DP, Kent AD (2014) Contribution of nitrogen fixation to first year *Miscanthus x giganteus*. *Global Change Biology Bioenergy*, **6**, 577–586.

Kim S, Dale BE, Ong RG (2012) An alternative approach to indirect land use change: allocating greenhouse gas effects among different uses of land. *Biomass and Bioenergy*, **46**, 447–452.

Kindt R, Coe R (2005) *Tree Diversity Analysis. A Manual and Software for Common Statistical Methods for Ecological and Biodiversity Studies*. World Agroforestry Centre (ICRAF), Nairobi.

Kirk JL, Beaudette LA, Hart M, Moutoglis P, Khironomos JN, Lee H, Trevors JT (2004) Methods of studying soil microbial diversity. *Journal of Microbiological Methods*, **58**, 169–188.

Klein DA, Paschke MW (2004) Filamentous fungi: the indeterminate lifestyle and microbial ecology. *Microbial Ecology*, **47**, 224–235.

Kunin V, Hugenholtz P (2010) PyroTagger: a fast, accurate pipeline for analysis of rRNA amplicon pyrosequence data. The Open Journal, Article 1. Available at: http://www.theopenjournal.org/toj_articles/1#5 (accessed 15 January 2015).

Legendre P, Gallagher ED (2001) Ecologically meaningful transformations for ordination of species data. *Oecologia*, **129**, 271–280.

Liang C, Jesus ED, Duncan DS, Jackson RD, Tiedje JM, Balser TC (2012) Soil microbial communities under model biofuel cropping systems in southern Wisconsin, USA: impact of crop species and soil properties. *Applied Soil Ecology*, **54**, 24–31.

Liu KL, Porras-Alfaro A, Kuske CR, Eichorst SA, Xie G (2012) Accurate, rapid taxonomic classification of fungal large-subunit rRNA genes. *Applied and Environmental Microbiology*, **78**, 1523–1533.

Lumini E, Orgiazzi A, Borriello R, Bonfante P, Bianciotto V (2010) Disclosing arbuscular mycorrhizal fungal biodiversity in soil through a land-use gradient using a pyrosequencing approach. *Environmental Microbiology*, **12**, 2165–2179.

Maherali H, Klironomos JN (2007) Influence of phylogeny on fungal community assembly and ecosystem functioning. *Science*, **316**, 1746–1748.

Mao YJ, Yannarell AC, Mackie RI (2011) Changes in N-transforming Archaea and Bacteria in soil during the establishment of bioenergy crops. *PLoS ONE*, **6**, e24750.

Mao YJ, Yannarell AC, Davis SC, Mackie RI (2013) Impact of different bioenergy crops on N-cycling bacterial and archaeal communities in soil. *Environmental Microbiology*, **15**, 928–942.

Murphy CA, Foster BL (2014) Soil properties and spatial processes influence bacterial metacommunities within a grassland restoration experiment. *Restoration Ecology*, **22**, 685–691.

Oates LG, Duncan DS, Robertson GP, Gelfand I, Miller N, Jackson RD (2015) Nitrous oxide emissions during establishment of eight alternative cellulosic bioenergy crops in the North Central United States. *Global Change Biology Bioenergy*, doi: 10.1111/gcbb.12268.

Oksanen J, Blanchet FG, Kindt R et al. (2012) vegan Community Ecology Package. Available at: http://CRAN.R-project.org/package=vegan (accessed 12 October 2012).

Penton CR, St Louis D, Cole JR et al. (2013) Fungal diversity in permafrost and tallgrass prairie soils under experimental warming conditions. *Applied and Environmental Microbiology*, **79**, 7063–7072.

Poly F, Monrozier LJ, Bally R (2001) Improvement in the RFLP procedure for studying the diversity of *nifH* genes in communities of nitrogen fixers in soil. *Research in Microbiology*, **152**, 95–103.

R Core Team (2012) *R: A language and environment for statistical computing*, Vienna, Austria, R Foundation for Statistical Computing. Available at: http://www.R-project.org/ (accessed 15 October 2012).

Roberts DW (2012) *labdsv: Ordination and Multivariate Analysis for Ecology*. Available at: http://CRAN.R-project.org/package=labdsv (accessed 15 October 2012).

Roesch LF, Fulthorpe RR, Riva A et al. (2007) Pyrosequencing enumerates and contrasts soil microbial diversity. *ISME Journal*, **1**, 283–290.

Schnoor TK, Lekberg Y, Rosendahl S, Olsson PA (2011) Mechanical soil disturbance as a determinant of arbuscular mycorrhizal fungal communities in semi-natural grassland. *Mycorrhiza*, **21**, 211–220.

Schroeder-Moreno MS, Greaver TL, Wang SX, Hu SJ, Rufty TW (2012) Mycorrhizal-mediated nitrogen acquisition in switchgrass under elevated temperatures and N enrichment. *Global Change Biology Bioenergy*, **4**, 266–276.

Swift MJ, Izac AMN, van Noordwijk M (2004) Biodiversity and ecosystem services in agricultural landscapes - are we asking the right questions? *Agriculture Ecosystems and Environment*, **104**, 113–134.

Tilman D, Hill J, Lehman C (2006) Carbon-negative biofuels from low-input high-diversity grassland biomass. *Science*, **314**, 1598–1600.

Tjepkema JD, Burris RH (1976) Nitrogenase activity associated with some Wisconsin prairie grasses. *Plant and Soil*, **45**, 81–94.

US-DOE (2011) *U.S. Billion-Ton Update: Biomass Supply for a Bioenergy and Bioproducts Industry*, Oak Ridge, TN, Oak Ridge National Laboratory. Available at: http://www1.eere.energy.gov/bioenergy/pdfs/billion_ton_update.pdf (accessed 15 January 2015).

Wang Q, Garrity GM, Tiedje JM, Cole JR (2007) Naive Bayesian classifier for rapid assignment of rRNA sequences into the new bacterial taxonomy. *Applied and Environmental Microbiology*, **73**, 5261–5267.

Wang Q, Quensen JF, Fish JA, Lee TK, Sun YN, Tiedje JM, Cole JR (2013) Ecological patterns of *nifH* genes in four terrestrial climatic zones explored with targeted metagenomics using FrameBot, a new informatics tool. *mBio*, **4**, e00592-00513.

Webster CR, Flaspohler DJ, Jackson RD, Meehan TD, Gratton C (2010) Diversity, productivity and landscape-level effects in North American grasslands managed for biomass production. *Biofuels*, **1**, 451–461.

Werling BP, Dickson TL, Issacs R et al. (2014) Perennial grasslands enhance biodiversity and multiple ecosystem services in bioenergy landscapes. *Proceedings of the National Academy of Sciences of the United States of America*, **111**, 1652–1657.

Wright L, Turhollow A (2010) Switchgrass selection as a 'model' bioenergy crop: a history of the process. *Biomass and Bioenergy*, **34**, 851–868.

Yao H, He Z, Wilson MJ, Campbell CD (2000) Microbial biomass and community structure in a sequence of soils with increasing fertility and changing land use. *Microbial Ecology*, **40**, 223–237.

Zan CS, Fyles JW, Girouard P, Samson RA (2001) Carbon sequestration in perennial bioenergy, annual corn and uncultivated systems in southern Quebec. *Agriculture Ecosystems and Environment*, **86**, 135–144.

Zehr JP, McReynolds LA (1989) Use of degenerate oligonucleotides for amplification of the *nifH* gene from the marine Cyanobacterium *Trichodesmium thiebautii*. *Applied and Environmental Microbiology*, **55**, 2522–2526.

19

Impact of land-use change to Jatropha bioenergy plantations on biomass and soil carbon stocks: a field study in Mali

JEROEN DEGERICKX[1], JOANA ALMEIDA[1], PIETER C.J. MOONEN[1], LEEN VERVOORT[1], BART MUYS[1] and WOUTER M.J. ACHTEN[2]

[1]Division Forest, Nature and Landscape, Katholieke Universiteit Leuven, Celestijnenlaan 200E Box 2411, 3001 Leuven, Belgium, [2]Institute for Environmental Planning and Land Use Planning, Université libre de Bruxelles, Avenue Franklin D. Roosevelt 50 CP 130/02, 1050 Brussels, Belgium

Abstract

Small-scale Jatropha cultivation and biodiesel production have the potential of contributing to local development, energy security, and greenhouse gas (GHG) mitigation. In recent years however, the GHG mitigation potential of biofuel crops is heavily disputed due to the occurrence of a carbon debt, caused by CO_2 emissions from biomass and soil after land-use change (LUC). Most published carbon footprint studies of Jatropha report modeled results based on a very limited database. In particular, little empirical data exist on the effects of Jatropha on biomass and soil C stocks. In this study, we used field data to quantify these C pools in three land uses in Mali, that is, Jatropha plantations, annual cropland, and fallow land, to estimate both the Jatropha C debt and its C sequestration potential. Four-year-old Jatropha plantations hold on average 2.3 Mg C ha^{-1} in their above- and belowground woody biomass, which is considerably lower compared to results from other regions. This can be explained by the adverse growing conditions and poor local management. No significant soil organic carbon (SOC) sequestration could be demonstrated after 4 years of cultivation. While the conversion of cropland to Jatropha does not entail significant C losses, the replacement of fallow land results in an average C debt of 34.7 Mg C ha^{-1}, mainly caused by biomass removal (73%). Retaining native savannah woodland trees on the field during LUC and improved crop management focusing on SOC conservation can play an important role in reducing Jatropha's C debt. Although planting Jatropha on degraded, carbon-poor cropland results in a limited C debt, the low biomass production, and seed yield attained on these lands reduce Jatropha's potential to sequester C and replace fossil fuels. Therefore, future research should mainly focus on increasing Jatropha's crop productivity in these degraded lands.

Keywords: allometry, biofuel, carbon debt, carbon sequestration, *Jatropha curcas* L., land conversion, root-to-shoot ratio, West Africa

Introduction

The current demand for reducing greenhouse gas (GHG) emissions, in combination with the depletion of fossil fuel reserves and the growing concern on energy security and independence (Verrastro & Ladislaw, 2007) led to a growing interest in the production of liquid biofuels. In this context, *Jatropha curcas* L., a tropical deciduous shrub, was claimed to provide high oil yields on degraded lands with minimal nutrient and management inputs, thereby avoiding competition with food production (Achten *et al.*, 2010a). However, more recent research has come to disprove these early claims (van

Correspondence: Prof Dr Bart Muys
e-mail: bart.muys@ees.kuleuven.be

Eijck *et al.*, 2014) and a large fraction of Jatropha initiatives failed because of low yields due to insufficient agronomic knowledge (Nielsen *et al.*, 2013; Singh *et al.*, 2014).

Despite this negative experience, small-scale Jatropha cultivation can still play an important role as a local energy source in low-income areas (e.g., Sahel region), thereby contributing to local development, energy security, and GHG mitigation (Achten *et al.*, 2010b; Nielsen *et al.*, 2013; Muys *et al.*, 2014). The latter can be attained through (i) C sequestration in Jatropha biomass and soil during cultivation and (ii) the production of biodiesel to replace fossil fuels (Van Rooijen, 2014). Besides the well-known environmental benefits, GHG mitigation can boost the economic viability of Jatropha projects through C trading mechanisms (Nielsen *et al.*, 2013; Van Rooijen, 2014).

In recent years however, the GHG mitigation potential of crop-based liquid biofuels has been heavily debated. In particular, land-use change (LUC) due to biofuel crop establishment may create initial losses in soil and biomass C stocks as a result of increased microbial decomposition and burning. This C debt can have a significant negative impact on the biofuel's GHG balance (Fargione et al., 2008). In the case of Jatropha, multiple studies have been made addressing this particular issue (Struijs, 2008; Bailis & Baka, 2010; Achten & Verchot, 2011; Bailis & McCarty, 2011; Romijn, 2011; Rasmussen et al., 2012; Achten et al., 2013). A wide variety of C debts and associated repayment times have been reported, the latter ranging from a few years up to multiple centuries. The repayment time depends (i) on the C debt created (i.e., the land cover which is replaced by Jatropha) and (ii) on the life cycle CO_2 reduction potential of the biofuel substituting fossil fuel (kg CO_2 ha^{-1} yr^{-1}), indicating a high dependency on local conditions. However, for both aspects data quality (measurements vs. modeled estimation) and assumptions (e.g., assumed yields, fertilizer use, and field emissions) also play an important role. Most studies conclude that GHG mitigation through Jatropha production can only be achieved when it is planted on degraded lands poor in C stocks (Achten & Verchot, 2011; Romijn, 2011). However, the accuracy of these earlier analyses can be questioned, as frequent use is made of default values and nonvalidated estimates of seed yield and C stocks, which are in turn based on little empirical data. This practice can give rise to significant errors in the analysis of Jatropha C debts, as the magnitude and dynamics of C stocks depend strongly on local biophysical conditions (Powers et al., 2011). In addition, assumptions are frequently made which have not been verified in the field (e.g., soil organic carbon (SOC) remaining constant upon LUC), adding more uncertainty to currently available estimates. Therefore, there is an urgent need for more empirical data on Jatropha C stocks compared to other LUs to verify the results reported by the studies mentioned above (Romijn, 2011; Rasmussen et al., 2012).

To answer this call for more empirical data, a field study was set up in Mali with the aim of quantifying soil and biomass C stocks in small-scale Jatropha plantations and comparing these with other LUs. Mali is one of the few sub-Saharan countries explicitly encouraging Jatropha cultivation in its policy, aiming for a 20% replacement of diesel by Jatropha oil by 2023 (Favretto et al., 2012). Whereas traditionally Jatropha was mainly grown as a living fence for local soap production, its cultivation was recently redirected toward small-scale plantations for local energy production. By 2011, this resulted in a total area of almost 5000-ha of Jatropha, mainly situated in the provinces of Koulikoro, Sikasso, and Kayes (Favretto et al., 2012). The gathered C stock data were used to estimate the C debt and associated repayment time of Jatropha-based biofuel and soap production in Mali.

Materials and methods

General setup and study area

The impact of LUC on biomass and soil C stocks was studied using the C stock change method (UNFCCC, 2009), in which C stocks prior to and after LUC are compared. As a monitoring study was practically unfeasible, we applied the ergodic principle, that is, presenting assumed changes over time by comparing different LU classes in space at one point in time. C stocks were measured during summer 2011 in 18 triplets of neighboring fields, each comparing Jatropha, cropland, and fallow LU, thereby assuming that all factors other than the effect of LU are constant within each triplet (spatially paired site design; Conteh, 1999). Sampling sites were equally divided over two distinct ecoregions in Mali: Koulikoro, in the central part of the country and Garalo, a smaller village in the Southern province of Sikasso (Fig. 1). Koulikoro is situated in the Sudanese agroecological zone, which is characterized by a semi-arid climate [mean annual temperature (MAT) of 27.6 °C and mean annual precipitation (MAP) of 815 mm; New_LocClim (FAO, Rome, Italy)], dry woodlands (Magin, 2011), and farming systems integrating sedentary livestock rearing with crop production (Coulibaly, 2003). Garalo, belonging to the North-Guinean zone, has a subhumid climate [MAT = 27.0 °C, MAP = 1142 mm; New_LocClim (FAO)] giving rise to a more lush savannah vegetation and a larger diversity of crops (Coulibaly, 2003). Highly degraded soils dominate the landscape in Garalo (Ferric and Plinthic Acrisol), whereas soils in Koulikoro are more productive due to the deposition of Saharan dust (Lixisol) (FAO, 2007). Within each ecoregion, a representative selection of Jatropha fields was made, taking into account various factors as plantation age, management factors (e.g., plant spacing, intercropping), soil conditions, and the presence of neighboring cropland and fallow land. Jatropha plantations were always part of an outgrowers production system managed by a private company (Koulikoro) or local NGO (Garalo).

Data collection

General information on the history and management of each field was gathered using a brief, semistructured interview with the field's owner. Exact field locations and surface areas were recorded using GPS.

Biomass carbon. Only long-term C pools, that is, perennial shrubs and trees, were included in the estimation of biomass C. To determine Jatropha biomass, an allometric equation was first derived from destructive measurements on a representative sample of 46 Jatropha trees originating from within the selected fields and five trees from an additional field in Koulikoro.

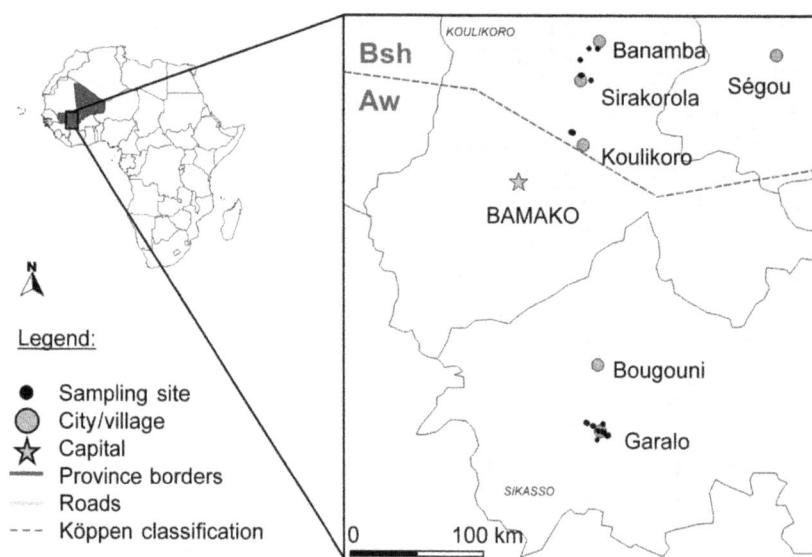

Fig. 1 Location of study sites in relation to Köppen-Geiger climate classification. Bsh, Hot steppe climate and Aw, Tropical savannah climate.

After measuring tree dimensions (i.e., basal stem diameter, tree height, crown diameter in two perpendicular directions, number and diameter of primary branches), the trees were cut down and their woody aboveground biomass (excluding leaves; wAGB) was measured fresh on the field. Subsequently, representative samples of stem and branches were taken, weighed, dried until constant weight (105 °C), and weighed again to calculate the total dry weight of wAGB per tree. Nonlinear regression analysis was used to find the most suitable allometric relation. Using the selected equation (see section Jatropha biomass, allometric relation and root-to-shoot ratio, under Results), Jatropha wAGB was then estimated in three square plots per field, each containing nine healthy and representative Jatropha trees, and finally expressed in Mg ha^{-1} using the plot's surface area. Allometric equations for other tree species and shrubs were obtained from the literature (see Box 1). In Jatropha fields and annual cropland, all mature trees and shrubs other than Jatropha were measured individually, whereas a nested sampling design was applied in fallow land, consisting of one 10 × 10 m plot in one 20 × 20 m plot. All trees with a stem diameter exceeding 6 cm were measured in the large plot, while other trees and shrubs were only appraised in the small plot.

Belowground biomass (BGB) was estimated using root-to-shoot ratios. A region-specific value was obtained for Jatropha through destructive measurements of 17 Jatropha trees. After measuring plant dimensions, these trees were uprooted and their dry BGB was determined in a similar way as described above for wAGB. For other species, the literature values were used, that is, 0.28 and 0.56 for trees in subtropical dry forest with more and <20 Mg AGB ha^{-1}, respectively, and 0.32 for scrubland in subtropical steppe (Eggleston *et al.*, 2006). The resulting biomass estimates were converted to C stocks in Mg ha^{-1} using C content data from the literature: 0.46 for Jatropha (based on Firdaus *et al.*, 2010; Torres *et al.*, 2011; Firdaus &

Box 1

Literature-based allometric equations for aboveground biomass

1) Shea tree (*Vitellaria paradoxa* C.F. Gaertner): (Nouvellet *et al.*, 2006)

$$AGB = ((a \times G) - b)) \times wD \qquad (1)$$

with AGB = aboveground biomass [Mg] of an individual tree, G = girth at breast height [m], wD = wood density (=0.85 Mg m^{-3}; Louppe, 1994); a = 2.4612 (DBH > 0.63 cm) or 0.6868 (DBH < 0.63 cm); b = 1.5130 (DBH > 0.63 cm) or 0.1314 (DBH < 0.63 cm).

2) Trees of dry tropical forest (generic): (UNFCCC, 2006)

$$AGB = \exp[-1.996 + (2.32 \times \ln(DBH))] \times 10^{-3} \quad (2)$$

with AGB = aboveground biomass [Mg]; DBH = diameter at breast height [cm].

3) Shrubs (generic): (UNFCCC, 2006)

$$AGB = \sum \left(\frac{\pi}{3} \times BA_i \times H \times wD \right) \qquad (3)$$

with AGB = aboveground biomass [Mg]; BA$_i$ = basal area of branch i [m²]; H = height of shrub [m]; wD = wood density (=0.62 Mg m^{-3}; UNFCCC, 2006).

Husni, 2012; Hellings *et al.*, 2012) and 0.50 for other tree species (Eggleston *et al.*, 2006).

Soil carbon. Four soil layers were sampled in each field, that is, 0–5, 5–10, 10–20, and 20–30 cm, for which both SOC concentration and bulk density were determined to calculate SOC stocks in Mg ha^{-1} (see Eqn 4).

$$SOC_{stock} = \frac{SOC}{100} \times BD \times \left(1 - \frac{G}{100}\right) \times d \times 10 \qquad (4)$$

with SOC_{stock} = SOC stock [Mg ha^{-1}]; SOC = SOC mass percentage [g C (100 g soil)$^{-1}$]; BD = bulk density [kg m^{-3}]; G = mass percentage of coarse fragments (>2 mm) [g (100 g soil)$^{-1}$] and d = depth of soil layer [m].

Jatropha fields were sampled most intensively to study the spatial variability of SOC (3 plots × 2 sampling locations per field; Fig. 2). In each Jatropha plot, sample A_1 was mixed with B_1 and A_2 with B_2, yielding two BD samples and two SOC samples for each soil layer per plot. In cropland, three samples were taken per field for both SOC and BD per depth. In each fallow plot (10 × 10 m), six SOC samples, situated on three transects, were taken. These samples were bulked per transect and depth, yielding three SOC samples and one BD sample per depth.

SOC samples were air-dried, passed through a 2-mm sieve, ground, and homogenized with a mortar, oven-dried at 60 °C and analyzed using the automated dry combustion method (Carlo Erba 1110 Elemental Analyzer, Carlo Erba Instruments, Rodano, Italy). As nitrogen levels are determined in the same analysis, these results were also used to calculate the C/N ratio. BD was determined using the gravimetric method, that is, drying samples with a fixed volume of 100 cm³ overnight (105 °C) and weighing them on a precision balance. These samples were then passed through a 2-mm sieve to calculate the mass fraction of gravel in the soil. Finally, soil texture was measured through laser diffraction analysis (Beckman Coulter – LS 13 320 Laser Diffraction Particle Size Analyzer, Beckman Coulter, Miami, FL, USA) and pH-H$_2$O was determined using an electrode (Van Reeuwijk, 2002) on one mixed sample per field.

Data analysis

Throughout this study, statistical analyses were conducted in SPSS 17.0 (IBM, Chicago, IL, USA), and a significance level (α) of 0.05 was used, unless stated otherwise. Whenever appropriate,

the data were lognormal-transformed to meet the criteria of parametric statistical tests. In general, differences between LUs, soil types, or ecoregions were assessed using ANOVA in combination with Tukey *post hoc* tests. To determine the impact of LUC on SOC using all gathered data, mixed ANOVA was used in which LU was included as a fixed factor and a unique field ID as a random factor, nested in LU to account for subsampling. This analysis was conducted in SAS 9.3 (SAS Institute Inc., Cary, NC, USA) using the MIXED procedure.

The total C debt was calculated as the difference between the total carbon stock (biomass + soil) of the previous land use and the total carbon stock of the Jatropha plantation at year 0 (Fargione *et al.*, 2008). The latter was approximated by subtracting the amount of newly sequestered carbon in Jatropha biomass and soil from the total carbon stock measured in the Jatropha plantation. The associated repayment time, that is, the time it takes before the initial C emissions are compensated through the substitution of fossil fuels by Jatropha biodiesel, is calculated by dividing the C debt by the yearly C reduction potential, which is in turn derived from the comparison of the global warming potential (GWP) of Jatropha-based biofuel with the GWP of the fossil fuel reference system. The GWP of both fuels are obtained from a life cycle analysis (LCA) conducted in Koulikoro (Almeida *et al.*, 2014).

Results

Description of fields

With the exception of one missing fallow land in Koulikoro, nine fields of each LU type were visited in each ecoregion. The Jatropha plantations under study are 3–5 years old and most frequently established on former cropland. In Koulikoro, Jatropha is always mixed with other crops and wide planting distances of 5 × 2 m are frequently used, whereas in Garalo intercropping is rare and smaller planting distances of 3 × 3 and 4 × 3 m are applied. Furthermore, Jatropha fields are generally ploughed once a year and receive no irrigation or pruning. Cropland most frequently consists of monocultures and is ploughed once a year. In both ecoregions, crops are mainly cultivated in agroforestry parkland systems, where some mature, widely spaced

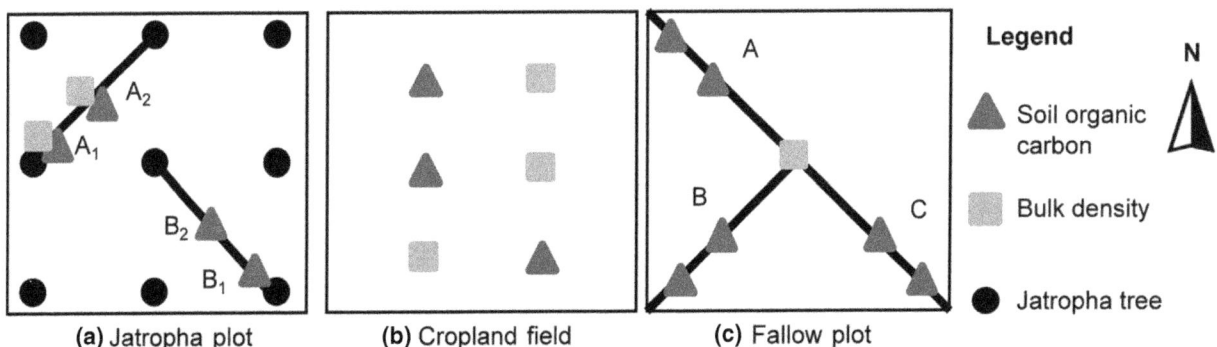

(a) Jatropha plot   **(b) Cropland field**   **(c) Fallow plot**

Legend

N

▲ Soil organic carbon

▨ Bulk density

● Jatropha tree

Fig. 2 Soil sampling locations per land-use type (letters represent sampling transects, while numbers refer to sampling locations).

trees (e.g., *Vitellaria paradoxa* C.F. Gaertner, *Parkia biglobosa* (Jacq.) R. Br. ex G. Don and *Mangifera indica* L.) are kept on the field. These provide nutrients to the crops and an extra income to the farmer through the selling of nonwood tree products such as mango fruit and Shea nuts. Major crops are corn, cotton, and sesame in Garalo and corn, sorghum and millet in Koulikoro. Fallow vegetation consists in both ecoregions of bushes combined with mature trees up to 15 m high. Detailed metadata for each field can be found in Table S1.1 in the Supporting information (Data S1). Examples of the three LUs are presented in Figure S1.1 in the Supporting information (Data S1).

Soil conditions

In both ecoregions, two soil types can be distinguished based on hierarchical cluster analysis: sandy vs. loamy soils in Koulikoro and gravel vs. nongravel soils in Garalo. The mean values of the clustering variables for these soil types are presented in Table 1. The loamy soils in Koulikoro closely resemble the nongravel soils in Garalo. The soil variables given in Table 1 were compared between the three LUs for each ecoregion separately using ANOVA analysis. No significant differences were found between the LUs over all triplets of fields (not shown). Although the similarity in soil conditions between individual fields within each triplet cannot be statistically assessed (only one measurement of soil texture available per field), this outcome does provide a good indication that field selection meets the criteria of a paired site sampling design.

Biomass carbon

Jatropha biomass, allometric relation, and root-to-shoot ratio. A summary of plant dimensions and biomass measurements of individual Jatropha trees is given in Table 2. Nonlinear regression analysis resulted in the crown area (in m²) to be selected as the best predicting variable for wAGB ($R^2 = 0.803$; see Eqn 5 and Fig. 3). The average root-to-shoot ratio for Jatropha amounts to 0.48 (Table 2).

$$wAGB = 0.897 \times CA^{1.244} \tag{5}$$

with wAGB = woody aboveground biomass in kg and CA = crown area in m².

Biomass carbon stocks in the different land uses. Mature trees, although low in abundance, represent the largest share of biomass C in all LU types (Fig. 4). On average, only 18.6% of the total biomass stock in a Jatropha plantation is in Jatropha trees. The partitioning of the biomass C stock among the different vegetation elements is similar in both ecoregions (not shown), with

Table 1 Mean values of edaphic variables for the soil types in both ecoregions (0–30 cm depth)

Ecoregion	Soil class	# Jatropha fields assessed	% Sand	% Silt	% Clay	% Gravel	pH	% C	% N	BD (g cm⁻³)
Koulikoro	Sandy	2	76.7 (7.4/6)	17.3 (5.8/6)	6.0 (2.1/6)	0.0 (0.0/6)	4.9 (0.4/6)	0.25 (0.11/24)	0.02 (0.01/24)	1.39 (0.05/20)
	Loamy	7	42.1 (7.00/20)	45.4 (5.7/20)	12.5 (3.5/20)	0.0 (0.0/20)	5.2 (0.5/20)	0.58 (0.25/73)	0.04 (0.02/73)	1.34 (0.07/72)
Garalo	Gravel	6	45.3 (11.4/18)	43.1 (7.2/18)	11.6 (4.9/18)	60.8 (14.3/18)	5.1 (0.4/18)	0.72 (0.20/72)	0.05 (0.01/72)	1.46 (0.09/60)
	Non-gravel	3	43.7 (12.5/9)	42.8 (6.4/9)	13.5 (7.7/9)	1.1 (2.2/9)	5.1 (0.2/9)	0.55 (0.19/36)	0.04 (0.01/36)	1.41 (0.1/30)

BD, bulk density.
Numbers in brackets represent the standard deviations and number of samples, respectively.

Table 2 Averages of measurements on individual Jatropha trees, grouped per ecoregion and soil type

Ecoregion	Soil type	Age (years)	Basal area (cm²)	Height (m)	# primary branches	Crown area (m²)	wAGB (kg)	BGB (kg)	R/S
Koulikoro	Loamy	3.45 (0.64/203)	91.98 (55.45/203)	1.70 (0.47/203)	4.16 (1.86/203)	2.97 (2.33/203)	2.55 (2.39/11)	2.14 (0.69/2)	0.46 (0.25/2)
	Sandy	3.50 (0.50/54)	127.57 (44.63/54)	1.80 (0.30/54)	5.69 (1.79/54)	3.43 (1.65/54)	2.67 (1.59/6)	2.35 (-/1)	0.59 (-/1)
Garalo	Gravel	4.55 (0.50/164)	120.25 (47.70/164)	1.89 (0.31/164)	4.14 (1.55/164)	3.09 (1.80/164)	3.58 (2.46/20)	1.57 (1.13/11)	0.44 (0.11/11)
	Non-gravel	4.00 (0.00/81)	99.43 (42.73/81)	1.74 (0.30/81)	4.26 (1.28/81)	2.57 (1.47/81)	3.28 (2.15/9)	2.69 (1.18/3)	0.61 (0.13/3)
All	–	3.90 (0.02/502)	106.25 (2.23/502)	1.78 (0.02/502)	4.33 (0.07/502)	3.00 (0.09/502)	3.16 (0.34/46)	1.88 (0.27/17)	0.48 (0.04/17)

wAGB, dry woody aboveground biomass per tree; BGB, dry belowground biomass per tree; R/S, root-to-shoot ratio. Numbers in brackets represent standard deviation and number of samples, respectively. 'All' stands for the total mean and standard deviation calculated according to stratified random sampling design.

the exception of the fraction of shrub biomass in fallow land being higher in Garalo (31.4%) compared to Koulikoro (11.4%).

LU has a pronounced effect on the biomass C stock in both ecoregions (Table 3). A significant difference was found between fallow and Jatropha on the one hand ($P = 0.026$ for Garalo and $P = 0.020$ for Koulikoro) and fallow and cropland on the other hand ($P = 0.004$ for Garalo and $P = 0.010$ for Koulikoro). Biomass C stocks in Jatropha plantations are not significantly different from those under annual cropland. This is explained by a similar presence of mature trees in both LUs. Depending on the density and dimensions of these scattered trees in the landscape, the variability in biomass C stocks within each LU is high, implying that the impact of LUC is highly variable as well (see section Total C stock, C debt and C repayment time).

Soil carbon

Soil organic carbon concentrations. SOC concentrations measured in Garalo generally show a logarithmic decrease with depth, being most pronounced in fallow land, followed by Jatropha and cropland (Fig. 5). In Koulikoro, cultivated soils are found to be more homogeneous and are more depleted in organic matter at the surface as compared to Garalo. The latter difference is only found significant for Jatropha ($P = 0.004$). SOC concentrations in fallow land are similar between the two ecoregions. Although SOC concentrations are higher under fallow compared to cropland and Jatropha in all soil layers (Fig. 5), the difference is found to be only significant in the upper 5 cm for Garalo and 10 cm for Koulikoro (see Table S2.1 in the Supporting Information, Data S2).

Soil carbon stocks in the different land uses. SOC stocks are found to follow the same trend as biomass C, that is, being largest under fallow and without significant differences between cropland and Jatropha (Table 3). The effect of LUC is primarily visible in the upper soil layers (Fig. 6).

Spatial variability. A paired t-test was conducted to look for significant differences in SOC between the two sampling locations within Jatropha plantations, that is, directly underneath the shrubs vs. in between the shrubs (Fig. 2). A significant difference is only found for the third soil layer (10–20 cm), where values are larger underneath the shrubs (4.72 Mg ha^{-1}) compared to between the shrubs (4.35 Mg ha^{-1}).

Finally, the within-field spatial variability of SOC, expressed by means of the coefficient of variation (CV), is compared to the between-field variability (Table 4).

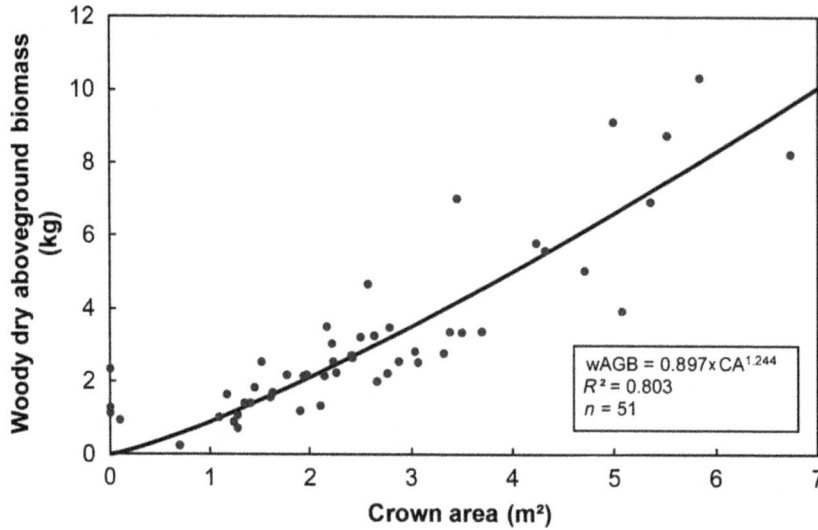

Fig. 3 Allometric relation for woody dry aboveground biomass of individual Jatropha trees based on their crown area.

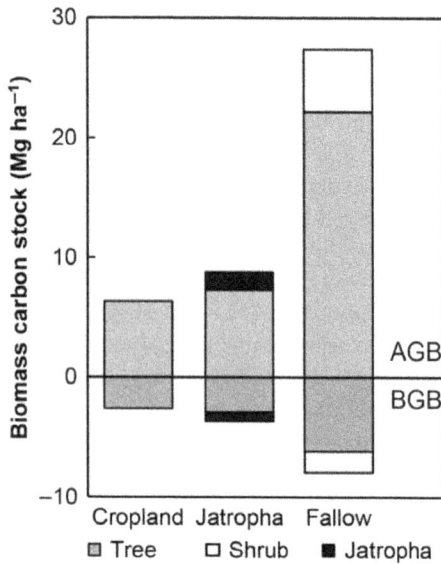

Fig. 4 Partitioning of total biomass carbon stock between aboveground and belowground biomass (AGB and BGB, respectively) and between the different vegetation elements for each land use type. BGB is read below the x-axis and AGB above it. The stacked bars represent the vegetation elements: trees, shrubs, and Jatropha.

Spatial variability is largest in fallow and lowest in Jatropha fields, but none of these differences are statistically significant. In all LUs, the within-field CV varies widely between the different fields, making it difficult to estimate the number of samples needed for an accurate estimation of SOC stock in a particular LU. The variability between different fields is the largest source of variation, exceeding the local within-field variability by a factor 2–3.

Total C stock, C debt, and C repayment time

Total C stock differs significantly between fallow land and cultivated land, that is, cropland and Jatropha (Fig. 7a). The same trends were found for the two ecoregions (Table 3), which are therefore displayed together. In cropland and Jatropha fields, most C is stored in the soil, while in fallow land biomass is the dominant C pool. By subtracting the current C stock in Jatropha plantations (at year 4) from the C stock in another LU, the so-called remaining C debt is calculated, that is, the fraction of the initial C debt (C debt at year 0 of the plantation's life cycle) that has not yet been compensated by C sequestration in the Jatropha plantation during the past 4 years (Fig. 7b). On average, this remaining C debt amounts to 32.4 and −3.1 Mg C ha^{-1} for the conversion of fallow land and cropland, respectively, the latter being not significantly different from zero. Based on the nonsignificant differences in SOC between cropland and Jatropha on the one hand and between the two sampling locations within Jatropha fields on the other hand, it can be assumed that SOC sequestration in a timeframe of 4 years is negligible, and consequently, the initial C debt can be approximated by the sum of the remaining C debt and the C stock in Jatropha biomass after 4 years. This results in an average initial C debt of 34.7 Mg C ha^{-1} for fallow land. As can be seen from Fig. 7a, this carbon debt can be mainly attributed to biomass removal prior to planting Jatropha (on average 73% of the total carbon debt is caused by the difference in biomass C content). It should be noted that standard errors of total C stocks are large, resulting in a large variability of C debts (Fig. 7b). For both LUCs under study, the C debt varies

Table 3 Average and standard deviation (within brackets) of Jatropha carbon stocks, total biomass carbon stocks, soil organic carbon stocks (SOC; 0–30 cm depth) and total carbon stocks grouped per ecoregion and land use. Significant differences between land uses per ecoregion are indicated using differing superscript letters

Ecoregion	Land use	Number of fields	Jatropha C (Mg ha⁻¹)	Total biomass C (Mg ha⁻¹)	SOC (Mg ha⁻¹)	Total C (Mg ha⁻¹)
Koulikoro	Cropland	9	–	7.97[a] (7.73)	17.12[a] (5.08)	25.09[a] (10.92)
	Jatropha	9	2.68 (1.57)	11.26[a] (7.65)	17.04[a] (5.90)	28.30[a] (12.67)
	Fallow	7	–	44.75[b] (41.37)	28.08[a] (16.06)	72.83[b] (44.15)
Garalo	Cropland	9	–	9.74[a] (6.28)	14.66[a] (6.49)	24.40[a] (10.93)
	Jatropha	9	2.01 (1.25)	13.70[a] (9.41)	13.77[a] (6.91)	27.47[a] (15.02)
	Fallow	8	–	27.00[b] (12.92)	20.61[a] (10.23)	47.61[b] (14.39)

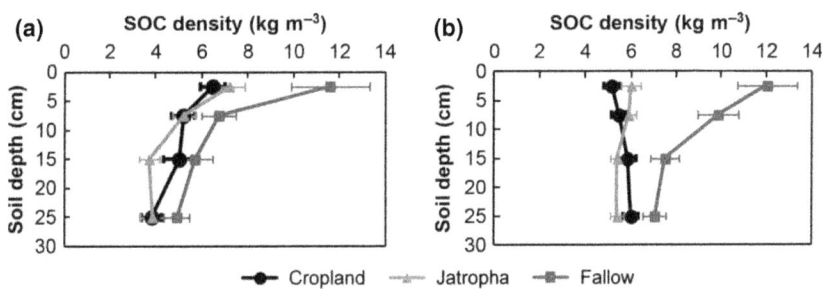

Fig. 5 Relation of soil organic carbon density with soil depth in cropland, Jatropha, and fallow for the Garalo (a) and Koulikoro (b) ecoregions. The error bars represent standard error of the mean.

from highly positive to highly negative, depending on the local situation. All extreme cases (outliers in Fig. 7b) can be explained by large differences in the presence of mature trees between the LUs.

The remaining C debt after 4 years of Jatropha cultivation can be further compensated through substitution of fossil fuels by the produced biodiesel. For this case study, an average biofuel C repayment rate of 0.09 Mg C ha⁻¹ yr⁻¹ was estimated based on Almeida et al. (2014), assuming a seed yield of 0.6 Mg ha⁻¹ yr⁻¹ (based on local observations). Hence, it would take on average 349 years of Jatropha cultivation and biodiesel production to repay the C debt created after fallow conversion. The calculated repayment time varies between 0 and 1278 years, depending on the C debt. Instead of energy production, Jatropha oil can be diverted to the cosmetic industry or small-scale soap production (Contran et al., 2013), a very attractive practice to smallholders for its simplicity and profitability. Based on the LCA model of Almeida et al. (2014), the ratio of materials stated in Contran et al. (2013) and assuming that the reaction is heated with fuel wood, the global warming potential (GWP) of Jatropha-based soap production in Koulikoro would amount to 1.2 kg CO_2 eq kg⁻¹ soap. The GWP of soaps present in ecoinvent v3 database (The Swiss Centre for Life Cycle Inventories, Switzerland) is on average 5.6 kg CO_2 eq kg⁻¹ soap. Hence, with soap production, the average C debt here reported

would be repaid within 256 years (range: 0–938 years, depending on C debt).

Discussion

Biomass carbon in Jatropha plantations

Allometric relations based on stem diameter are most frequently used in the literature to estimate the aboveground biomass of Jatropha (Ghezehei et al., 2009; Achten et al., 2010c; Firdaus & Husni, 2012; Hellings et al., 2012; Bayen et al., 2015). However, due to the specific tree architecture of Jatropha, that is, branching close to the soil surface, stem diameter is often difficult to measure. In this study, crown area was found to be the best alternative to predict wAGB. The use of this predictor variable should, however, be restricted to cases where there is no pruning and canopy closure is not yet reached, as these factors highly influence crown dimensions. This shows that the allometric relation to be used for Jatropha biomass estimation should be both location and management specific. The potential sources of error mentioned above can be partly avoided in future allometric relations using both stem and crown diameter simultaneously.

The average root-to-shoot ratio observed in this study (0.48) is higher compared to the value of 0.32 reported by Hellings et al. (2012) in similar climatic conditions

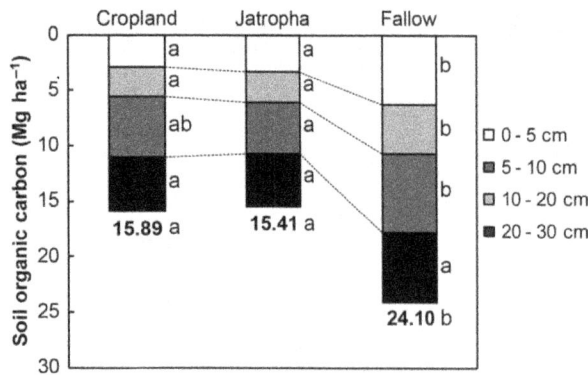

Fig. 6 Average differences in soil organic carbon stocks between the three land uses for each soil layer. Significant differences between land uses are indicated for each soil layer using letters: a soil layer marked with 'a' differs significantly from the same layer in another land use marked with 'b', but not from 'a' or 'ab'; bold numbers represent the total soil carbon stock.

(Northern Tanzania), but agrees well with the value of 0.51 found by Torres *et al.* (2011) for a humid climate in Brazil, both for a similar plant age. Hence, caution should be exercised when using any of these values as a default root-to-shoot ratio for Jatropha in future studies, as this plant characteristic is not only affected by climate, but also by local soil conditions (cf. Table 2: largest root biomass found in stone-free, coarse-textured soils). As most manual measurements of BGB were conducted in plantations on gravel soils, the average root-to-shoot ratio was likely to be underestimated.

Woody biomass stocks for 4-year-old Jatropha plantations found in this study (on average 5.04 Mg ha^{-1}) agree well with the average value of 3.9 Mg ha^{-1} reported by a study under similar environmental conditions in Burkina Faso (Bayen *et al.*, 2015), but are at the lower end of the range between 9 and 28 Mg ha^{-1} given in the literature for various other locations and planting densities (Reinhardt *et al.*, 2007; Firdaus *et al.*, 2010;

Table 4 Coefficient of variation (CV) of soil organic carbon stocks within and between fields

	Land use	Within field CV (%)					Between field CV (%)
		Mean	Standard deviation	Number of fields	Minimum	Maximum	
Koulikoro	Cropland	12.13	13.51	9	0.87	44.06	29.67
	Jatropha	10.27	6.36	9	0.82	19.41	34.60
	Fallow	21.40	12.97	7	8.75	43.20	53.46
Garalo	Cropland	22.59	17.17	9	3.54	62.74	44.27
	Jatropha	16.63	13.77	9	1.54	38.72	50.18
	Fallow	25.76	20.79	8	4.54	62.09	50.08

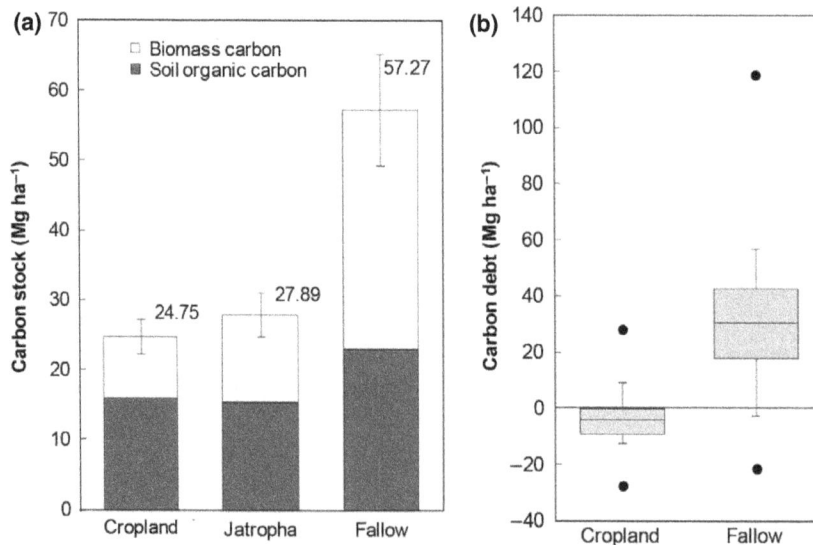

Fig. 7 (a) Average biomass carbon and soil carbon (0–30 cm depth) stocks per land use type. The error bars represent the standard error of the total carbon stocks. (b) Differences in total carbon stocks between Jatropha on the one hand and cropland and fallow on the other hand, or Jatropha carbon debts, presented as boxplots. Large dots represent outliers.

Bailis & McCarty, 2011; Torres et al., 2011; Firdaus & Husni, 2012; Wani et al., 2012), mainly owing to the relatively low amount of rainfall, poor soil conditions, and lack of management in the sites at hand. It should be noted that plant mortality, although frequently observed (on average 30% in Garalo, mainly due to termite activity – data not shown), was not taken into account in the calculation of Jatropha biomass. Due to this simplification, the biomass results reported here represent the achievable biomass under current management practices and likely overestimate reality. In addition, leaf biomass was not considered in this study due to a lack of data. According to Bayen et al. (2015), leaf biomass represents on average 9% of the total AGB in Jatropha plantations.

Soil carbon in Jatropha plantations

The average SOC stock of 15.4 Mg ha^{-1} in the top 30 cm soil profile of Jatropha plantations is lower compared to the value of 28.0 Mg ha^{-1} reported for intensively managed Jatropha plantations in Burkina Faso (0–20 cm). This difference can be due to multiple factors, including climate (Jobbagy & Jackson, 2000), soil (Walker & Desanker, 2004; Takimoto et al., 2008), and management (mainly fertilization and tillage). Within our study, soil texture and gravel content explained most of the observed variability in SOC content between the different sites in Koulikoro and Garalo, respectively (Table S2.2, Supporting Information Data S2).

In general, SOC densities found in this study for cropland and fallow (respectively, 16 and 22 Mg ha^{-1}) agree well with the range of 10–30 Mg ha^{-1} reported in similar environmental conditions (Tschakert et al., 2004; Woomer et al., 2004; Vagen et al., 2005; Takimoto et al., 2008; Saiz et al., 2012; Baumert et al., 2014), but are slightly lower than the IPCC default values for a tropical dry climate and low activity clay soils (20 and 35 Mg ha^{-1}, respectively; Eggleston et al.,, 2006). The logarithmic relation between SOC and soil depth found in this study is confirmed by Jobbagy & Jackson (2000) and Walker & Desanker (2004) for various ecosystems around the globe.

Land-use change impact and carbon sequestration by Jatropha plantations

Unlike C emissions from biomass, which are concentrated on the moment of land clearing, soil C emissions triggered by LUC can continue for multiple years due to the slow process of mineralization. This implies that, as the moment of LUC, two opposite C fluxes are simultaneously occurring in the Jatropha plantations: (i) continuous carbon emission from soil due to LUC and (ii)

building up of newly sequestered C in Jatropha woody biomass and soil (through litterfall and root decay). As only total C stocks were measured at year 0 and year 4, there is no way to strictly separate or quantify either of both fluxes (Conteh, 1999). Despite this drawback, some qualitative conclusions can still be made. The C debt created by converting cropland to Jatropha is generally low and is compensated within 4 years of Jatropha cultivation through C sequestration in Jatropha biomass. There is no significant SOC sequestration taking place within the first 4 years after Jatropha establishment, as there are no differences found in SOC content between Jatropha vs. cropland nor between inter-row and within-row locations in Jatropha plantations. This concurs with the findings of Baumert et al. (2014) in Burkina Faso, who used a similar paired sites approach on 4-year-old Jatropha plantations, supplemented by ^{13}C isotope measurements. However, multiple monitoring studies have demonstrated the positive effect of Jatropha cultivation on several soil properties, including SOC concentrations (Ogunwole et al., 2008; Wani et al., 2012; Srivastava et al., 2014). In addition, Baumert et al. (2014) found a significantly larger SOC stock in 15- to 20-year-old Jatropha living fences compared to surrounding cropland. Converting cropland to Jatropha thus may have a positive effect on SOC in the long term, but further monitoring is required to confirm this trend for our case. Despite the negligible SOC sequestration estimated in our case study, SOC should not be disregarded from future C sequestration assessments of Jatropha. The high share of SOC in the total ecosystem C stock (38–64%, which agrees well with the range of reported values for West African savannah systems, i.e. 30–90%; Tschakert et al., 2004; Bationo et al., 2007; Takimoto et al., 2008) highlights the importance of this C pool and stresses the need for good crop management practices (Lal, 2004) to avoid the loss of SOC during cultivation.

Converting fallow land to Jatropha has a clear negative impact on C stocks, especially biomass. Due to the protection of some tree species, such as *Vitellaria paradoxa* C.F. Gaertner, not all biomass is removed upon LUC. These few mature trees still make up the largest fraction of biomass after 4 years of Jatropha cultivation (Fig. 4), which clearly shows their benefits from a GHG mitigation perspective. In addition to biomass C, on average 8 Mg SOC ha^{-1} (34%) is lost, which is at the lower end of the 20–60% range that is reported in literature for the conversion of natural land to cropland in similar conditions (Elberling et al., 2003; Walker & Desanker, 2004; Vagen et al., 2005). The calculated total C debt of 34.7 Mg C ha^{-1} is in line with the estimations of Achten et al. (2013) for the conversion of scrubland in semi-arid regions (24–28 Mg C ha^{-1}). Although being at the lower end of the wide range found for various biofuel crops in various ecosystems (0–940 Mg C ha^{-1}, Fargione *et al.*, 2008), it still represents a considerable environmental impact, as can be seen from the high repayment times, implying that the production of Jatropha-based biofuel (and soap) on fallow land under current practices in Mali is unsustainable. Rasmussen *et al.* (2012) found similar high repayment times (187–966 years) for a case study in Mozambique.

One could conclude that Jatropha plantations should only be established in degraded ecosystems with low initial biomass and soil C stocks, as is also recommended by, for example, Achten & Verchot (2011) and Romijn (2011). However, the initial C stocks in soil and biomass are not the only factors that should be considered. Oil yields on degraded lands are often low, giving rise to low repayment rates and hence long repayment times. Low yields incentivize farmers to shift Jatropha to more productive lands, containing more C, and thus giving rise to higher C debts. This trend may cause competition with food production and additional indirect LUCs, which again increase the C debt (Achten & Verchot, 2011). Hence, there is a need for more agronomical research aiming at stabilizing and optimizing Jatropha yields on degraded lands (Muys *et al.*, 2014). Still, in regions such as the Sahel, where rainfall is erratic, significant annual yield variations are expected, causing C repayment rates to be highly variable from 1 year to another.

In addition to repayment through substitution of fossil fuels, the remaining C debt at year 4 can also be partially repaid by additional biomass growth in the Jatropha plantations (until the average biomass C stock of a rotation is reached; Achten *et al.*, 2013). This aspect is, however, not included in the calculation of the repayment time and most likely led to a slight overestimation of the latter. Furthermore, the calculation of C debt and associated repayment time neglects the fate of the C stocks in the biomass and soil of Jatropha plantations. While the C sequestered in biomass will be in principle released after the rotation ends, the evolution of SOC is unknown. This is an important factor to the repayment time because a trend of SOC sequestration may speed it up whilst a trend of loss will postpone it. Due to the lack of long-term chronosequences, it is not possible to infer from the data here presented whether there is sequestration or loss of SOC throughout the lifetime of a Jatropha plantation. The literature data are also contradictory in this matter (e.g., Rasmussen *et al.*, 2012; Baumert *et al.*, 2014).

Finally, the repayment of the C debt is based on the assumption that there is 100% substitution of the fossil fuel in question. However, it is not always the case. It can be argued that in Mali the availability of a liquid fuel in a rural setting may instead add to the energy which is already consumed, given that the energy demand is increasing rapidly in this part of the world (CIA, 2014). In this case, the actual repayment time would be even larger compared to the results reported here. Alternatively, Jatropha oil or biodiesel can replace fuel wood or charcoal, which are the most common fuels in the region, particularly in rural areas (Dasappa, 2011). These fuels are obtained with negligible energy input. In case, they are taken from sustainably managed woodland they are fully renewable and truly C neutral. In such case, the repayment would not exist.

Concluding remarks

Unlike many previous repayment time studies, this study is completely based on field data, which means that the analysis takes into account local specificities which can strongly influence the results and are often missed by modeling approaches (in this case: the dominant effect of retaining mature trees on the C debt). Our C stock data can therefore serve as valuable input for local Jatropha biofuel policy (Witcover *et al.*, 2013), Jatropha sustainability and C sequestration assessments (van Eijck *et al.*, 2014) and for estimating benefits from selling Jatropha-based C credits. Despite the large potential of semi-arid ecosystems to sequester SOC, C stock data in these regions remain particularly scarce (Saiz *et al.*, 2012). Our empirical database might therefore be used in a broader sense, for example, for the calibration and validation of local LUC and SOC models (e.g., RothC, DayCent). However, the results presented here cannot be generalized without caution, as C dynamics are known to be highly dependent on environmental characteristics and local management factors (Powers *et al.*, 2011).

The spatially paired site design applied in this study only results in an approximation of the C dynamics under Jatropha. Monitoring studies using a stock change approach with a timespan of more than 5 years should be conducted on Jatropha plantations to further assess its biomass and soil C sequestration potential, as data on plantations older than 5 years are particularly scarce for this biofuel tree (Rasmussen *et al.*, 2012). In addition, there is a need for more detailed studies that quantify the amount of C lost during LUC, for example, using the eddy covariance technique (Zenone *et al.*, 2012). Finally, future studies aiming at assessing the effect of LU on SOC are advised to not only determine total SOC stocks, but also to look at the different fractions of SOC (particulate organic matter (OM) vs. stable OM; fractions of humic acid, fulvic acid, and humin), as this can provide valuable information regarding the quantity of newly sequestered SOC (Guimarães *et al.*, 2013).

The high repayment times associated with the conversion of fallow land corroborate previous concerns regarding the mitigation potential of Jatropha cultivation and biofuel production (e.g., Rasmussen *et al.*, 2012; Achten *et al.*, 2013). In this paper, we present an empirical dataset to support these claims. It is, however, important to realize that Jatropha cultivation and the associated LUC can have various other environmental, economic and social effects, either positive or negative (Achten & Verchot, 2011). Research has pointed out positive effects on the level of increased erosion control (Reubens *et al.*, 2011) and, on the societal side, empowerment of rural communities involved in smallholder projects (van Eijck *et al.*, 2014). Negative issues pertain mostly to failure in secure access to food and land as well as economic unviability (Skutch *et al.*, 2011; van Eijck *et al.*, 2014). Hence, this study should be seen as part of a larger complex story and should be complemented with a more holistic study in which all these other impacts are included.

Acknowledgements

This study was part of the 'Jatrophability Mali' project, financed by Agricultural Research for Development – Dimension of the European Research Area (ERA-ARD), the Belgian Development Cooperation (BTC) and the Royal Museum for Central Africa. The paper was prepared in the framework of the KLIMOS project on Sustainable Development (Acropolis funding under VLIR/ARES/DGD). J. Almeida holds a doctoral grant from the Foundation for Science and Technology – Portugal. The authors greatly acknowledge the local project partners in Mali, Institut d'Economie Rural du Mali, Mali Folkecenter, Fondation Mali Biocarburant and MaliBiocarburant Société Anonyme (MBSA) for their support and cooperation during the field campaign. The field campaign was also made possible by the VLIR-UOS IRO travel grants of J. Degerickx and L. Vervoort.

References

Achten WMJ, Verchot LV (2011) Implications of biodiesel-induced land-use changes for CO_2 emissions: case studies in tropical America, Africa and Southeast Asia. *Ecology and Society*, **16**, 14–52.

Achten WMJ, Nielsen LR, Aerts R et al. (2010a) Towards domestication of *Jatropha curcas*. *Biofuels*, **1**, 91–107.

Achten WMJ, Maes WH, Aerts R et al. (2010b) Jatropha: from global hype to local opportunity. *Journal of Arid Environments*, **74**, 164–165.

Achten WMJ, Maes WH, Reubens B, Mathijs E, Singh VP, Verchot L, Muys B (2010c) Biomass production and allocation in *Jatropha curcas* L. seedlings under different levels of drought stress. *Biomass and Bioenergy*, **34**, 667–676.

Achten WMJ, Trabucco A, Maes WH et al. (2013) Global greenhouse gas implications of land conversion to biofuel crop cultivation in arid and semi-arid lands – Lessons learned from Jatropha. *Journal of Arid Environments*, **98**, 135–145.

Almeida J, Moonen PCJ, Soto I, Achten WMJ, Muys B (2014) Effect of farming system and yield in the life cycle assessment of Jatropha-based bioenergy in Mali. *Energy for Sustainable Development*, **23**, 258–265.

Bailis RE, Baka JE (2010) Greenhouse gas emissions and land use change from *Jatropha curcas*-based jet fuel in Brazil. *Environmental Science & Technology*, **44**, 8684–8691.

Bailis R, McCarty H (2011) Carbon impacts of direct land use change in semiarid woodlands converted to biofuel plantations in India and Brazil. *GCB Bioenergy*, **3**, 449–460.

Bationo A, Kihara J, Vanlauwe B, Waswa B, Kimetu J (2007) Soil organic carbon dynamics, functions and management in West African agro-ecosystems. *Agricultural Systems*, **94**, 13–25.

Baumert S, Khamzina A, Vlek PLG (2014) Soil organic carbon sequestration in *Jatropha curcas* systems in Burkina Faso. *Land Degradation & Development*. doi:10.1002/ldr.2310.

Bayen P, Bognounou F, Lykke AM, Ouédraogo M, Thiombiano A (2015) The use of biomass production and allometric models to estimate carbon sequestration of *Jatropha curcas* L. plantations in western Burkina Faso. *Environment, Development and Sustainability*. doi:10.1007/s10668-015-9631-4.

CIA (2014) The world factbook. Available at: www.cia.gov/library/publications/the-world-factbook/ (accessed 20 May 2014).

Conteh A (1999) *Estimation of Changes in Soil Carbon Due to Changed Land Use*. National Carbon Accounting System, Technical Report No. 2. Webbnet Land Resource Services Pty Ltd., Canberra, Australia.

Contran N, Chessa L, Lubino M, Bellavite D, Roggero PP, Enne G (2013) State-of-the-art of the *Jatropha curcas* productive chain: from sowing to biodiesel and by-products. *Industrial Crops and Products*, **42**, 202–215.

Coulibaly A (2003) Country pasture/forage resource profiles - Mali. Available at: http://www.fao.org/ag/agp/AGPC/doc/Counprof/Mali/mali.htm (accessed 29 May 2014).

Dasappa S (2011) Potential of biomass energy for electricity generation in sub-Saharan Africa. *Energy for Sustainable Development*, **15**, 203–213.

van Eijck J, Romijn H, Balkema A, Faaij A (2014) Global experience with Jatropha cultivation for bioenergy: an assessment of socio-economic and environmental aspects. *Renewable and Sustainable Energy Reviews*, **32**, 869–889.

Eggleston HS, Buendia L, Miwa K, Ngara T, Tanabe K (2006) *IPCC Guidelines for National Greenhouse Gas Inventories*. IGES, Tokyo, Japan.

Elberling B, Touré A, Rasmussen K (2003) Changes in soil organic matter following groundnut–millet cropping at three locations in semi-arid Senegal, West Africa. *Agriculture, Ecosystems & Environment*, **96**, 37–47.

FAO (2007) Digital Soil Map of the World. Available at: http://www.fao.org/geonetwork/srv/en/metadata.show?id=14116 (accessed 25 May 2014).

Fargione J, Hill J, Tilman D, Polasky S, Hawthorne P (2008) Land clearing and the biofuel carbon debt. *Science*, **319**, 1235–1238.

Favretto N, Stringer LC, Dougill AJ (2012) Policy and institutional frameworks for the promotion of sustainable biofuels in Mali. Paper No. 35. Sustainability Research Institute, University of Leeds, UK.

Firdaus MS, Husni MHA (2012) Planting *Jatropha curcas* on constrained land: emission and effects from land use change. *The Scientific World Journal*, **2012**, 1–7.

Firdaus MS, Hanif AHM, Safiee ASS, Ismail MR (2010) Carbon sequestration potential in soil and biomass of *Jatropha curcas*. In: *19th World Congress of Soil Science, Soil Solutions for a Changing World* (eds Gilkes RJ, Prakongkep N), pp. 62–65. IUSS, Brisbane, Qld, Australia.

Ghezehei SB, Annandale JG, Everson CS (2009) Shoot allometry of *Jatropha curcas*. *Southern Forests: a Journal of Forest Science*, **71**, 279–286.

Guimarães DV, Gonzaga MIS, da Silva TO, da Silva TL, da Silva Dias N, Matias MIS (2013) Soil organic matter pools and carbon fractions in soil under different land uses. *Soil and Tillage Research*, **126**, 177–182.

Hellings BF, Romijn HA, Franken YJ (2012) *Carbon Storage in Jatropha curcas Trees in Northern Tanzania*. FACT foundation, Eindhoven, the Netherlands.

Jobbagy EG, Jackson RB (2000) The vertical distribution of soil organic carbon and its relation to climate and vegetation. *Ecological Applications*, **10**, 423–436.

Lal R (2004) Soil carbon sequestration to mitigate climate change. *Geoderma*, **123**, 1–22.

Louppe D (1994) *Le karité en Côte d'Ivoire. Projet de promotion et de developpement des exportations agricoles (PPDEA)*. CIRAD-Föret, Montpellier, France.

Magin C (2011) Western Africa: Stretching from Senegal through Niger - Afrotropics (AT0722). Available at: http://www.worldwildlife.org/ecoregions/at0722 (accessed 20 May 2014).

Muys B, Norgrove L, Alamirew T et al. (2014) Integrating mitigation and adaptation into development: the case of *Jatropha curcas* in sub-Saharan Africa. *GCB Bioenergy*, **6**, 169–171.

Nielsen F, Raghavan K, deJongh J, Huffman D (2013) *Jatropha for Local Development - after the Hype*. Hivos, Den Haag, The Netherlands.

Nouvellet Y, Kassambara A, Besse F (2006) Le parc à karités au Mali?: inventaire, volume, houppier et production fruitière. *Bois et Forêts des Tropiques*, **287**, 5–20.

Ogunwole JO, Chaudhary DR, Ghosh A, Daudu CK, Chikara J, Patolia JS (2008) Contribution of *Jatropha curcas* to soil quality improvement in a degraded Indian entisol. *Acta Agriculturae Scandinavica, Section B - Plant Soil Science*, **58**, 245–251.

Powers JS, Corre MD, Twine TE, Veldkamp E (2011) Geographic bias of field observations of soil carbon stocks with tropical land-use changes precludes spatial extrapolation. *Proceedings of the National Academy of Sciences of the United States of America*, **108**, 6318–6322.

Rasmussen LV, Rasmussen K, Bech Bruun T (2012) Impacts of Jatropha-based biodiesel production on above and below-ground carbon stocks: a case study from Mozambique. *Energy Policy*, **51**, 728–736.

Reinhardt G, Gärtner S, Rettenmaier N, Münch J, von Falkenstein E (2007) *Screening Life Cycle Assessment of Jatropha Biodiesel*. IFEU, Heidelberg, Germany.

Reubens B, Achten WMJ, Maes WH, Danjon F, Aerts R, Poesen J, Muys B (2011) More than biofuel? *Jatropha curcas* root system symmetry and potential for soil erosion control. *Journal of Arid Environments*, **75**, 201–205.

Romijn HA (2011) Land clearing and greenhouse gas emissions from Jatropha biofuels on African Miombo Woodlands. *Energy Policy*, **39**, 5751–5762.

Saiz G, Bird MI, Domingues T *et al.* (2012) Variation in soil carbon stocks and their determinants across a precipitation gradient in West Africa. *Global Change Biology*, **18**, 2676.

Singh K, Singh B, Verma SK, Patra DD (2014) *Jatropha curcas*: a ten year story from hope to despair. *Renewable and Sustainable Energy Reviews*, **35**, 356–360.

Skutch M, de los Rios E, Solis S *et al.* (2011) Jatropha in Mexico: environmental and social impacts of an incipient biofuel program. *Ecology and Society*, **16**, 11.

Srivastava P, Sharma YK, Singh N (2014) Soil carbon sequestration potential of *Jatropha curcas* L. growing in varying soil conditions. *Ecological Engineering*, **68**, 155–166.

Struijs J (2008) *Shinda Shinda - Option for Sustainable Bioenergy: a Jatropha Case Study*. RIVM, Bilthoven, the Netherlands.

Takimoto A, Nair PKR, Nair VD (2008) Carbon stock and sequestration potential of traditional and improved agroforestry systems in the West African Sahel. *Agriculture, Ecosystems & Environment*, **125**, 159–166.

Torres CMME, Jacovine LAG, Toledo D de P, Soares CPB, Ribeiro SC, Martins MC (2011) Biomass and carbon stock in *Jatropha curcas* L. *CERNE*, **17**, 353–359.

Tschakert P, Khouma M, Sène M (2004) Biophysical potential for soil carbon sequestration in agricultural systems of the Old Peanut Basin of Senegal. *Journal of Arid Environments*, **59**, 511–533.

UNFCCC (2006) Revised simplified baseline and monitoring methodologies for selected small- scale afforestation and reforestation project activities under the clean development mechanism. Available at: http://cdm.unfccc.int/filestorage/C/D/M/CDMWF_AM _A3II6AX6KGW5GBB7M6AI98UD3W59X4/EB28_repan18_AR%20SSC0001_ver03.pdf?t=aWV8bjZjczA4fDBHY8Lsr_K0NZ19_1zcqo6_ (accessed 25 May 2014).

UNFCCC (2009) Approved afforestation and reforestation baseline methodology AR-AM0002 "Restoration of degraded lands through afforestation/reforestation." Available at: https://cdm.unfccc.int/methodologies/DB/6ZZXJUKK49WK-LID7ZH8FG3B S9WTCCH/view.html (accessed 25 May 2014).

Vagen T-G, Lal R, Singh BR (2005) Soil carbon sequestration in sub-Saharan Africa: a review. *Land Degradation & Development*, **16**, 53–71.

Van Reeuwijk LP (2002) *Procedures for Soil Analysis*. International Soil Reference and Information Centre, Wageningen, the Netherlands.

Van Rooijen L (2014) Pioneering in marginal fields: Jatropha for carbon credits and restoring degraded land in eastern Indonesia. *Sustainability*, **6**, 2223–2247.

Verrastro F, Ladislaw S (2007) Providing energy security in an interdependent world. *The Washington Quarterly*, **30**, 95–104.

Walker SM, Desanker PV (2004) The impact of land use on soil carbon in Miombo woodlands of Malawi. *Forest Ecology and Management*, **203**, 345–360.

Wani SP, Chander G, Sahrawat KL, Srinivasa Rao C, Raghvendra G, Susanna P, Pavani M (2012) Carbon sequestration and land rehabilitation through *Jatropha curcas* (L.) plantation in degraded lands. *Agriculture, Ecosystems & Environment*, **161**, 112–120.

Witcover J, Yeh S, Sperling D (2013) Policy options to address global land use change from biofuels. *Energy Policy*, **56**, 63–74.

Woomer P, Tieszen L, Tappan G, Touré A, Sall M (2004) Land use change and terrestrial carbon stocks in Senegal. *Journal of Arid Environments*, **59**, 625–642.

Zenone T, Gelfand I, Chen J, Hamilton SK, Robertson GP (2013) From set-aside grassland to annual and perennial cellulosic biofuel crops: effects of land use change on carbon balance. *Agricultural and Forest Meteorology*, **182–183**, 1–12.

PERMISSIONS

All chapters in this book were first published in GCB BIOENERGY, by John Wiley & Sons Ltd.; hereby published with permission under the Creative Commons Attribution License or equivalent. Every chapter published in this book has been scrutinized by our experts. Their significance has been extensively debated. The topics covered herein carry significant findings which will fuel the growth of the discipline. They may even be implemented as practical applications or may be referred to as a beginning point for another development.

The contributors of this book come from diverse backgrounds, making this book a truly international effort. This book will bring forth new frontiers with its revolutionizing research information and detailed analysis of the nascent developments around the world.

We would like to thank all the contributing authors for lending their expertise to make the book truly unique. They have played a crucial role in the development of this book. Without their invaluable contributions this book wouldn't have been possible. They have made vital efforts to compile up to date information on the varied aspects of this subject to make this book a valuable addition to the collection of many professionals and students.

This book was conceptualized with the vision of imparting up-to-date information and advanced data in this field. To ensure the same, a matchless editorial board was set up. Every individual on the board went through rigorous rounds of assessment to prove their worth. After which they invested a large part of their time researching and compiling the most relevant data for our readers.

The editorial board has been involved in producing this book since its inception. They have spent rigorous hours researching and exploring the diverse topics which have resulted in the successful publishing of this book. They have passed on their knowledge of decades through this book. To expedite this challenging task, the publisher supported the team at every step. A small team of assistant editors was also appointed to further simplify the editing procedure and attain best results for the readers.

Apart from the editorial board, the designing team has also invested a significant amount of their time in understanding the subject and creating the most relevant covers. They scrutinized every image to scout for the most suitable representation of the subject and create an appropriate cover for the book.

The publishing team has been an ardent support to the editorial, designing and production team. Their endless efforts to recruit the best for this project, has resulted in the accomplishment of this book. They are a veteran in the field of academics and their pool of knowledge is as vast as their experience in printing. Their expertise and guidance has proved useful at every step. Their uncompromising quality standards have made this book an exceptional effort. Their encouragement from time to time has been an inspiration for everyone.

The publisher and the editorial board hope that this book will prove to be a valuable piece of knowledge for researchers, students, practitioners and scholars across the globe.

LIST OF CONTRIBUTORS

Ani L Baral and Chris Malins
The International Council on Clean Transportation, 1225 Eye St., NW Suite 900, Washington, DC 20005, USA

Nicolai David Jablonowski, Moritz Nabel, Simone Krafft and Ulrich schurr
Forschungszentrum Jülich GmbH, Institute of Bio- and Geosciences, IBG-2: Plant Sciences, 52428 Jülich, Germany

Tobias Kollmann
Forschungszentrum Jülich GmbH, Institute of Bio- and Geosciences, IBG-2: Plant Sciences, 52428 Jülich, Germany
Project Management Jülich, Renewable Energies – EEN, 52425 Jülich, Germany

Tatjana Damm and Holger Klose
RWTH Aachen University, Institute of Botany and Molecular Genetics IBMG, Worringerweg 3, 52074 Aachen, Germany,

Michael Müller, Marc Bläsing and Sören Seebold
Forschungszentrum Jülich GmbH, Institute of Energy and Climate Research, IEK-2: Microstructure and Properties of Materials, 52428 Jülich, Germany

Isabel Kuperjans and Markus Dahmen
FH Aachen, Aachen University of Applied Sciences, Institut NOWUM-Energy, Heinrich-Mussmann-Str. 1, 52428 Jülich, Germany

Susanne Frydendal-Nielsen and Uffe Jørgensen
Department of Agroecology, Aarhus University, Blichers Alle 20, Tjele, Denmark

Maibritt Hjorth
Department of Engineering, Aarhus University, Hangøvej 2, 8200, Aarhus N, Denmark

Claus Felby
Department of Geosciences and Natural Resource Management, University of Copenhagen, Rolighedsvej 23, 1958 Frederiksberg C, Denmark

René gislum
Department of Agroecology, Aarhus University, Forsøgsvej 1, 4200 Slagelse, Denmark

Eric K. Anderson, Dokyoung Lee and Thomas B. Voigt
Department of Crop Sciences, University of Illinois at Urbana-Champaign, Urbana, IL 61801, USA

Damian J. Allen
Mendel BioEnergy Seeds, Mendel Biotechnology Inc., 3935 Point Eden Way, Hayward, CA 94545, USA,
Department of Agronomy, Purdue University, West Lafayette, IN 47907, USA

Roger T. Koide
Department of Biology, Brigham Young University, Provo, UT 84602, USA

Binh T. Nguyen and Patrick J. Drohan
Department of Ecosystem Science and Management, The Pennsylvania State University, University Park, PA 16802, USA

R. Howard Skinner, Curtis J. Dell and Paul R. Adler
Pasture Systems and Watershed Management Research Unit, USDA-ARS, University Park, PA 16802, USA

Matthew S. Peoples
Department of Plant Science, The Pennsylvania State University, University Park, PA 16802, USA

Eric G. Mbonimpa
Department of Systems Engineering and Management, Air Force Institute of Technology, WPAFB, OH, USA

Sandeep Kumar and Rajesh Chintala
Plant Science Department, South Dakota State University, Brookings, SD, USA

Vance N. Owens
North Central Regional Sun Grant Center, South Dakota State University, Brookings, SD, USA

Heidi L. Sieverding and James J. Stone
Department of Civil and Environmental Engineering, South Dakota School of Mines and Technology, Rapid City, SD, USA

Deepak Rajagopal
Institute of the Environment and Sustainability, University of California, Los Angeles, CA, USA
School of Public and Environmental Affairs, Indiana University, Los Angeles, CA, USA

Colin W. Murphy
Institute of Transportation Studies and Energy Institute, University of California, 1605 Tilia St. Suite 100, Davis, CA 95616, USA,

Alissa Kendall
Department of Civil and Environmental Engineering, University of California, 1 Shields Ave., Davis, CA 95616, USA

Marcos Siqueira Neto, Brigitte J. Feigl and Carlos C. Cerri
Centro de Energia Nuclear na Agricultura, Universidade de São Paulo (CENA/USP), Av. Centenário, 303, Piracicaba, SP, Brazil

Marcelo V. Galdos
Laboratório Nacional de Ciência e Tecnologia do Bioetanol (CTBE), R. Giuseppe Máxiom Scalfaro, 10.000, 13083-970 Campinas, SP, Brazil

Carlos E. P. Cerri
Escola Superior de Agricultura Luiz de Queiroz, Universidade de São Paulo (ESALQ/USP), Av. Pádua Dias, 11, Piracicaba, SP, Brazil

Lawrence G. Oates, David S. Duncan and Randall D. Jackson
DOE-Great Lakes Bioenergy Research Center & Department of Agronomy, University of Wisconsin–Madison, Madison, WI 53706, USA

Ilya Gelfand, Neville Millar and G. Philip Robertson
DOE-Great Lakes Bioenergy Research Center & W.K. Kellogg Biological Station, Michigan State University, Hickory Corners, MI 49060, USA
Department of Plant, Soil and Microbial Sciences, Michigan State University, East Lansing, MI 48824, USA

Patricia Y. Oikawa
Department of Environmental Science, Policy and Management, University of California, Berkeley, CA 94702, USA

G. Darrel Jenerette and David A. Grantz
Department of Botany and Plant Sciences, University of California, Riverside, CA 92521, USA

Ilhan Ozturk
Faculty of Economics and Administrative Sciences, Cag University, 33800 Mersin, Turkey

Juan Sesmero and Xin Sun
Agricultural Economics, Purdue University, 403 West State Street, West Lafayette, IN 47907, USA

Judith A. Verstegen and Floor Van Der Hilst
Faculty of Geosciences, Copernicus Institute for Sustainable Development, Utrecht University, Heidelberglaan 2, 3584 CS, Utrecht, The Netherlands

Geert Woltjer
LEI, Wageningen University & Research Centre, Alexanderveld 5, 2502 LS Den Haag, The Netherlands

Derek Karssenberg and Steven M. De Jong
Department of Physical Geography, Faculty of Geosciences, Utrecht University, Heidelberglaan 2, 3584 CS, Utrecht, The Netherlands

André P. C. Faaij
Energy and Sustainability Research Institute Groningen, University of Groningen, Blauwborgje 6, Groningen, The Netherlands

Chunwu Zhu, Xi Xu, Jianguo Zhu and Gang Liu
State Key Laboratory of Soil and Sustainable Agriculture, Institute of Soil Science, Chinese Academy of Sciences, NO.71 East Beijing Road, Nanjing 210008, China

Dan Wang
International Center for Ecology, Meteorology and Environment, School of Applied Meteorology, Nanjing University of Information Science and Technology, Nanjing 210044, China

Saman Seneweera
Centre for Systems Biology, University of Southern Queensland, Toowoomba, Qld 4350, Australia

Richard G. Bennett and Gary Ogden
Ecosystem Sciences, CSIRO, Underwood Avenue, Floreat, WA 6014, Australia

Daniel Mendham
Ecosystem Sciences, CSIRO, College Road, Sandy Bay, Tas. 7025, Australia

John Bartle
Department of Parks and Wildlife, Kensington, WA, Australia

Thomas Buchholz
Spatial Informatics Group, LLC, 3248 Northampton Ct., Pleasanton, CA 94588, USA
Gund Institute for Ecological Economics, University of Vermont, 617 Main Street Burlington, Vermont 05405, USA

Matthew D. Hurteau
Department of Ecosystem Science and Management, The Pennsylvania State University, Bigler Road, State College, PA 16803, USA

John Gunn
Spatial Informatics Group, Natural Assets Laboratory, 11 Pond Shore Drive, Cumberland, ME 04021, USA

David Saah
Spatial Informatics Group, LLC, 3248 Northampton Ct., Pleasanton, CA 94588, USA
Department of Environmental Science, University of San Francisco, 2130 Fulton Street, San Francisco, CA 94117, USA

John F. Quensen and James M. Tiedje
Center for Microbial Ecology and DOE Great Lakes Bioenergy Research Center, Michigan State University, 540 Plant and Soil Sciences Building, East Lansing, MI 48824-1325, USA

Ederson Da C. Jesus
Center for Microbial Ecology and DOE Great Lakes Bioenergy Research Center, Michigan State University, 540 Plant and Soil Sciences Building, East Lansing, MI 48824-1325, USA
Embrapa Agrobiologia, BR 465, km 7, Seropedica, Rio de Janeiro 23890-000, Brazil

Randall D. Jackson and Teresa C. Balser
Department of Agronomy and DOE Great Lakes Bioenergy Research Center, University of Wisconsin-Madison, 1575 Linden Drive, Madison, WI 53706, USA

Chao Liang
Department of Agronomy and DOE Great Lakes Bioenergy Research Center, University of Wisconsin-Madison, 1575 Linden Drive, Madison, WI 53706, USA
State Key Laboratory of Forest and Soil Ecology, Institute of Applied
Ecology, Chinese Academy of Sciences, Shenyang 110164, China

Endang Susilawati
Center for Microbial Ecology and DOE Great Lakes Bioenergy Research Center, Michigan State University, 540 Plant and Soil Sciences Building, East Lansing, MI 48824-1325, USA
Department of Chemical Engineering and Applied Chemistry, University of Toronto, Toronto, ON M5S 3E5, Canada

Jeroen Degerickx, Joana Almeida, Pieter C. J. Moonen, Leen Vervoort and Bart Muys
Division Forest, Nature and Landscape, Katholieke Universiteit Leuven, Celestijnenlaan 200E Box 2411, 3001 Leuven, Belgium

Wouter M.J. Achten
Institute for Environmental Planning and Land Use Planning, Université libre de Bruxelles, Avenue Franklin D. Roosevelt 50 CP 130/02, 1050 Brussels, Belgium

Index

www.ingramcontent.com/pod-product-compliance
Lightning Source LLC
Chambersburg PA
CBHW082050190326
41458CB00010B/3495